Global Geological Record of Lake Basins is the first in a short series of books containing concise geological summaries of lake basin deposits. Lakes occur in a wide variety of climatic and tectonic settings – they range in size from ponds to lacustrine giants such as the Caspian Sea, and their depths range from millimeters on playa flats to the vast depths of Lake Baikal. Lacustrine deposits are also widespread throughout the geological record and geoscientists worldwide have cooperated to bring together the information available on a global spectrum of lacustrine deposits – looking in particular at paleoenvironmental aspects. The compilation has resulted from IGCP 219 – Comparative Lacustrine Sedimentology in Space and Time – which has given impetus to the new field of limnogeology.

The sequence of sediments deposited in lakes forms an archive of high resolution records of the dynamics in continental paleoenvironments. Lake sequences are also increasingly important for paleogeographic reconstructions and in strategic exploration for hydrocarbons, rare elements, salts and other economic resources. This series of books will assess for every geological time window, key lacustrine sequences worldwide, in the form of short summaries of specific deposits. For the first time a common language will be defined to describe these deposits. Ancient and modern case studies illustrate the plethora of potential subfacies in lakes. The information will allow correlation with marine deposits, define paleo-rainfall patterns and help calibrate paleoclimate parameters. The information is also being integrated onto a computer database.

T0207041

Global Geological Record of
Lake Basins
Volume 1

World and Regional Geology Series
Series Editors: M. R. A. Thomson and J. A. Reinemund

This series comprises monographs and reference book on studies of world and regional geology. Topics will include comprehensive studies of key geological regions and the results of some recent IGCP projects.

Geological Evolution of Antarctica – M. R. A. Thomson, J. A. Crame & J. W. Thomson (eds.)

Permo-Triassic Events in the Eastern Tethys – W. C. Sweet, Yang Zunyi, J. M. Dickins & Yin Hongfu (eds.)

The Jurassic of the Circum-Pacific – G. E. G. Westermann (ed.)

Global Geological Record of Lake Basins vol. 1 – E. Gierlowski-Kordesch & K. Kelts (eds.)

Paleocommunities: A Case Study from the Silurian and Lower Devonian A. J. Boucot & J. D. Lawson (eds.)

Earth's Glacial Record – M. Deynoux, J. M. G. Miller, E. W. Domack, N. Eyles, I. Fairchild & G. M. Young (eds.)

Global Geological Record of Lake Basins
Volume 1

EDITED BY

E. GIERLOWSKI-KORDESCH

Department of Geological Sciences
Ohio University

AND

K. KELTS

Limnological Research Center
University of Minnesota

CAMBRIDGE
UNIVERSITY PRESS

CAMBRIDGE UNIVERSITY PRESS
Cambridge, New York, Melbourne, Madrid, Cape Town, Singapore, São Paulo

Cambridge University Press
The Edinburgh Building, Cambridge CB2 2RU, UK

Published in the United States of America by Cambridge University Press, New York

www.cambridge.org
Information on this title: www.cambridge.org/9780521414524

First published 1994
This digitally printed first paperback version 2006

A catalogue record for this publication is available from the British Library

Library of Congress Cataloguing in Publication data

Gierlowski-Kordesch, E. (Elizabeth)
Global geological record of lake basins / E. Gierlowski-Kordesch
and K. Kelts. vol. 1
  p. cm.
ISBN 0-521-41452-0
1. Paleolimnology. 2. Sedimentary basins. I. Kelts, K. (Kerry)
II. Title.
QE39.5.P3G54 1994
551.48′2–dc20 93-28336 CIP

ISBN-13 978-0-521-41452-4 hardback
ISBN-10 0-521-41452-0 hardback

ISBN-13 978-0-521-03168-4 paperback
ISBN-10 0-521-03168-0 paperback

Contents

viii   Contents

Contents

Contributors

E. ABBATE
Dipartimento Scienze della Terra, via La Pira 4, 50121 Firenza, Italy

A. M. ALONSO ZARZA
Dpto Petrologia y Geoquímica, Facultad C. C. Geológicas, Universidad Complutense, E-28040 Madrid, Spain

P. ANADON
Institut de Ciènces de la Terra (J. Almera), CSIC, c. Marti i Franqués s.n., E-08028, Barcelona, Spain

M. ANGELES GARCIA DEL CURA
Institute of Economic Geology, Consejo Superior de Invastigaciones Cientificas, E-28040 Madrid, Spain

P. ARLHAC
Laboratoire de Géologique structurale et appliquée, Université de Provence, case postale 28, F-13331 Marseille cedex-3, France

I. ARMENTEROS
Department of Geology, University of Salamanca, E-37071 Salamanca, Spain

M. E. ARRIBAS
Departamento Petrologia y Geoquimica, Facultad C C Geológicas Universidad Complutense, E-28040 Madrid, Spain

T. ASTIN
Postgraduate Research Institute of Sedimentology, University of Reading, Reading, Berks RG6 2AB, UK

M. W. BINFORD
Graduate School of Design, Harvard University, Cambridge, MA 02138, USA

P. BLOT
Laboratoire de Géologie, Université de Reims, F-51100 Reims, France

M. BRENNER
Florida Museum of Natural History, Gainesville, Florida 32611, USA

M. E. BROOKFIELD
Land Resource Science, Guelph University, Guelph, Ontario NIG 2W1, Canada

P. BRUNI
Dipartimento Scienze della Terra, via La Pira 4, 50121 Firenza, Italy

L. A. BUATOIS
Facultad de Ciencias Naturales e Instituto Miguel Lillo, Universidad Nacional de Tucumán, Casilla de correo 1, 4000 San Miguel de Tucumán, Argentina

L. CABRERA
Dpto Geología Dinàmica, Geofisica i Paleontologia, Facultad de Geologia, Zona Universitaria de Pedralbes, E-08028 Barcelona, Spain

J. P. CALVO
Departamento Petrologia y Geoquimica, Facultad CC Geológicas, Universidad Complutense, E-28040 Madrid, Spain

T. E. CERLING
Department of Geology and Geophysics, University of Utah, Salt Lake City, Utah 84112, USA

S. CÉSARI
Faculted de Ciencias Exactas y Naturales, Universidad de Buenos Aires, Dpto Geoloía, Pabellón II, Cuidad Universitaria, 1428 Buenos Aires, Argentina

J-P. COLIN
Esso Rep, 213 Cours Victor Hugo, F-33323 Bègles, France

A. CORROCHANO
Department of Geology, University of Salamanca, E-37071 Salamanca, Spain

T. CSERNY
Hungarian Geological Institute, Népstadion út. 14, H-1142 Budapest, Hungary

D. R. CURREY
Dept Geography, University of Utah, Salt Lake City, UT 84112, USA

J. H. CURTIS
Dept Geology, University of Florida, Gainesville, FL 32611, USA

C. J. DABRIO
Dpto Estratigrafía, Facultad de Ciencia Geológicas, Universidad Complutense, E-28040 Madrid, Spain

G. DAM
Geological Survey of Greenland, Øster Voldgade 10, DK-1350 Copenhagen K, Denmark

G. M. S. DONGOL
Geological Survey, His Majesty's Government of Nepal, Katmandu, Nepal

K. T. DORSEY
Kinnetic Laboratories Inc., Santa Cruz, CA 95060, USA

R. F. DUBIEL
US Geological Survey, MS 919 Box 25046 DFC, Denver, CO80225, USA

C. DUPUIS
Geologie fondamentale et appliquée, Faculté Polytechnique, rue de
Houdain, B-7000 Mons, Belgium

E. ELIZAGA
Dpto Geologicas, ITGE, Rios Rosas 23, E-28003 Madrid, Spain

R ERTUS
Laboratoire de géochimie des roches sedimentaires, UA723, Université de
Paris-sud, F-91405 Orsay Cedex, France

M. FLOQUET
Centre des Sciences de la Terre, Université de Bourgogne, 6 Bd Gabriel,
F-21000 Dijon, France

E. GIERLOWSKI-KORDESCH
Department of Geological Sciences, Ohio University, Athens, OH 45701-
2979, USA

J. C. GÓMEZ FERNÁNDEZ
Dpto Estratigrafía, Facultad de Ciencia Geológicas, Universidad
Complutense, E-28040 Madrid, Spain

M. A. GONZÁLEZ
CONCEIT, CC 289, Sucursal 13 (B), 1413 Buenos Aires, Argentina

P. J. W. GORE
Department of Geology, DeKalb College, 555 North Indian Creek Drive,
Clarkston, Georgia 30021, USA

N. HAUSCHKE
Department of Geology and Palaeontology, University of Münster,
Corrensstraße 24, D-48149 Münster, Germany

ROGER HIGGS
Sedimentological Consultant, Moreton Cottage, Beer Road, Seaton,
Devon EX12 2PR, UK

A. HIGUERA-GUNDY
Florida Museum of Natural History, University of Florida, Gainesville,
FL 32611, USA

D. A. HODELL
Dept Geology, University of Florida, Gainesville, FL 32611, USA

M. HOYOS
National Museum of Natural Sciences, Consejo Superior de
Investigaciones Cientificas, E-28006 Madrid, Spain

G. A. JONES
Woods Hole Oceanographic Institution, Wood Hole, MA 02543, USA

C. Z. KAAYA
Geology Department, University of Dar es Salaam, PO Box 35052,
Tanzania

K. KELTS
Limnological Research Center, University of Minnesota, Pillsbury Hall,
Minneapolis, MN 55455-0219, USA

T. KREUSER
Geologisches Institut, Universität Köln, Zülpicherstr. 49. D-5000 Köln 1,
Germany

P. KUMAR MAULIK
Geological Studies Unit, Indian Statistical Institute, 203 Barrackpore
Trunk Road, Calcutta, 700 035, India

M. LAMY AU ROUSSEAU
Centre des Sciences de la Terre, Université de Bourgogne, 6 Bd Gabriel,
F-21100 Dijon, France

A. LESLIE
Marine Geology, British Geological Survey, Murchison House, West
Mains Road, Edinburgh EH9 3LA, UK

C. O. LIMARINO
Faculted de Ciencias Exactas y Naturales, Universidad de Buenos Aires,
Dpto Geología, Pabellón II, Cuidad Universitaria, 1428 Buenos Aires,
Argentina

J. MARSHALL
Dept Geology, University of Southampton, Southampton S09 5NH, UK

A. MARTÍN-SERRANO
Instituto Tecnológico Geominero de España, Rios Rosas 23, E-28003
Madrid, Spain

M. MARZO
Dpto Geología Dinàmica, Geofisica i Paleontologia, Facultad de
Geologia, Zona Universitaria de Pedralbes, E-08028 Barcelona, Spain

K. MASTALERZ
Institute of Geological Sciences, University of Wroclaw, Cybulskiego 30,
PL 50-205 Wroclaw, Poland

R. MEDIAVILLA
Instituto Tecnológia Geominero de España, Cristobal Bordiú 235,
E-28003 Madrid, Spain

A. MELENDEZ
Facultad de Ciencias Departamento Geologia, Universidad de Zaragoza,
E-50009 Zaragoza, Spain

N. MELENDEZ
Departamento Estratigrafia, Facultad C C Geológicas, Universidad
Complutense, E-28040 Madrid, Spain

H. MULLINS
Department of Geology, Syracuse University, Syracuse, NY 13244, USA

A. MUÑOZ
Facultad de Ciencias, Departamento de Geologia, Universidad de
Zaragoza, E-50009 Zaragoza, Spain

M. NORMATI
Centre des Sciences de la Terre, Université de Bourgogne, 6 Bd Gabriel,
F-21000 Dijon, France

D. NURY
Laboratoire de Géologique structurale et appliquée, Université de
Provence, case postale 28, F-12221 Marseille cedex 3, France

S. ORDOÑEZ
Dpto Petrologia y Geoquímica, Facultad C. C. Geológicas, Universidad
Complutense, E-28040 Madrid, Spain

C. OVIATT
Department of Geology, Kansas State University, Manhattan, KS 66506,
USA

R. B. OWEN
Dept Geography, Hong Kong Baptist College, 224 Waterloo Road,
Kowloon, Hong Kong

G. PARDO
Faculted de Ciencias, Departmento de Geología, Universidad de
Zaragoza, E-50009 Zaragoza, Spain

J. PARNELL
Department of Geology, Queen's University, Belfast BT7 1NN, UK

A. PÉREZ
Facultad de Ciencias, Departamento de Geologia, Universidad de
Zaragoza, E-50009 Zaragoza, Spain

N. H. PLATT
Geo-Prakla, Ltd., Schlumberger House, Buckinghamshire Gate, Gatwick Airport, West Sussex RH6 0NZ, UK

E. RAMOS-GUERRERO
Departamento Geologia Dinàmica, Geofisica i Paleontologia, Facultad de Geologia, Zona Universitaria de Pedralbes, E-08028 Barcelona, Spain

R. W. RENAUT
Department of Geological Sciences, University of Saskatchewan, Saskatoon, Saskatchewan S7N 0W0, Canada

E. I. ROBBINS
US Geological Survey, Reston, VA 22092, USA

D. ROGERS
Lomond Associates, 48 West Regent Street, Glasgow G2 2RA, UK

D. K. RUDRA
Geological Studies Unit, Indian Statistical Institute, 203 Barrackpore Trunk Road, Calcutta 700 035, India

D. SACK
Dept Geography, University of Wisconsin, Madison, WI 53706, USA

M. SAGRI
Dipartamento Scienze della Terra, via La Pira 4, I-50121 Firenze, Italy

J. SALOMON
Centre des Sciences de la Terre et URA CNRS no 157, Université de Bourgogne, 6 Bd Gabriel, 21000 Dijon, France

J. I. SANTISTEBAN
Instituto Tecnológico Geominero, Cristobal Bordiú 35, E-28003 Madrid, Spain

T. SCHLÜTER
Department of Geology, University of Makerere, PO Box 7062, Kampala, Uganda

B. SHUKLA
Department of Geology, Queen's University, Belfast BT7 1NN, UK

R. SIMON-COINÇON
Laboratoire de Géographie Physique, CNRS, F-92195 Meudon, France

V. SKOCEK
Czech Geological Institute, Malostr., nam. 19, 118 21 Prague 1, Czech Republic

C. P. SLADEN
Exploration Overseas Sedimentology, British Petroleum plc, Britannic House, Moor Lane, London EC2Y 9BU, UK

P. N. SOUTHGATE
Division of Continental Geology, Bureau of Mineral Resources, PO Box 378, Canberra 2601, ACT, Australia

D. STEAD
Dept Geological Sciences, University of Saskatchewan, Saskatoon, Saskatchewan, S7N OWO, Canada

L. STEMMERIK
Geological Survey of Greenland, Øster Voldgade 10, DK-1350 Copenhagen K, Denmark

J. T. TELLER
Department of Geological Sciences, University of Manitoba, Winnipeg, Manitoba, R3T 2N2, Canada

D. A. TEXTORIS
Department of Geology, University of North Carolina, Chapel Hill, North Carolina 27599, USA

P. A. THAYER
Department of Earth Sciences, University of North Carolina at Wilmington, Wilmington, NC 28403, USA

M. THIRY
Ecole des Mines de Paris, 35 rue St Honoré, F-77305 Fontainebleau, France

N. TOUTIN-MORIN
Université d'Orléans, URA au CNRS no 1366, BP 6759, F-45067 Orléans Cedex 2, France

M. E. TUCKER
Department of Geological Sciences, University of Durham, South Road, Durham DH1 3LE, UK

J. P. VADOT
Centre des Sciences de la Terre et URA CNRS no 157, Université de Bourgogne, 6 Bd Gabriel, 21000 Dijon, France

B. VALERO GARCÉS
Facultad de Ciencias, Departamento de Geologia, Universidad de Zaragoza, E-50009 Zaragoza, Spain

J. VILLENA
Faculted de Ciencias, Departmento de Geología, Universidad de Zaragoza, E-50009 Zaragoza, Spain

J. K. WARREN
School of Applied Geology, Curtin University GPO Box U1987, Perth WA 6001, Australia

S. D. WEEDMAN
Geosciences Department, Pennsylvania State University, University Park, PA 16802, USA

R. W. WELLER
Department of Geology, Syracuse University, Syracuse, NY 13244, USA

D. WIRRMANN
Agreement ORSTROM, Universidad Mayor de San Andrés, La Paz, Bolivia

R. YURETICH
Department of Geology and Geography, University of Massachusetts, Amherst, Massachusetts 01003, USA

Preface

This volume series aims to provide concise summaries of information available on a global spectrum of lacustrine deposits, especially in terms of their paleoenvironmental interpretations. It is the outgrowth of the International Geological Correlation Program, Project 219 (Comparative Lacustrine Sedimentology in Space and Time) to understand the significance of lake deposits in the geological record. Initiated in 1984, IGCP-219 gave impetus to a new field called LIMNOGEOLOGY.

Numerous colleagues cooperated in this compilation effort, which has evolved through annual meetings and numerous special publications, listed below. Following the momentum, a successor project IGCP-324 (Global Paleoenvironmental Archives in Lacustrine Systems) continues to coordinate research on ancient lake records.

Lake sequences are of current interest as the high-resolution archives of changes in continental paleoenvironments. They are also becoming a focus of paleogeography as well as strategic exploration objects for hydrocarbons, rare elements, salts, ores, diatomite and other valuable economic resources. Considerable effort remains to develop consistent sedimentary models of diverse lacustrine systems and subfacies based on modern and ancient case studies, such as in this volume.

The IGCP-219 survey of lake basins worldwide shows distinct patterns in space and time which are linked to major phases of continental break-up, stretching, rifting and shearing. Lake basins are much more widespread in geological history than any textbook suggests. There are many ancient lacustrine giants with Caspian Sea proportions. Regional climatic patterns, coupled with tectonics, favored large lakes during certain intervals, such as the Permo-Carboniferous of Gondwanaland, the Lower Cretaceous of West Africa and Brazil, and the Pliocene of the Great Basin, USA. Paleocene–Eocene lake basins are widespread in China and in North America, forming as the result of transtensional tectonics. Opening of the North Atlantic Ocean produced a series of Upper Triassic–Lower Jurassic lacustrine rift basins along its margins.

Our goal in this volume series is to place the occurrence, geometry and type of lacustrine deposits within their geological context, with examples from each geological period. The IGP-219 strategy proposed five theme areas as a way of organizing the geological record

of lake basins and highlighting research opportunities. These provide the framework for this volume.

(1) Modern systems

Lakes are dynamic systems. We require the study of modern lakes to develop sedimentation models and environmental criteria. From ancient lakes we have only sediments; from existing lakes we can study production, sedimentation and preservation of the system components and the full range of biotic/abiotic interactions. Water levels, sedimentation and composition may change rapidly. In some cases, we must trace freshwater, brackish, saline or hypersaline facies over only cm-scale sequences, with incomplete sedimentological, biotic, or mineralogical evidence.

Coring and seismic profiling of modern lakes have provided valuable two- and three-dimensional views of facies architecture. Pollution studies from modern lakes have added insight on processes in freshwater systems. More information is needed from unusual lacustrine environments such as alkaline lakes, crater lakes or salt lakes. Their deposits are actually common in the geological record.

Models of hydrological evolution and mass balance should distinguish (a) lakes mainly receiving water from direct precipitation; (b) lakes dominated by drainage runoff; and (c) lakes controlled by groundwater systems. New analytical techniques seek to define ancient salinity, composition and pH from calcareous components of the geological record. We lack environmental range charts for lacustrine biota, sediment structures, carbonates and trace elements.

(2) Cenozoic lacustrine basins

Our present-day view of lake basins is dominated by high-latitude glacial lakes and low-latitude arid-zone lakes. The Tertiary record, however, holds evidence of widespread, vast lakes representing a spectrum of types. These lake deposits offer special opportunities to study closely the interaction of tectonics and sedimentation. The occurrence of lacustrine deposits within well-defined structural settings, such as rifts, pull-aparts, cratonic sags or craters provide opportunities to follow sedimentation, paleo-climate and tectonic histories.

Quaternary lake deposits provide the best continuous record of regional climate dynamics; a record that can be used to inspire, test or calibrate, global climatic models. Long sediment records drilled from Cenozoic lakes will aid land–sea telecorrelations. We need more studies to critically evaluate solar, orbital or tectonic rhythms reflected in lacustrine sequences. Further progress is needed on criteria for interpreting lacustrine subfacies from core material. Concomitant research themes concern diagenesis of clays, zeolites, silica, organic matter, sulfides and carbonates in lacustrine settings. The database of undeformed Cenozoic lacustrine deposits can be used, for example, to identify sites for future studies of paleoprecipitation of paleoaltitude using stable isotope sedimentology.

(3) Mesozoic lacustrine basins

The Mesozoic record of lake basins is more fragmentary but extended deposits document the early history of supercontinent rifting. Some lie beneath thick passive margin cover. The IGCP project highlighted (a) Triassic–Jurassic rift lakes of the North Atlantic margins; (b) the Eromango and Otway systems of Australia; (c) large-scale rift-associated deposits from the Lower Cretaceous, South Atlantic margin and central African paleo-lake province; and (d) Late Cretaceous lacustrine giants of China and central Asia. Interpretation of Mesozoic sequences are hampered by burial, tectonic overprints and less precise knowledge of paleogeography, geometry and basin morphology. Some challenging problems include paleoenvironment of basinal calcareous deposits, paleoecology of lacustrine faunal, floral and trace-fossil assemblages, as well as intra- and extrabasinal correlations.

(4) Paleozoic/Precambrian lacustrine basins

The intracontinental environmental history of Pangaea and Gondwana are reflected in widespread, but disjointed deposits preserved from Paleozoic lake basins. Research was encouraged to identify various lacustrine facies in Karoo deposits of Africa and similar sequences in Australia. Much progress has been made on the sedimentological interpretation of Devonian (Orcadian) and Carboniferous (Dinantian) lake deposits of Europe, some of which may source major oil fields.

Little remains of the Precambrian sedimentary record. Precambrian margins have been accreted, fragmented, subducted and reconfigured through time. We can perhaps postulate that undeformed Precambrian rift deposits are, in fact, lacustrine. Such deposits may derive from short-lived, aborted rifts from continental interiors, similar to thick mid-Cretaceous rift deposits in central Africa. A central problem is to define criteria which distinguish among Precambrian lacustrine and Precambrian marine sediments, in the absence of fossils. Progress has been made in Australia, northern Africa and North America using both sedimentological arguments and evidence from strontium isotopes. Several Precambrian deposits show characteristics of playa settings.

(5) Resources in lacustrine basins

Lacustrine basins may hold among the most prolific hydrocarbon reserves. Progress has been made in the evaluation of lacustrine basins in terms of their source potential but exploration models are lacking. Lacustrine salt deposits are mined in many areas of the world and include strategic trace salts (lithium, boron, arsenids, rare earths) as well as the common Na-, Mg- and K-salts. Some geochemical arguments support models of alkaline lakes as the host environments for the deposition of Precambrian synsedimentary sulfide ores. As an example of non-metallic resources, Tertiary lake sequences in southern Asia contain thick diatomites that are economically important as substitutes for asbestos fibers in housing insulation materials.

The over 70 contributions in this volume provide a taste of the richness and diversity of the global geological record from lake basins. Through the auspices of IGCP, a global community of geoscientists are discovering the potential for limnogeological research to unlock Earth secrets of ancient climates, non-marine environments, ecosystems and resources. As the effort progresses, we will see a striving towards coherent research and views of the plethora of ancient lakes.

We are pleased to acknowledge the continuing help and guidance of the Cambridge University Press editors, especially Catherine Flack. Beatrice Schwertfeger helped greatly with the IGCP organization. The task of collecting and reviewing such a large number of manuscripts is awesome, slow and only possible with the help of numerous dedicated reviewers and contributors. The voluntary task spread out from 1989. The compilation also suffered delays from lack of funding, which finally arrived from the Petroleum Research Fund of the American Chemical Society (24200-GB2). Clearly the support and encouragement of the International Geological Correlation Program was instrumental.

Elizabeth Gierlowski-Kordesch
Kerry Kelts
1993

Major IGCP 219/324 symposia

1985 Kraków, Poland – Comparative Lacustrine Sedimentology (International Association of Sedimentologists Regional Meeting)

1985 London, England – Lacustrine Petroleum Source Rocks (Geological Society of London Meeting)

1986 Canberra, Australia – Sedimentology of Lacustrine Giants (12th International Sedimentological Congress)

1987 Kehrsiten, Switzerland – The Phanerozoic Lake Record (IGCP 219 Workshop)

1987 Strasbourg, France – Paleoclimatic and Environmental Reconstructions (European Union of Geosciences IV)

1988 Beijing, China – Commparative Lacustrine Sedimentology (International Association of Sedimentologists Special Meeting)

1988 Barcelona, Spain – Lacustrine Facies Models in Rift Systems (IGCP 219 Workshop/Field Seminar)

1988 Wroclaw, Poland – Non-marine Permian Rotliegendes Lacustrine Basins (IGCP 219 Workshop/Field Seminar)

1989 Strasbourg, France – Isotope Sedimentology of Lacustrine Deposits (European Union of Geosciences V)

1989 Washington, D.C. – Lacustrine Deposits as Records of Paleoenvironments (28th International Geological Congress)

1990 Orkney Isles, Scotland – The Orcadian Basin, a Devonian Lacustrine Giant (IGCP 219 Field Seminar)

1990 Nottingham, England – Lacustrine Sedimentation (13th International Sedimentological Congress)

1990 Lake Tahoe, California – Large Lakes and their Stratigraphic Record (Geological Society of America Penrose Conference)

1991 Ankara, Turkey – Neogene Lacustrine Strike-Slip Basins (IGCP 324 Annual Meeting)

1991 Saskatoon, Canada – Sedimentary and Paleolimnological Records of Saline Lakes

1991 Beijing, China – Quaternary Lakes and Global Change (XIII Int. Congress, International Union of Quaternary Research)

1992 Madrid, Spain – Geochemical Signals in Lacustrine Sequences (IGCP 324 Annual Meeting)

1992 Salamanca, Spain – Lacustrine Sedimentation and Facies Analysis (III Congreso Geológico de España)

1992 Kyoto, Japan – Climatic vs. Tectonic Controls on Lake Sedimentation (29th International Geological Congress)

1993 Marrakesh, Morocco – Lacustrine Sedimentation (International Association of Sedimentologists Regional Meeting)

1993 University Park, USA – Climatic and Tectonic Rhythms in Lake Deposits (SEPM Annual Mid-year Meeting)

IGCP 219/324 publications

1988 *Lacustrine Petroleum Source Rocks*, eds. A. Fleet, K. Kelts & M. Talbot, Geological Society of London Special Publication No. 40, Blackwell, Oxford, 391 pp.

Twenty-eight contributions dealing with general lacustrine systems, paleoenvironmental indicators, organic geochemistry and 15 case-studies of lacustrine basins.

1989 *The Phanerozoic Record of Lacustrine Basins and their Environmental Signals*, eds. M. Talbot & K. Kelts *Palaeogeography, Palaeoclimatology, Palaeoecology*, vol. 70, no. 1–3, 304 pp.

Twenty-two contributions dealing with facies models, stable isotope sedimentology, environmental sequence analysis and resource/regional studies.

1991 *Lacustrine Facies Analysis*, eds. P. Anadón, L. Cabrera & K. Kelts, International Association of Sedimentologists (IAS) Special Publication No. 13, Blackwell, Oxford, 318 pp.

Fifteen contributions dealing with sedimentation and tectonics, natural resources, modern processes in East African lakes, rhythmic stratigraphy and geochemistry, and organic remains.

1992 *Chinese Lacustrine Basins*, eds. K. Kelts, Yang Sheng & P. Anadón, Special Publication of China Earth Science, vol. 3–4, Intergeos, The Netherlands.

Twelve contributions on lacustrine basins and sediments in China.

1993 *Sedimentology and Geochemistry of Modern and Ancient Saline Lakes*, eds. R.W. Renaut & W.M. Last, SEPM Special Publication No. 50. (In press.)

Twenty-six contributions on various aspects of saline lakes and their sedimentation records.

1993 *Sedimentary and Paleolimnological Records of Saline Lakes*, ed. M. S. Evans, *Journal of Paleolimnology*, vol. 8, no. 2, pp. 97–169.

Six papers covering aspects of the biology, chemistry, and climatic record of saline lakes.

1994 *Climatic and Tectonic Rhythms in Lake Deposits*. Special Issue. *Journal of Paleolimnology*, vol. 11, in press.

Seven papers on lake deposits and the signatures of climatic and tectonic overprints.

Introduction

ELIZABETH GIERLOWSKI-KORDESCH[1] AND KERRY KELTS[2]

[1] Department of Geological Sciences, Ohio University, Athens, OH 45701-2979 USA
[2] Limnological Research Center, University of Minnesota, Pillsbury Hall, Minneapolis, MN 55455-0219 USA

Limnogeology

A. Forel (1892–1902) helped establish the scientific discipline of limnology with his integrated study of biology, chemistry, circulation and sedimentation in modern Lake Geneva. Subsequently, the study of modern lakes in Europe evolved largely separate from geology. Lake deposits were not a central feature of mountain belts. This contrasts with the integrated development of marine sciences where questions of the modern ocean and sediments were linked by a common tradition and logistics. Classic texts with the limnological viewpoint include: F. Forel (1901), F. Ruttner (1963), G.E. Hutchinson (1957), D. Frey (1974), R. Wetzel (1983) and F. Taub (1984). Strakhov (1970) and Pia (1933) are among a few dealing with lithology of modern lake deposits.

The geology of ancient calcareous lake deposits was already an important theme in Lyell's *Principles of Geology* (1830), and marl and varve lake deposits were key elements of the Ice Age hypothesis. Limnogeology was, however, given its greatest boost with the exploration of the American West; the US Geological Surveys of the Great Basin and their discovery of widespread Tertiary and Quaternary non-marine outcrops (e.g. Davis, 1882; Russell, 1885). In his masterpiece monograph on Pleistocene Lake Bonneville, G.K. Gilbert (1890) laid out the principles of facies analyses of lacustrine shoreline deposits and delta formation. He used lake deposits to show the vast extent of late Pleistocene freshwater lakes as evidence of glacial-age climate changes, to show the catastrophic demise of the large lake and to argue isostatic tectonic rebound from the loading by a water body. Subfacies analogs were derived from modern Lake Michigan, and calcareous deposits were used to reconstruct and even predict lake-level curves. This tradition and style led to a broad geological interest in lacustrine deposits and processes through the early 1900s in North America (e.g. Davis, 1901; Grabau, 1924; Bradley, 1929). The Eocene Green River Formation became the schoolbook example of a lacustrine giant with immense oil shale reserves. Major lake deposits were eventually recognized as components of Triassic, Jurassic, Cretaceous and Tertiary basins (Feth, 1964; Picard & High, 1972).

System

A lake may be defined as an inland body of standing water occupying a depression in the Earth's crust. As such, lakes can exhibit a wide range of possible settings, sizes, chemistries, concentrations and morphologies. The Greek *limne* at the root of limnology and limnogeology, means marsh, lake or pool and by implication has been generally applied to freshwater environments. Thirteen of the world's 40 largest lakes, and innumerable smaller ones are, however, without outlets and commonly quite saline. Of the largest lakes, only 20 are deeper than 400 m but these hold most of the world's fresh lake water (e.g. 23,000 km^3 in Lake Baikal).

Essential conditions for a lake are simply a topographical depression, and a hydrological balance (input–output) that is adequate to support surface water (Gilbert, 1890). The hydrological balance, in turn, is a function of the prevailing climate which is influenced by latitude-dependent zonal winds, geography, altitude, continentality and orbital parameters (Fig. 1; Street-Perrott & Harrison, 1984; Kutzbach & Street-Perrott, 1985). Global moisture and lake distribution, therefore, can be radically altered by changing plate positions (Parrish & Barron, 1986). Depending on the balance of input versus evaporation, a lake may be considered hydrologically open or closed. An open system is characterized by relatively stable shorelines, and a limited residence time for solutes. Open systems may also be linked in cascading chains, such as the Pleistocene pluvial lakes of Panamint Valley, California (Eugster & Hardie, 1978).

The source of water for lakes is precipitation. Rainwater composition is a function of distance to oceans, wind strengths and evolution of the vapor. Near oceans, rain may contain 60 ppm TDS (Total Dissolved Solids) with seawater ratios. Continental interiors may have less than 6 ppm TDS, or much more, of a very different ionic ratio than seawater (Eugster & Hardie, 1978). Thus the starting point for rainwater to a basin may differ. Isotopic compositions are directly affected. A lake with drinking water (e.g. 600 ppm TDS) may show oxygen isotopic values of extreme evaporation that are the same as hypersaline brine in another region.

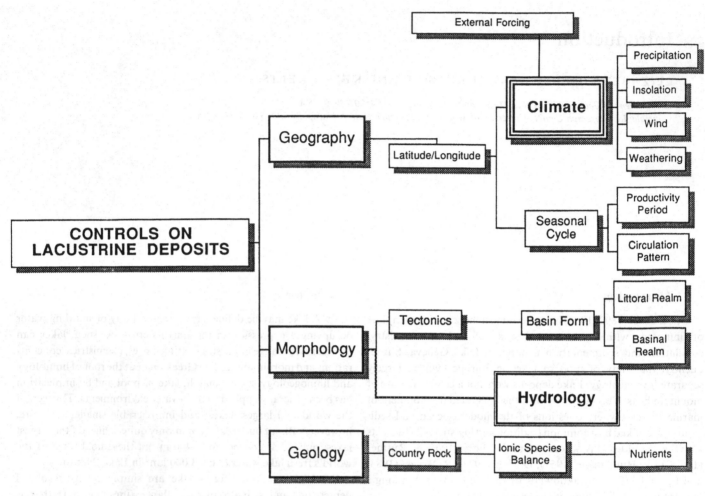

Fig. 1. General controls on lacustrine sedimentation (after Glenn & Kelts, 1991).

Lake water solute concentrations may range over five orders of magnitude, from dilute, monsoonal rainwater (e.g. 10 ppm) to viscous chloride brines of 500,000 mg/kg.

One main difference between marine and lacustrine systems is that the initial ionic balance of lake waters is not constant. Salinity and composition are not the same. Mobilization of elements from the drainage basin geology by weathering determines the initial ionic ratios of most lakes fed by surface water (Eugster & Hardie, 1978). This fingerprint is already present in the early dilute to freshwater phases but becomes particularly important as concentration increases. In closed basins, the concentration of brines may fluctuate widely, but the chemical type will remain relatively uniform. As brines evolve, a main control is exerted by the calcium/carbonate ionic ratio and the early precipitation of calcium carbonates (calcite, aragonite; Eugster & Hardie, 1978). If the calcium ion concentration is less than that of the carbonate ion, it may be depleted before significant brine concentration is reached, and thus reduce the buffering capacity of a lake. Without buffering components, lakes may have pH ranges from less than 1 in some sulphuric acid-rich Japanese crater lakes to more than 11.0 in alkaline brines of several East African Rift lakes. There are, of course, concomitant

effects in the corresponding ecosystems. Eugster & Hardie (1978) provided a chemical classification based on solute species from modern lakes.

Solute chemistry is but one variable used to measure variations between the ocean and lakes. Table 1 is a schematic attempt to summarize some aspects in comparisons of lacustrine versus marine environments. Major differences are due to the environmental sensitivity of lakes to their regional setting and the accompanying biological diversity. This is in contrast to the marine environment, where foraminifera and calcareous nannoplankton are the most abundant basinal, calcareous components; there are few planktonic calcareous fossils in lake deposits. Most fine-grained lacustrine basinal micrite is an inorganic precipitate, induced by biological or physical-chemical processes (Kelts & Hsü, 1978). Along marginal areas, oncolite, bioherm and stromatolitic carbonates are more prevalent than in marine environments. Charophyte chalks are mostly lacustrine. Benthic ostracods, molluscs and gastropods may form significant beds. Fish, molluscs, ostracods, gastropods and other limnic fauna are particularly subject to endemism. There are very few systematic paleoecological studies of past non-marine environments. Picard & High (1972, 1981) concluded that there are

Table 1 *Contrasting aspects for lacustrine versus marine environments of deposition*

Aspect	Lacustrine	Marine
Aqueous reservoir	Limited, variable	Immense, uniform
Chemistry	Highly variable, ionic species function of drainage basin geology and climate	Uniform Na-Cl
Salinity	Highly variable, 10^1–10^5 mg TDS/1	Uniform around 35‰
pH	Variable 1.5–11.0	8.3 in surface waters 7.7 in bottom waters
Size	Highly variable, 1 up to 80,000 km³ today	Immense
Sediment rates	0.1–2 m/ty rapid	0.001–0.35 m/ty modest
Tectonics	Event basins, sag or rifts + fault control	Sea floor spreading. Continental margin subsidence
Geodynamics	Includes altitude variations, drainage capture, sudden changes	Sea level, epirogeny; slower changes
Climate control	Zonal latitude dependent	
Climate change	Immediate, drastic response; level changes, and composition: tens of years	Long-term response; 1000s of years
Residence time	1–1000 yr	1000 yr +
Cycles	Annual, sun-spot, short-term climate, Milankovitch	Long-term climate, paleooceanography, Milankovitch
Tide	No tides, seasonal level variations	Tidal-dominated
Organic matter	Algae/bacteria; land plants. Type I common	Marine algae, or land plant. Type II & II
Productivity	Very high, high nutrient	Modest, upwelling zones
Preservation potential	High with high sed. rates. Anoxia, low sulphate common	Requires high sed. rates, or anoxia
Bacteria/algae	Special adaptations. Photochemotrophs	Marine
Paleontology		
Silica microfossils	Diatoms dominant since Eocene, sponges	Diatoms since late Cretaceous, radiolaria
Calc. microfossils	Rare pelagic calcareous	Forams, Nannofossils dominant
Benthic microfossils	Ostracods. Local endemism	Forams. Worldwide index
Dinoflagellates	Abundant, but few cysts preserved	Cysts preserved
Macrofossils	Micromammals, reptiles, fish (kills), insects, chironomids, non-marine molluscs	Marine invertebrates
Littoral/shelf	No corals; molluscs, stromatolites common, algal bioherms, charophyte chalks	Coral reefs. calc. algae, molluscs, subtidal, marine chara rare
Offshore	Macrofossils scarce	
Bioturbation	Worms, insects, vertebrates. No tiering. Few deep burrowers	Numerous burrowers, common tiering
Facies		
Evaporites	Derived, evaporitic concentration variable types, reworked, thin, fractionated basins	Marine, fixed sequence, giants may be kms thick
Carbonates	No barrier reefs, no calcareous plankton oozes, mostly chemical, dolomite common Stromatolites, algal bioherms	Mostly biogenic, calcite dominant
Oolites	Saline and brackish lakes	Turbulent margins
Silica	Abiotic chert common	Biogenic chert common
Basinal	Anoxia common	Anoxia not common
Deltas	Short-term, rapid variance response to level changes	Long-term stability
Turbidites	Common in dilute waters	Rare events
Sands	Fan-delta complexes, alluvial	Clastic shorelines, beach
Transgression/regression	Very short period	Long period phenomena
Stratigraphy	Rapid facies change laterally and vertically	Walthers' law, transitional
Life span	< 1 мa is old, up to 35 мa	1–100 мa
Biomarkers	Botryococcane, bacterial common	

(After Kelts, 1988.)

few criteria which by themselves are sufficient to indicate lacustrine deposits. A combination of positive and negative evidence is needed. Hardie (1984) weighs sedimentological and geochemical signatures that separate marine and non-marine evaporites.

Origin of lake basins

Lakes are formed by numerous mechanisms that have been classified by Hutchinson (1957) after Davis (1882) into 76 main types. For geological purposes we may simplify this scheme by grouping them into just three categories: event basins, paralic basins and tectonic basins.

Event lake basins are formed by short-term processes and thus less likely to preserve thick lake deposits in the geological record. These include meteorite and volcanic craters, landslide and other dams, glacial lakes, kettle-holes, karst sink-holes, groundwater lakes and river meanders. Paralic lake basins include cut-off marine embayments and shoreline depressions which are controlled by sea level fluctuations. Tectonic lake basins are most commonly encountered in the geological record. These may be broad regional sags, foreland deeps, orogenic-collapse basins, or rifts and strike-slip basins. Pull-apart basins and continental rift separations have the maximum preservation potential because of their rapid subsidence and long histories (Biddle & Christie-Blick, 1985). Ancient lacustrine basin deposits often record the conditions during the first stages of continental break-up.

Multi-dimensional lacustrine lattice

Various classifications of lake basin deposits as well as descriptions of sedimentologic aspects of ancient lake systems are spread over various sources (e.g. Bradley, 1929; Anderson & Kirkland, 1960; Picard & High, 1972, 1981; Schuiling 1977; Eugster & Hardie, 1978; Hardie, Smoot, & Eugster, 1978; Matter & Tucker, 1978; Dean & Fouch, 1983; Eugster & Kelts, 1983; Hardie, 1984; Hsü & Kelts, 1984; Allen & Collinson, 1986; Frostick *et al.*, 1986; Kelts, 1988; Fleet, Kelts, & Talbot, 1988; Talbot & Kelts, 1989; Katz, 1990; Glenn & Kelts, 1991; Platt & Wright, 1991; Anadón, Cabrera & Kelts, 1991). An all encompassing sedimentary model for lacustrine depositional systems is difficult to build because the spectrum of lake types, both modern and ancient, is indeed quite broad, because Nature paints in shades of grey. Commonly, classification of lake deposits are based on features of modern lakes which are not directly seen or measured in outcrop, such as whether a system is hydrologically open versus closed. Furthermore, subfacies (subenvironments) have been defined directly from specific geomorphological models, such as saline lake/playa versus deep, freshwater, turbidite-basin lake. These are fundamental aspects of a given lake, but this limits our ability to deal with the dynamic sequence changes found in many systems. Discussion is needed on lake facies development in terms of a continuum of processes, and to reduce the number of special types. Ideally, the set of defining parameters should be applied to all lakes and deposits, and derived

from features that can be described at the outcrop level. What elements do all lakes have in common?

At the outcrop level, sedimentology descriptions determine the dominance of clastic versus chemical or biological processes. Factors which can be estimated include: (1) the energy of deposition and hydraulic features, (2) the stability of the system including probable lake level fluctuations, (3) geomorphological gradients, (4) evidence of biotic/abiotic interactions and (5) concentration and compositional information (e.g. index minerals or structures). For example, the presence of fine-grained aragonitic or dolomitic muds may indicate elevated Mg/Ca ratios, suggesting the evolution of brines; trona or sodium carbonate pseudomorphs indicate the dominance of continental alkaline brines. Each subenvironment can be examined by these criteria.

We can view lacustrine subenvironments as fundamental subdivisions in almost every lake, albeit with different distribution patterns or degree of development. We should not fall prey to stereotypes, for instance, equating lake water concentrations with lake levels, or forgetting that both playas and deep, freshwater lakes may expose huge, low-gradient mudflats. Parameters such as changing residence time and mass balances, better define sedimentation than merely whether a lake was open or closed. As lakes evolve, thresholds are crossed which change the character of sediments. For example, *Artemia salina* thrive between 7–29% TDS, leaving characteristic fecal pellet piles. With increasing concentration, the density and viscosity of water changes with concomitant effects on the distribution of suspended sediment. Numerous primary parameters are interconnected, yet require independent evaluation.

High productivity leads to anoxia and formation of laminations which may be characteristic of either shallow saline or deep freshwater lakes. We suggest an approach stemming from the geomorphological subdivisions: supralittoral, eulittoral, intralittoral, sublittoral, slope and basinal (or pelagic) realms (Fig. 2). Fundamental to lake shorelines is the dominance of seasonal changes in contrast to lunar tidal effects for marine environments. Annual fluctuations in levels may alternately expose and submerge narrow to extensive aprons (eulittoral). These are commonly subjected to strong chemical and redox gradients and diagenetic potential. On low gradients, the extent of the eulittoral zone may be tens of kilometers. Examples include large playas, the Caspian Sea, Great Salt Lakes or even Lake Michigan. Alluvial fan deltas or stream facies interfinger with various littoral subfacies. Each lake type may have beaches, platform, lags, shoal, lagoons, deltas, marsh deposits and mudflats. Carbonate benches, bars, spits or coquinas are common in sublittoral regions. Intralittoral describes the nearshore zone which is permanently subaqueous on an annual scale. Pelagic sedimentation into quiet realms at many depths or salinities will lead to basinal sediments which are seasonally laminated, unless perturbed. Turbid suspensions from inflowing waters may be carried long distances as turbidity currents in dilute lakes, but overflow and settle out from low-salinity wedges over denser waters.

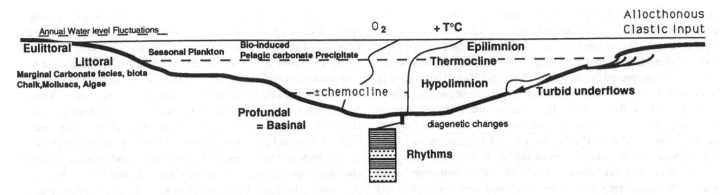

Fig. 2. Major depositional regimes of lake systems ((cf. Hutchinson, 1957); Kelts, 1988; Glenn & Kelts, 1991).

A combination of salt pan evaporites with ephemeral, cracked floodplain muds, salt casts and dunes are considered index features for a saline playa, but might also occur along the lagoons of a major brackish lake. Stromatolites are no longer automatic criteria for high solute concentrations. We encourage researchers to explore new ways to view sedimentary features in lacustrine systems in terms of a continuum within an n-dimensional lattice of different processes: clastic, chemical, biological activity, biota, gradient, depth, energy, water composition, water concentration, residence time, productivity, anoxia, stratification, solar cycles or wind. Each is linked in various ways to the climatic and tectonic regimes.

Time windows – tectonic factor

The geological record of major lacustrine basins with hydrocarbon potential shows a preference for certain periods with a favorable combination of climate, paleogeography and structural setting (Smith, 1991). Paleogeographic reconstructions (Ziegler et al., 1985) are the starting point of basin analysis. From the Paleozoic, widespread Devonian dark lacustrine source beds are associated with continental, semi-arid basins near the locus of an opening Iapetus Ocean. Shallow Carboniferous lakes of Britain persisted in humid climate, with abundant land-derived plant matter. The configuration of Gondwanaland in the Permian provides the setting for widespread, fluvial lacustrine, giant lowland lake deposits in a taiga-like, cool-temperate climate. These include the Permian Gippsland and Cooper Basin area of Australia (Powell, 1986) and Karoo of southern Africa as well as the Parana oil shale basin of South America. The Permian hypersaline Zechstein of Europe and the Delaware Basin of Texas behaved as giant saline lakes in arid climates. Triassic lacustrine basins are common along the east coast concomitant with North Atlantic rifting. Many show playa characteristics (Demicco & Gierlowski-Kordesch, 1986). Manspeizer (1988) attributes these partly to rift-associated orographic rain shadows. Africa has Upper Jurassic semi-arid lake basins, whereas Jurassic fluvial lacustrine floodplain facies occur in Eromango Basin, Australia (Powell, 1986) consistent with its more temperate latitude. Lower Cretaceous rift lake basins occur parallel with Gondwana break-up along South Atlantic margins (Brice et

al., 1982) and the interior Central African Rift (Fairhead, 1986) as well as in the Otway Basin Australia (Smith, 1991). The giant Songliao inland lake in Middle Cretaceous, China (Yang & Gao, 1985) was a dominantly humid zone feature as were the lake basins of the Bohai region. Eocene and Oligocene lacustrine source rocks of China are more associated with arid climates. The Eocene giant Uinta Complex, USA was in a subtropical, semi-arid environment with strong seasonality similar to the conditions expected in other Great Basin Tertiary lakes, or the Tertiary Qaidan Basin of interior China. More tropical wooded conditions surrounded lakes of Miocene Sumatra and Thailand (Gibling et al., 1985). Studies of Pleistocene evaporite lakes (Hardie, 1984; Warren, 1986) and East African rift lakes (Rosendahl, 1986) have provided important paleoclimatic information as well as facies models for such basins in the past.

Climate influence

Lacustrine sedimentary sequences are important as archives of continental and global climate history. High rates of accumulation and the absence of bioturbation suggest that sediments from some anoxic lake basins may preserve particularly refined records of past climate. The key is learning to read the criteria, and to correlate trends over long distances.

It is now well-established that mid- to low-latitude lacustrine basins can be sensitive and reliable recorders of even quite subtle climatic changes. Continental-scale compilations of lake-level variations during the Late Quaternary (Street-Perrott & Roberts, 1983; Street-Perrott & Harrison, 1984, 1985), demonstrate that precipitation/evaporation changes have often been synchronous on a continental or intercontinental scale, and confirm that they represent regional expressions of global climatic events. Both the scale and rate of these climatically-induced changes have generally been sufficient to override tectonic influences on lacustrine sedimentation.

Tropical and temperate lakes reflect fundamental differences in their climatic forcing which are transferred to their sedimentation patterns. Whereas the changes in physical conditions directly control the biota of higher latitude lakes (temperature, ice, solar), tropical lakes show less physical variation, and tend to reflect the controls imparted by the biotic interactions. Cyclic changes in the level of tropical lakes do, however, reflect isolation forcing of low latitude climate (Kutzbach & Street-Perrott, 1985). Milankovitch-type cycles, suspected by Bradley (1929) and Van Houten (1962), have now been identified in ancient lacustrine sequences (Olsen, 1986; Fischer & Roberts, 1991). Furthermore, although there is obviously a tectonic bias in the distribution of lakes over the non-glacial areas of the earth, the nature and occurrence of inland waters bears a clear relationship to global climatic belts.

Lake deposits potentially carry paleoclimatic information on two scales. Medium-term variations (millenial-scale) are probably related to astronomic forcing of global climate. Such variations are reflected in changes in the character of lacustrine sediments on a lamina-by-lamina to tens of meters scale.

A broader-scale record of global climatic belts and their spatial relationship to the major continental plates, is reflected in the nature and distribution of water bodies occupying continental depressions. Long-term changes in the climate of the basins, due to plate movement, or displacement or disruption of climatic belts (e.g. due to uplift of topographic barriers), will be reflected in sedimentary sequences that are hundreds to thousands of meters thick and span millions of years.

Lacustrine life cycles

Lake deposits often display rhythmic sedimentation on different scales, from sub-millimeter to kilometers. In Fig. 3, hypothetical depositional sequences portray millenial-scale rhythmic or cyclic development of lacustrine facies in the geological record (after Glenn & Kelts, 1991). Each represents vertical successions caused by variations in climate, water level, chemistry and clastic input.

Siliciclastic lake models commonly depict laminated silt or mud coarsening up to sand, leading eventually to deltaic sands and gravels. Such straightforward successions are rarely observed. Lakes seem more likely to simply silt up rather than become infilled by the inward migration of marginal clastic facies. (See reviews in Picard & High, 1972, 1981; Allen & Collinson, 1986; Cohen, 1989.)

Closed lake basins occur at all latitudes and are ideal settings for the development of transgressive-regressive cycles because such basins respond very sensitively to variations in evaporation, rainfall and alluvial input, and their shorelines thus tend to move rapidly and frequently. Rates of draw-down have commonly been so rapid that a progression from dilute and open to shallow and eventually hypersaline facies may be preserved in a few centimeters of pelagic carbonates. Alternatively, thick carbonate bioherms marking paleoshorelines, for example in the Green River Formation, may represent only a few tens or hundreds of years. Sedimentation cycles produced by repetitive inflow and desiccation in such settings may commonly contain unconformities produced by evaporite dissolution and fluvial erosion following evaporative draw-downs.

Open lakes with permanent outlets tend to have stable shorelines. Sedimentation in such lakes may be dominated by clastic input or, if protected from such inputs, by chemical and biochemical sedimentation. Cores from temperate perialpine settings provide a good example of proglacial to chemical sedimentation (Hsü & Kelts, 1984). The lower 30 m sequence from Lake Zürich is dominated by fine-grained glacial sediments characterized by laminated clay-silt varves deposited during freeze-thaw conditions, respectively. These overlie older glacial tills with dropstones. Sedimentation rates for clastic varves are usually very high, typically centimeters per year. Following glacial retreat, as meltwater clastic influx subsides with glacier retreat, new vegetation also hinders soil erosion so that chemically-precipitated lacustrine chalk and marl facies become dominant. These may be laminated or varved and very organic-rich if basinal areas are nearly anoxic.

Organic carbon contents provide one sensitive monitor of changing lake conditions. Whereas organic carbon-rich pelagic sediments are a typical feature of many modern dilute lakes due to seasonal productivity blooms and water column stratification, algal productivity and organic carbon preservation may reach a maximum in the brackish to saline brines of shallow closed lakes experiencing evaporative draw-down (Kelts, 1988). Cyanobacterial mats bordering ephemeral lakes may also contribute significant organic matter to sediments. Little is known about the significance of chemotrophic bacterial mats in deeper water.

Accumulation rates vary drastically both within and among different facies as illustrated in Fig. 3. Estimates of the temporal periodicity combine evidence from sections with greatly differing sedimentation rates. It would be incorrect, for example, to assume that the glacial–interglacial cycle, as represented in Fig. 3C, represents a long cold period followed by a relatively brief warm period of interglacial chemical sedimentation, as it would likewise be erroneous to assume on the basis of the relative thicknesses in Fig. 3A that deep lake phases uniformly alternate with shallow/playa phases.

GGLAB concept

The Global Geological Record of Lake Basins (GGLAB) encompasses more than a collection of papers. Research on lake deposits has been basin-specific, with a diversity of approaches, criteria and classifications. Participants of IGCP 219 and 324 foresaw the need for a central database of ancient lacustrine deposits. Data from this worldwide compilation could be used to recognize depositional patterns through time and to identify unique sequences for paleoenvironmental analyses. The first step towards coherency is to provide an inventory forum on state-of-knowledge of lacustrine sequences from different perspectives, as a means to encourage more comprehensive syntheses. This book is the first in a series of publications in which lake scientists can publish summaries of past and present research on ancient lake basins.

We envisage long-term projects linked with a growing global inventory of lake basins and their deposits. These include formulation of subfacies models for lacustrine depositional patterns or, for example, identification of evolutionary patterns in non-marine biota and ecosystems. Some sequences will aid correlations among

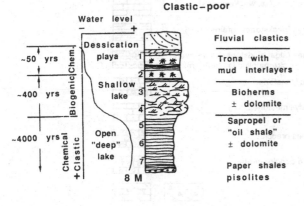

**A CLOSED SALT LAKE SYSTEM
BRACKISH TO ALKALINE LAKE**
Clastic–poor

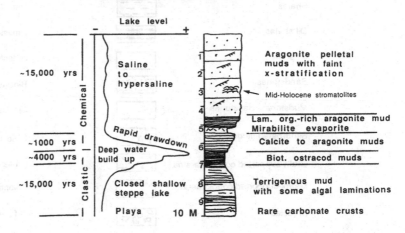

**B CLOSED TO OPEN SYSTEM
FRESH TO SALINE LAKE**

] 1 meter

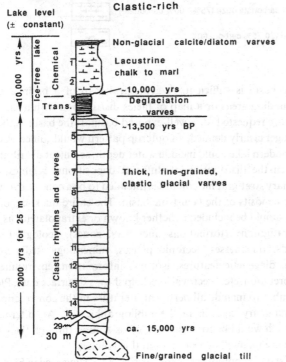

**C OPEN SYSTEM
DEEP GLACIAL/INTERGLACIAL LAKE**
Clastic-rich

Fig. 3. Hypothetical comparative development for three types of millennial-scale lacustrine rhythms driven by climatic change; emphasis on abrupt changes in lake levels, sedimentation, and accumulation (after Glenn and Kelts, 1991). A: Closed, shallow, clastic-poor salt system in which a perennial alkaline lake dries out. B: Sequence stratigraphy from closed basin filling stage, through open-system overflow and flushing stage, followed by abrupt down-draw to a perennial hypersaline lake.
C: Deglaciation sequence from open-system proglacial lake to interglacial chalk-facies lakes.

1982; Parrish & Barron, 1986). New tools applied to sequence analyses, such as stable isotope sedimentology, will help define rate, source and magnitude of paleoenvironmental change.

The preparation of this volume has begun to encourage researchers to view their work on a global scale, and to seek a common sedimentological language as a basis on which to compare non-marine deposits. The step beyond is aimed at a computer compatible GGLAB database. An open invitation is extended to all lake scientists to participate in this global collection of data. This database will be interactive; paleogeographic maps showing locations of near contemporaneous lake basins will serve as windows to reveal underlying data on each individual lake deposit. Contributions covering lake deposits within specific 'time slices' are requested. The GGLAB computer database also includes data from lake basins in addition to those from GGLAB volume series. Literature research is underway to incorporate little-studied lake sequences. This is accompanied by the compilation of a bibliography of lake deposits which currently encompasses over 3000 references.

The database follows the 1986 Geological Society of America, DNAG convention for the geologic time scale. Basin references are indexed by a temporal and a spatial code. Time is located by the Era code (PC,Pz,Mz,Cz,Q) followed by a period abbreviation (Mio, Plio, K,J-Tr, Dev, etc). The world is divided into eight major regions (EUR, NAMER, SAMER, ASIA, AFR, ME, AUS, CHINA) plus the polar and oceanic regions. Each basin is thus

non-marine and marine deposits, and others will help diagnose the extent of control on sedimentation by tectonic or climatic parameters. A major use of the database will center on the reconstruction of regional paleoclimate (paleoprecipitation, paleohydrologic) patterns for specific paleogeographic configurations. Because the present distribution of lakes is so clearly related to global rainfall patterns, such a compilation will have considerable potential for testing and refining reconstructions of atmospheric circulation under varying continental configurations (e.g. Parrish & Curtis,

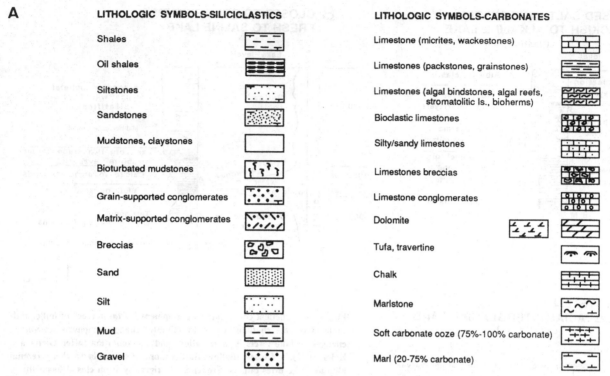

A

LITHOLOGIC SYMBOLS-SILICICLASTICS

Shales

Oil shales

Siltstones

Sandstones

Mudstones, claystones

Bioturbated mudstones

Grain-supported conglomerates

Matrix-supported conglomerates

Breccias

Sand

Silt

Mud

Gravel

LITHOLOGIC SYMBOLS-CARBONATES

Limestone (micrites, wackestones)

Limestones (packstones, grainstones)

Limestones (algal bindstones, algal reefs, stromatolitic ls., bioherms)

Bioclastic limestones

Silty/sandy limestones

Limestones breccias

Limestone conglomerates

Dolomite

Tufa, travertine

Chalk

Marlstone

Soft carbonate ooze (75%-100% carbonate)

Marl (20-75% carbonate)

Fig. 4 (*see next page for caption*)

quickly located by a code comprising the major region and a modifier for the country, state or subregion (e.g. E, N, S, W) following postal conventions. For example the Green River Fm. is indexed as Cz, Eo and NAMER,WY.

GGLAB guidelines for future editions

The spectrum of modern lake types in the world reflects the wide variability possible in the geological record. Unlike marine deposits, lake deposits are extremely variable in facies patterns, geochemistry and depositional histories (Table 1). A simple database with numbered categories cannot begin to tabulate the spectrum of possibilities in sediment type, chemical signatures, faunal and floral combinations and other factors. GGLAB is thus conceived as a datafile in graphic and text formats, so that information on a specified topic can be found in context with other geological data. The format for GGLAB contributions is thus very open. In general, each contribution should be a concise summary of the information available. Five main components should be included: (1) a locality map, (2) a summary stratigraphic column, (3) detailed lithologic columns of the lacustrine sequences, (4) a bibliographic list, and (5) a short explanatory text on paleoenvironments.

The locality map should include a clear outline map of the lacustrine basin or lake. At least one longitude and one latitude measurement is requested. Also, the latitude and longitude of the approximate midpoint of the lake or lake deposit is needed for exact positioning on a paleogeographic map. Scale and north arrow are indispensable. The type of map for an ancient lake deposit depends on surface exposure. If the data is subsurface, then a general outline

of the basin is sufficient. A geological map of a basin and its surrounding areas or a map of facies distribution within the lake basin are requested for a well-exposed basin. If the basin outline is no longer clearly defined, an outcrop pattern would suffice. A map of a modern lake could include water depth, sediment distribution, or even the distribution of different depositional regimes. The summary stratigraphic column is intended to be a concise overview of the deposits of the lacustrine basin. The entire thickness of the basin should be included (whether known or extrapolated), as well as stratigraphy (formations, members, zones), geological time periods, thicknesses, tectonic phases, mineralogy, sedimentary cycles, diagenetic features, isotopic information, environmental interpretation, etc. Electrical logs help show subsurface data. Please remember to include all pertinent literature in the construction of this summary column in the bibliographic list. An alternative approach would be to illustrate a cross-section through the basin. There are many good examples in this volume.

A list of symbols of rock type, fossils, etc. is suggested in Figs. 4A, B, C and abbreviations in Fig. 5. A standard legend throughout the book would save space and lend uniformity to lake sequence representations and facilitate comparisons. Most contributors in this volume chose their own legends for the sake of convenience. We encourage future contributors to consider using the GGLAB legend when drawing stratigraphic and lithologic columns.

Age determinations on lake deposits are often difficult and quite inexact. The dating method used for a particular lake sequence can be paleontologic, isotopic, paleomagnetic, lithostratigraphic or a combination of methods. It is important to establish the reliability of a dating method when correlating non-marine (and marine)

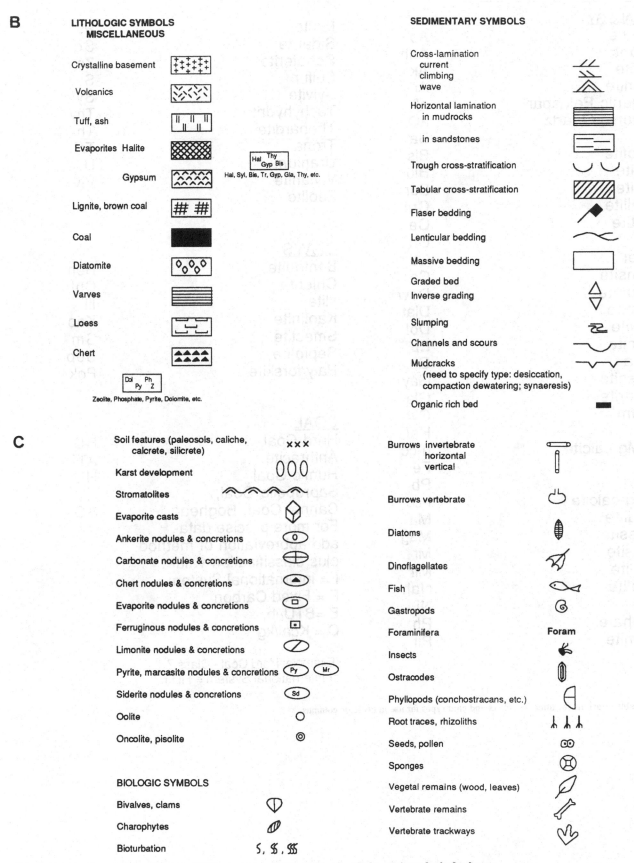

B

**LITHOLOGIC SYMBOLS
MISCELLANEOUS**

Crystalline basement

Volcanics

Tuff, ash

Evaporites Halite

Gypsum

Hal, Syl, Bis, Tr, Gyp, Gla, Thy, etc.

Lignite, brown coal

Coal

Diatomite

Varves

Loess

Chert

Zeolite, Phosphate, Pyrite, Dolomite, etc.

SEDIMENTARY SYMBOLS

Cross-lamination
    current
    climbing
    wave

Horizontal lamination
    in mudrocks

    in sandstones

Trough cross-stratification

Tabular cross-stratification

Flaser bedding

Lenticular bedding

Massive bedding

Graded bed
Inverse grading

Slumping

Channels and scours

Mudcracks
    (need to specify type: desiccation,
    compaction dewatering; synaeresis)

Organic rich bed

C

Soil features (paleosols, caliche,
    calcrete, silicrete)

Karst development

Stromatolites

Evaporite casts

Ankerite nodules & concretions

Carbonate nodules & concretions

Chert nodules & concretions

Evaporite nodules & concretions

Ferruginous nodules & concretions

Limonite nodules & concretions

Pyrite, marcasite nodules & concretions

Siderite nodules & concretions

Oolite

Oncolite, pisolite

BIOLOGIC SYMBOLS

Bivalves, clams

Charophytes

Bioturbation

Burrows invertebrate
    horizontal
    vertical

Burrows vertebrate

Diatoms

Dinoflagellates

Fish

Gastropods

Foraminifera   **Foram**

Insects

Ostracodes

Phyllopods (conchostracans, etc.)

Root traces, rhizoliths

Seeds, pollen

Sponges

Vegetal remains (wood, leaves)

Vertebrate remains

Vertebrate trackways

Fig. 4. Suggested symbols for lithology (A/B), structures (B/C), and fossils (C) for use in geological columns.

MINERALOGY

Analcime	Ac
Anhydrite	Anh
Ankerite	Ak
Aragonite	Ar
authigenic Feldspar	aF
authigenic Quartz	aQ
Barite	Ba
Bischofite	Bis
Bloedite	Blo
Burkeite	Bur
Carnallite	Crn
Celestite	Ce
Chert	Ch
Copper	Cu
Corrensite	Cor
Dawsonite	Daw
Diatomite	Diat
Dolomite	Dol
Epsomite	Ep
Galena	Gl
Gaylussite	Gay
Glauberite	Gla
Gypsum	Gyp
Halite	Hal
high Mg-calcite	h-cc
Iron	Fe
Lead	Pb
low Mg-calcite	l-cc
Magadiite	Ma
Magnesite	Mag
Marcasite	Mr
Mirabilite	Mir
Nahcolite	Nah
Nickel	Ni
Phosphate	Ph
Pirssonite	Pir

Pyrite	Py
Siderite	Sd
Sphalerite	Zn
Sulfur	S
Sylvite	Syl
Tachyhydrite	Thy
Thenardite	The
Trona	Tr
Uranium	U
Vivianite	Viv
Zeolite	Z

CLAYS

Bentonite	Ben
Chlorite	Chl
Illite	Ill
Kaolinite	Kao
Smectite	Sm
Sepiolite	Sep
Palygorskite	Pgk

COAL

Hard Coal	HD
Anthracite	AT
Humic Coal	HU
Sapropelic Coal,	
Cannel Coal, Boghead	SO

For more precise data-
add abbreviation of method
plus classification value:
I = International System
F = Fixed Carbon
B = BTU/lb.
C = Kcal/kg

(Example: Hard Coal - Class 7
of International System = HD I.7)

Fig. 5. Suggested abbreviations for minerals, clays and coal types for use in geologic columns.

sediments over long distances for paleogeographic purposes. Please include a reliability index: (A) complete biostratigraphy with fossil zonal indices of both the top and bottom time division present; (B) some biostratigraphic information with some zonal fossils; (C) stratigraphic interpolation – zonally useful fossils present in underlying and/or overlying strata; (D) geologic inferences – unit can be correlated lithologically with other dated units; (E) radiometric determination with about $+5\%$ error; (F) secondary guesswork – for example, from regional correlations; and (G) guesswork.

The detailed lithologic columns should be dedicated to the detailed representation of the lacustrine sequences found within a particular basin. For ancient basins, detailed geological columns should include rock type, grain size, sedimentary structures, diagenetic structure, unusual mineralogies, and fossils. We strongly recommend use of a double column, one for lithology and one for sedimentary structures, with width of columns dependent on grain size. Comments on paleoenvironmental interpretation, sedimentary cycles and tectonic phases should be kept separate from the data. For modern lake contributions, core data, seismic sections, or some other important aspect, such as fauna or chemistry, can be included here.

The bibliographic list should contain all relevant references on the stratigraphy, paleontology, tectonics, geochemistry and sedimentology of the lake deposit. Modern lake references should contain summary information on biology, sediments and chemistry. Any interpretation should have a literature citation (interpretations have been known to change with time).

The text itself should be a short and concise summary (average length 4–6 pages) of information on facies descriptions and interpretations, paleontology, dating procedures, tectonic situation of the lake basin, isotopic data or other relevant information. Various examples are found in this volume. Again, each contribution will be unique because the amount and kind of information for each lake deposit or lake sediment are dependent on the lake type as well as the progress of the scientific investigations. As a guide to possible research opportunities, as well as important information to consider while compiling a GGLAB contribution, refer to the original IGCP 219 Global Lacustrine Inventory Questionnaire (Fig. 6A). Explanatory notes to the Questionnaire and the Geological Society of America DNAG time scale are given in Figs. 6B and 6C.

Further details on the preparation of GGLAB contributions may be obtained from the editors.

Patterns in this volume

This volume is organized with the first seven papers as initial syntheses of process studies or regions. Sladen's analysis of lacustrine facies and hydrocarbons provides a guide to the importance of a differentiated understanding of lacustrine basin subfacies in exploration strategies. Cerling explains aspects of chemical diversity. The Greenland synthesis of Dam and Stemmerik illustrates a typical example of multiple lacustrine episodes recurring in a tectonically active region. Four papers treat combined basinal

stratigraphy respectively of Mesozoic–Cenozoic Zaïre, southern France and Duero, Spain. Most of the contributions follow the strategy of the database and summarize a lacustrine basin deposit within a convenient geologic time interval and one of the eight global regions. The chronostratigraphic control on the deposits varies widely. The reader should keep in mind that typical sedimentation rates for lacustrine basins are 0.1 to 1 mm/yr, and a 1000 m section rarely represents more than a million years. Because climate variations may change a lake from fresh to hypersaline within a few thousand years (a few cm of sediment), we must be cautious in the interpretation of whole geological stages in terms of the occurrence of a few meters of lacustrine evaporites.

Surprisingly, a large number of the reported deposits include playa sedimentation. Lacustrine dolomite is common. Partly, this may reflect criteria and confidence for recognition of lacustrine deposits. Pseudomorphs after trona, and other sediment features of alkaline lakes, are among the most conclusive evidence of lacustrine deposition in Precambrian and lower Paleozoic basins which lack diagnostic fossil assemblages.

Carboniferous to Permian deposits are characteristically rich in references to shallow freshwater facies, abundant land-plant debris, reworked detrital material and non-marine limestone.

Triassic–Jurassic examples tend to congregate around the North Atlantic margins related to incipient rifting and multiple deep and shallow lake episodes concomitant with changing continental climates. Playa-alluvial stages are common and diagnostic, along with thinly laminated, organic carbon-rich deepwater calcareous shales.

The four Early Cretaceous examples are from the Iberian peninsula and display variable compositions and deep and shallow water facies. These tend to reflect fossil-rich, subtropical to tropical conditions which are related to equable climate, Tethyan ocean circulation and the incipient opening of the South Atlantic. The Late Cretaceous examples tend to represent paludal carbonate sequences along a broad, stable platform margin to the Tethyan and Alpine sea.

Four Paleocene–Eocene examples are from southern Europe for a time characterized by more arid episodes with evaporites and carbonates. Concurrently, the Bighorn Basin in Wyoming, USA, was characterized by an extensive freshwater, paludal carbonate system, leading to vast coal resources.

Oligocene examples are diverse. Varied lacustrine carbonate, chert and organic carbon-rich facies occur in Somalia. Oil shales are important in Northern Ireland. Thin calcareous beds, some evaporitic, occur in the Swiss foreland basin, and thick rhythmic lacustrine limestones are characteristic of several Oligocene basin deposits of Spain.

Seven of the Miocene–Pliocene contributions deal with southern Europe, whereas only one is from northwest India and illustrates mainly clastic deposition in an intermontane basin. Neogene deposits of Iberia occur in strike-slip basins and comprise varied chemical sediments, some rhythmically bedded, which include evaporites, oil shales, dolomite, aragonitic paper-shales, lignitic limestones and even phosphates.

The nine examples from the Quaternary are spatially and

A

IGCP PROJECT-219   GLOBAL LACUSTRINE DEPOSITS INVENTORY

DEPOSIT:BasInName: _____

1. Country_____ 2. Geogr. Reg._____ 3. _____ 4. my? _____

5. Formation,member,Group_____6. Thickness_____m

7. Geogr.Locat._____ 8. Long._____Lat._____

9. Dating Method_____Reliability A B C D E F G. Ref?___
10. Index Fauna_____
11 .Index Flora_____
12. BIOTA:

13. Tectonic Setting: Rift___Strike Slip___Fault___Craton___Glacial___Volc___ Meteor___Other_____
14. Est.Max +Min size of paleolake: km_____ AreaKm^2_____ depth m____
15. 0pen_____ or Closed _____ basin? Varied? _____ Est.Salinity ranges? _____
16. Fresh___ Brackish___Saline___Penesaline___Hypersaline___Dry___ (Alk___)
17. Facies: Terrig-Dominated___Carbonate-Dominated___Evaporite-Dominated___ Cyclic___Other?___
    Playa____ Paludal____ Paralic____ Shallow water___ Deep water___Cyclic___ Other?_____
18. Environmentel Index features:
19.Clastics (C, S, M, T) Carbonate (L, D, S, Marl) Evap (G, H, K, ...) Biogen (P, C, Corg, Dia, Cq)

20. Min/Biog._____
_____

21. Sed. Struct:_____
_____

22. Facies Types in Outcrops ___ and/or Cores___: brief description.

    Marginal_____

    Basinal_____

23.Diagenetic Features_____

24. Isotopic Info.ref._____

25. 0rganic Geochem _____

26. Economic Potenial._____
27. Outcrop Condition: Undeform__ Tilted__ Folded__ Thrust__ Sheared__ High.Deform__ Meta_____
28. Lithology of Source area: Xtln___ Acid Pluton___ Mafic Volc___ Carb___ Sh+SS___? _____

29.Ref. #1_____

_____

Ref.#2_____

_____

Ref.#3_____

                                            Address_____
                                            _____

Reporter_____   _____Tel:_____

Fig. 6A. Questionnaire for initial survey by IGCP 219 to collect information for global inventory of lake deposits.

B

Information on IGCP-219 INVENTORY Questionnaire –GLOBAL LACUSTRINE DEPOSITS

The main purpose of the data sheet is to collect infomation on major lacustrine deposits in a compact format (one page) for entry into a computer data base. This will be used for an IGCP-219 Monograph to assess a global view of the significance of lacustrine deposits. Of course lacustrine systems are dynamic and the information cannot be complete. Choices and flexibility are necessary. Ideally, fill in the items that apply to a deposit of your interest and use the other items as a source of thought simulation. Several forms can be used for the same deposit if different aspects (facies changes, age ranges, different researchers, unconformities, etc.) wish to be stressed. The data bank is created with the program Microsoft File for an Apple Macintosh with Hyperdrive, and can be easily modified to improve content.

1. Country with the International Posal Code (See Listing attached)
2. Broad Geographic Regions: 1) ANT-Antartica 2) AUS-Australia 3) CAN-Canada 4) CHI-China 5) EUR Europe toTurkey 6) IND-India tolran 7) NAF-Northern Africa and Mid-east 8) SAF- Southern Africa 9) SAM- Latin America S+C 10) SEA-South East Asia 11) RUS- FSU 12) USA Oceania to nearest large region.
3.+ 4. Ages should include Epoch(s) and Stage(s) with absolute age (my) range according to DNAG Geol.Soc. Amer. Time Scale (Geology, Sept. 1983). (See Listing Attached)
5. Formation name with Group and Member if available.
6. Thickness estimated of lacustrine facies only if possible to separate from fluvial and alluvial deposits.
7. Location information for others to quickly find on a 1:1 mio. scale map.
8. Longitude and Latitude can be related to an approximate mid-point of the deposit.
9. Dating method: paleontologlcal(ostracod,palynol), paleomagnetic, isotope,...

Reliability   A) Complete Biostratigraphy with Fossil zonal indices of both top and bottom of time   division present.

                B) Some Biostratigraphic information with some zonal Fossils.
                C) Stratigraphic Interpolation--Zonally useful fossils present in underlying and or overlying strata (eventually marine deposits)
                D) Geologic Inferrence--Unit can be correlated lithologically with others dated
                E) Radiometric Determination with about $\pm$ 5% error.
                F) Secondary Information---eg. From regional compilations
                G) Guesswork   GIVE REFERENCES

12. Common groups of fossils occurrences, other than index - of ecological importance.
13. Strike slip includes pull-apart. Fault for normal or thrust fault-caused basins. Craton refers to warps or sags, or general subsidence eg. L. Victoria or Green River Formation.
14. Deposit of major interest are those with around 1000 + km^2.
15. For closed basins, salinities fluctuate rapidly. Index minerals and biota help locate the range.
16. Major deposits may include several.
19. Circle appropriate categorie(s) for Background Sediment: (C-coarse, S-sandy, M-Mudstone, Clayey, T-Till, L-Limestone, D-Dolomite, Marl, S-Siderite.
    Evaporites (G-Gypsum, Anhydrite, H-Halite, K-potassium chlorides,)ior
    Biogene (C-Coal, P-Peat, Corg - oil-shale, Dia-diatomites, Cq-coquina)
20. What authigenic mineral or biogenic components are present that narrow the possible environments of deposition. eg: zeolites, carbonate phase, Phos, Pyrite, Chert, Sulphur, Trona, Sepiolite, etc.
    What fauna/flora occurs that is an environmental index?
21. What sediment structures occur that indicate environments, Gilbert deltas, Flood sands, desiccation cracks, debris flows, turbidites, ripples, dunes, x-lam, varves.
22. Sedimentological evidence visible from outcrop or cores (note) - margln-oolites, stromatolites, pisolites, bioherms, roots, channels, etc.
    Basin - Homog.muds, laminated varves, current reworking, turbidites, index layers, bullseye, etc.
23. Diagenetic cements, concretions, breccias, etc.
24. Are there stable or radiogenic isotope analyses for carbonates, or organic matter? What value ranges? <u>Reference?</u>
25. Have there been analyses of organic matter? Aquatic or torrestrial? Pyrolysis? Bio Markers? Reference?
26. Economic occurrences of hydrocarbon (coal, oil shale, etc.) mineral salts (which)? Diatomite? or other? either exploited or suspected.
27. We would like to locate areas that present good field excursion possibilities for lacustrine basin analyses.
28. We suggest you use formats according to the attached listing.
- additional informaion on stratigraphy, fossils, organic matter, etc. can be added to the reverse side. Simple index maps and cross or vertical sections would greatly enhance the information.

Fig. 6B. Explanatory notes to original questionnaire.

C

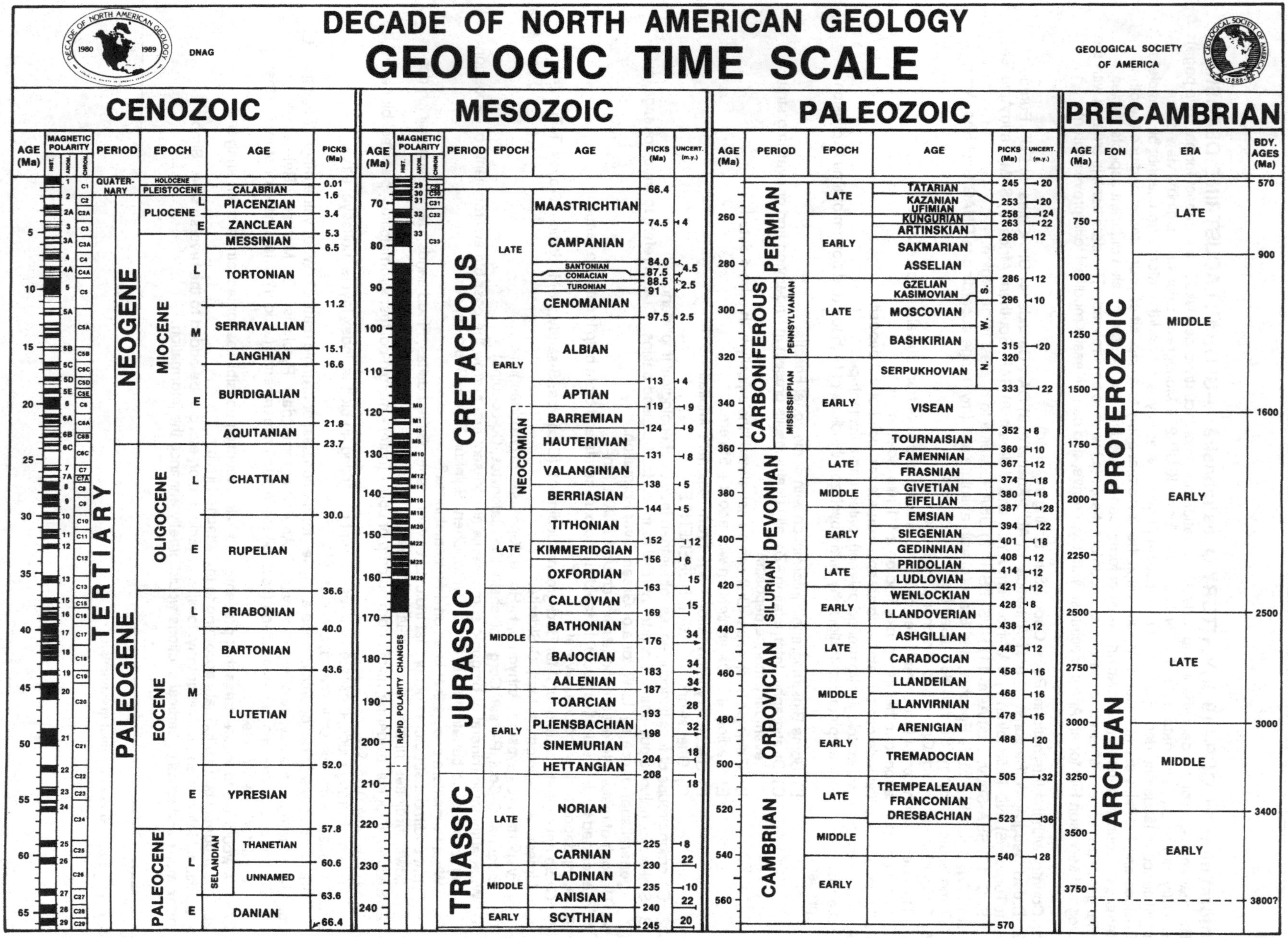

DECADE OF NORTH AMERICAN GEOLOGY
GEOLOGIC TIME SCALE

DNAG 1980 1989

GEOLOGICAL SOCIETY OF AMERICA

CENOZOIC

AGE (Ma)	MAGNETIC POLARITY	PERIOD	EPOCH	AGE	PICKS (Ma)
1	C1	QUATER-NARY	HOLOCENE		0.01
2	C2		PLEISTOCENE	CALABRIAN	1.6
2A	C2A		PLIOCENE L	PIACENZIAN	3.4
5	C3		E	ZANCLEAN	5.3
	3A			MESSINIAN	6.5
	C3A				
	4				
	C4				
10	4A		MIOCENE L	TORTONIAN	
	C4A				11.2
	5				
	C5				
	5A				
15	C5A		M	SERRAVALLIAN	
	5B C5B				15.1
	5C C5C			LANGHIAN	16.6
	5D C5D				
	5E C5E		E	BURDIGALIAN	
20	6 C6				
	6A C6A				21.8
	6B C6B			AQUITANIAN	23.7
	6C C6C				
25	7 C7		L	CHATTIAN	
	7A C7A				
	8 C8				
	9 C9				30.0
30	10 C10		E	RUPELIAN	
	11 C11				
	12 C12				
35	13 C13				36.6
	15 C15		L	PRIABONIAN	
	16 C16				
40	17 C17				40.0
	18 C18			BARTONIAN	43.6
	19 C19				
45	20 C20		M	LUTETIAN	
	21 C21				
50					52.0
	22 C22				
	23 C23		E	YPRESIAN	
55	24 C24				57.8
	25 C25			THANETIAN	
60	26 C26			UNNAMED	60.6
	27 C27				63.6
65	28 C28		E	DANIAN	
	29 C29				66.4

Periods: NEOGENE, TERTIARY, PALEOGENE. Epochs: OLIGOCENE, EOCENE, PALEOCENE. SELANDIAN.

MESOZOIC

AGE (Ma)	MAGNETIC POLARITY	PERIOD	EPOCH	AGE	PICKS (Ma)	UNCERT. (m.y.)
	29 C29				66.4	
70	30 C30			MAASTRICHTIAN		
	31 C31				74.5	4
80	32 C32		LATE	CAMPANIAN		
	33 C33			SANTONIAN	84.0	4.5
				CONIACIAN	87.5	
90				TURONIAN	88.5	2.5
					91	
				CENOMANIAN		
100					97.5	2.5
			EARLY	ALBIAN		
110						
				APTIAN	113	4
120	M0				119	9
	M1			BARREMIAN	124	9
	M3			HAUTERIVIAN		
	M5				131	
130	M10			VALANGINIAN		
	M12				138	5
140	M14			BERRIASIAN		
	M16				144	5
	M18					
150	M20			TITHONIAN	152	12
	M22		LATE	KIMMERIDGIAN	156	6
	M25			OXFORDIAN		
160	M29				163	15
				CALLOVIAN	169	15
170				BATHONIAN		
			MIDDLE		176	34
180				BAJOCIAN	183	34
				AALENIAN	187	34
190				TOARCIAN	193	28
			EARLY	PLIENSBACHIAN	198	32
200				SINEMURIAN	204	18
210				HETTANGIAN	208	18
220			LATE	NORIAN		18
					225	8
230				CARNIAN	230	22
			MIDDLE	LADINIAN	235	10
240				ANISIAN	240	22
			EARLY	SCYTHIAN	245	20

Periods: CRETACEOUS, JURASSIC, TRIASSIC. NEOCOMIAN. RAPID POLARITY CHANGES.

PALEOZOIC

AGE (Ma)	PERIOD	EPOCH	AGE	PICKS (Ma)	UNCERT. (m.y.)
			TATARIAN	245	20
		LATE	KAZANIAN	253	20
260	PERMIAN		UFIMIAN		
			KUNGURIAN	258	24
			ARTINSKIAN	263	22
280		EARLY	SAKMARIAN	268	12
			ASSELIAN		
			GZELIAN	286	12
300		LATE	KASIMOVIAN	296	10
			MOSCOVIAN		
320	CARBONIFEROUS		BASHKIRIAN	315	20
			SERPUKHOVIAN	320	
340		EARLY	VISEAN	333	22
			TOURNAISIAN	352	8
360				360	10
		LATE	FAMENNIAN	367	12
			FRASNIAN	374	18
380	DEVONIAN	MIDDLE	GIVETIAN	380	18
			EIFELIAN	387	28
			EMSIAN	394	22
400		EARLY	SIEGENIAN	401	18
			GEDINNIAN	408	12
		LATE	PRIDOLIAN	414	12
420	SILURIAN		LUDLOVIAN	421	12
		EARLY	WENLOCKIAN	428	8
440			LLANDOVERIAN	438	12
		LATE	ASHGILLIAN	448	12
460			CARADOCIAN	458	16
	ORDOVICIAN	MIDDLE	LLANDEILAN	468	16
			LLANVIRNIAN	478	16
480		EARLY	ARENIGIAN	488	20
500			TREMADOCIAN		
		LATE	TREMPEALEAUAN	505	32
520			FRANCONIAN		
			DRESBACHIAN	523	36
	CAMBRIAN	MIDDLE			
540				540	28
		EARLY			
560				560	
				570	

Pennsylvanian, Mississippian. N.W.S., N.S.

PRECAMBRIAN

AGE (Ma)	EON	ERA	BDY. AGES (Ma)
			570
750		LATE	
			900
1000	PROTEROZOIC		
1250		MIDDLE	1000
1500			
			1600
1750		EARLY	
2000			
2250			
2500			2500
2750		LATE	
3000	ARCHEAN		3000
3250		MIDDLE	3400
3500			
3750		EARLY	3800?

Fig. 6C. Decade of North American Geology 1983, Geologic Time Scale, Geological Society of America, Boulder, CO 80301, USA.

sedimentologically diverse. Two are from saline examples of the East African half-grabens. Lake Agassiz was an immense proglacial lake of the North American ice sheet, with vast deposits of fine gray mud and subtle shorelines. These were coincident with the transition of Lake Bonneville from a vast freshwater lake to the modern prennial shallow salt lake. Stromatolitic carbonates are a feature of lakes on the limestone substrates of southern France. South America is represented by two examples of salars or playa systems whereas Katmandu Basin fill is characterized by the erosion products of a rising Himalayan chain.

The eight Holocene examples illustrate a variety of saline-playa types including the Coorong of Australia, and several from British Columbia. Lake Titicaca is a brackish deep water lake with rapid level changes whereas Lake Balaton fluctuates from a nearly desiccated mud playa to a very shallow freshwater chalk system. The Finger Lakes provide examples of typical sequence stratigraphy in deep periglacial lakes.

The contributions which make up this Volume 1 of GGLAB represents only a preliminary and limited sampling of the numbers of lacustrine basins. It is premature for time-interval syntheses although some patterns are emerging. Examples range from Precambrian to Recent, but they do not include many of the vast lacustrine deposits of China, Africa, Russia and former Soviet Asia, or South America. At this stage in our understanding, we chose to have the geoscientists present individual basin studies with differing styles, detail and emphasis. These provide examples to encourage further progress, innovative approaches and compilations from other regions. Only with a broad global coverage will temporal and spatial patterns emerge with clarity.

What next

As part of the current efforts of IGCP-324, Global Paleoenvironmental Archives in Lake Systems (GLOPALS) focus on criteria and methodology applied to lacustrine sequences in order to understand the high-resolution response of our natural system to external forcing throughout the geological record.

Understanding of the lacustrine record will require more process studies to improve interpretations of geochemical and biological signatures in terms of paleoclimate or subfacies development in lake basins. We foresee increased applications of geochemical stratigraphy (isotopes of carbon, oxygen, strontium and others) to define paleoenvironment from lake basin records as archives of global changes.

The study of lacustrine sequences will continue to challenge the limits of time resolution for non-marine systems. Much progress is needed to refine the correlation of lake deposits within time windows. Improved time-synchronous patterns of regional climates can lead to validation of global models of past climate change or tectonic position.

We have not yet developed common criteria for lacustrine facies and subfacies recognition similar in scope to marine deposits. Serious terminology problems exist with regards to diverse lithological facies; what is meant by offstep/onlap sequences in lacustrine

system; how do we understand shoreline desiccation processes; what are the possibilities for pseudomorphs and crystalline molds in lacustrine sequences; what kind of special sedimentary structures identify subfacies, for example, marginal playa deposits or basin margin floodplain? Do we even have proper terminologies for the subfacies in the important zone where lake waters fluctuate on an annual scale, covering the marginal upper littoral apron with seasonal flooding? Comparative sedimentology in modern lake basins remains central for interpreting the past records. The study of modern lakes thus complements the interests of Quaternary paleolimnologists or paleoecologists seeking to understand environmental dynamics from lacustrine deposits.

Lakes are sensitive pieces of the Earth system which require an increasingly multidisciplinary and global approach to understand the importance of their high resolution records. Because lakes represent a spectrum, and change rapidly in response to environmental pressures, it is necessary to view them in a unified lattice which considers geological settings, chronology, salinity concentration, depth, chemistry and biota as parts of system trends and patterns.

A major goal is to tie non-marine paleoenvironment records closer to coeval marine history by comparing the reconstructed paleogeography within time intervals. A key step is to merge the land record with the global paleogeographic database (Ziegler et al., 1985). In order to achieve these results it will be necessary to stimulate the application of new techniques for dating, stable isotope geochemistry, paleoaltimetry, paleomagnetism and even global paleogeography and paleoceanography. Future volumes of GGLAB will reflect the progress and include more regional syntheses. It is hoped that our large gaps from the basins of Asia, South America and Africa can soon be rectified.

Acknowledgements

Portions of this introduction derive from collaboration with our friends Mike Talbot, Blas Valero, Craig Glenn, Lluis Cabrera, and Pere Anadón. Discussions on GGLAB have been integral to all of the IGCP-219 meetings, thanks to the IGCP participants. Special thanks to Fred Ziegler, Dave Rowley and staff of the Paleogeographic Atlas Project for encouragement with the GGLAB database structure. Beatrice Schwertfeger aided with illustrations.

Bibliography

Allen, P.A. & Collinson, J.D., 1986. Lakes. In *Sedimentary Environments and Facies*, ed. H.G. Reading, 2nd ed., pp. 63–94. Blackwell, Oxford.

Anadón, P., Cabrera, L. & Kelts, K. (eds.), 1991. *Lacustrine Facies Analysis*. Int. Assoc. Sediment. (IAS) Spec. Publ. No. 13. Blackwell, Oxford.

Anderson, R.Y. & Kirkland, D.W., 1960. Origin, varves, and cycles of Jurassic Todilto Formation, New Mexico. *Amer. Assoc. Petrol. Geol. Bull.*, **44**, 37–52.

Biddle, K.T. & Christie-Blick, N., 1985. *Strike-slip Deformation, Basin Formation, and Sedimentation*. Soc. Econ. Paleontol. Mineral. Spec. Publ., 37.

Bradley, W.H. 1929. The varves and climate of the Green River epoch. *U.S. Geol. Surv. Prof. Pap.*, **158**, 87–110.

Brice, S., Cochran, M.D., Pardo, G. & Edwards, A.D., 1982. Tectonics and sedimentation of the South Atlantic rift sequence, Cabinda, Angola. In *Studies in Continental Margin Geology*. ed. J.S. Watkins & C.L. Drake, pp. 5–18. Am. Assoc. Pet. Geol. Mem., 34.

Cohen, A.S., 1989. Facies relationships and sedimentation in large rift lakes and implications for hydrocarbon exploration: examples from Lake Turkana and Tanganyika. *Palaeogeogr., Palaeoclimatol., Palaeoecol., 70*, 65–80.

Davis, C.A., 1901. A second contribution to the natural history of marl. *J. Geol., 8*, 491–506.

Davis, W.M., 1882. On the classification of lake basins. *Proc. Boston Soc. Nat. Hist., 21*, 315–81.

Dean, W.E. & Fouch, T.D., 1983. Lacustrine environment. In *Carbonate Depositional Environments*, ed. P. Scholle, D. Bebout & C. Moore, pp. 97–130. Amer. Assoc. Petrol. Geol. Memoir 33.

Demicco, R.V. & Gierlowski-Kordesch, E., 1986. Facies sequences of a semi-arid closed basin: the Lower Jurassic East Berlin Formation of the Hartford Basin, New England, USA. *Sedimentology, 33*(1), 107–18.

Eugster, H. & Hardie, L.A., 1978. Saline Lakes. In *Lakes, Chemistry, Geology, Physics*, ed. A. Lerman, pp. 239–93. Springer, NY.

Eugster, H.P. & Kelts, K. 1983. Lacustrine chemical sediments. In *Chemical Sediments and Geomorphology*, ed. A. Goudie & K. Pye, pp. 321–68. Academic Press, London.

Fairhead, J.D. 1986. Geophysical controls on sedimentation within the African Rift systems. In *Sedimentation in the African Rifts*, ed. L.E. Frostick, R.W. Renaut, I. Reid & J.J. Tiercelin, pp. 19–27. Geol. Soc. Spec. Publ., 25.

Feth, J.H., 1964. *Review and Annotated Bibliography of Ancient Lake Deposits (Precambrian to Pleistocene) in the Western States*. U.S. Geol. Survey Bull. 1080.

Fischer, A.G. & Roberts, L.T. 1991. Cyclicity in the Green River Formation (Lacustrine Eocene) of Wyoming. *J. Sed. Petrol., 61*(7), 1146–54.

Fleet, A.J., Kelts, K. & Talbot, M.R. (Eds.), 1988. *Lacustrine Petroleum Source Rocks*. Geol. Soc. Spec. Publ., 40.

Forel, F.A., 1901. *Handbuch der Seenkunde*. Allgemeine Limnologie, Stuttgart.

Frey, D.G., 1974. Paleolimnology. *Mitt. Int. Ver. Limnol., 20*, 95–123.

Frostick, L.E., Renaut, R.W., Reid, I. & Tiercelin, J.-J. (Eds.), 1986. *Sedimentation in the African Rifts*. Geol. Soc. Spec. Publ., 25.

Gibling, M.R., Charn, T., Wutti, U. Theerapongs, T. & Mungkorn, H., 1985. Oil shale sedimentology and geochemistry in Cenozoic Mae Sot Basin, Thailand. *Bull. Amer. Assoc. Petrol. Geol. 69*, 767–80.

Gilbert, G.K., 1890. *Lake Bonneville*. U.S. Geol. Surv. Monograph 1.

Glenn, C. & Kelts, K., 1991. Sedimentary rhythms in lake deposits. In *Cycles and Events in Stratigraphy*, ed. G. Einsele, W. Ricken & A. Seilacher, pp. 188–221. Springer Verlag, Berlin.

Grabau, A.W., 1924. *Principles of Stratigraphy*. 1960 Dover Editions, NY.

Hardie, L.A., 1984. Evaporites: marine or non-marine. *Amer. J. Sci., 234*, 193–240.

Hardie, L.A., Smoot, J.P. & Eugster, H.P., 1978. Saline lakes and their deposits: a sedimentological approach. In *Modern and Ancient Lake Sediments*, ed. A. Matter & M.E. Tucker, pp. 7–41. Spec. Publs. Int. Assoc. Sediment 2.

Hsü, K. & Kelts, K. (Eds.), 1984. Quaternary geology of Lake Zurich: an interdisciplinary investigation by deep lake drilling. *Contrib. Sedimentol., 13*, 1–210.

Hutchinson, G.E., 1957. *A Treatise on Limnology*. I, *Geography, Physics and Chemistry*. John. Wiley & Sons, NY.

Katz, B.J., 1990. *Lacustrine Basin Exploration: Case Studies and Modern Analogs*. Amer. Assoc. Petrol. Geol. Memoir 50.

Kelts, K., 1988. Environments of deposition of lacustrine petroleum source rocks: an introduction. In *Lacustrine Petroleum Source Rocks*, ed. A. Fleet, K. Kelts & M. Talbot, pp. 3–26. Geol. Soc. Spec. Publ., 40.

Kelts, K.R. & Hsü, K.J., 1978. Freshwater carbonate sedimentation. In *Lakes – Chemistry, Geology, Physics*, ed. A. Lerman, pp. 295–323. Springer, NY.

Kutzbach, J.E. & Street-Perrott, F.A., 1985. Milankovitch forcing of fluctuations in the level of tropical lakes from 18 to 0 kyr B.P. *Nature, 317*, 130–4.

Lyell, Ch., 1830. *Principles of Geology*. Vol. 1. J. Murray, London,

Manspeizer, W. (Ed.), 1988. *Triassic Jurassic Rifting, Continental Break-up and the Origin of the Opening of the Atlantic Ocean and Passive Margins*. Elsevier, Amsterdam.

Matter, A. & Tucker, M.E. (Eds.), 1978. *Modern and Ancient Lake Sediments*. Spec. Publs. Int. Assoc. Sediment. 2.

Olsen, P.E., 1986. A 40-million-year lake record of early Mesozoic orbital climatic forcing. *Science, 234*, 842–8.

Olsen, P.E., 1990. Tectonic, climatic, and biotic modulation of lacustrine ecosystems – Examples from Newark Supergroup of Eastern North America. In *Lacustrine Basin Exploration – Case Studies and Modern Analogs*, ed. B. Katz, pp. 209–24. Amer. Assoc. Petrol. Geol. Mem. 50.

Parrish, J.T. & Barron, E.J., 1986. Paleoclimates and economic geology. *Soc. Econ. Paleontol. Mineral. Short Course, 18*, 1–162.

Parrish, J.T. & Curtis, R.L., 1982. Atmospheric circulation, upwelling, organic-rich rocks in the Mesozoic and Cenozoic eras. *Palaeogeogr., Palaeoclimatol., Palaeoecol., 40*, 31–66.

Pia, J., 1933. *Kohlensäure und Kalk. Die Binnengewässer 13*. Schweizerbartsche Verlag. vii.

Picard, M.D. & High, L.R., 1972. Criteria for recognizing lacustrine rocks. In *Recognition of Ancient Sedimentary Environments*, ed. J.K. Rigby & W.K. Hamblin, pp. 108–45. Spec. Publ. Soc. Econ. Pal. Min. 16.

Picard, M.D. & High, L., 1981. Physical stratigraphy of ancient lacustrine deposits. In *Recent and Ancient Non-Marine Depositional Environments: Models for Exploration*, ed. F.G. Ethridge *et al.*, pp. 233–59. Spec. Publ. Soc. Econ. Pal. Min., 31.

Platt, N.H. & Wright, V.P., 1991. *Lacustrine carbonates: facies models, facies distributions and hydrocarbon aspects*, pp. 57–74. Spec. Publs. Int. Assoc. Sediment. 13.

Powell, T.G., 1986. Petroleum geochemistry and depositional setting of lacustrine source rocks. *Mar. Pet. Geol., 3*, 200–19.

Rosendahl, B.R., 1987. Architecture of continental rifts with special reference to East Africa. *Ann. Rev. Earth Planet. Sci., 15*, 445–503.

Russel, I.C., 1885. *Geological History of Lake Lahontan, a Quaternary Lake of Northwestern Nevada*. U.S. Geol. Surv. Monograph 11.

Ruttner, F., 1963. *Fundamentals of Limnology*. (Transl. D. Frey & F. Fry.) Univ. Toronto Press.

Schuiling, R.D., 1977. Source and composition of lake sediments. In *Interactions Between Sediments and Freshwater*, ed. H.L. Golterman, pp. 12–18. Junk, The Hague.

Smith, M. 1991. Lacustrine Oil Shale in the Geologic Record. In *Lacustrine Basin Exploration – Case Studies and Modern Analogs*, ed. B. Katz, pp. 43–60. Amer. Assoc. Petrol. Geol. Mem. 50.

Strakhov, N.M., 1970. *Principles of Lithogenesis*. Plenum. NY.

Street-Perrott, F.A. & Harrison, S.P., 1984. Temporal variations in lake levels since 30,000 yr B.P. — an index of the global hydrological cycle. *Geophys. Monogr., 29*, 118–29.

Street-Perrott, F.A. & Harrison, S.P., 1985. Lake levels and climate reconstructions. In *Paleoclimate Analysis and Modeling*, ed. A.D. Hecht, pp. 291–304. Wiley, New York, N.Y.

Street-Perrott, F.A. & Roberts, N., 1983. Fluctuations in closed basin lakes as indicators of past atmospheric circulation patterns. In *Variations in the Global Water Budget*, ed. F.A. Street-Perrott, M. Beran & R.A.S. Ratcliffe, pp. 331–45. Reidel, Dordrecht.

Talbot, M.R. & Kelts, K. (Eds.), 1989. Phanerozoic Record of Lacustrine Basins and their Environmental Signals. *Paleogeogr., Paleoclimatol., Paleoecology, 70* (1–3).

Taub. F.B. (Ed.), 1984. *Lakes and Reservoirs. Ecosystems of the World*, Nr. 23. Elsevier, Amsterdam.

Van Houten, F.B., 1962. Cyclic sedimentation and the origin of analcime-rich Upper Triassic Lockatong Formation, west-central New Jersey and adjacent Pennsylvania. *Am. J. Sci.*, **260**, 561–76.

Warren, J.K., 1986. Perspectives: Shallow-water evaporitic environments and their source rock potential. *J. Sed. Petrol.* **56**(3), 442–54.

Wetzel, R., 1983. *Limnology*. 2nd. ed., Saunders College Publ. Philadelphia.

Yang, W., Li, Y. & Gao, R., 1985. Formation and evolution of non-marine petroleum in Songliao Basin, China. *Amer. Assoc. Petrol. Geol. Bull.*, **69**, 1112–22.

Ziegler, A.M., Rowley, D.B., Lottes, A.L., Sahagian, D.L. Hulver, M.L. & Gierlowski, T.C., 1985. Paleogeographic Interpretation: with an example from the Mid-Cretaceous. *Ann. Rev. Earth. Planet. Sci.*, **13**, 385–425.

Selected Topics

Key elements during the search for hydrocarbons in lake systems

CHRIS P. SLADEN

British Petroleum plc, Britannic House, Moor Lane, London, EC2Y 9BU, UK

Introduction

Objectives of paper

The principal objective of this contribution is to highlight those aspects of lake basins and lake sequences necessary to successfully carry out hydrocarbon exploration in the subsurface. Techniques that can aid and improve hydrocarbon exploration in lake basins are also emphasized.

Many of the techniques commonly used to explore marine clastic and carbonate sequences have very limited application in lacustrine settings because most major controls on lacustrine sedimentation and lacustrine sequences differ from those operating in the marine realm. For example, changes in global eustacy tend to be relatively gradual and the actual change in sea-level may only be a few tens of meters. In many closed lake basins without an outflow to the sea, global eustacy is of little consequence. In contrast however, changes in lake levels of a few hundred meters may occur in only a few tens of thousands of years due to small shifts in climate or perhaps a slight alteration in basin configuration creating an outflow. Applying techniques such as seismic stratigraphy in lake basins requires an understanding of these rather different 'cause and effect' relationships. In addition, exploration philosophy has to be altered to match the very different combinations of reservoirs, seals, source-rocks and hydrocarbons that are common in lake basins.

During the course of this contribution, we will examine some of the methods that can be used to classify lake basins, understand their evolution, identify and model the lake sequences, and then the systems tracts and finally the hydrocarbon reservoir-source-seal characteristics within the sequences.

Data base available

Two problems are inherent in the data base and hence worthy of mention:

(1) Only in the last 2 years or so have many western oil companies contemplated widespread exploration in lake basins and many basins are still in the early phases of exploration. Consequently, the state-of-the-art in lacustrine exploration is generally less advanced, and published information is scarcer than for many other geological settings.

In addition, a large proportion of the hydrocarbons produced to date from lacustrine basins has been from within China (*c.* one billion barrels per year, Gorst, 1986). Relative to most other major hydrocarbon provinces (e.g. Alaska, the Middle East, or the North Sea), these basins are poorly understood by the industry at large outside of China. Many have not yet been explored by the more advanced industry techniques.

(2) Much of the research conducted on lakes during the past hundred years has concentrated on glacial lakes, and/or lake chemistry. Comparatively little has been done on the fundamental controls on ancient lake systems, particularly where, why and how they have developed through geological time. These features are naturally crucial to effective hydrocarbon exploration and only in the last few years have many of these questions been seriously attacked. Published data have until very recently been quite scarce.

Lake basin classification, occurrence and evolution: basic principles and philosophy during hydrocarbon exploration

'Tectonic' lakes and floodplain lakes

Two major categories of lake can be identified. These are, (a) lakes that owe their origin to the form of a tectonically induced depression, typically an extensional or flexural basin. These 'tectonic' lakes are usually long-lasting features, often existing for more than 1 Ma, and sometimes > 10 Ma; and (b) lakes that occur within part of the alluvial milieu in between river channels and distributaries. These 'floodplain' lakes are typically temporary features lasting only a few hundred thousand years, often much less. Glacial lakes are very rarely important in hydrocarbon exploration and are not considered in this contribution.

Of greatest importance in hydrocarbon terms are the 'tectonic' lakes. These can develop a hydrocarbon system whereby potential source-rocks may accumulate in low-energy anoxic parts of the lakes and, during burial, hydrocarbons may be generated and become reservoired in other lake facies such as the deltas and fans that feed into the lakes, as well as in other older and younger sequences. In this contribution 'tectonic' lakes will henceforth simply be termed lakes.

Examples of lake sequences rich in hydrocarbons

Examples of significant hydrocarbons that were generated from lake sediments occur in:

(1) many onshore extensional basins in China, for example numerous Tertiary basins within the North China rift system, and the Cretaceous Songliao basin in northeast China (e.g. Xu Shice & Wang Hengjian, 1981; Ma Li *et al.*, 1982; Zha Quanheng, 1984; Powell, 1986);

(2) extensional basins in southern Thailand, for example the Tertiary Phitsanulok basin, and, further south, the Kamphaeng Saen and Suphan Buri basins (e.g. Flint *et al.*, 1988; Burri, 1989; Chinbunchorn, Pradidtan & Sattayarak, 1989; O'Leary & Hill, 1989);

(3) extensional basins formed along the Brazilian and W. African margins at the start of South Atlantic rifting, for example the Campos, Reconcavo and Sergipe-Alagoas basins (Brice, Kelts & Arthur, 1980; Bertani & Carozzi, 1985);

(4) extensional basins in Sudan (e.g. Schull, 1988; Derksen & McLean-Hodgson, 1988);

(5) Tertiary flexural basins in the western USA, for example the Unita Basin, Utah (e.g. Lucas & Drexler, 1975; Swain, 1981);

(6) flexural basins in northwest China, for example Junggar during the Permian, Tarim during the Mid Jurassic, Qaidam during the Oligo-Miocene (e.g. Powell, 1986; Watson *et al.*, 1987; Taner, Kamen-Kaye & Meyerhoff, 1988).

(7) extensional basins in Central Sumatra, Indonesia (e.g. Derksen & McLean-Hodgson, 1988; Atkinson, 1989).

Until now, the most prolific areas for hydrocarbon discoveries and production which are linked to development of lacustrine hydrocarbon source-rocks appear to be Central Sumatra and northeast China. When taken together, basins in these areas have contributed over 20 billion barrels of hydrocarbon reserves.

Evolution of lakes in extensional/transtensional basins

The evolution of lakes in extensional (rift) systems can usually be divided into three phases (cf. Watson *et al.*, 1987), which reflect the development, acme and then decline in rifting (Fig. 1):

(1) 'Early': during this initial stage when faults are beginning to have a topographic expression, sedimentation may keep pace with subsidence. The rocks are characteristically a clastic alluvial fill which is lithologically very variable. Any lake sites that develop tend to be shallow and temporary, and have insignificant source potential. In the subsurface, their position is very difficult to predict. The alluvial clastics may be intermixed with shallow lake clastics or even carbonates.

The thickness of the early rift fill will vary depending on the speed at which major fault(s) begin to exert influence and control subsidence. The thickness of early alluvial syn-rift fill is normally less than 500 m and typically less than 200 m. In extreme cases, and particularly in transtensional (strike-slip) settings where subsidence rates can be extremely rapid relative to sedimentation rates, it may only be a few meters thick.

(2) 'Middle': during this stage, subsidence outpaces sedimentation as major fault activity reaches its peak and basin margin faults dominate basin evolution. Whilst subsidence outpaces sedimentation, stable long-lived lake sites may develop if a suitable topographic hole develops and is maintained. It is during this stage of syn-rift activity that significant thicknesses of lacustrine mudrock are most likely to develop which can act as both source-rock and seal. Marginal to the lakes, various fans, deltas and shallow lake sequences may accumulate as potential reservoirs.

(3) 'Late': during this stage, sedimentation begins to outpace subsidence as fault activity dies out. The lake is gradually infilled, becoming smaller and shallower and eventually it is eliminated. As a consequence, facies typically revert to alluvial. Parts of the succession may still accumulate hydrocarbon source-rocks, whilst other parts may contain significant reservoirs in alluvial sequences.

Repetition of the stages described above is a characteristic of many extensional basins. If extension is pulsed, this is reflected in the abundance and type of lake sediments and alluvium. Thus in the areas most prone to lake development, the sequences may switch repeatedly from deep lake to shallow lake and alluvial, and then vice versa (e.g. see Zhou Guangjia, 1981). In many transtensional (strike-slip) basins, extension and sedimentation may be more rapid than dip-slip basins. Evolutionary features in the sedimentary record may therefore be strongly condensed or in places appear to be missing.

Evolution of lakes in flexural basins

In flexural basins that develop due to loading of the crust, the evolution of lakes is also often characterized by three phases similar to extensional basins (Watson *et al.*, 1987). These represent basin initiation, the development and acme of thrusting, loading and basin subsidence, and then the decline and end of thrusting and loading:

(1) 'Early': during this stage of basin initiation, initial collision, compression and development of a mountain belt is occurring. The basin-fill is characteristically lithologically variable alluvium. Any lake sites that develop tend to be shallow and temporary, and have insignificant source potential.

(2) 'Middle': during this stage, the mountain belt grows and the loading causes the basin to experience maximum subsidence. If subsidence outpaces sedimentation, a stable lake site may develop. Many lakes in this setting are areally extremely large, perhaps 5000 km² or more. The lateral movement

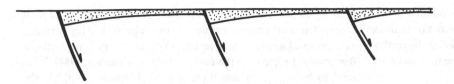

① **RIFT INITIATION CHARACTERISED BY WIDESPREAD ALLUVIAL DEPOSITION.**

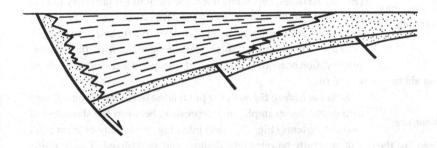

② **SUBSIDENCE OUTPACES SEDIMENTATION AS EXTENSION REACHES A PEAK. MOST LIKELY PERIOD FOR LONG–LIVED LAKES,GIVEN SUITABLE TECTONICALLY CLOSED BASIN. MOST LIKELY PERIOD FOR ACCUMULATION OF THICK LACUSTRINE MUDSTONES CAPABLE OF BEING HYDROCARBON SOURCE-ROCKS.**

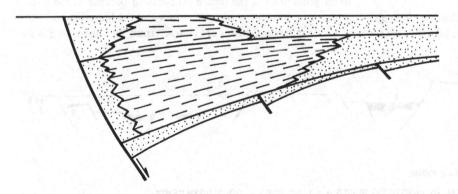

③ **CESSATION OF RIFTING. AS SUBSIDENCE DECREASES LAKE SITES ARE INFILLED AND ELIMINATED BY FLUVIAL SYSTEMS.**

IN FLEXURAL BASINS, SIMILAR STAGES DEVELOP RELATED TO BASIN INITIATION, THE ACME OF THRUSTING , BASIN LOADING AND SUBSIDENCE, AND THEN THE DECLINE AND CESSATION OF THRUSTING AND LOADING. MODIFIED FROM WATSON ET. AL. 1987.

Fig. 1. Characteristic stages in the development of lake sequences in an extensional (rift) setting. Repetition of these stages is common when extension is pulsed.

characteristic of flexural basin depocenters continually shifts the facies belts and this tends to inhibit the formation of thick vertically monotonous sequences of lake facies. Nevertheless, this is the period during which lacustrine mudrocks are most likely to accumulate.

(3) 'Late': during this stage, collision, compression and growth of the thrust belt diminishes and ends. Flexural subsidence decreases. The basin is gradually infilled and the lake site(s) eliminated. The fill characteristically returns to alluvium.

Repetition of the stages described above can be observed in many flexural basins, especially where there are renewed periods of thrusting, compression and collision (e.g. Steidtmann, McGee & Middleton, 1983). In parts of the basin close to the thrust belt, uplift and cannibalization of the basin sequences by the thrust front may result in many evolutionary features being condensed, missing, or subsequently being eroded.

Identifying and modelling lacustrine sequences as an aid to hydrocarbon exploration

Fundamentals of locating lake sequences in the subsurface

The position of lake sites is primarily determined by the relationship between tectonic subsidence and sedimentation. Of principal importance are the major active fault systems at the time. To locate a lake site in the subsurface, it is essential to be able to recognize these faults and their geometric relationships to each other in order to determine which faults were controlling subsidence patterns. Gravity data provide a first-pass assessment on whether a long-lived closed basin capable of containing a lake exists in the subsurface. Seismic data is then clearly important as this allows isopach maps highlighting the active faults to be prepared for individual sequences of the basin fill.

Lakes in half-graben tend to be 'tucked-up' close to the major controlling faults (Fig. 2). In many instances, the lake may extend right up to the major faults, in places touching the footwall fault scarp. The bathymetric profile of the lake is usually strongly asymmetrical and deep lake facies can develop very close (within a few hundred meters) to the fault scarp (e.g. Crossley, 1984); lakes tend to be deepest where there is a single major fault. On the opposite side of the rift, where the lake extends up the more gently dipping hangingwall, it gradually shoals. The lake shoreline is often controlled by the position of scarps of antithetic faults, or sometimes by the position of other synthetic faults. A typical isopach map for a lacustrine sequence in a half-graben, prepared from seismic and well data, is shown in Fig. 3.

Lakes in symmetrical graben are normally concentrated in the topographically low closed areas, occupying central parts of the rift (Fig. 2). At times, depending upon fluctuations in lake levels and the position of outflow points along the rift system, lakes may occupy almost the entire graben and spill over onto rift shoulders (Fig. 2). However, in this situation the lake sequences with the greatest preservation potential are those in the deeper, more central portions of the rift.

Lakes outside of the main rift but related to rift development may also occur, for example, in a depression between the shoulders of two rift systems (Fig. 2). These lakes may be areally extensive but will generally be relatively shallow and short-lived. The resulting sequences have poor preservation potential as they are likely to be eroded to form parts of the sedimentary fill in nearby basins.

Lakes in strike-slip basins are usually in the central parts of rhomb-shaped graben and/or close to the faults with large throws (e.g. see Nilsen & McLaughlin, 1985; Cabrera, Roca & Santanach, 1988; Anadon *et al.*, 1989). Phases of extension and local shortening may be repeated (and be occurring at the same time), and rotation of fault blocks may alter the positions of lake sites. In strike-slip basins, it is particularly difficult to reconstruct fault geometries and basin geometries at the time a lake existed because of the lateral motions of different parts of the basin and the hinterlands.

Lakes in flexural basins are characterized by a migrating depo-

1. IN BETWEEN RIFT SHOULDERS e.g. L. VICTORIA, KENYA

2. HALF GRABEN, EXTENT OF LAKE CONTROLLED BY ANTITHETIC FAULTS e.g. L. BOGORIA, L. SOLAI BOTH KENYA

3. IN CENTRAL PART OF RIFT e.g. L. BARINGO, L. MAGADI BOTH KENYA

4. FULL TO SPILL, OCCUPYING MOST OF RIFT e.g. L. MALAWI, MALAWI

5. PERCHED ABOVE MAIN RIFT, EXTENT OF LAKE CONTROLLED BY SYNTHETIC FAULTS, e.g. L. OLBOLASSAT, KENYA

6. ON EDGE OF RIFT, EXTENT OF LAKE CONTROLLED BY ANTITHETIC FAULTS e.g. L. CHONGOLE, MALAWI

N.B. SEDIMENTS IN LAKES 1, 5 & 6 HAVE RELATIVELY LOW PRESERVATION POTENTIAL

Fig. 2. Lakes in rift systems can develop in a variety of locations. Various present-day lakes in East Africa highlight this variety.

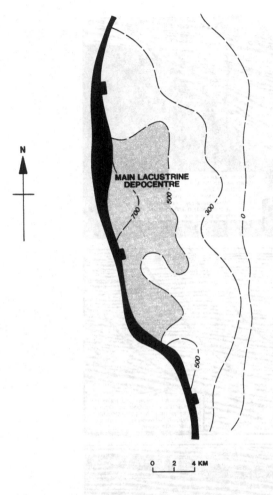

N

MAIN LACUSTRINE
DEPOCENTRE

0   2   4 KM

(THIS EXAMPLE IS BASED UPON SEISMIC & WELL DATA FROM A BASIN IN THAILAND. ISOPACHS ARE IN METRES).

Fig. 3. Typical isopachs for a half-graben lake sequence. The basin possesses a closed depocentre close to the major fault where subsidence is most likely to outpace sedimentation and be conducive to develop a lake.

center that moves ahead of the propagating thrust front (e.g. see Steidtmann *et al.*, 1983). The bathymetric profile is typically asymmetric – deep water lying close to the growing thrust fronts, the lake shoaling toward the peripheral bulge. Because the depocentre is migrating laterally, the water depth changes with time at any one point, characteristically from shallow to deep then to shallow (cf. many basins described by Lucas & Drexler, 1975; Yuretich, 1989). As a result, maintenance of a long-lived lake site is usually less common than in rifts and many lakes in flexural basins never develop significant deep-lake facies (see Steidtmann *et al.*, 1983; Yuretich *et al.*, 1984; Yuretich, 1989). In order to aid location of lacustrine sequences, modelling must incorporate well-constrained and carefully prepared palinspastic and isopach maps.

Lakes on passive margins normally fail to develop because the tectonic regime is unsuitable to creating long-lived 'closed holes' capable of maintaining a lake. Many shallow, temporary, lakes and lagoons (often of great areal extent) may exist on coastal and deltaic plains, but individual lakes have a short lifespan (normally < 100,000 years) and the thicknesses of the resulting lake sequences are relatively small.

Seismic sequences, sequence boundaries and seismic facies

As noted earlier, it is important to show on isopach maps which tectonic elements were controlling subsidence for each lake sequence. The distribution of various seismic attributes can be added to these maps to help identify and predict the extent and type of lake sediments present.

Key attributes of lake sediments on seismic sections are a high degree of continuity and conformity of reflections (Fig. 4). Lateral continuity of reflections develops because sedimentation over large parts of many lakes is dominated by rather uniform suspension fallout. In thick sequences, the frequency of reflections may vary considerably. Monotonous mudstone/siltstone sequences may possess few lithological contrasts and be almost reflection-free on seismic sections. However, there are often changes in organic or calcareous content, or the presence of evaporites, biogenic oozes and thin turbidites that cause significant velocity variation, creating laterally extensive reflections with variable amplitudes. This can create a distinctive 'stripey' seismic facies (Fig. 4).

The presence of clinoforms on seismic sections can be used to help identify a lake site because they indicate deltas and fans building into a standing-water body. The direction of progradation and position of clinoforms in a sequence can be used as an indicator of both the position of a lake site, and that of potential alluvial and deltaic reservoirs (Fig. 4).

The effects of rising and falling lake level are seen in a variety of seismic configurations. Rising lake level is often seen as onlap in all directions as the lake expands. Such onlap is typically gentle across the hangingwall or peripheral bulge (e.g. across deltas and low energy shorelines) and stronger on the footwall or thrust belt side (e.g. across the surface of alluvial and lacustrine fans). Mapping the direction of onlaps can thus provide an indicator of the direction towards an area of more-permanent lake.

Most sequence boundaries in lake systems are a response to tectonic change. Pulses of accelerated rifting or compression may create new lakes, and either eliminate or re-establish old ones. Truncations in the basin (lake) margin stratigraphy are common features. The deep lake environment is least likely to show any breaks in deposition and it is here (as with marine sequences), that one should expect to find the correlative conformities.

Changing climate, whilst not usually documented in sequence stratigraphy as being a cause of seismic sequence boundaries, frequently appears to be capable of generating a seismic sequence boundary in lake deposits. This is because small climatic changes can have such dramatic effects on lake level, size, depth and consequently, deposition and erosion around the lake. For example, in the last 20,000 years the level of the Dead Sea has fallen about 200 m, creating a lake lowstand (Manspeizer, 1985). This is a result of warmer temperatures and lower rainfall. Runoff has decreased, and evaporation has increased. Marginal alluvial and lacustrine

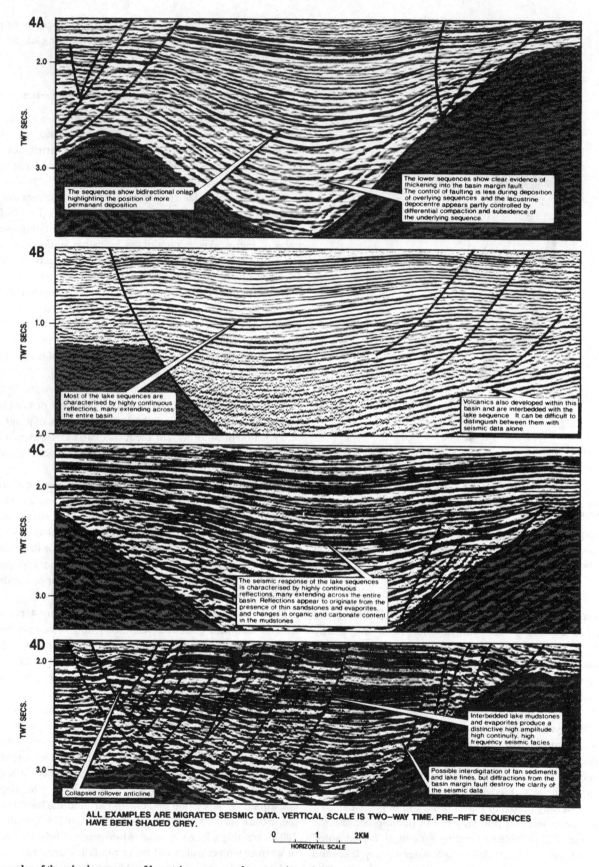

4A

The sequences show bidirectional onlap highlighting the position of more permanant deposition

The lower sequences show clear evidence of thickening into the basin margin fault The control of faulting is less during deposition of overlying sequences and the lacustrine depocentre appears partly controlled by differential compaction and subsidence of the underlying sequence.

4B

Most of the lake sequences are characterised by highly continuous reflections, many extending across the entire basin

Volcanics also developed within this basin and are interbedded with the lake sequence It can be difficult to distinguish between them with seismic data alone

4C

The seismic response of the lake sequences is characterised by highly continuous reflections, many extending across the entire basin Reflections appear to originate from the presence of thin sandstones and evaporites, and changes in organic and carbonate content in the mudstones

4D

Interbedded lake mudstones and evaporites produce a distinctive high amplitude, high continuity, high frequency seismic facies

Collapsed rollover anticline.

Possible interdigitation of fan sediments and lake fines, but diffractions from the basin margin fault destroy the clarity of the seismic data

ALL EXAMPLES ARE MIGRATED SEISMIC DATA. VERTICAL SCALE IS TWO–WAY TIME. PRE–RIFT SEQUENCES HAVE BEEN SHADED GREY.

0   1   2KM
HORIZONTAL SCALE

Fig. 4. Examples of the seismic response of lacustrine sequences from a variety of rift settings.

fans, lacustrine deltas and shallow lake deposits have been severely, and widely eroded (truncated) and the earlier sequences have been strongly dissected as erosion has cut down to the new base level. Sediment previously held in feeder channels and wadis, has been transported further out toward the basin center. Consequently, the alluvial and lacustrine fans have prograded out over the older sequences. Sedimentation in the deeper parts of the lakes has continued uninterrupted although deposition of varved lake marls and lake carbonates has given way to that of evaporites, due to the increased evaporation. A converse climatic shift toward a more humid climate, would raise lake level and cause widespread onlap of lake sediments onto the marginal alluvium.

In Lake Tanganyika, Africa, climate changes have also caused dramatic changes in lake levels capable of producing sequence boundaries in the geological record. About 14,000 years ago, the lake level was c. 300 m lower than at present (Haberyan & Hecky, 1987). Truncation of older sequences, and onlap by younger sequences should be observed, similar to that seen in the subsurface in Lake Turkana (Johnson et al., 1987).

Many sequence boundaries in the lacustrine record can be interpreted as a result of climatic changes, of a similar style to those observed in the Dead Sea. For example, in sequences such as the Early Tertiary in China, palynological study reveals marked changes in climatic conditions and lake chemistry that coincide with mappable seismic sequences (unpublished BP reports).

Clinoforms associated with lacustrine delta systems may be extremely gentle sigmoidal forms, often with vertically compressed topset/foreset/bottomset relationships where the deltas are building into very shallow lakes (e.g. early Pliocene deltas in the Ridge Basin, California; Wood & Link, 1987). Frequently these clinoforms will be unresolvable on seismic. An absence of clinoforms is thus not indicative of the absence of a lacustrine delta. Fans developing where alluvial fans build into a lake can also develop topset, foreset and bottomset relationships. On seismic sections, the clinoforms will always be difficult to see because the fans typically develop adjacent to faulted terrains. Consequently seismic diffractions from the faults tend to obscure the clinoforms.

Maps of seismic interval velocity data may also be used to detect, or infer the presence of thick lake mudstones. Overpressuring is common due to very rapid deposition rates impeding porewater expulsion (mudrock deposition rates can be in excess of 1 mm/year and may reach 5 mm/year; Johnson et al., 1987). Organic maturation and clay-mineral dehydration reactions may also contribute to overpressuring. Many Chinese basins have overpressured portions including parts of the Bohai and the South Yellow Sea, and in the Uinta Basin, USA (e.g. Lucas & Drexler, 1975). These overpressured areas may have relatively low seismic interval velocities.

Understanding controls on lake facies to improve facies prediction

For accurate facies prediction in lacustrine basins, and hence accurate exploration models, it is necessary to understand controls on the basin in terms of:

(1) tectonic evolution/subsidence rate;
(2) basin size and shape;
(3) catchment areas and their geology, topography and bedrock composition:
(4) relationships to temperate or tropical climatic belts, and humid vs. arid climates;
(5) regional paleogeography, proximity to oceans, elevation;
(6) lake hydrology, that is, open vs. closed, fresh vs. saline, deep vs. shallow.

Refer to Table 1 for further details.

The major controls on accumulation of a thick lacustrine sequence of mudstones that may be a good hydrocarbon source-rock appear to be:

(1) Tectonics. Formation of a large, 'closed hole' is required, accompanied by prolonged periods of subsidence outpacing or equalling sedimentation (cf. Blair & Bilodeau, 1988).
(2) Climate. Subtropical to tropical humid climates are preferable. Warm temperatures encourage organic matter to flourish in the lakes; humid climates enable vegetation to flourish in the hinterland, providing additional detrital organic input. Although humid climates may cause considerable sediment input to the lake basin, they also provide continual recharge of nutrients into the lakes, and help inhibit hypersalinity.

Whilst high salinities may aid organic preservation due to salinity stratification, the advent of hypersalinity normally causes organic sterility and consequently no source-rock potential (cf. Powell, 1986).
(3) Nature of clastic input. Hinterlands conducive to providing mostly silt and mud-grade sediment result in lakes dominated by fine-grained sequences more likely to preserve organic matter. The optimum hinterland composition appears to be limestones because these produce virtually no clastics, and usually liberate only small amounts of clay and silt-grade sediment during weathering. Clastic sedimentation rates are thus relatively low and the lakes are not polluted with sediment that inhibits organic production.

Paleotopographic analysis as an aid to hydrocarbon exploration in lacustrine systems

Paleotopographic analysis is simply a form of paleogeographic study that attempts to reconstruct the lake basin, its catchment, and the hinterland topography and geology as an aid to understanding and predicting sedimentation patterns. It helps in predicting the various facies and lithologies and hence contributes to a better assessment of hydrocarbon potential. It requires the following:

(1) Careful mapping and isopaching of the lacustrine sequence of interest.
(2) Identification of active syn-depositional faults and removal of subsequent structural modifications. Unequivocal removal of structural modifications in flexural basin thrusted terrains may be particularly difficult.

Table 1. *Summary of some of the main controls on lake facies, and common effects that occur*

Control	Effects on lake facies	Control	Effects on lake facies
1. Structural evolution	Long-lived lake sites, and thick lake sequences are more likely where subsidence outpaces sedimentation for long periods (1–10 Ma). In half graben, these areas are often close to major faults. Strike-slip basins are ideal lake sites, frequently forming a topographic 'hole', fault bounded on all sides. In symmetrical graben, lake sites may develop in topographically-closed lows usually in the rift axis. Lakes may develop in between rift shoulders or perched high on footwall blocks but these sequences have low preservation potential. In flexural basins, lakes commonly develop in front of the growing thrust front in basins characterised by a laterally migrating depocentre. Lakes may also develop in thrust sheet top basins but have low preservation potential.	6. Humid or arid climates	Humid climates promote a rich vegetation around the lake, hence there may be considerable detrital organic input together with greater runoff and a high clastic input which inhibits and swamps in-situ organic production. Periods of aridity, within otherwise humid climates may lead to dramatic falls in lake level, or even complete drying out of lakes. Around the lake margins there may be significant erosion and reworking of lake sequences.
2. Basin size, lake size and shape	Large, or complex basins often have one or more lake sites. Many rifts and flexural basins are elongated; the lakes in them are similarly elongate and major river-dominated deltas commonly enter axially from the ends, whilst small fans enter laterally from the sides. Shoreline and carbonate facies are usually best developed along the sides. Large lakes are more conducive towards wave activity, and consequently shoreline facies may be higher energy with wave reworking of fans and deltas producing sandy linear clastic shorelines. Large lakes have more chances of a quiet portion with little clastic input, and also have a greater area for organic productivity, both conducive to accumulation of organic-rich facies. Note however that large lakes are not necessarily deep (or small lakes shallow) and consequently the development of water stratification and bottom water anoxia which aids organic preservation, cannot be simply judged from lake size (see also 10).	7. Proximity to oceans, elevation	Proximity to oceans usually ameliorates the climate, which can inhibit turnover and increase organic production and preservation. Low lying areas close to oceans may experience marine influxes promoting salinity stratification. Later annexation, and evaporation of the 'fossil' seawater body, may lead to widespread/thick evaporites.
		8. Open or closed lakes	Open lakes have an outflow and hence throughput of water. Stable stratification can be hard to maintain unless the lake is very deep. The arrangement of facies belts in open lakes is different to closed lakes. Closed lakes more commonly undergo evaporative concentration leading to evaporite precipitation, creating lakes with unusual water chemistries that alter organic production rates, and develop unusual, often endemic faunas and floras.
3. Catchment area and topography	Relatively large catchments have relatively higher runoff. Large catchments are typical of rivers draining axially whilst small catchments characterise rivers draining transversely. If relief is strong it increases the potential of transporting coarse clastics and may locally affect the climate in the vicinity of the basin.	9. Fresh or saline lakes, water chemistry	Alkaline and saline lakes encourage carbonate facies. At high salinities anoxic decomposition processes are inhibited, and the low sulphate concentrations typical in lake waters result in little sulphate to oxidize organic matter; good organic preservation may thus occur in saline lakes (over half the world's lakes are brackish or saline today). Low sulphate concentrations mean that iron sulphides (pyrite, etc) are scarce; siderite is the common iron mineral in lake sediments. The initial porewater composition and hence early diagenesis will be determined by the salinity/alkalinity of the lake waters.
4. Catchment bedrock, composition and geology	When weathered, granites, schists and gneisses produce mostly sand. Metapelites and mudrocks produce mostly mud. Limestones mostly dissolve releasing a little mud and silt. Limestone terrains are thus conducive towards silt/mud dominated lake systems, and hence mudstone sequences that could be potential hydrocarbon source-rocks.	10. Deep or shallow lakes	The maximum lake depth is controlled by the spill point out of the basin. The high deposition rates common in lakes can rapidly alter lake depth. Deep lakes are more likely to maintain a stable, stratified water column leading to greater organic preservation. Turbidites and gravity flows may also be more common. Shallow lakes are more likely to have aerated substrates and experience turnover; marginal/shoreline facies may be more widespread. Lakes do not need to be

Table 1 (*cont.*)

Control	Effects on lake facies	Control	Effects on lake facies
5. Temperate or tropical climates	Temperate climates are more conducive to seasonal over-turning, destroying stratification, oxygenating bottom waters and inhibiting organic matter preservation. In tropical climates, the higher temperatures are more conducive to year-round production of in-situ organic matter.	11. Volcanic activity	deep to produce organic matter; the greatest organic production is between 0.1 and 20m. Extruded lavas seek topographic lows; they may displace or eliminate lake sites. Hydrothermal springs may introduce considerable dissolved ions, affecting water chemistry and hence sediment composition, organic characteristics and content.

(3) Reconstruction of the likely elevation of sediment source areas, their composition, modification by erosion, and catchment area.

(4) Seismic facies analysis.

(5) An understanding of the numerous controls on the lake systems.

Lacustrine hydrocarbon fairways, systems tracts, reservoirs, seals and source rocks

Previous exploration results

Previous exploration results present a confusing picture of lacustrine hydrocarbon reserves and distributions. Partly this is because in many areas exploration has concentrated either upon structures at the edges of potential lake basins or on structural highs and shallow fault segments. The result is that the sequence stratigraphy, facies and full hydrocarbon potential of many potential lake basins remains poorly known. Whether or not a lacustrine sequence is present frequently remains unknown as the deep parts of many basins remain undrilled.

Discovery histories in lacustrine basins simply reflect the fact that most hydrocarbons are reservoired in large structural closures. Consequently the addition of new reserves is small once the major structures have been drilled and represents either smaller structures or stratigraphic traps. The number of fields discovered may increase steadily but the addition to total basin reserves is minor.

Approaches to exploration differ onshore and offshore due to differences in the cost of wells and the minimum size of field that can be economically developed. This further distorts our knowledge of lacustrine hydrocarbons and their distribution. Most wells offshore are located upon large, relatively shallow structures, typically fault blocks, rollovers and drapes.

The majority of lacustrine basins explored to date are onshore (e.g. in China, Thailand, Brazil, USA and Sudan), consequently the data base is inherently biased. Onshore, the lower drilling and field development costs enable much greater trap coverage and a much closer well spacing. Consequently, large numbers of small stratigraphic traps (< 10 MM barrels of oil) may be discovered (e.g. North China Basin).

Common stratigraphic trapping mechanisms (refer Fig. 5) appear to be:

(1) Basinward pinchout of lacustrine reservoir/basinward appearance of seal, for example lacustrine fans, deltas, shoreline and shallow lake facies that have undergone structural tilting and reversal (e.g. Fuller Field, Redwash Field, Altamont-Bluebell Field and Shaunghezhen Field; Osmond *et al.*, 1968; Lucas & Drexler, 1975; Ma Li *et al.*, 1982; Ray, 1982).

(2) Basinward appearance of reservoirs such as shallow water carbonates and axial lacustrine turbidites (e.g. Gaosheng Field, Liangjialou Field; Ma Li *et al.*, 1982).

(3) Unconformities (up-dip truncation) within lacustrine sequences. These tend to be best developed around basin margins. They are commonly related to tilting, or to changes in lake level (e.g. Jinjiazhuang Field, Ma Li *et al.*, 1982).

(4) Simple onlap caused by expansion of the area of lake deposition and submergence of marginal and shallow lake reservoir facies by fine-grained lake sediments (e.g. Shanjiasi Field and Zhongshi Field, China; Hu *et al.*, 1984).

(5) Buried-hill traps are a type of unconformity trap often created when lakes expand, drowning pre-existing 'basement' topography. This may cause onlap and drape of the basement topography with lacustrine fines enabling hydrocarbons to become trapped in basement reservoirs (e.g. Renqiu Field, Nanmeng Field, Balizhuang Field and Wei 11–1 Field; Zhai Guangming & Zha Quanheng, 1982; Fei Qi & Wang Xie-Pei, 1984; Zhang Qiming & Su Houxi, 1989).

Hydrocarbon characteristics

Lacustrine-derived hydrocarbons frequently have a very distinctive set of characteristics originating from the types of organic matter that accumulates in lakes.

Organic matter in lake sediments can be divided into that produced in the lake (most notably algae) and that growing around the lake and in the hinterlands, which is carried into the lake by rivers and during floods (typically plant debris). Not surprisingly lakes developing in hot, very arid climates tend to have source intervals characterized by a high proportion of algae, whereas those lakes developing in hot, humid climates are characterized by a mixture of algae and also land-derived detrital organic input from heavily vegetated hinterlands.

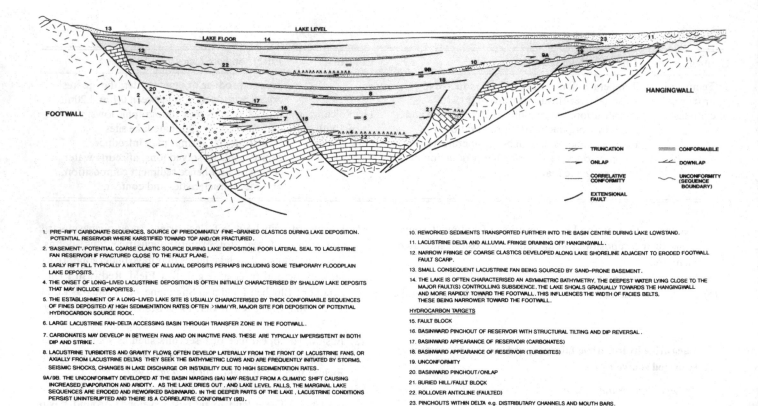

1. PRE-RIFT CARBONATE SEQUENCES. SOURCE OF PREDOMINATLY FINE-GRAINED CLASTICS DURING LAKE DEPOSITION. POTENTIAL RESERVOIR WHERE KARSTIFIED TOWARD TOP AND/OR FRACTURED.

2. 'BASEMENT'. POTENTIAL COARSE CLASTIC SOURCE DURING LAKE DEPOSITION. POOR LATERAL SEAL TO LACUSTRINE FAN RESERVOIR IF FRACTURED CLOSE TO THE FAULT PLANE.

3. EARLY RIFT FILL TYPICALLY A MIXTURE OF ALLUVIAL DEPOSITS PERHAPS INCLUDING SOME TEMPORARY FLOODPLAIN LAKE DEPOSITS.

4. THE ONSET OF LONG-LIVED LACUSTRINE DEPOSITION IS OFTEN INITIALLY CHARACTERISED BY SHALLOW LAKE DEPOSITS THAT MAY INCLUDE EVAPORITES.

5. THE ESTABLISHMENT OF A LONG-LIVED LAKE SITE IS USUALLY CHARACTERISED BY THICK CONFORMABLE SEQUENCES OF FINES DEPOSITED AT HIGH SEDIMENTATION RATES OFTEN >1MM/YR. MAJOR SITE FOR DEPOSITION OF POTENTIAL HYDROCARBON SOURCE ROCK.

6. LARGE LACUSTRINE FAN-DELTA ACCESSING BASIN THROUGH TRANSFER ZONE IN THE FOOTWALL.

7. CARBONATES MAY DEVELOP IN BETWEEN FANS AND ON INACTIVE FANS. THESE ARE TYPICALLY IMPERSISTENT IN BOTH DIP AND STRIKE.

8. LACUSTRINE TURBIDITES AND GRAVITY FLOWS OFTEN DEVELOP LATERALLY FROM THE FRONT OF LACUSTRINE FANS, OR AXIALLY FROM LACUSTRINE DELTAS THEY SEEK THE BATHYMETRIC LOWS AND ARE FREQUENTLY INITIATED BY STORMS. SEISMIC SHOCKS, CHANGES IN LAKE DISCHARGE OR INSTABILITY DUE TO HIGH SEDIMENTATION RATES.

9A/9B. THE UNCONFORMITY DEVELOPED AT THE BASIN MARGINS (9A) MAY RESULT FROM A CLIMATIC SHIFT CAUSING INCREASED EVAPORATION AND ARIDITY. AS THE LAKE DRIES OUT, AND LAKE LEVEL FALLS, THE MARGINAL LAKE SEQUENCES ARE ERODED AND REWORKED BASINWARD. IN THE DEEPER PARTS OF THE LAKE, LACUSTRINE CONDITIONS PERSIST UNINTERUPTED AND THERE IS A CORRELATIVE CONFORMITY (9B).

10. REWORKED SEDIMENTS TRANSPORTED FURTHER INTO THE BASIN CENTRE DURING LAKE LOWSTAND.

11. LACUSTRINE DELTA AND ALLUVIAL FRINGE DRAINING OFF HANGINGWALL.

12. NARROW FRINGE OF COARSE CLASTICS DEVELOPED ALONG LAKE SHORELINE ADJACENT TO ERODED FOOTWALL FAULT SCARP.

13. SMALL CONSEQUENT LACUSTRINE FAN BEING SOURCED BY SAND-PRONE BASEMENT.

14. THE LAKE IS OFTEN CHARACTERISED AN ASYMMETRIC BATHYMETRY, THE DEEPEST WATER LYING CLOSE TO THE MAJOR FAULT(S) CONTROLLING SUBSIDENCE. THE LAKE SHOALS GRADUALLY TOWARDS THE HANGINGWALL AND MORE RAPIDLY TOWARD THE FOOTWALL. THIS INFLUENCES THE WIDTH OF FACIES BELTS. THESE BEING NARROWER TOWARD THE FOOTWALL.

HYDROCARBON TARGETS

15. FAULT BLOCK

16. BASINWARD PINCHOUT OF RESERVOIR WITH STRUCTURAL TILTING AND DIP REVERSAL.

17. BASINWARD APPEARANCE OF RESERVOIR (CARBONATES)

18. BASINWARD APPEARANCE OF RESERVOIR (TURBIDITES)

19. UNCONFORMITY

20. BASINWARD PINCHOUT/ONLAP

21. BURIED HILL/FAULT BLOCK

22. ROLLOVER ANTICLINE (FAULTED)

23. PINCHOUTS WITHIN DELTA e.g. DISTRIBUTARY CHANNELS AND MOUTH BARS.

Fig. 5. Schematic diagram illustrating many typical facies, sequence relationships and hydrocarbon targets found in lake sequences in an evolving non-marine half-graben (compiled from numerous examples worldwide).

Most lacustrine source-rocks of commercial significance appear to have accumulated in areas of hot, humid climates. Consequently, there are mixtures of algae and plant debris. The source-rocks are normally strongly oil-prone (e.g. see Powell, 1986).

Given suitable subsidence/sedimentation rates, lakes often accumulate vast thicknesses of potential source-rock (Table 2a, b). Analyses frequently show that much of this is of rather mediocre quality, interspersed with organically rich horizons. These variations can occur both on a millimeter scale, such as with individual laminations, and on a much larger scale, for example the development of oil-shale sequences perhaps 5–40 m thick within otherwise poor source mudstone sequences.

Mediocre quality source sequences may be many hundreds or thousands of meters thick, with total organic carbon (TOC) contents of 0.5–1.5% and production capabilities (S_2 values) of 3–10 kg/t. The low organic content is often due to dilution of organic matter by the high clastic sedimentation rates common in lakes. Oil-shale sequences have considerable higher TOC contents, typically c. 20%, and production capabilities of up to 100 kg/t or more. They usually have very high contents of algal organic matter. Unlike most source-rocks which are grey or black in color, lacustrine oil-shales are characteristically light brown or brown in color (e.g. Henstridge & Missen, 1982). Many oil shale outcrops have been mined in the past and oil has been directly distilled from the shale.

Lacustrine source-rocks tend to produce oils with high wax contents. Wax contents of 10–35% are common (Powell, 1986) and result in high viscosities and notably high pour points. Many oils have pour points of 30 °C or higher due to the wax content and are solid at room temperature.

Another typical characteristic of lacustrine hydrocarbons is low sulfur contents in oils, and low hydrogen sulfide contents in gases. This results from the low sulfur content of lake waters and lake sediments. As a further consequence, anoxic lacustrine shales are typically rich in siderite ($FeCO_3$) instead of pyrite (FeS_2). A few instances of sulfur-rich oils are known from basins that contained hypersaline lake systems (see Powell, 1986).

Lacustrine hydrocarbon fairways and their relative richness

Hydrocarbon fairways, in the context of this contribution are taken to be those parts of a basin with a common set of reservoir, source, seal and hydrocarbon characteristics. Fairways in extensional and flexural lake basins can frequently be subdivided into three segments, related to the depositional systems and systems tracts that developed, (i) along the basin axis, (ii) from the footwall or thrust belt, and (iii) from the hangingwall or peripheral bulge (Fig. 6).

The structural configuration of the basin is a prime control on the relative prospectivity of fairways. In a simple single half-graben or flexural basin, where source-rocks occur toward the basin axis, the

Table 2A. *Thickness of rift fills, and their lacustrine mudstone sequences in some rift basins in China. Many of these mudstones are potential hydrocarbon source rocks*

Basin	Area (km²)	Max. thickness of fill (m)	Lacustrine mudstones (m)
Songliao	260,000	7,000	800
Jizhong	25,000	7,000	1,000
Jianghan	25,000	10,000	1,000
Bohai	10,000	7,000	2,000
Biyang	1,000	8,000	1,900
Changwei	880	7,000	900
Baise	830	5,000	1,200
Damintun	800	7,000	1,000
Nan-4	450	4,000	600

(Data compiled from numerous sources.)

Table 2B. *Typical thicknesses of hydrocarbon reservoirs within rift lake systems*

Reservoir	Typical thickness (m)
Meander channel	4–15
Distributary (meandering)	3–9
Distributary (straight)	2–6
Channel mouth bar	0.5–3
Shoreline	1–5
Delta front (top)	2
Delta front (base)	1
Interdistributary	1–2
Turbidite/gravity flow	3–5
Oolite shoal	0.5–4
Bioherm	0.5–3
Fan complex	70 +

(Data assimilated from numerous sources covering numerous lake basins both onshore and offshore China.)

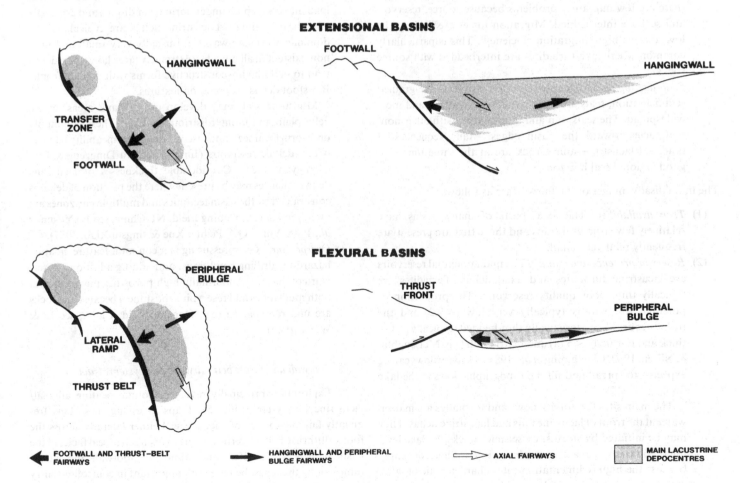

Fig. 6. Simplified view of lacustrine hydrocarbon fairways in extensional and flexural basins.

asymmetry of the basin results in the bulk of hydrocarbons first filling axial traps and then migrating up-dip toward the hangingwall or peripheral bulge. On the footwall or thrust belt side, the drainage area of mature source-rocks is usually small, and can be almost negligible if sand-rich fans are well-developed and occupy most of the potential drainage area.

Axial fairways and systems tracts

These occur in the axial parts of basins that have attracted fine-grained potential hydrocarbon source-rocks and seals. The hydrocarbon 'charge' usually originates from this area.

The main advantages of the fairway are as follows:

(1) *Source presence.* Thick sequences of potential lacustrine source-rocks commonly develop in these depocenters. In long-lived lake sites, where there is prolonged subsidence, significant thicknesses of fine-grained clastics may accumulate, often 500 m or more (Table 2).

(2) *Source maturity and migration distances.* Source-rocks are more likely to become mature due to the greater subsidence and burial occurring in the axis. The distances from hydrocarbon source 'kitchens' to traps are usually very short and there are few migration problems because source, reservoir and seal are interbedded. Migration losses are likely to be low, that is, high 'migration efficiency'. This is particularly common where lake turbidites are interbedded with source rocks.

(3) *Seal development.* Because the axis attracts fine-grained sedimentation, seals are usually thickest, uniform and most widespread. The seals thin and interdigitate with other non-seal facies toward the basin edges. Many sequences of potential lacustrine source-rock are at the same time also good regional seal horizons.

The main disadvantages of the fairway are as follows:

(1) *Trap availability.* The axial parts of many basins have relatively few structural traps and those that are present are frequently relatively small.

(2) *Reservoir presence and quality.* The main potential reservoirs are lacustrine turbidites and axial deltas. Turbidites are typically thin, poor quality reservoirs. The proportion of potential reservoir is typically very low (<0.3) and the turbidites are characteristically thin-bedded (typically <1 m thick and normally <5 m) and matrix rich (e.g. McLaughlin & Nilsen, 1982; Changming *et al.*, 1984). Gravity flows can be expected to spread and fill the topographic lows in the lake floor.

The main sites for gravity flows and turbidites are in deep water at the front of lacustine fans and lacustrine deltas. They may be initiated by storms, by seismic shocks, by lake level fluctuations, by increases in discharge into the lake, or simply because the high sedimentation rates characteristic of lakes creates oversteepening and unstable pore pressure regimes

that result in slumping and slope failure. In a few instances, fan systems analogous to submarine fans may be built at the base of very large delta-front slopes where there are powerful rivers, for example Miocene fans in the Bekes Basin, Hungary (Phillips *et al.*, 1989).

Deltas commonly develop at the ends of lakes in elongate basins where rivers have a large catchment and drain axially. Deltas may often also develop where rivers drain the hangingwall of a half-graben or peripheral bulge in a flexural basin (e.g. Cohen, 1989). The principal reservoirs are deposits of distributary channels and mouth bars. The proportion of reservoir is typically low (<0.5) and sandbodies thin (<10 m). The sandbodies have poorly connected ribbon shapes (shoestrings), typically only 100–1000 m wide and separated by intraformational seals comprising overbank floodplain and shallow lake fines (Yinan Qui *et al.*, 1987). Crevasse splay deposits also form potential reservoirs but are usually poor quality reservoirs, typically only 1–3 m thick and incorporating considerable fines.

Deltas are major sites of deposition with very high sedimentation rates (often 1 mm/year or more, e.g. Frostick & Reid, 1986; Johnson *et al.*, 1987). Delta distributaries are also responsible for introducing much of the suspended sediment load into lakes, both under normal conditions and especially after heavy rainfall. Lacustrine deltas are typically river-dominated because waves are usually very small and tides non-existent in all but a few extremely large lakes. The deltas tend to build highly constructive forms with arcuate fronts. Birdsfoot deltas are common features.

Mudstone seals may develop in overbank areas on the delta plain and in interdistributary bays. These commonly only form local seals and discontinuous permeability barriers within deltaic reservoirs (Jin Yusun, Liu Dingzeng & Luo Changyan, 1985). Considerable thicknesses of lacustrine delta sequences may be present where the position of deltas is constrained by the basin tectonics and multiple pay zones are often present (e.g. Daqing Field, NE China, see Jin Yusun *et al.*, 1985; Yinan Qui, Peihua Xue & Jingsiu Xiao, 1987).

(3) *Overpressure.* Overpressuring is a common feature creating hazardous drilling conditions, and adding additional reservoir complexity. Abnormally high porosities may occur (in both mudstones and reservoir sandstones) because the rocks are undercompacted (e.g. Dagang Field, China, see Li & Wang, 1985).

Footwall and thrust belt fairways and systems tracts

Exploration principally revolves around locating alluvial/lacustrine fan systems (Fig. 5). Traps involving these fans frequently fail due to lack of effective seal either laterally across the footwall/thrust-belt, or vertically due to insufficient seal facies in the basin sequences. Transfer zones along the footwall and lateral ramps along the thrust belt are both important in controlling entry points of clastics into lake basins, and the positions of fans.

Consequently they are also focal points for hydrocarbons migrating out of the basin axis.

The main advantages of the fairway are as follows:

(1) *Migration distances and source maturity*. Migration distances from nearby 'deep lake' source kitchens are short when source-rocks developed in the adjacent basin axis. Coarse fan sediments may interfinger directly with source facies.

(2) *Reservoir quality*. Reservoirs may occur where fans deposit a thick pile of sandstones and conglomerates. In rifts, these reservoirs are commonly located along major fault scarps, with sediment typically entering transverse to the rift axis. Fan positions may be controlled by antecedent drainage systems, transfer zones or transverse faults. In flexural basins where fans commonly develop off the thrust belt, positions of the larger fans may be controlled by lateral ramps and thrust tip-lines (e.g. Hirst & Nichols, 1986; Nichols, 1987a, b).

(3) Fan sequences can attain great thicknesses (100–1500 m) with individual sandstone/conglomerate reservoir bodies 10–100 m thick. Whilst entire fans may comprise potential reservoir, they are characterized by extremely rapid basin-ward net-to-gross changes. In proximal areas, the proportion of potential reservoir may approach 1.0 but decrease rapidly to < 0.2 in under 2 km (e.g. see Sneh, 1979; Manspeizer, 1985; Nichols, 1987a, b). Most of the fan may have net to gross of between 0.4 and 0.8. At the same time that the proportion of reservoir decreases, sandbody thickness and vertical connectivity usually decrease markedly as beds wedge out into lake fines.

Shoreline facies around fans tend to form an extremely narrow strip which is difficult to locate in the subsurface. In between fans, and on inactive fans, shallow water biogenic carbonates can develop (e.g. Cohen, 1989). These are laterally impersistent in both dip and strike section (typically only a few kilometers), and only a few meters thick. Carbonates are usually low porosity/low permeability reservoirs due to their predominantly fine-grained, often algal micritic nature. Calcareous silt, and shell blankets and coquinas of gastropods and bivalves also form potential reservoirs (see Bertani & Carozzi, 1985). Meteoric/vadose diagenesis, giving rise to carbonate dissolution, is desirable in most lacustrine carbonates to enhance reservoir quality.

Disadvantages of the fairway include the following:

(1) *Seal*. Top seal on lacustrine fans is typically poor because the fans lie close to the basin edge where the facies are usually coarse-grained. It is unlikely that any intraformational seals will develop over an entire fan and extend over the fan apex up to the fault scarps and uplands from which the fans emerge. Any partial seals are likely to be extremely thin (< 10 m) and be broken by faults of small throw or cut out by channels. Good seals can develop, typically in sequences overlying the fans if lake level rises, the lake expands, choking and finally drowning the fans.

(2) *Drainage area*. The potential drainage area of lake source-

rocks is frequently small due to the asymmetrical structural configuration and in some instances, much of the potential drainage area may actually be occupied by lacustrine fan reservoir (cf. Fig. 5).

Hangingwall and peripheral bulge fairways and systems tracts

The comparatively low depositional slopes result in the position of shoreline and shallow lake facies belts moving large distances with only small changes in subsidence rate/water level. Consequently the sequences may contain reservoir and seal facies that are more extensively interdigitated than elsewhere in the basin.

Advantages of the fairway can be summarized as:

(1) *Source drainage*. Due to the asymmetry of many rifts and flexural basins, the hangingwall or peripheral bulge usually receives the bulk of the hydrocarbon charge from source-rocks in the basin axis.

(2) *Trap availability*. Numerous stratigraphic traps may exist (refer Fig. 5). Factors contributing to success of these traps are threefold. First, the strongly interdigitated nature of reservoirs and seals (contrast this to the difficulties of sealing on footwall and thrust belt fairways). Secondly, stratigraphic traps frequently develop throughout deposition of the sequences because slight changes in lake level may shift facies belts large distances. Many of these traps are available to receive a hydrocarbon charge from source-rocks toward the basin axis, as soon as these source-rocks begin to generate hydrocarbons. Thirdly, the traps may receive a relatively large hydrocarbon charge due to the asymmetry of the basin.

The main disadvantages of the fairway are:

(1) *Migration pathways*. Whilst there may be a large hydrocarbon charge directed towards the hangingwall or peripheral bulge, the distance laterally between traps and source kitchens is relatively large. The gentle dips characteristic of the hangingwall or peripheral bulge are not favorable toward focusing of hydrocarbons. Migration pathways therefore tend to be broad and losses of hydrocarbons during migration are often high. To better understand the fairways, it is particularly important to carefully model the structure of migration pathways throughout their burial history.

Seal potential increases basinward as the facies become finer-grained. However, no regional seals are likely because of the strong interbedding of lithologies (and facies); this adds further complexity to migration pathways.

(2) *Reservoir quality*. Shoreline and shallow lake facies are typically low energy as the depositional slope is very low and there are normally only small waves with limited fetch and no tidal currents. Many reservoir sandstones are thus very fine-grained and argillaceous. The depth to which the substrate is reworked and scoured is small, often only a meter or two. As a consequence 'offshore' bars and sand banks that might

form good quality reservoirs in a marine environment are minor or absent, and the shoreface zone is narrow, often with a rapid basinward transition to low energy silt and clay deposition (e.g. Nir, 1986).

Non-reservoir sequences are commonplace because the shoreline and shallow lake facies on the hangingwall or peripheral bulge may be entirely mud and silt. Frequent lithological variations result from small lake level fluctuations shifting facies belts large distances, for example variably colored muds, thin-bedded evaporites, paleosols and carbonaceous beds. Carbonates often develop in arid/semi-arid tropical/sub-tropical lakes where there is little clastic input.

Thick shoreline sandstone reservoirs are rare because the low depositional slopes ensure that small fluctuations in lake level cause the shoreline to move considerable distances laterally. Reservoirs may be very thin (1–2 m), but laterally extensive sheet-like bodies. The precise position of a shoreline at any one time is particularly difficult to locate, and most shoreline and shallow lake reservoir sandstones will be too thin or lack diagnostic criteria, to be recognizable on seismic sections.

Acknowledgements

I would like to thank many friends at BP who have helped develop the ideas and methods presented here, especially Tony Hayward, If Price, Richard Herbert, John Pooler and John Hurst. Jon Bellamy very kindly reviewed the manuscript, and suggested numerous improvements.

This contribution is published by permission of, and with assistance from, BP Exploration.

Bibliography

Anadon, P., Cabrera, L., Julia, R., Roca, E. & Rosell, L. 1989. Lacustrine oil-shale basins in Tertiary grabens from NE Spain (Western European Rift System). *Palaeogeogr., Palaeoclimatol., Palaeoecol.*, **70**, 7–28.

Atkinson, C.M., 1989. Coal and oil shale in Tertiary intermontane basins of Indonesia and eastern Australia. *International Symp. on Intermontane Basins: Geology and Resources, Chang Mai Univ., Chang Mai, Thailand*, Proceedings, pp. 77–88.

Bertani, R.T. & Carozzi, A.V., 1985. Lagoa Feia Formation (Lower Cretaceous), Campos Basin, offshore Brazil: rift valley stage lacustrine carbonate reservoirs – I & II. *J. Petrol. Geol.*, **8**, 37–58, 199–220.

Blair, T.C. & Bilodeau, W., 1988. Development of tectonic cyclothems in rift, pull-apart, and foreland basins: sedimentary response to episodic tectonism. *Geology*, **16**, 517–20.

Brice, S.E., Kelts, K.R. & Arthur, M.A., 1980. Lower Cretaceous lacustrine source beds from early rifting phases of south Atlantic. *Bull. Am. Assoc. Petrol. Geol.*, **64**, 680–1.

Burri, P., 1989. Hydrocarbon potential of Tertiary intermontane basins in Thailand. *International Symp. on Intermontane Basins: Geology and Resources, Chang Mai Univ., Chang Mai, Thailand*, Proceedings, p. 3–12.

Cabrera, Ll., Roca, E. & Santanach, P., 1988. Basin formation at the end of a strike-slip fault: the Cerdanya Basin (eastern Pyrenees). *J. Geol. Soc., London*, **145**, 261–8.

Changming, C., Jiakuan, H., Jingshan, C. & Xingyou, T., 1984. Depositional models of Tertiary rift basins, eastern China, and their application to petroleum prediction. *Sedim. Geol.*, **40**, 73–88.

Chinbunchorn, N., Pradidtan, S. & Sattayarak, N., 1989. Petroleum potential of Tertiary intermontane basins in Thailand. *International Symp. on Intermontane Basins: Geology and Resources, Chang Mai Univ., Chang Mai, Thailand*, Proceedings, pp. 29–42.

Cohen, A.S., 1989. Facies relationships and sedimentation in large rift lakes and implications for hydrocarbon exploration: examples from Lakes Turkana and Tanganyika. *Palaeogeogr., Palaeoclimat. Palaeoecol.*, **70**, 65–80.

Crossley, R., 1984. Controls on sedimentation in the Malawi rift valley, central Africa. *Sedim. Geol.*, **40**, 33–50.

Derksen, S.J. & McLean-Hodgson, J., 1988. Hydrocarbon potential and structural style of continental rifts: examples from East Africa and Southeast Asia. Seventh Offshore Southeast Asia Conference, Singapore, Proceedings, pp. 120–34.

Fei Qi & Wang Xie-Pei, 1984. Significant role of structural fractures in Renqui buried-hill oil field in eastern China. *Bull. Am. Assoc. Petrol. Geol.*, **68**, 971–82.

Flint, S., Stewart, D.J., Hyde, T., Gevers, E.C.A., Dubrule, O.R.F. & Van Riessen, E.D., 1988. Aspects of reservoir geology and production behavior of Sirikit oil field, Thailand: an integrated study using well and 3-D seismic data. *Bull. Am. Assoc. Petrol. Geol.*, **72**, 1254–69.

Frostick, L.E. & Reid, I., 1986. Evolution and sedimentary character of lake deltas fed by ephemeral rivers in the Turkana basin, northern Kenya. *Geol. Soc. London Spec. Publ.*, **25**, 113–25.

Gorst, I., 1986. Onshore key to increased output. *Petrol. Economist*, 451–7.

Haberyan, K.A. & Hecky, R.E., 1987. The Late Pleistocene and Holocene stratigraphy and palaeolimnology of Lakes Kivu and Tanganyika. *Palaeogeogr., Palaeoclimat., Palaeoecol.*, **61**, 169–97.

Henstridge, D.A. & Missen, D.D., 1982. Geology of oil shale deposits within the Narrows Graben, Queensland, Australia. *Bull. Am. Assoc. Petrol. Geol.*, **66.**, 719–31.

Hirst, J.P.P. & Nichols, G.J., 1986. Thrust tectonic controls on Miocene alluvial distribution patterns, southern Pyrenees. *Internat. Assoc. Sedimentologists*, Publ. No. 8, pp. 153–64.

Hu, J., Fan, C., Zhang, J., Liu, S., Xu, S. & Tong, X., 1984. *Stratigraphic-lithologic oil and gas pools in continental basins, China*. Beijing Petrol. Geol. Symp.

Jin Yusin, Liu Dingzeng & Luo Changyan, 1985. Development of Daqing oil field by water flooding. *J. Petrol. Tech.*, **37**, 269–74.

Johnson, T.C., Halfman, J.D., Rosendahl, B.R. & Lister, G.S., 1987. Climatic and tectonic effects on sedimentation in a rift-valley lake: evidence from high-resolution seismic profiles, Lake Turkana, Kenya. *Geol. Soc. Am. Bull.*, **98**, 439–47.

Li, S. & Wang, D., 1985. Relationship between diagenesis and porosity of sandstones on the southern and northern flanks of Bei Dagang structural zone. *Oil and Gas in China*, **1**, 21–35.

Lucas, P.T. & Drexler, J.M., 1975. Altamont-Bluebell: a major fractured and overpressured stratigraphic trap, Uinta Basin, Utah. *Rocky Mountain Assoc, Symposium, Deep Drilling Frontiers in the central Rocky Mountains, Denver, Colorado*, Proceedings, pp. 265–73.

Ma Li, Ge Taisheng, Zhao Xueping, Zie Taijun, Ge Rong & Dand Zhenrong, 1982. Oil basins and subtle traps in the eastern part of China. *Am. Assoc. Petrol. Geol. Memoir*, **32**, 287–315.

Manspeizer, W., 1985. The Dead Sea rift: impact of climate and tectonism on Pleistocene and Holocene sedimentation. *Soc. Econ. Paleon. Mineral.*, Spec. Publ., **37**, 143–58.

McLaughlin, R.J. & Nilsen, T.H., 1982. Neogene non-marine sedimentation and tectonics in small pull-apart basins of the San Andreas fault system, Sonoma County, California. *Sedimentology*, **29**, 865–76.

Nichols, G.J., 1987a. Structural controls on fluvial distributary systems – the Luna system, northern Spain. *Soc. Econ. Paleon. Mineral.*, Spec. Publ., **39**, 269–77.

Nichols, G.J., 1987b. Syntectonic alluvial fan sedimentation, southern Pyrenees. *Geol. Mag.*, **124**, 121–33.

Nilsen, T.H. & McLaughlin, R.J., 1985. Comparison of tectonic framework and depositional patterns of the Hornelen strike-slip basin of Norway and the Ridge and Little Sulphur Creek strike-slip basins of California. *Soc. Econ. Paleon. Mineral.*, Spec. Publ., **37**, 79–103.

Nir, Y., 1986. *Recent Sediments of Lake Kinneret (Tiberias), Israel*. Geol. Surv. Israel, Report No. GS1/18/86.

O'Leary, J. & Hill, G.S., 1989. Tertiary basin development in the southern central plains, Thailand. *International Symp. on Intermontane Basins: Geology and Resources, Chang Mai Univ., Chang Mai, Thailand*, Proceedings, pp. 254–64.

Osmond, J.C., Locke, R., Dille, A.C., Praetorius, W. & Wilkins, J.G., 1968. Natural gas in Uinta Basin, Utah. *Am. Assoc. Petrol. Geol. Memoir*, **9**, 174–98.

Phillips, R.L., Berczi, I., Mattick, R.E. & Rumpler, J., 1989. Lacustrine-deltaic processes and sedimentation in a deep lake basin, southeast Hungary. *Twenty-eighth Intern. Geol. Congress*, Abstracts 2, p. 605.

Powell, T.G., 1986. Petroleum geochemistry and depositional setting of lacustrine source rocks. *Marine Petrol. Geol.*, **3**, 200–19.

Ray, R.R., 1982. Seismic stratigraphic interpretation of the Fort Union Formation, western Wind River basin: example of subtle trap exploration in a nonmarine sequence. *Am. Assoc. Petrol. Geol. Memoir*, **32**, 169–80.

Schull, T.J., 1988. Rift basins of interior Sudan: petroleum exploration and discovery. *Bull Am. Assoc. Petrol. Geol.*, **72**, 1128–42.

Sneh, A., 1979. Late Pleistocene fan-deltas along the Dead Sea rift. *J. Sed. Petrol.*, **49**, 541–52.

Steidtmann, J.R., McGee, L.C. & Middleton, L.T. 1983. Laramide sedimentation, folding, and faulting in the southern Wind River Range, Wyoming. *Rocky Mountain Assoc. Geol. Symposium, Foreland Basins and Uplifts, Denver, Colorado*, Proceedings, pp. 161–7.

Swain, F.M., 1981. Petroleum in continental facies. In *Petroleum Geology in China*, ed. J.F. Mason, pp. 5–25. Pennwell Books, Tulsa, Oklahoma.

Taner, I., Kamen-Kaye, M. & Meyerhoff, A.A., 1988. Petroleum in the Junggar Basin, northwestern China. *J. Southeast Asian Earth Sci.*, **2**, 163–74.

Watson, M.P., Hayward, A.B., Parkinson, D.N. & Zhang, Z.M., 1987. Plate tectonic history, basin development, and petroleum source rock deposition onshore China. *Marine Petrol. Geol.*, **4**, 205–25.

Wood M.F. & Link, M.H., 1987. The non-marine Late Miocene to Pliocene Ridge Route Formation and its depositional framework. In *Sedimentary Facies, Tectonic Relations, and Hydrocarbon Significance in Ridge Basin, California*, ed. M.H. Link, pp. 5–20. Soc. Econ. Paleon. Mineral. Pacific Guidebook 51.

Xu Shice & Wang, Hengjian, 1981. Deltaic deposits of a large lake basin. In *Petroleum Geology in China*, ed. J.F. Mason, pp. 202–13. Pennwell Books, Tulsa, Oklahoma.

Yinan Qiu, Peihua Xue & Jingsiu Xiao, 1987. Fluvial sandstone bodies as hydrocarbon reservoirs in lake basins. *Soc. Econ. Paleon. Mineral.*, Spec. Publ., **39**, 329–42.

Yuretich, R.F., 1989. Paleocene lakes of the central Rocky Mountains, western United States. *Palaeogeogr., Palaeoclimat., Palaeoecol.*, **70**, 53–63.

Yuretich, R.F., Hickey, L.J., Gregson, B.P. & Hsia, Y.L., 1984. Lacustrine deposits in the Paleocene Fort Union Formation, northern Bighorn Basin, Montana. *J. Sed. Petrol.*, **54**, 836–52.

Zha Quanheng, 1984. Jizhong depression, China – its geologic framework, evolutionary history, and distribution of hydrocarbons. *Bull. Am. Assoc. Petrol. Geol.*, **68**, 983–92.

Zhai Guangming & Zha Quanheng, 1982. Buried-hill oil and gas pools in the North China Basin. *Am. Assoc, Petrol. Geol. Memoir.*, **32**, 317–35.

Zhang Qiming & Su Houxi, 1989. Hydrocarbon accumulations in the Beibuwan (Beibu Gulf) Basin. *China Earth Sciences*, **1**, 31–41.

Zhou Guangjia, 1981. Character of organic matter in source rocks of continental origin and its maturation and evolution. In *Petroleum Geology in China*, ed. J.F. Mason, pp. 26–47. Pennwell Books, Tulsa, Oklahoma.

East Greenland lacustrine complexes

GREGERS DAM AND LARS STEMMERIK

Geological Survey of Greenland, Øster Voldgade 10, DK-1350 Copenhagen K, Denmark

Introduction

The sedimentary basins between 70° and 74° N in East Greenland have a long non-marine history spanning the Middle Devonian to earliest Jurassic, almost 185 Ma, with the exception of a period of 13–15 Ma in the Late Permian and earliest Triassic. During this time the structural style in the region changed from regional extension and deposition in a broad basin during the Devonian to rifting and deposition in narrow tilted half-grabens during the Carboniferous and Early Permian (e.g. Surlyk, 1990; Surlyk *et al.*, 1981, 1986; Stemmerik *et al.*, 1993). The post-Permian basin history is controlled mainly by thermal subsidence. Thus, within the same area, it is possible to study lacustrine complexes formed during very different structural conditions. Furthermore, the northwards drift of the Laurasian continent produced a gradual change in climate from humid tropical, over arid subtropical to humid temperate during the Devonian–earliest Jurassic, with an overall trend towards cooler climates.

Most of the lacustrine successions exhibit one of two orders of pronounced cyclicity. This cyclicity can possibly be related to cyclic variations in precipitation controlled by orbital forcing on climate in most cases, but also penecontemporaneous tectonics and eustacy seem to have been governing factors.

Middle–Upper Devonian lacustrine complexes

The Middle–Upper Devonian succession includes more than 8 km of continental clastic and volcanic rocks (Friend *et al.*, 1983; Olsen & Larsen, 1993). The vast majority of these sediments were deposited in fluviatile and aeolian environments, but lacustrine complexes were developed both during the Middle and Upper Devonian (Figs 1, 2).

Orbital forcing on lacustrine cyclicity; Wimans Bjerg Formation (Famennian)

The playa-lake deposits of the Wimans Bjerg Formation contain playa mudflat, shallow ephemeral lake and ephemeral stream facies (Olsen, 1994).

(1) *Playa mudflat facies* consists of brecciated and apparently

massive silty mudstones. The brecciated mudstones are characterized by a patchy fabric riddled by superimposed desiccation cracks. Close examination of the massive mudstones indicates the presence of submillimeter scale cracks. Dolomite crystal rhombs and nodules of dolomite and quartz abound in the massive and brecciated mudstones. The brecciated fabric of the mudstones is probably due to repeated wetting and drying of clays and the brecciated and apparently massive mudstones are interpreted as playa mudflat deposits.

(2) *Shallow ephemeral lake facies* consists of laminated siltstones. The laminated siltstones are composed of centimeter-scale thick beds of siltstone or silty very fine sandstone with sharp and commonly erosive bases. Individual beds commonly exhibit wave-ripple lamination, parallel lamination, cross-lamination and parallel-draped lamination. Desiccation cracks commonly occur at the top of the beds. The laminated mudstones were deposited in shallow lakes experiencing rapid lake level fluctuations.

(3) *Ephemeral stream facies* consists of rare channel-shaped siltstone and sandstones that exhibit internal desiccation cracks. The siltstones and sandstones are interpreted as ephemeral stream deposits.

The Wimans Bjerg Formation exhibits a pronounced cyclicity (Olsen, 1993). Individual cycles consist of a unit dominated by ephemeral lake deposits and a unit exclusively composed of playa mudflat deposits. A higher order cyclicity is also observed in the variation in thickness of the ephemeral lake siltstones. The cyclicity indicates variations in environmental conditions between periods with frequent ephemeral lake expansions and contractions, and periods exclusively characterized by playa mudflats. The higher order of cyclicity indicates long-term variations in the periods of punctuated ephemeral lake conditions. The cyclic variations are in both cases explained by cyclic variations in precipitation controlled by orbital forcing on climate (precession and eccentricity cycles) (Olsen, 1993).

Uppermost Devonian–Lower Permian lacustrine complexes

The uppermost Devonian–Lower Permian succession is up to 3 km thick in outcrop (Stemmerik *et al.*, 1991). It consists of continental clastics that were deposited in a system of rotated half-

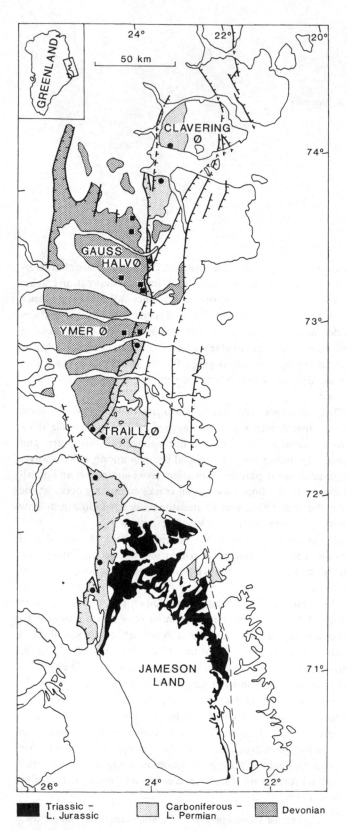

Triassic –
L. Jurassic

Carboniferous –
L. Permian

Devonian

Fig. 1. Simplified map of central East Greenland showing distribution of the Upper Paleozoic and Triassic–Lower Jurassic sediments. Major Upper Paleozoic lacustrine complexes marked: Devonian (squares) and Upper Carboniferous (circles).

grabens (Surlyk *et al.*, 1986). Lacustrine complexes are mainly confined to the western, down-faulted parts of the half-grabens (Christiansen *et al.*, 1990; Piasecki *et al.*, 1990; Stemmerik *et al.*, 1990). During the Early Permian temporary lakes also formed in more widespread floodplains (Fig. 2).

Anoxic lakes; Upper Carboniferous (Westphalian)

The Upper Carboniferous part of the uppermost Devonian–Lower Permian syn-rift package includes several lacustrine intervals along the western down-faulted margin (Christiansen *et al.*, 1990; Piasecki *et al.*, 1990; Stemmerik *et al.*, 1990) (Fig. 1). These intervals have a proposed areal extent of 150–500 km². The lacustrine deposits consists of hypolimnic and epilimnic facies (see Fig. 4).

(1) *Hypolimnic facies* is dominated by black laminated shale with a moderately high content of micaceous silt and fine-grained, windblown, quartz sand. The lamination is defined by couplets of mica-rich and organic-rich laminae. Associated with these laminated shales are non-laminated shales, mudstones with algae and shaly limestones with abundant ostracods.

The shales have a TOC (total organic carbon) content in the range 2–10% and 80–90% of the organic material is amorphous kerogen believed to be degraded algal material. Geochemically these shales have HI (hydrogen index) values in the range 300–800 and OI (oxygen index) values below 35.

The hypolimnic shales were deposited in the deep anoxic areas of the lakes, well below wave base.

(2) *Epilimnic facies* consists of grey silty shales and fine-grained sandstones, and includes abundant macroscopic plant fragments. The shales are laminated or micro cross-laminated. Individual laminae are commonly graded and have erosive bases.

The shales have TOC values in the range 0.5–7% and are dominated by coaly fragments and woody material. HI values are less than 80 and OI values are in the range 10–40.

Associated with the shales are thin, sandy wave cross-laminated and hummocky cross-stratified siltstones, and low-angle cross-bedded fine to medium-grained sandstones.

These sediments were deposited in the shallow oxic parts of the lakes. Their organic content indicates a significant contribution of river-transported material during deposition and most likely the shales represent pro-delta deposits in front of small lacustrine deltas. This is confirmed by their association with fluvial sandstones in the upper part of the lacustrine successions.

The Westphalian lacustrine successions are asymmetrical with the apparent deepest facies directly overlying coarse-grained alluvial deposits (Fig. 3). The successions are showing a gradual coarsening-upward trend and are terminated by sharply based fluvial sandstones, suggesting a structural control on the formation of the lakes (Stemmerik *et al.*, 1990).

Middle Triassic–Lower Jurassic lacustrine complexes

The Middle Triassic–Lower Jurassic lacustrine succession is up to 2 km in outcrop and is restricted to the southern end of the

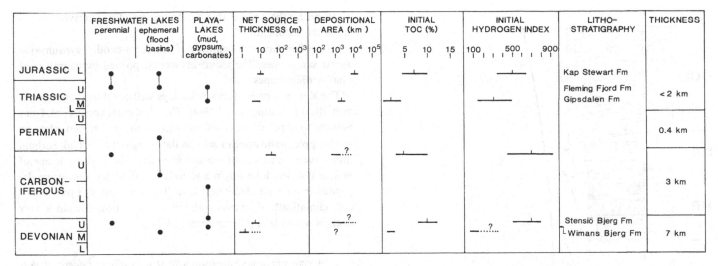

Fig. 2. Summary of lacustrine complexes known in East Greenland.

exposed part of the East Greenland rift system (Surlyk *et al.*, 1981) (Figs 1, 2, 4). Minor tectonic rejuvenation of the border faults and a major drop in relative sea-level isolated this basin during the Scythian. Bedded gypsum-bearing playa-lake complexes marked the beginning of the sedimentation in this lacustrine basin during (?) Anisian time (Clemmensen, 1978a, b, 1979, 1980a, b) (Fig. 4). They were gradually succeeded by Middle–Late Triassic playa mudstones, lacustrine carbonates and floodplain deposits indicative of a more humid climate (Bromley & Asgaard, 1979; Clemmensen, 1980a, b). By latest Triassic, Rhaetian, time the climate became warm and humid and the perennial wave and storm-dominated lacustrine basin of the Kap Stewart Formation developed. The marked change from semi-arid to humid conditions reflects a slow northwards drift of the Laurasian continent (cf. Smith *et al.*, 1981). In the Early Pliensbachian the lacustrine complex was transgressed and a shallow marine embayment was formed, probably reflecting the important eustatic sea-level rise during the Late Triassic–Early Jurassic (cf. Hallam, 1988).

Cyclic aeolian, sabkha and shallow-lake deposits; Gipsdalen Formation ((?) Anisian)

The basal Kolledalen Member of the Gipsdalen Formation consists of shallow lake, sabkha, aeolian and alluvial-fan facies (Clemmensen, 1978a) (Fig. 5).

(1) *Shallow lake facies* consists of red sandy micaceous, locally gypsum-bearing siltstones. The siltstones were deposited in a low-energy aquatic and evaporitic environment, probably a shallow lake.

(2) *Sabkha facies* consists of predominantly very fine to fine light-yellowish, reddish or rarely variegated sandstone with gypsum nodules. The sandstone generally lacks sedimentary structures but horizontal or cross-laminations do occur. Sandy lenses or streaks, or thin discontinuous sandy horizons in a matrix of darker finer-grained sandstones, are not uncommon. The sandstones were deposited in a warm evaporitic environment. The sandy lenses and

thin, discontinuous sandy horizons are thought to represent adhesion ripples. The parallel laminated sandstones represent aeolian sheet sands and the cross-laminated sandstones shallow water deposits. Thus, it appears that the sediments were deposited in an environment that alternated between aeolian and shallow water, and it is suggested that the depositional environment was a sabkha.

(3) *Aeolian dune facies* consists of fine to medium-grained, well-sorted, quartz-rich or gypsum-rich sandstones. In the quartz-rich sandstones, gypsum occurs frequently as small nodules or as a thin secondary crust seen on weathered surfaces. The sandstones are interpreted as aeolian dune deposits formed along the edge of a sabkha.

(4) *Alluvial fan facies* consists of structureless or occasionally cross-bedded pebbly sheet or channel sandstones, frequently arranged in fining-upward successions. The sandstones were deposited on a lower alluvial fan during periods of sheetflooding.

The overlying Kap Seaforth Member consists of sabkha flat-marginal desert lake, marginal desert lake, expanded sabkha lake and aeolian facies (Clemmensen, 1978a).

(1) *Sabkha flat-marginal desert lake facies* consists of parallel laminated greyish-red or greenish mudstone. Wave-rippled silty and sandy laminae frequently occur. Casts of halite crystals and mud-cracked surfaces are locally present. The depositional environment was one of a sabkha flat or marginal areas of a shallow desert lake.

(2) *Marginal desert lake facies* consists of green, fine-grained sandstones closely interbedded with thin variegated, irregular to horizontally laminated silty mudstones with gypsum nodules and layers. The sandstones display well-preserved wave-ripples, an abundance of halite crystal casts and incipient desiccation cracks. These sediments denote the marginal areas of a shallow desert lake.

(3) *Sabkha lake facies* consists of red gypsiferous fine-grained silty sandstones. The sandstones are structureless or irregular bedded. Wave ripples and horizons with incipient wavy cross-lamination frequently occur. Bedding planes often display desiccation cracks. These sediments were deposited in a sabkha

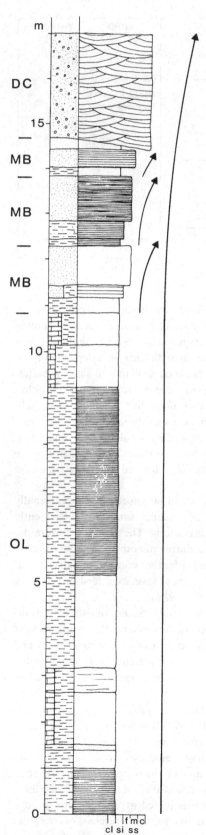

Fig. 3. Detailed sedimentological log of Upper Carboniferous lacustrine sediments in southwestern Traill. OL: Open lacustrine; MB: Mouth bar; DC: Distributary channel.

environment that frequently was flooded, resulting in wave-working of the sediment.

(4) *Aeolian dune facies* consists of red, cross-bedded gypsum-rich quartz sandstones. The sediments were deposited by migration of small aeolian dunes.

The Kap Seaforth Member displays well-developed cyclic sedimentation (Clemmensen, 1978a). The ideal cycle consists of, from bottom to top: cross-bedded aeolian sandstone, structureless or wave-rippled sandstone of sabkha flat-marginal desert lake origin, thin wave-rippled sandstone and irregularly bedded mudstone of marginal desert lake origin and parallel laminated mudstone deposited in an expanded desert lake. The sedimentary cycles delineate climatically determined shoreline oscillations within a very shallow desert lake (Clemmensen, 1978a).

Cyclic playa-lake deposits with stromatolites; Edderfugledal Member (basal Flemming Fjord Formation) (?Ladinian)

The Edderfugledal Member represents open lake deposition (Clemmensen, 1978b, 1980a, b) (Fig. 6). The lower part of the Edderfugledal Member contains open lacustrine, dolomitic shoreline and stromatolitic shoreline facies, representing a quasi-saline stage of the lake.

(1) *Open lacustrine facies* consists of green, parallel laminated siliciclastic mudstones. The mudstones are rich in organic matter and show no evidence of bioturbation. They were deposited in a shallow offshore environment under reducing conditions during periods of high lake-level.

(2) *Dolomitic shoreface facies* comprises a varied assemblage of laminated siltstone with dolomicritic laminae, wave-rippled sandstones and structureless lime mudstones composed of clotted dolomicrite with algal filament mounds. The structureless lime mudstone probably formed in a shallow water environment characterized by algal blooms. The laminated siltstone and wave-rippled sandstone were probably dolomitized by evaporitic pumping in a marginal carbonate mudflat environment experiencing frequent subaerial exposure. These facies formed during times of low lake level.

(3) *Stromatolitic shoreface facies* includes laterally-linked hemispheroidal and cumulate stromatolitic structures which consist of millimeter-scale couplets of dark grey calcite laminae and yellow-brown dolomite laminae. They formed under wave or current activity in the shallow nearshore environment, during periods of relative lake stability. Flat pebble conglomerates containing peloids and intraclasts of stromatolitic limestone and oolitic limestone frequently occur associated with the stromatolitic facies. They were deposited in a nearshore or beach environment by wave-reworking of collapsed stromatolites or desiccated mudflats.

The upper part of the Edderfugledal Member contains the previous three facies together with an additional two facies representing fringing sand and mud flats (Fig. 7).

(4) *Sand flat facies* consists of wave-rippled sheet sandstones interbedded with thin desiccation-cracked mudstones, cross-stratified oolitic calcarenites and thin coquinas. These sediments were

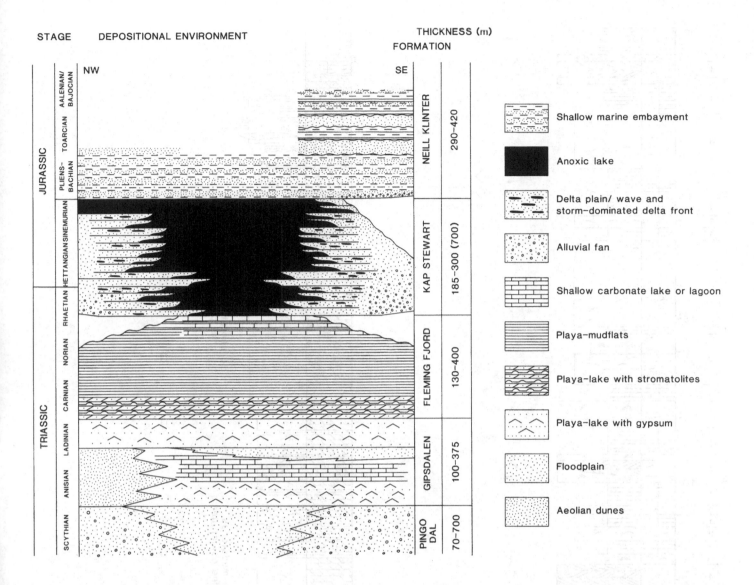

STAGE   DEPOSITIONAL ENVIRONMENT

THICKNESS (m)
FORMATION

Shallow marine embayment

Anoxic lake

Delta plain/ wave and storm-dominated delta front

Alluvial fan

Shallow carbonate lake or lagoon

Playa-mudflats

Playa-lake with stromatolites

Playa-lake with gypsum

Floodplain

Aeolian dunes

Fig. 4. Stratigraphy and depositional environments of the Middle Triassic–Lower Jurassic succession in the Jameson Land Basin. From Dam & Surlyk (1993).

deposited on a shallow shoreline sandflat subject to intermittent wave activity and periodic exposure.

(5) *Alluvial mudflat facies* consists of mudstones containing abundant desiccation cracks. They are intensively bioturbated and contain small calcrete nodules. They were deposited on an alluvial mudflat.

Climatic changes resulting in lake-level fluctuations are suggested as the cause of the well-developed cyclicity observed in the Edderfugledal lake (Clemmensen, 1978b). Moreover, there is evidence to suggest that the lake evolved from fairly saline to nearly fresh water (Clemmensen, 1978b; Bromley & Asgaard, 1979). This change is linked with an overall climatic change from arid to more humid during the Triassic period in East Greenland.

Cycles in a large wave- and storm-dominated anoxic lake; Kap Stewart Formation (Rhaetian-Sinemurian)

The Kap Stewart Formation represents a large, wave- and storm-dominated, anoxic lake (Dam & Christiansen, 1990; Dam, 1991; Dam & Surlyk, 1992, 1993). The Kap Stewart Formation contains open lacustrine, delta front, delta plain, alluvial plain and delta abandonment facies (Fig. 8).

(1) *Open lacustrine facies* consists of black, parallel laminated kaolinitic mudstones, in places with siderite concretions and beds of coarse-grained sandstones with symmetrical ripples. The mudstone typically contain 2–10% TOC. The HI ranges from approximately 100 to more than 700. The organic matter is a mixture of freshwater algal (*Botryococcus*) and higher plant remains. The mudstones were deposited in a stratified lake with anoxic bottom waters during periods of high lake level.

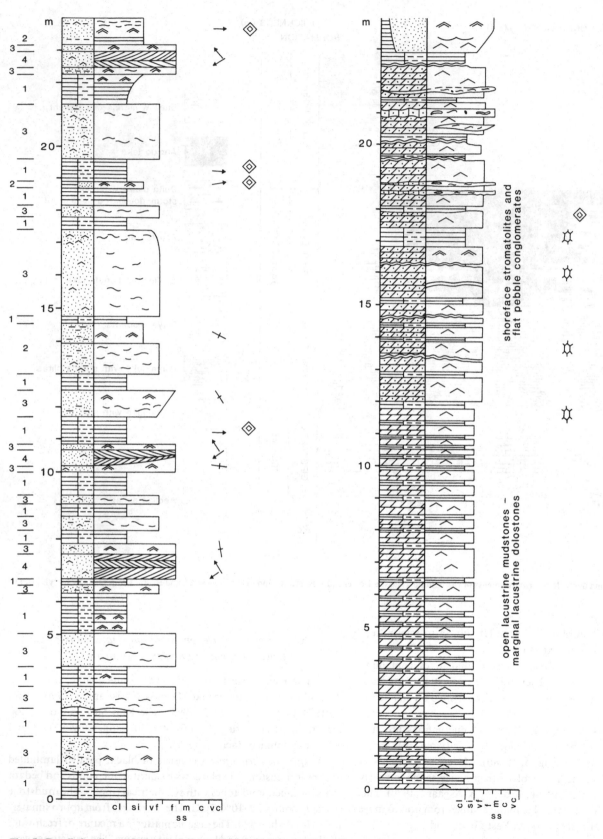

Fig. 5. Detailed sedimentological log of interbedded aeolian, sabkha and shallow-lake deposits from the Middle Triassic Gipsdalen Formation in Jameson Land. From Clemmensen (1978a).

Fig. 6. Detailed sedimentological log of playa-lake deposits with stromatolites from the basal part of the Middle Triassic Edderfugledal Member in Jameson Land. From Clemmensen (1978b).

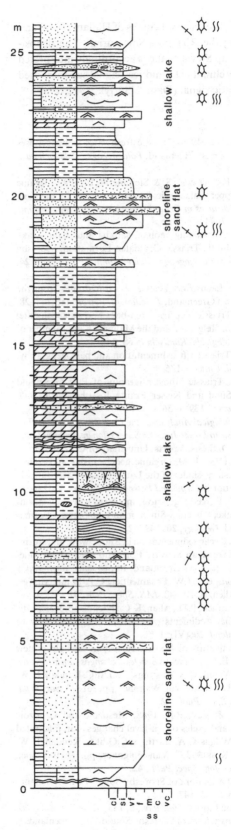

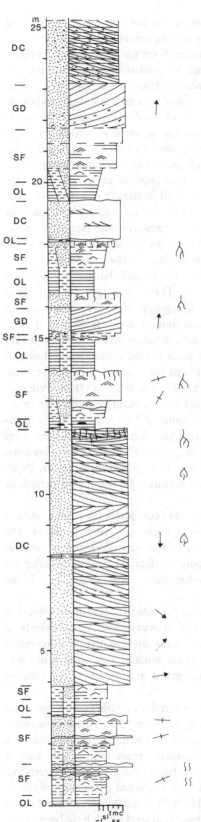

Fig. 7. Detailed sedimentological log of the shallow lacustrine Pingel Dal Beds of the upper part of the Edderfugledal Member. From Clemmensen (1978b).

Fig. 8. Detailed sedimentological log of wave and storm-dominated anoxic lake deposits from the Upper Triassic–Lower Jurassic Kap Stewart Formation in Jameson Land. OL: Open lacustrine; GD: Gilbert delta (terminal lobe); DC: Distributary channel.

(2) *Delta front facies* consists of heterolithic sediments and sandstones, including deposits of the shoreface, terminal lobe and distributary channels. The lower shoreface deposits include wave- and storm-dominated interbedded mudstones, siltstones and very fine-grained sandstones showing incipient wave-ripple lamination. Upper shoreface deposits consist of parallel laminated, wave ripple cross-laminated, hummocky and swaley cross-stratified sandstones. The shoreface deposits are commonly overlain by single foreset beds composed of medium to very coarse-grained sandstone. The foreset beds are interpreted as terminal lobe deposits.

The shoreface and terminal lobe deposits are overlain erosively by an upwards-fining sandstone unit, initiated by a conglomerate of logs, mudstone and coal intraclasts, and quartzite pebbles followed by very coarse pebbly to medium-grained parallel laminated, cross-bedded and cross-laminated sandstone. The fining-upward sandstone units are interpreted as distributary channel deposits.

(3) *Delta plain facies* consists of interbedded distributary channel and interdistributary deposits. The distributary channel facies consists of single, sheet-like, fining-upward pebbly conglomerates and sandstones. The interdistributary deposits consist of non-laminated to laminated kaolinitic mudstones interbedded with thin sheet sandstones. Abundant plant fossils, drifted logs, allochthonous coal seams and rootlet beds are interbedded with the mudstones. Harris (1926, 1931a, b, 1932a, b, 1935, 1937) and Pederson & Lund (1980) recorded a very diverse floral assemblage from the interdistributary deposits, dominated by cycadophytes, ginkgophytes, conifers, pteridosperms and ferns. The mudstones were deposited from suspension in protected lacustrine or lagoonal environments on the delta plain. The sheet sandstones were deposited from ephemeral hyperconcentrated flows during periods of intense overbank flooding.

(4) *Alluvial plain facies* consists of conglomeratic sheet sandstone bodies arranged in multistorey fining-upward successions. The conglomeratic sheet sandstone bodies were deposited in low sinuosity to braided rivers sweeping an alluvial plain, representing the more proximal, fluvially-dominated part of the deltaic distributaries.

(5) *Delta abandonment facies* consists of sheets of wave-worked shoreface deposits (as described above) and coal seams overlying rootlet beds. The thin coal seams and rootlet beds represent compressed autochthonous plant remains deposited above seat earths when the backswamp spread over the delta front during abandonment.

Wave ripple data show that water depth during deposition of the delta front sands was generally less than 15 m. This shallow-water depth and the evidence for heavy wave action is incompatible with long-term anoxia in a lake covering more than 12,000 km². A sequence stratigraphic interpretation of the succession suggests that the deltaic sheet sandstones were deposited during forced regressions (Dam & Surlyk, 1992, 1993). The lake thus experienced a large number of high-frequency changes in level which have been related to Milankovitch-type climatic cycles. The high-frequency cycles of the Kap Stewart Formation can be grouped in a number of low-frequency cycles. These cycles seem to show the same trend and

number of fluctuations as the sea-level curves of Hallam (1988) and Haq *et al.* (1988) during the Rhaetian–Sinemurian. The depositional evolution of the Kap Stewart lake is accordingly interpreted to show a high-frequency climatically-induced signal superimposed on a low-frequency eustatic signal (Dam & Surlyk, 1992, 1993).

Bibliography

Bromley, R.G. & Asgaard, U., 1979. Triassic freshwater ichnocoenoses from Carlsberg Fjord, East Greenland. *Palaeogeogr., Palaeoclimat., Palaeoecol.,* **28,** 39–80.

Christiansen, F.G., Olsen, H., Piasecki, S. & Stemmerik, L., 1990. Organic geochemistry of Upper Palaeozoic lacustrine shales in the East Greenland Basin. *Advances in Organic Geochemistry 1989. Organic Geochemistry,* **16,** 287–94.

Clemmensen, L.B., 1978a. Alternating aeolian, sabkha and shallow-lake deposits from the Middle Triassic Gipsdalen Formation, Scoresby Land, East Greenland. *Palaeogeogr., Palaeoclimat., Palaeoecol.,* **24,** 111–35.

Clemmenson, L.B., 1978b. Lacustrine facies and stromatolites from the Middle Triassic of East Greenland, *J. Sediment. Petrol.,* **48,** 1111–28.

Clemmensen, L.B., 1979. Triassic lacustrine red-beds and palaeoclimate: the 'Buntsandstein' of Helgoland and the Malmros Klint Member of East Greenland. *Geologische Rundschau,* **68,** 748–74.

Clemmensen, L.B., 1980a. Triassic rift sedimentation and paleogeography. *Bull. Grønlands Geol. Unders.,* **136,** 1–72.

Clemmensen, L.B., 1980b. Triassic lithostratigraphy of East Greenland between Scoresby Sund and Kejser Frantz Josephs Fjord. *Bull. Grønlands Geol. Unders.,* **139,** 1–56.

Dam, G., 1991. *A Sedimentological Analysis of the Continental and Shallow Marine Upper Triassic to Lower Jurassic Succession in Jameson Land, East Greenland.* Ph.D. thesis, 6 parts, University of Copenhagen.

Dam, G. & Christiansen, F.G., 1990. Organic geochemistry and source potential of the lacustrine shales of the Late Triassic–Early Jurassic Kap Stewart Formation. *Marine and Petroleum Geology,* **7,** 428–43.

Dam, G. & Surlyk, F. 1992. Forced regressions in a large wave and storm-dominated anoxic lake, Rhaetian-Sinemurian Kap Stewart Formation, East Greenland. *Geology,* **20,** 749–52.

Dam, G. & Surlyk, F. 1993. Cyclic sedimentation in a large wave and storm-dominated anoxic lake; Kap Stewart Formation (Rhaetian–Sinemurian), Jameson Land, East Greenland. In *Sequence stratigraphy and facies associations,* ed. H.W. Posamentier, C.P. Sommerhayes, B.U. Haq & G.P. Allen, p. 419–48. *IAS. Spec. Publ.,* **18.**

Friend, P.F., Alexander-Marrack, P.D., Allen, K.C., Nicholson, J. & Yeats, A.K., 1983. Devonian sediments of East Greenland. Review of results. *Meddr Grønland,* **206**(VI), 1–96.

Hallam, A.V. 1988. A reevaluation of Jurassic eustasy in the light of new data and the revised Exxon curve. In *Sea-level changes – an integrated approach,* ed. C.K. Wilgus, C.A. Hastings, C.G.St.C. Kendall, H.W. Posamentier, C.A. Ross & J.C. Van Wagoner, pp. 261–73. *Soc. Econ. Paleont. Mineralog. Spec. Publ.,* **42.**

Haq, B.U., Hardenbol, J. & Vail, P.R. 1988. Mesozoic and Cenozoic chronostratigraphy and cycles of sea-level changes – an integrated approach, ed. C.K. Wilgus, C.A. Hastings, C.G.St.C. Kendall, H.W. Posamentier, C.A. Ross & J.C. Van Wagoner, pp. 71–108. *Soc,. Econ. Paleont. Mineralog. Spec. Publ.,* **42.**

Harris, T.M., 1926. The Rhaetic flora of Scoresby Sound East Greenland. *Meddr Grønland,* **68**(2), 45–147.

Harris, T.M., 1931a. Rhaetic Floras. *Biol. Rev.,* **6,** 133–62.

Harris, T.M., 1931b. The fossil flora of Scoresby Sound East Greenland. 1. Cryptogams. *Meddr Grønland,* **85**(2), 1–102.

Harris, T.M., 1932a. The fossil flora of Scoresby Sound East Greenland. 2. Descriptions of seed plants *incertae sedis* together with a discussion of certain cycadophyte cuticles. *Meddr Grønland,* **85**(3), 1–112.

Harris, T.M., 1932b. The fossil flora of Scoresby Sound East Greenland. 3. Caytoniales and Bennittitales. *Meddr Grønland*, **85**(5), 1–133.

Harris, T.M., 1935. The fossil flora of Scoresby Sound East Greenland. 4. Ginkgoales, Coniferales, Lycopodiales and isolated fructifications. *Meddr Grønland*, **112**(1), 1–176.

Harris, T.M., 1937. The fossil flora of Scoresby Sound East Greenland. 5. Stratigraphic relations of the plant beds. *Meddr Grønland*, **112**(2), 1–114.

Harris, T.M. 1946. Liassic and Rhaetic plants collected in 1936–38 from East Greenland. *Meddr Grønland*, **114**(9), 1–41.

Olsen, H. 1994. Orbital forcing on continental depositional systems – lacustrine and fluvial cyclicity in the Devonian of East Greenland. In *Orbital Forcing and Cyclic Sequences*. ed. P.L. de Boer & D.G. Smith, *IAS. Spec. Publ.* (In press.)

Olsen, H. & Larsen, P.-H., 1993. Lithostratigraphy of the continental Devonian sediments in North-East Greenland. *Bull. Grønlands geol. Unders.*, **165**, 1–108.

Pedersen, K.R. & Lund, J.J., 1980. Palynology of the plant-bearing Rhaetian to Hettangian Kap Stewart Formation, Scoresby Sund, East Greenland. *Rev. Palaeobot. Palynol.*, **31**, 1–69.

Piasecki, S., Christiansen, F.G. & Stemmerik, L., 1990. Depositional history of a Late Carboniferous organic-rich lacustrine shale from East Greenland. *Bull. Can. Petrol. Geol.*, **38**, 273–87.

Smith, A.G., Hurley, A.M. & Briden, J.C., 1981. *Phanerozoic Paleocontinental World Maps*. Cambridge University Press, Cambridge.

Stemmerik, L., Christiansen, F.G. & Piasecki, S., 1990. Carboniferous lacustrine shales in East Greenland – an additional source rock in the northern North Atlantic? In *Lacustrine exploration case studies and modern analogues*, ed. B. Katz, pp. 277–86. *Mem. Am. Ass. Petrol. Geol.* **50**.

Stemmerik, L., Christiansen, F.G., Piasecki, S., Jordt, B., Marcussen, C. & Nøhr-Hansen, H., 1993. Depositional history and petroleum geology of the Carboniferous to Cretaceous sediments in the northern part of East Greenland. In *Arctic geology and petroleum potential*, ed. T.O. Vorren, E. Bergsager, Ø.A. Dahl-Stannes, E. Holter, B. Johansen, E. Lie & T.B. Lund., pp. 67–87. NPD, Spec. Publ., **2**.

Stemmerik, L., Vigran, J.O. & Piasecki, S., 1991. Dating of late Paleozoic rifting events in the North Atlantic: New biostratigraphic data from the uppermost Devonian and Carboniferous of East Greenland. *Geology*, **19**, 218–21.

Surlyk, F., 1977. Mesozoic faulting in East Greenland. In *Fault tectonics in NW Europe*. ed. R.T.C. Frost & A.J. Dikkers, (eds.) *Geologie en Mijnbouw*, **56**, 311–27.

Surlyk, F., 1978. Jurassic basin evolution of East Greenland. *Nature*, **274**(5667), 130–3.

Surlyk, F., 1990. Timing, style and sedimentary evolution of Late Palaeozoic-Mesozoic extensional basins of East Greenland. In *Tectonic events responsible for Britain's oil and gas reserves*, ed. R.F.P. Hardman & J. Brooks. *Geol. Soc. Lond. Spec. Publ.*, **55**, 107–25.

Surlyk. F., Clemmensen, L.B. & Larsen, H.C., 1981. Post-Paleozoic evolution of the East Greenland continental margin. In *Geology of the North Atlantic Borderlands*. ed. J.W. Kerr, & A.J. Ferguson, *Mem. Can. Soc. Petrol*, **7**, 611–645.

Surlyk, F., Hurst, J.M., Piasecki, S., Rolle, F., Scholle, P.A., Stemmerik, L. & Thomsen, E., 1986. The Permian of the western margin of the Greenland Sea – A future exploration target. In *Future Petroleum Provinces of the World*. ed. M.T. Halbouty, *Am. Ass. Petrol. Geol. Mem.*, **40**, 629–59.

Chemistry of closed basin lake waters: a comparison between African Rift Valley and some central North American rivers and lakes

T.E. CERLING

Department of Geology and Geophysics, University of Utah, Salt Lake City, Utah 84112, USA

The chemistry of closed-basin lakes is determined by the composition of inflowing waters and the chemical reactions taking place as the water evaporates. The composition of inflowing waters is determined by bedrock geology. Therefore the primary minerals formed in lakes and preserved in lake sediments is directly related to the bedrock of the drainage basin and the subsequent evolution of the basin waters. Because certain minerals, such as gypsum, can form as primary or as secondary minerals it is extremely important to know if such a mineral is an indicator of the original water chemistry or of some later diagenetic mechanism. In this contribution I examine and compare the chemistry of river and lake waters from two very different bedrock geologies. The bedrock in the East African Rift is predominantly volcanic, whereas that in north central North America is of mixed sedimentary rocks, including glacial till, that contain a lot of shale.

Weathering of volcanic rocks produces cations (mostly Na^+ and Ca^{+2}) and anions (HCO_3^-) because silicates are broken down by silicate hydrolysis reactions. Sedimentary rocks, on the other hand, have minerals such as calcite, gypsum and halite which readily dissolve and pyrite which is easily oxidized. Waters draining shales have an additional cation exchange reaction which releases Na^+ into the water while Ca^{+2} is absorbed by clay minerals (Cerling *et al.*, 1989, 1990). Thus, the latter waters are often high in Na^+ and SO_2^{-2}, with HCO_3^- and Cl^- often being subordinate in concentration to the other anions. Therefore, the water chemistry of rivers that drain exclusively volcanic, limestone, or shale terrains are different (Fig. 1). Because the first minerals to form in evaporating solutions are salts, the ratios of various cations and anions are extremely important in determining the evolutionary pathway of an evaporating brine. This important concept, that of chemical divides, was developed by Hardie & Eugster (1970) to generalize the mass balance relationships of evaporating waters using the approach of Garrels & MacKenzie (1967) (see also Drever, 1988, for discussion).

The first mineral to form is generally calcite by the reaction:

$$CA^{+2} + 2HCO_3^- = CaCO_3 + H_2O$$

The ratio $[m(HCO_3^-)/2m(Ca^{+2})]$ is extremely important because either the Ca^{+2} ion or the HCO_3^- ion will be consumed first. If Ca^{+2}

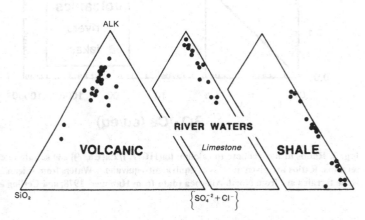

Fig. 1. Water chemistry of rivers draining exclusively volcanic (left), limestone (center), or shale (right) terrains normalized to the molar concentrations of dissolved SiO_2^-, HCO_3^-, and $[SO_4^{-2} + Cl^-]$.

is consumed first, the water will have an excess of alkalinity (HCO_3^-) after calcite precipitation. A second extremely important chemical divide is that which forms gypsum:

$$Ca^{+2} + SO_4^{-2} + 2H_2O = CaSO_4 \cdot 2H_2O$$

Thus the ratio $[m(SO_4^{-2})/m(Ca^{+2})]$ is also important because it also determines if the waters will tend to be depleted in Ca^{+2} or in SO_4^{-2}.

In the preceding paragraphs I showed that waters draining volcanic terrains should have high $m(HCO_3^-)/2m(Ca^{+2})$ ratios (units of eq/eq) and should therefore evolve along a pathway to Na-HCO_3 alkaline waters. Waters from shale terrains, on the other hand, should have high $m(SO_4^{-2})/m(Ca^{+2})$ ratios and should therefore evolve into Na-SO_4 waters. Fig. 2 shows the ratios and the evolution of waters from volcanic terrains in East Africa and waters from shale terrains in North America. This figure shows that the composition of rivers and lakes with these two different bedrock sources have vastly different chemical evolutionary pathways and composition.

These different evolutionary pathways of waters means that the primary minerals that can form in the lakes are very different: calcite, tri-octahedral smectite, analcime, and various sodium bicarbonate-carbonate minerals form in lakes where volcanic

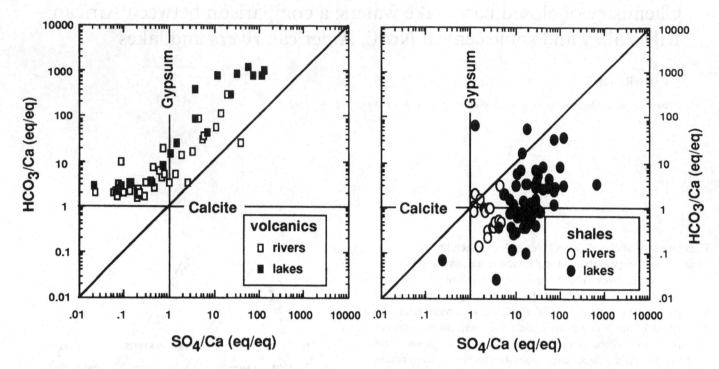

Fig. 2. Ratios of bicarbonate to calcium [$m(HCO_3^-)/2m(Ca^{+2})$] and sulfate to calcium [$m(SO_4^{+2})/m(Ca^{+2})$] in rivers and lakes from either volcanic or shale terrains. Ratios are given in units of equivalent/equivalent. Waters from volcanic terrains are from East Africa (Cerling, 1979, and unpublished); waters from shale terrains are from North America (data from Hammer, 1978, and Cerling *et al.*, 1989).

terrains predominate, whereas the sulfate suite of minerals is very important in shale terrains. Significantly, all of the East African lakes studied are undersaturated with respect to gypsum (Cerling, 1979), whereas virtually all of the North American lakes from shale terrains are at or slightly supersaturated with respect to gypsum. This has important consequences when interpreting the meaning of minerals observed in lacustrine deposits. For example, the mineral gypsum, which occurs as a secondary mineral in East African lake sediments such as in the Turkana Basin, cannot be considered to be indicative of evaporative conditions in terrains where $m(HCO_3^-)/2m(Ca^{+2})$ ratios are greater than 1! More likely, it forms by secondary oxidation of sulfide minerals as they are exposed during weathering processes. The dominant mechanism for sulfate loss in modern East African lakes is by sulfate reduction and the formation of iron sulfide minerals. For the example, of East African lake sediments such as those in the Turkana Basin, these considerations of chemical evolution of waters have important consequences on the interpretation of mineral assemblages in lake sediments. For those sediments, the presence of gypsum indicates sulfide oxidation during weathering rather than highly evaporated waters at the time

the sediments were deposited in the paleo-lake. Therefore, it is extremely important to understand the evolutionary pathway of an evolving water system in the study of paleolimnology.

Bibliography

Cerling, T.E., 1979. Paleochemistry of Plio-Pleistocene Lake Turkana, Kenya. *Palaeogeogr., Palaeoclimatol., Palaeoecol.*, **27**, 247–85.

Cerling, T.E., Pederson, B.L. & Von Damm, K.L., 1989. Sodium-calcium ion exchange in the weathering of shale: implications for global weathering budgets. *Geology*, **17**, 552–4.

Cerling, T.E., Pederson, B.L. & Von Damm, K.L., 1990. Reply to comment on: sodium-calcium ion exchange in the weathering of shale: implications for global weathering budgets. *Geology*, **18**, 63–5.

Drever, J.I., 1988. *Geochemistry of Natural Waters*. Prentice-Hall, Englewood Cliffs.

Garrels, R.M. & MacKenzie, F.T., 1967. Origin of the chemical compositions of some springs and lakes. *Equilibrium Concepts in Natural Water Systems*, pp. 222–42. Am. Chem. Soc. Adv. Chem. Ser. 67.

Hammer, U.T., 1978. The saline lakes of Saskatchewan. III. Chemical characterization. *Internat. Rev. Gesamt. Hydrobiol.*, **63**, 311–35.

Hardie, L.A. & Eugster, H.P. 1970. The evolution of closed-basin brines. *Mineral. Soc. Am. Spec. Pub.*, **3**, 273–90.

Mesozoic–Cenozoic lacustrine sediments of the Zaïre Interior Basin

J.-P. COLIN

Esso Rep, 213 Cours Victor Hugo, F-33323, Bègles, France

Thick fluvial and lacustrine sediments of Middle Jurassic to Eocene age occupy most of the superficy of the Zaïre Interior Basin (Fig. 1). The following succession is recognized.

Lualaba Group

Stanleyville Formation

Lithology

Basal conglomerates unconformably overlying Permo-Triassic Karroo sediments, overlain by alternating shales and sandstones, bituminous at many levels, and especially in the lower half. Thickness of 460 m in the type-area, around Kisangani. Based on lithology, the Stanleyville Formation has been divided into 14 horizons (Fig. 2).

Paleontology

Ostracodes Grékoff (1957) subdivided the Stanleyville Formation into nine ostracode zones, each defined by the first appearance of one or more ostracode species. These are, from bottom to top (Fig. 2):

Zone 1 (= horizon 2D): *Darwinula leguminella, Darwinula oblonga* and *Theriosynoecum duboisi.*

Zone 2 (= horizons 3 to 6A): *Theriosynoecum lualabaensis, T. passaui, Klieana?* sp. and *Reconcavona? longaensis.*

Zone 3 (= horizons 6B to 7A): *Theriosynoecum longaensis, T. bitufoensis.*

Zone 4 (= horizon 7F): *Theriosynoecum roberti.*

Zone 5 (= horizon 8): *Theriosynoecum mushopensis.*

Zone 6 (= horizons 9A to 9B): *Theriosynoecum kanianiaensis.*

Zone 7 (= horizons 9C to 12): *Theriosynoecum litokoensis.*

Zone 8 (= horizon 13): *Theriosynoecum cornutum, Reconcavona? paupera, Metacypris?* 12040 and *Salvadoriella* 12089.

Zone 9 (= horizon 14): *Reconcavona? liloensis.*

Phyllopods Defretin-Lefranc (1967) subdivided the series using phyllopods:

Horizon 2: *Polygrapta biaroensis, Bairdestheria caheni.*

Horizons 3–4: *Paleolimnadiopsis lombardi, Palaeestheria passaui.*

Horizon 5: *Glyptoasmussia corneti.*

Horizons 6–10: *Pseudoasmussia duboisi, Bairdestheria evrardi.*

Horizons 11–14: *Bairdestheria lualabaensis.*

Palynology The following forms were reported (Bose, 1977; Maheshwari et al., 1977b): *Cerebropollenites mesozoicus, Classopollis* sp., *Corrugatisporites* sp., *Cyathidites* sp., *Eucomiidites* sp., *Inaperturopollenites* cf. *indicus, Laevigatosporites ovatus, Laricodites desquamatus, Laricodites magnus, Perinopollenites alatoides, Perotriletes reticulatus, Piceapollenites, Psilasporites* cf. *marcidus* and *Schizosporites parvus.*

Fishes Fossil fishes occur at several levels of the Stanleyville and have been studied in detail by de Saint-Seine (1953, 1955). The most common species in horizon 2 (Lime Fine) is *Ophiopsis lepersonnei* which has strong Middle Jurassic affinities, although originally dated as Kimmeridgian. Other reported species are *Austropleuropholis lombardi, Lualabea henryi, L. lerichei, Lepidotes congolensis, Catervariolus hornemanni, C. passaui, Lombardia decorata, Signeuxella preumonti, Pleuropholis lannoyi, P. jamotti, Parapleuropholis olbrechtsi, P. koreni, Ligulella sluysi, L. fourmarieri, Majokia brasseuri, Macrosemius maeseni, Songanella callida, Pholidophorus aequatorialis, Paraclupavus caheni* and *Ihybodus songaensis.*

Pelecypods Studied by Cox (1953, 1960), they include the following species (Cox, 1960): *Mytilus?* sp. (horizons 9 or 10), *Laevitrigonia* sp. (horizons 13–14), *Quenstedtia? lomaniensis* (horizon 9), *Pleuromya? ruikiensis* (horizons 12–13), *Cercomya? undosa* (horizons 11 or 13), *Plectomya subrugosa* (horizons 6 or 7) and *Pleuromya? elambana.*

Gastropods Cox (1960) reported an indeterminate species, hypothetically assigned to the genus *Natica* (horizon 12).

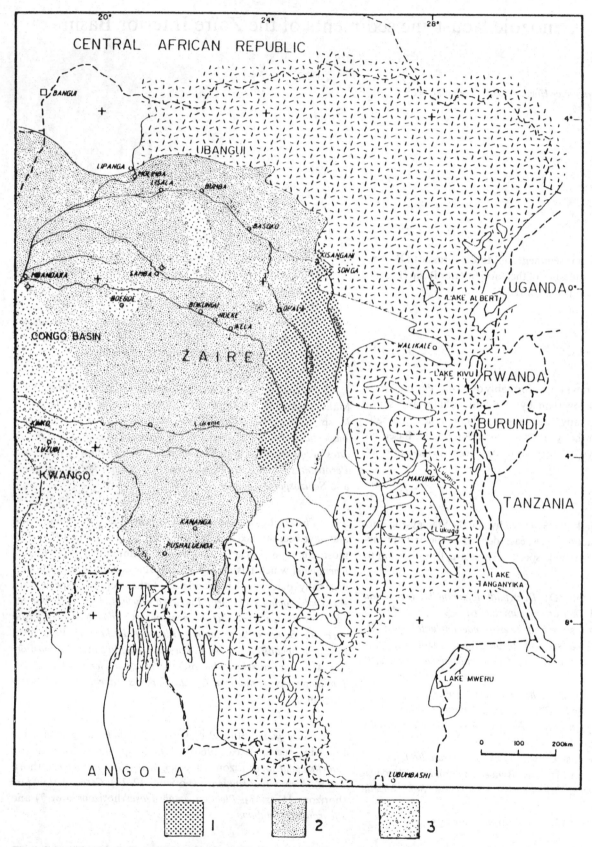

Fig. 1. Schematic geological map of the Zaïre Interior Basin showing the distribution of Jurassic to Tertiary lacustrine deposits. 1: Stanleyville Fm.; 2: Loia Fm.; 3: Kwango Group.

SERIES AND FORMATIONS			LEVELS	OSTRACODES ZONES	AGE
		GRES POLYMORPHES			EOCENE
KWANGO	N'SELE	ARKOSIC AND CONGLOMERATIC SANDSTONES			MAASTRICHTIAN SENONIAN
KWANGO	INZIA	KITARI - KIMBAU	5		TURONIAN LATE CENOMANIAN
KWANGO	INZIA	SANDSTONE	4		TURONIAN LATE CENOMANIAN
KWANGO	INZIA	BUMBA SHALE	3		TURONIAN LATE CENOMANIAN
KWANGO	INZIA	SANDSTONE	2		TURONIAN LATE CENOMANIAN
KWANGO	INZIA	LUTSHIMA SHALE	1		TURONIAN LATE CENOMANIAN
LUALABA	LOIA S.L.	BOKUNGU	16	11	EARLY CENOMANIAN ALBIAN LATE APTIAN
LUALABA	LOIA S.L.	YAKOKO - IKELA (LOIA S.S.)	15	10	EARLY CENOMANIAN ALBIAN LATE APTIAN
LUALABA	STANLEYVILLE	LILO	14	9	MIDDLE JURASSIC
LUALABA	STANLEYVILLE	RUIKI	13	8	MIDDLE JURASSIC
LUALABA	STANLEYVILLE	Km 108	12		MIDDLE JURASSIC
LUALABA	STANLEYVILLE		11	7	MIDDLE JURASSIC
LUALABA	STANLEYVILLE	USENGWE (ASSENGWE)	10	7	MIDDLE JURASSIC
LUALABA	STANLEYVILLE		9	7	MIDDLE JURASSIC
LUALABA	STANLEYVILLE		8	6 / 5	MIDDLE JURASSIC
LUALABA	STANLEYVILLE		7	4	MIDDLE JURASSIC
LUALABA	STANLEYVILLE	LOSO COMPLEX	6	3	MIDDLE JURASSIC
LUALABA	STANLEYVILLE		5		MIDDLE JURASSIC
LUALABA	STANLEYVILLE	MINJARO , MEKOMBI , KEWE	4	2	MIDDLE JURASSIC
LUALABA	STANLEYVILLE		3		MIDDLE JURASSIC
LUALABA	STANLEYVILLE	LIME FINE , SONGA LIMESTONE	2	1	MIDDLE JURASSIC
LUALABA	STANLEYVILLE	FALLS SANDSTONE AND CONGLOMERATE	1		MIDDLE JURASSIC

Fig. 2. Jurassic–Tertiary lithostratigraphy of the Zaire Interior Basin.

Age

Originally considered as Upper Jurassic (Kimmeridgian) on fish and pelecypods (*Plectomya subrugosa*) evidence, the Stanleyville Formation is now considered as Middle Jurassic (Aalenian-Bathonian) and correlated with the M'Vone Formation of Gabon and the Aliança Formation of the Reconcavo-Tucano Basin in Brazil based on both palynological and micropaleontological evidence. The ostracode fauna, dominated by species of the genera *Theriosynoecum* and *Darwinula*, and the absence of the genus *Cypridea*, is comparable with the ostracodes known from the Aliança Formation of Brazil (Krömmelbein & Weber, 1971). Palynological assemblages show strong affinities with those of the M'Vone Series of Gabon (Jardiné, 1974).

Environment

Marine influence is suggested by fishes and pelecypods especially near the base (horizon 2 = Songa), but all the ostracodes are lacustrine and no marine microplankton was found in palynological preparations. According to Defretin-Lefranc (1967), the presence of numerous phyllopods at many levels is a clear indication of a hot and humid climate with seasonal dry periods.

Loia Formation (level 15 and 16)

Distribution

Widely outcropping in the Interior Basin. Also known in the Samba-1 well in the north-central part of the Basin (Cahen *et al.*, 1959b).

Lithology

Red and green sandstones with few streaks of bituminous shales (180–300 m), in the lower part (Yakoko and Ikela = level 15 = Loia Formation *sensu stricto*); red shales and sandstones (110–300 m) in the upper part (level 16 = Bokungu Formation).

Paleontology

Ostracodes   In Grékoff's (1957) zonal scheme, zone 10 corresponded to the Loia Formation. It is characterized by the following species: *Cypridea elisabethaensis, C. maringaensis, C. yakokoensis, Ilyocypris lomamiensis, I. minor, Metacypris polita, M. pustulosa* and *Pattersoncypris rotunda*.

Phyllopods   Marlière (1950) and Defretin-Lefranc (1967) reported the following characteristic species: *Bairdestheria kasaiensis, Pseudoasmussia banduensis, P. dekeseensis* in the Loia *sensu stricto Asmussia dekeseensis, A. ubangiensis* and *Euestheria sambaensis* in the Bokungu Formation.

Palynology   The Loia Formation *sensu stricto* has yielded the following palynomorphs (Bose, 1977; Maheshwari *et al.*, 1977a, b; Colin, unpub. results): *Afropollis jardinus, Elaterosporites castelaini, E. klaszi, Sofrepites legouxi, Galeocorna protensa* and *Steevesipollenites binodosus*.

Fishes   *Casieroides yamangaensis, Charobnius longicaudatus, Clupavus yamangaensis* and *Hyaelobatis, Leptolepis minor* in the Loia Formation *sensu stricto; Ceratodus* sp., *Hybodus molimbaensis, Lepidotus* sp., *Mawsonia ubangiensis, Paralepidosteus praecursor, Prororhyza molimbaensis* and *Stromerichthys* sp. in the Bokungu Formation (Casier, 1961, 1969a).

Pelecypods   *Anomia?* sp. and Unionacea in the Loia Formation *sensu stricto* (Cox, 1960).

Age

Palynology clearly indicates an Albian to Lower Cenomanian age.

Kwango Group

Known from the southwest part of the Interior Basin. It is divided into the Inzia Formation (120–210 m) and the N'sele Formation (105 m).

Inzia Formation

Lithology

The lithological succession is as follows:

Kitari-Kimbau complex, greenish to reddish fossiliferous shales (40 m).
Red sandstones, locally with pebbles and cherts (50–100 m).
Bumba Shales, red shales and sandstones (25 m).
Friable red sandstones (15–40 m).
Lutshima Shales, red, calcareous shales and red sandstones (25 m).

Paleontology

Ostracodes   In the Kitari-Kimbau complex toward the top of the Inzia Formation, Grékoff (1960) reported from various localities *Afrocythere?* 536, *Cypridea kitariensis, Darwinula kwangoenis, Ilyocypris compressa, I. luzubiensis* and *Metacypris* K3099.

Phyllopods   In the Kitari-Kimbau complex, Defretin-Lefranc (1967) reported *Bairdestheria kitariensis* and *Pseudoestheria lepersonnei*. Marlière (1950) also reported *Estheria lerichei* in the Kwango.

Fishes   *Rhipis moorseli* is reported from the Bumba Shales (Saint-Seine, 1953).

N'sele Formation

Lithology

Fine-grained red sandstones locally silicified (up to 105 m).

Paleontology

Ostracodes In the Kinko Makaw Horizon, at the base of the N'sele Formation, Grékoff (1960) reported *Dolerocypris kinkoensis* and *Paracypria makawaensis*.

Fishes Decertids and *Diplomystus* are reported in the Kipala horizon near the base on the N'sele (Casier, 1965).

Age

Based on the fish fauna from the Bumba Shale, the Kwango Formation is interpreted by Saint-Seine (1953) as Cenomanian–Turonian in age.

Grès polymorphes

Distribution

Mainly in the southern part of the Basin; particularly well developed in the Kwango Basin. Known as 'Séries des Plateaux Bateke' in the western part of the Basin.

Lithology

Basal conglomerates unconformable on the N'sele Formation, overlain by sandstones, often silicified.

Paleontology

Ostracodes *Cypris farnhami*, *C. lerichei*, *Erpetocypris* sp., *Gomphocythere?* sp., *Oncocypria?* sp. and *Stenocypris? bunzaensis* (Grékoff, 1958).

Gastropods *Planorbus fontainasi* and *Pyrgophysa cayeni* (Leriche, 1927).

Charophyta *Chara rauwi* and *C. saleei* (Polinard, 1931–1932).

Age

Interpreted as Eocene, this is based essentially on charophyta.

Bibliography

Bose, M.N., 1977. A palynological reconnaissance of the Mesozoic sediments of Zaire. *Ann. Mines Géol. Tunis*, **28**, 65–9.

Cahen, L., 1954. *Géologie du Congo Belge*. Vaillant-Carmanne, Liège.

Cahen, L., Ferrand, J.J., Haarsma, M.J.H.F., Lepersonne, J. & Verbeek, T., 1959a. Description du sondage de Dekese. *Ann. Mus. royal Congo Belge*, **34**, 1–115.

Cahen, L., Ferrand, J.J., Haarsma, M.J.H.F., Lepersonne, J. & Verbeek, T., 1959b. Description du sondage de Samba. *Ann. Mus. royal Congo Belge*, **29**, 1–210.

Casier, E., 1961. Matériaux pour la faune ichtyologique éocrétacique du Congo (résultats scientifiques, paléontologie). *Ann. Mus. royal Afrique Centrale*, **39**, 1–61.

Casier. E., 1965. Poissons fossiles de la Série du Kwango (Congo). *Ann. Mus. royal Afrique Centrale*, **50**.

Casier, E., 1969a. Addenda aux connaissances sur la faune ichtyologique de la série de Bokungu (Congo). *Ann. Mus. royal Afrique Centrale*, **62**, 1–17.

Casier, E., 1969b. Sur les conditions de dépot de quelques unes des formations mésozoiques du bassin du Congo. *Ann. Mus. royal Afrique Centrale*, **62**, 29–47.

Cox, L.R., 1953. Lamellibranchs from the Lualaba beds of the Belgian Congo. *Rev. Zool. Botan. afric.*, **47**, 99–107.

Cox, L.R., 1960. Further mollusca from the Lualaba beds of the Belgian Congo. *Ann. Mus. royal Congo Belge*, **37**, 1–15.

Defretin-Lefranc, S., 1967. Etudes des phyllopodes du bassin du Congo. *Ann. Mus. royal Afrique Centrale*, **56**, 1–122.

Grékoff, N., 1957. Ostracodes du bassin du Congo. I. Jurassique supérieur et Crétacé inférieur du Nord du bassin. *Ann. Mus. royal Congo Belge*, **29**, 1–87.

Grékoff, N., 1958. Ostracodes du bassin du Congo. III. Tertiaire. *Ann. Mus. royal Congo Belge*, **22**, 1–36.

Grékoff, N., 1960. Ostracodes du bassin du Congo. II. Crétacé. *Ann. Mus. royal Congo Belge*, **35**, 1–70.

Jamotte, A., 1947. Découverte au Katanga de l'horizon à ostracodes et poissons de l'étage du Lualaba. *Bull. Inst. royal colon. Belge*, **18** (1), 296–301.

Jardiné, S., 1974. Microflores des formations du Gabon attribuées au Karroo. *Rev. Palaeobot. Palynol.*, **17**, 75–112.

Krömmelbein, K. & Weber, R., 1971. Ostracoden des Nordost-Brasilianischen 'Wealden'. *Beih. geol. Jb.*, **115**, 1–93.

Lepersonne, J., 1977. Structure géologique du bassin intérieur du Zaïre. *Bull. Acad. royal Belgique*, **5** (63), 941–65.

Leriche, M., 1910. Sur les premiers poissons fossiles rencontrés au Congo Belge dans le système du Lualaba. *Comptes rendus Acad. sciences Paris*, **151**, 840–1.

Leriche, M., 1911. Les poissons des couches du Lualaba (Congo Belge). *Rev. Zool. afric.*, **1**, 190–7.

Leriche, M., 1913. Les Entomostracés des couches de Lualaba (Congo Belge). *Rev. Zool. afric.*, **3** (1), 1–11.

Leriche, M., 1927. Les fossiles des 'grès polymorphes' (couches du Lubilash) aux confins du Congo et de l'Angola. *Rev. Zool. afric.*, **15**, 408.

Lombard, A.L., 1967. Géologie des parties Nord (Ubangi) et Est (bassins du Lualalaba-Lomani) de la cuvette centrale congolaise (République Démocratique du Congo). *Bull. Soc. Belge Géol. Paléontol. Hydrol.*, **75** (1), 49–67.

Maheshwari, H.K., Bose, M.N. & Kumaran, K.P.N., 1977a. Mesozoic sporae dispersae from Zaire. II. The Loia and Bokungu groups in Samba borehole. *Ann. Mus. royal Afrique Centrale*, **80**, 1–43.

Maheshwari, H.K., Bose, M.N. & Kumaran, K.P.N., 1977b. Mesozoic sporae dispersae from Zaire. III. Some miospores from the Stanleyville group. *Ann. Mus. royal Afrique Centrale*, **80**, 49–60.

Marlière, R., 1948a. Ostracodes et phyllopodes du système du Karroo au Congo Belge. *Ann. Mus. Congo Belge*, **2**, 1–61.

Marlière, R., 1948b. Paléontologie du Karoo au Congo Belge: ostracodes et phyllopodes. *Bull. Soc. Belge Géol. Paléontol. Hydrol.*, **57**, 329–30.

Marlière, R., 1948c. Conclusions relatives à l'étude des ostracodes et phyllopodes du système du Karroo au Congo Belge. *Ann. Soc. géol. Belgique*, **71**, 260–2.

Marlière, R., 1950. Ostracodes et phyllopodes du système du Karroo au Congo Belge et les régions avoisinantes. *Ann. Mus. Congo Belge*, **6**, 1–43.

Marlière, R., 1955. Sur l'age de quelques phyllopodes et ostracodes mésozoiques du Congo Belge. *Congr. nat. Sci., Comptes rendus*, **3** (1), 24–5.

Marlière, R., 1956. Sur quelques Entomostracés de la cuvette Congolaise. *Bull. Soc. Belge Géol. Paléontol. Hydrol.*, **65**, 45–52.

Polinard, E., 1931–32. Découverte de gisements fossilifères d'eau douce sur les versants de la Lubudi au Katanga méridional. *Ann. Soc. Géol. Belgique*, **4**, 63–81.

Saint-Seine, P. de, 1953. Poissons de la cuvette congolaise. *Comptes rendus Soc. géol. France*, **16**, 343–5.

Saint-Seine, P. de, 1954. Les poissons des Schistes bitumineux de l'étage de Stanleyville (Congo Belge). *Comptes rendus Soc. géol. France*, **17**, 331–2.

Saint-Seine, P. de, 1955. Poissons fossiles de l'étage de Stanleyville (Congo Belge). 1ère partie: La faune des argilites et schistes bitumineux. *Ann. Mus. royal Afrique Centrale*, **44**, 1–52.

Saint-Seine, P. de & Casier, E., 1962. Poissons fossiles des couches de Stanleyville (Congo). 2ème partie: La faune marine des calcaires de Songa. *Ann. Mus. royal Afrique Centrale*, **44**, 1–52.

Ancient lacustrine basins in the southwestern French Massif Central

RÉGINE SIMON-COINÇON

Laboratoire de Géographie Physique, C.N.R.S., Meudon, F-92195 and Ecole Nationale des Mines de Paris, 33, Rue St.-Honoré, Fontainebleau Cedex, F-77305, France

Introduction

Erosional remnants of Tertiary fluvial and lacustrine deposits are exposed along the southwestern portion of the French Massif Central (Fig. 1). These rocks overlie crystalline basement and/or Jurassic bedrock within tectonic grabens. These grabens are either elongated sags on the edge of strike–slip faults (cf. Villefranche de Rouergue fault) or basins in front of reverse faults. Six basins are described here: St. Flour-Le Malzieu, Bramelou-Rodez, Varen, Asprières-Compolibat, St. Santin-Maurs, and Fumel Basins.

St Flour-Le Malzieu Basin

This basin is aligned along a strike–slip fault. Subsidence along this fault during the Eocene–Miocene allowed thick deposits of silt, mud, gravel, and sandy carbonate lenses to accumulate. The carbonate lenses contain Rupelian mammals and reptilian fauna (Fig. 2).

The sedimentary sequences (Figs 2 and 3) include the following from bottom to top. Red kaolinitic sandstones and claystones are interpreted as ferruginous soils developed on floodplain sediments. These rocks are overlain by green claystones containing smectite and carbonate concretions. Next, limestones with molluscan fossils are interpreted as lacustrine. The carbonates are followed by a unit containing interstratified lenses composed of pebbles, sand and clay, which are interpreted as deposits of a braided alluvial system. Some of these lenses have been silicified and contain plant impressions. At the top of the sequence, Miocene and Pliocene sandstones interbed with volcanic flows from which radiometric and paleomagnetic information have been determined. The fossil flora and fauna of these sandstones have been assigned to the Pikermi biozone; these sandstones are interpreted as fluvial and paludal sediments.

Bramelou-Rodez Basin

The bottom half of the sedimentary sequences of this basin complex (Figs 2 and 4) is interpreted as alluvial fan to floodplain deposits. The top portion is composed of limestones containing *Limnea* and *Potamides*, which are interpreted as saline lacustrine deposits. Their age is determined as Eocene to Oligocene.

Varen Basin

This basin is located on the southern end of the Villefranche de Rouergue Fault. The stratigraphic section from bottom to top (Figs 2 and 4) follows. Ferruginous conglomerates of Paleocene? age form the basal unit. Next, a thin kaolinitic red claystone with ferruginous concretions has been determined as Late Eocene (Auversian) in age. This is followed by calcareous deposits interpreted as calcretes and other exposed lacustrine carbonates, which are dated as Upper Eocene (Ludian) with charophytes and mammalian fossils. Claystones overlie the limestones and are interpreted as sheetflood deposits. Miocene–Pliocene units composed of fining-upward sequences containing sandstones to siltstones to claystones top the Varen Basin sequence. These sequences contain a basal erosional contact and are interpreted as a meandering river complex with point bars. The siltstones contain plant impressions and may represent backswamp or floodplain sediments.

Asprieres-Compolibat Basin

This basin is composed of several small half-grabens along the Villefranche de Rouergue Fault. Three stratigraphic units are present (Figs 2 and 5). (1) Gravels and sandy clays (which can be cemented by calcite) contain ferruginous pisolites and mammalian fossils of Auversian (Late Eocene) age. These deposits are interpreted as fluviatile in nature. (2) Conglomerates of Marinesian (Middle Eocene) age are interpreted as possible alluvial fan deposits associated with tectonic activity. (3) Limestones and marls are Ludian (Upper Eocene) in age, as determined by mammalian fossils. These rocks are interpreted as continental limestones deposited in a lacustrine–fluvial complex associated with marshes.

St Santin-Maurs Basin

This basin is positioned on the main lineament of the Grand Sillon Houiller. The stratigraphic sequence contains the following

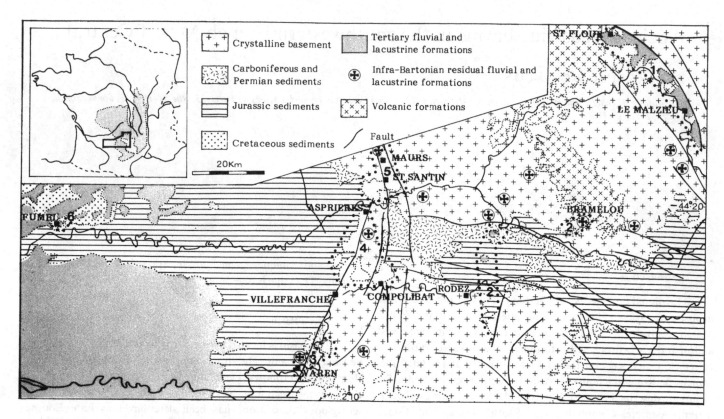

Fig. 1. Tertiary basins in the southwestern Massif Central. Numbers refer to the location of lithologic columns similarly numbered in Fig. 2.

units (Figs 2 and 6). Ferruginous conglomerates and mudstones (or unlithified gravels and pebbles) of Paleocene age compose the basal unit of the St Santin-Maurs Basin. There is a great variability in the amounts of clay, silt and sand particles within this unit. The next unit above is composed of cross-laminated limestones topped by sandy clay with gravels at St Santin and at Maurs the unit contains only the sandy clay with gravels. This second unit of the sequence is interpreted as fluviatile in nature. Overlying this unit at Maurs, a sandy claystone with conglomerates occurs; it is interpreted as alluvial fan deposits associated with tectonic activity. The following unit contains a thick sequence of limestone and marls at St Santin interpreted as lacustrine and fluviatile in origin. Mammalian fossils date these rocks as Ludian (Upper Eocene) in age. At Maurs, the limestones correlate with sandy claystones and conglomerates. The top of the stratigraphic sequence contains limestones and sandy clay which are supposedly Oligocene in age and interpreted as a lacustrine–fluvial complex.

Fumel Basin

This basin is located in the western-most portion of the study area (Fig. 1); sedimentation in this basin was continuous from the Early Tertiary to the Oligocene. The sedimentary sequence is as follows (Figs 2 and 6). Kaolinitic claystones with lignite (Thanetian in age) form the basal unit. These are overlain by kaolinitic

sandstones interpreted as deposits of a deltaic complex. The upper portions of these sandstones contain siliceous and ferruginous duricrusts probably formed during an arid event. A thin layer of sandstones and gritstones of deltaic to fluvial origin form the next unit. The top of the sedimentary sequence is composed of limestones of Rupelian age, as determined by mammalian fossils of the Montalban zone. These limestones are interpreted as lacustrine in origin. This entire sequence is interpreted as a prograding deltaic–fluviatile paleoenvironment on the western edge of a flexure.

Climate

Climatic change and tectonic activity are recorded in the rock sequences of these basins (Figs 3–7). The lower siliciclastic deposits of the Paleocene to early Eocene record a moist tropical climate and pulsed tectonic activity. A calm tectonic period followed during the middle Eocene with monsoonal climate and a marked dry season (evidence includes a change in clay minerals and the exposure of calcareous deposits). The onset of aridity is marked by the dominantly calcareous deposits of the Oligocene, which were deposited during episodic floods with intervening dry periods in endoreic (closed) environments. During the Miocene and Pliocene, a colder climate and increased precipitation occurred along the continued uplift of the Massif Central, as evidenced by dominantly siliciclastic fluviatile deposition.

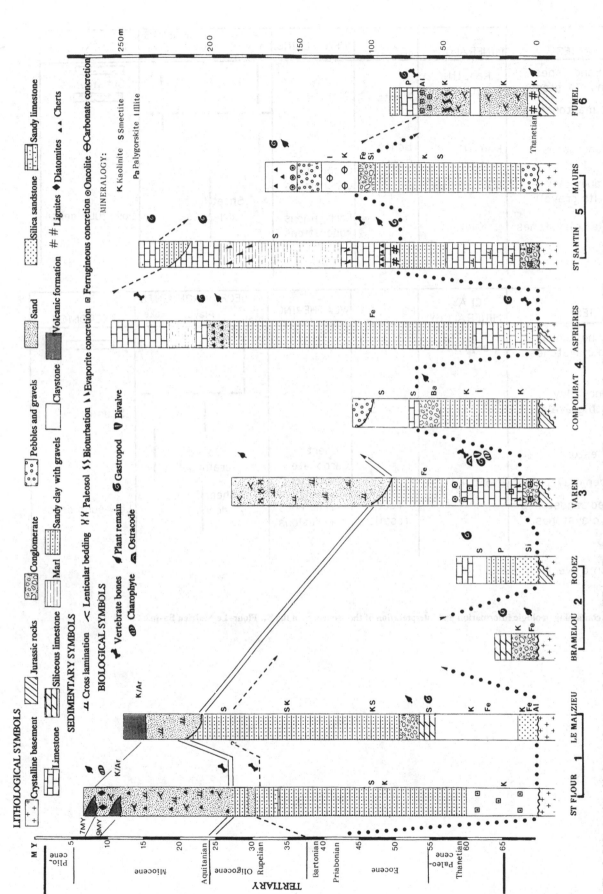

Fig. 2. Lithostratigraphic logs and age correlation among the different Tertiary basins shown in Fig. 1. Correlation line key: Late Miocene (single line); Aquitanian (double line); Lower Oligocene (dashed line); Priabonian (heavy dots).

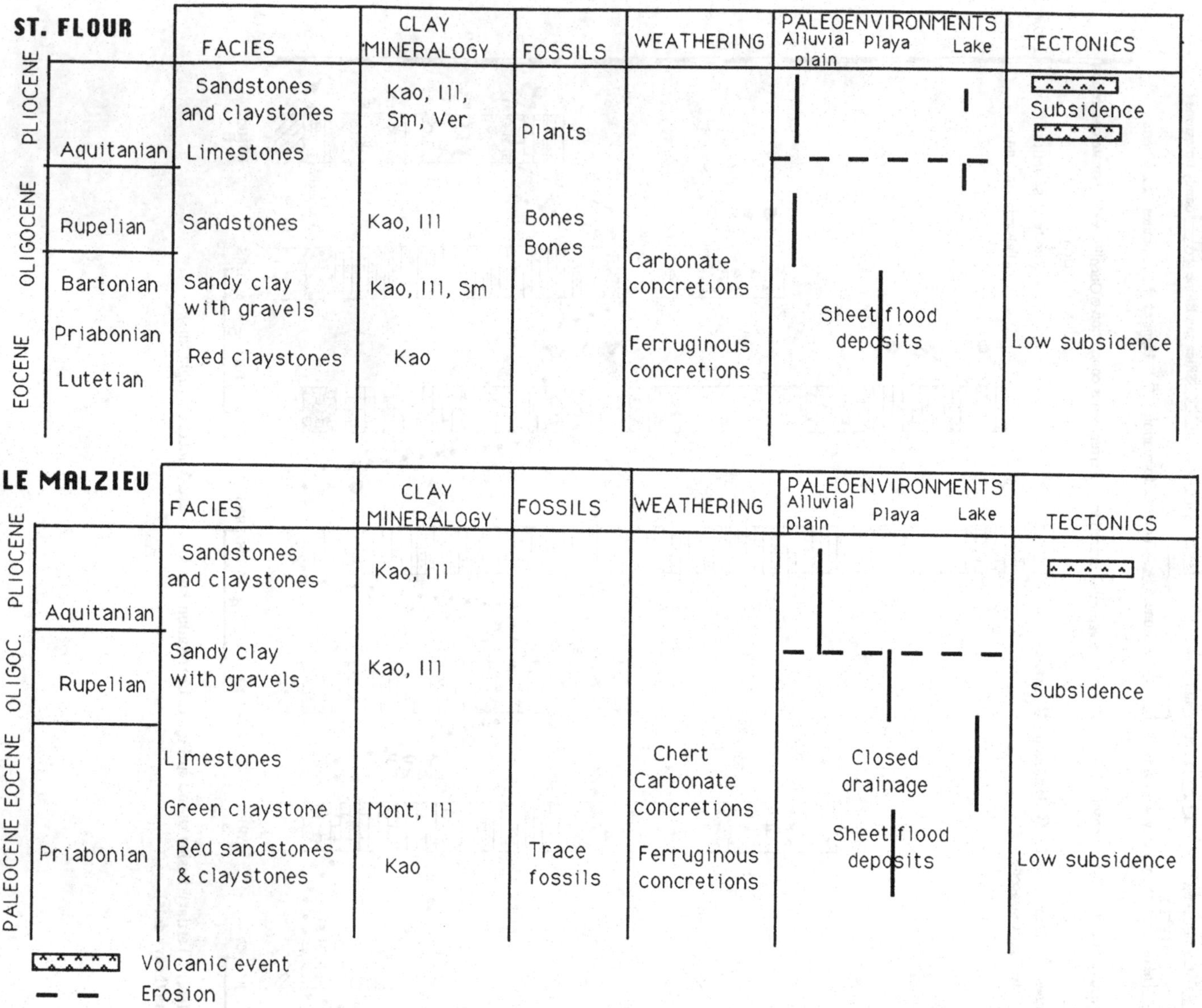

ST. FLOUR

		FACIES	CLAY MINERALOGY	FOSSILS	WEATHERING	PALEOENVIRONMENTS Alluvial plain / Playa / Lake			TECTONICS
PLIOCENE	Aquitanian	Sandstones and claystones	Kao, Ill, Sm, Ver	Plants					Subsidence
		Limestones							
OLIGOCENE	Rupelian	Sandstones	Kao, Ill	Bones Bones					
	Bartonian	Sandy clay with gravels	Kao, Ill, Sm		Carbonate concretions	Sheet flood deposits			Low subsidence
EOCENE	Priabonian								
	Lutetian	Red claystones	Kao		Ferruginous concretions				

LE MALZIEU

		FACIES	CLAY MINERALOGY	FOSSILS	WEATHERING	PALEOENVIRONMENTS Alluvial plain / Playa / Lake			TECTONICS
PLIOCENE	Aquitanian	Sandstones and claystones	Kao, Ill						
OLIGOC.	Rupelian	Sandy clay with gravels	Kao, Ill						Subsidence
EOCENE		Limestones			Chert Carbonate concretions	Closed drainage			
PALEOCENE	Priabonian	Green claystone	Mont, Ill			Sheet flood deposits			Low subsidence
		Red sandstones & claystones	Kao	Trace fossils	Ferruginous concretions				

⬡⬡⬡⬡⬡ Volcanic event

— — Erosion

Fig. 3. Summary tables containing geologic information and interpretation of the sequences in the St. Flour–Le Malzieu Basin.

BRAMELOU

PALEOCENE EOCENE OLIGOC.	FACIES	CLAY MINERALOGY	FOSSILS	WEATHERING	PALEOENVIRONMENTS Alluvial plain	Marsh	Lake	TECTONICS
Bartonian	Silicified limestones & claystones		Gastropods	Siliceous duricrust		Closed environments		Subsidence
Ypresian	Conglomerates and sandstones	Kao	Plants	Lateritic				

RODEZ

		FACIES	CLAY MINERALOGY	FOSSILS	WEATHERING	PALEOENVIRONMENTS Alluvial plain	Marsh	Lake	TECTONICS
OLIGOCENE	Priabonian	Limestones	Sm	Gastropods			Closed environments		Low subsidence
		Claystone	Sm						
EOCENE	Bartonian	Sandy clay with gravels	Sm, Kao		Calcitization				
		White, well sorted sandstones	Kao		Siliceous duricrust				Subsidence

VAREN

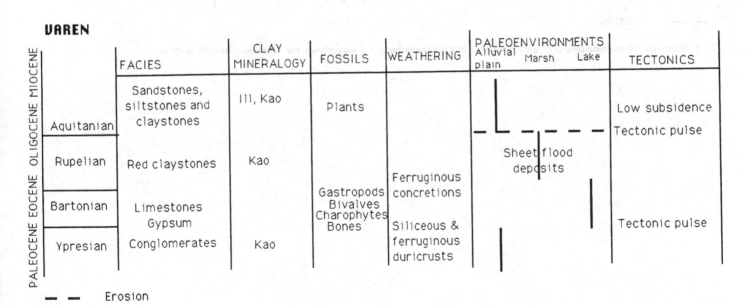

		FACIES	CLAY MINERALOGY	FOSSILS	WEATHERING	PALEOENVIRONMENTS Alluvial plain	Marsh	Lake	TECTONICS
PALEOCENE EOCENE OLIGOCENE MIOCENE	Aquitanian	Sandstones, siltstones and claystones	Ill, Kao	Plants					Low subsidence / Tectonic pulse
	Rupelian	Red claystones	Kao			Sheet flood deposits			
	Bartonian	Limestones Gypsum		Gastropods Bivalves Charophytes Bones	Ferruginous concretions				Tectonic pulse
	Ypresian	Conglomerates	Kao		Siliceous & ferruginous duricrusts				

– – Erosion

Fig. 4. Summary tables containing geologic information and interpretation of the sequences in Bramelou–Rodez and Varen Basins.

ASPRIERES

		FACIES	CLAY MINERALOGY	FOSSILS	WEATHERING	PALEOENVIRONMENTS Alluvial plain / Marsh / Lake	TECTONICS
MIOCENE		Limestones	Sm	Bones			Low subsidence
OLIGOCENE	Aquitanian	Marls	Sm				
		Limestones	Sm	Gastropods Plants			
	Rupelian	Well sorted sandstones	Kao, Ill, Sm		Early siliceous duricrusts		
		Sandy clay with gravels				Sheet flood deposits	Low subsidence
EOCENE	Bartonian	Limestones					
		Sandstones	Kao	Bones Gastropods	Calcitization		Tectonic pulses

COMPOLIBAT

		FACIES	CLAY MINERALOGY	FOSSILS	WEATHERING	PALEOENVIRONMENTS Alluvial plain / Marsh / Lake	TECTONICS
EOCENE	Bartonian	Sandstones and conglomerates					
		Marls	Sm				
	Priabonian	Limestones	Sm, Ill		Ferruginous concretions		Tectonic pulse
PALEOCENE		Sandy clay with gravels	Kao, Ill	Plants	Lateritic paleosols	Sheet flood deposits	High subsidence

– – – Erosion

Fig. 5. Summary tables containing geologic information and interpretation of the sequences in the Asprieres–Compolibat Basin.

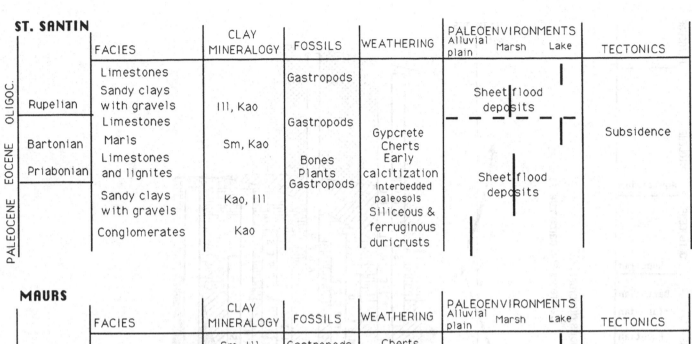

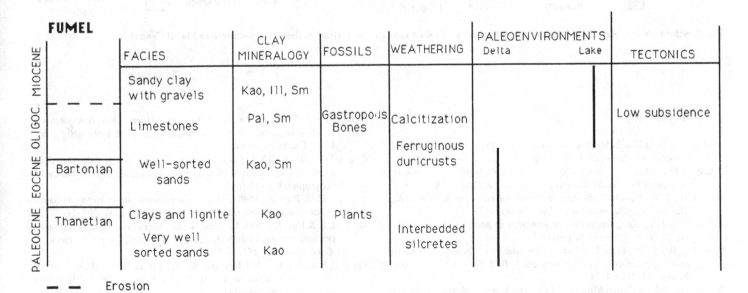

Fig. 6. Summary tables containing geologic information and interpretation of the sequences in the Saint Santin–Maurs and Fumel Basins.

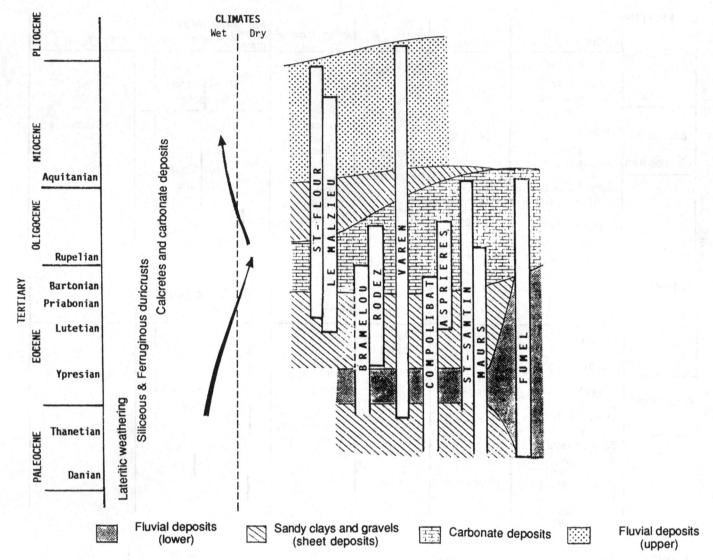

Fig. 7. Schematic interpretation of the climatic influences on the main facies of the Tertiary basins in the southwestern Massif Central.

Bibliography

Archanjo, J.D., 1988. *Le Sidérolithique du Quercy Blanc (France)*. Doctoral Dissertation, Univ. de Strasbourg.

Astruc, J.G., 1988. Le Paléokarst quercynois au Paléogène – altérations et sédimentations associées. *Documents du B.R.G.M.*, **133**, 1–135.

Bellon, H., 1971. *Datations absolues de laves d'Auvergne par la méthode K.A.* Thesis, 3rd cycle, petrology, Univ. Orsay, v. 1.

Billaud, Y., 1982. *Les paragénèses phosphatées du paléokarst des phosphorites du Quercy*. Thesis, 3rd cycle, Univ. Lyon.

Boisse de Black, Y., 1933. Le Détroit de Rodez et ses bordures cristallines (étude géologique et morphologique). *Bull. Serv. Carte géologique France*, **34** (188), 1–310.

Boule, M., 1986. le Cantel Miocène. *Bull. Serv. Carte géologique France*, 3rd Ser., **16**, 213–48.

Brousse, R. & Elhai, 1969. Age tiglien des diatomites de Murat (Cantal, France). Pollens et spores. (Unpublished manuscript.)

Coinçon, R., 1972. *La bordure occidentale de la Margeride de Neussargues à Saint-Alban: étude géomorphologique* Thesis, 3rd cycle, geography, Univ. Clermont-Ferrand.

Coque-Delhuile, B., 1978. *Les formations superficielles des plateaux de la Margeride occidentale: étude géomorphologique*. Thesis, 3rd cycle, geography, Univ. Paris I.

Depape, G. & Rey, R., 1949. Florule mio-pliocène des environs de Saint-Flour (Cantal). *Rev. Haute Auvergne (Aurillac)*, **31**, 208–14.

Durand, S. & Rey, R., 1963. Les formations à végétaux de Joursac (Cantal) peuvent être datées du Villafranchien par l'analyse pollinique. *C.R. Acad. Sci. Paris*, **257**, 2692–3.

Durand, S. & Rey, R., 1964. Le dépôt de diatomite de Sainte-Reine (Cantal) débute au Pliocène supérieur et permet de déceler les traces du refroidissement pretiglien. *C.R. Acad. Sci. Paris*, **259**, 1978–80.

Godard, A., Coinçon, R., & Valadas, B., 1978. *Carte géomorphologique de Saugues avec notice*. Centre Géographie R.G., Univ. Paris I.

Goer de Herve, A., 1972. La planèze de Saint-Flour (massif volcanique du

Cantal, France). *Ann. scientifiques de l'Univ. Clermont-Ferrand*, 2 volumes.

Goer de Herve, A., Tempier, P. & Conçon, R., 1981. *Notice explicative de la carte géologique de France au 1/50 000*. Feuille de Saint-Flour.

Jodot, P. & Rey, R., 1956. Observations stratigraphiques et malacologiques sur les bassins lacustres de Saint-Alban-sur-Limagnole (Lozère) et de Massiac (Cantal). *Bull. Soc. Géol. France*, **6**, 937–68.

Lavocat, R. Michel, R. & Rey, R., 1949. Age des dépôts sédimentaires des environs de Saint-Flour (Cantal). *C.R. Acad. Sci. Paris*, **228**, 191–2.

Marty, R., 1905. L'Oligocène du Peuch d'Alzou près de Bozouls (Aveyron). *Bull. Soc. Géol. France*, 3rd Ser., **5**, 560–4.

Marty, R., 1908. Flore miocène de Joursac. *Rev. Haute Auvergne (Aurillac) et Librairie Baillères, Paris*, pp. 1–91.

Muratet, B., 1983. *Géodynamique du paléogène continental en Quercy-Rouergue: analyse de la sédimentation polycyclique des bassins d'Asprières (Aveyron), Maurs (Cantal) et Varen (Tarn-et-Garonne)*. Thesis, 3rd cycle, Univ. Toulouse.

Muratet, B., Crochet, J.Y., Hartenberger, J.L., Sige, B., Sudre, J. & Vianey-Liaud, M., 1985. Nouveaux gisements à mammifères de l'Eocène supérieur et apport à la chronologie des épisodes sédimentaires et tectoniques à la bordure sud-ouest du Massif central. *Bur. Rech. Géol. Min., Géol de la France*, no. 3, 271–86.

Muratet, B., Feist, M., Hartenberger, J.L., Sige, B. & Vianey-Liaud, M., 1982. Un gisement fluvio-lacustre à vertébrés et charophites d'âge éocène terminal de la bordure orientale du Quercy. Implications sur la tectonique tertiaire du sud-ouest du Massif central. *C.R. Acad. Sci. Paris*, **294**(2), 123–6.

Rey, R., 1949. Stratigraphie des bassins tertiaires de Saint-Alban et du Malzieu (Lozère) de Saint-Flour et de Neussargues (Cantal). *C.R. Acad. Sci. Paris*, **237**, 325–6.

Rey, R., 1965. Deux gisements à plantes du flanc est du massif volcanique du Cantal: Sainte-Reine et Joursac. *Bull. Soc. Géol. et Min. de Bretagne (Rennes)* (1962–3; March, 1965), 211–73.

Simon-Coinçon, R., 1984a. Les dépôts sédimentaires tertiaires sur le socle à l'est de la faille de Villefranche de Rouergue (Massif central français). *C.R. Acad. Sci. Paris*, 2nd Ser., **298**(13), 563–8.

Simon-Coinçon, R., 1984b. Plateaux et bassins due socle du Rouergue au nord de Rieupeyroux (Aveyron). Paléogéographie et sédimentation tertiaire. *Rev. géographique des Pyrénées et du Sud-Ouest (Toulouse)*, **55** (fasc. 1), 57–69.

Simon-Coinçon, R., 1989. *Le rôle des paléoaltérations et des paléoformes dans les socles: l'exemple du Rouergue (Massif central francais)*. Doctoral dissertation, Univ. Paris I, Mémoires Sciences de la Terre, Ecole Nationale Supérieur des Mines de Paris, no. 9.

Simon-Coinçon, R., Goer de Herve, A., & Ginsburg, L., 1981. Première découverte d'un fragment de Rhinocérotidé dans les 'sables à chailles' du bassin de Saint-Flour (Cantal). *Rev. Sci. Nat. d'Auvergne (Clermont-Ferrand)*, **47**, 31–42.

Thévenin, A., 1903. Etude géologique de la bordure sud-ouest du Massif central. *Bull. Serv. Carte géologique France*, **14**(95), 1–203.

Virol, F., 1987. *Le contact Massif central/Bassin Aquitain au niveau du lot moyen et du Célé. Enseignements fournis par les formations superficielles d'âge secondaire et tertiaire en matière d'évolution géomorphologique*. Doctoral dissertation, Univ. Paris I. 300 p.

Lacustrine record in the continental Tertiary Duero Basin (northern Spain)

I. ARMENTEROS AND A. CORROCHANO

Department of Geology, University of Salamanca, E-37071 Salamanca, Spain

The Duero Basin is an intraplate basin in the northwestern part of the Iberian Peninsula (Fig. 1A) of Late Cretaceous or Early Paleogene age. The origin and evolution of the Duero Basin were controlled by successive reactivation phases of Late Hercynian faults during the Alpine orogeny. The basin fill comprises exclusively continental deposits. The sedimentary record of the Tertiary Duero Basin can be divided by age into three phases: Upper Cretaceous–Paleocene, Eocene–Lower Miocene, and Lower Miocene–Upper Miocene. The deposits are interpreted as marginal fluvial fringes and central lacustrine environments. The areas of maximum lacustrine accumulation were related to the maximal axis of subsidence next to the eastern margin of the Duero Basin (Fig. 1B, C). The space–time distribution of Tertiary lacustrine lithosomes is also a reflection of the differential degree of subsidence in the basin. Climatic factors have also exerted an important influence, not only because of latitudinal changes of the Iberian Peninsula throughout the Tertiary, but also because of more localized climatic variations.

The Cretaceous–Paleocene phase has only been recognized in the western margin of the Duero Basin and comprises alluvial sideroli-tic sediments with a thick layer of silcrete at the top (Fig. 2A). Lacustrine systems are recognized in the other two cycles (Fig. 2B, C).

Within the Eocene–Lower Miocene phase (Fig. 2B) along the eastern border of the basin, two large lacustrine subenvironments have been recognized: (1) fluvio-lacustrine fringes and (2) shallow inner lakes. Both are characterized by carbonate and siliciclastic facies associations. Throughout the Eocene–Lower Miocene phase, sedimentation occurred in shallow areas with fluctuating margins adjacent to basin edges (Fig. 3A).

The alluvial-lacustrine sequences, within the Lower Miocene–Upper Miocene phase, start with fluvial red-beds, followed by evaporitic facies (sometimes absent) and end with carbonates (Fig. 2C). Within the lacustrine systems, four subenvironments can be differentiated: (1) fluvio-lacustrine fringes, (2) mud flats, (3) a marginal lacustrine zone and (4) a shallow inner zone. The main lacustrine episode (Upper Aragonian) represents the highest degree of closed-basin conditions in this phase (Fig. 3C). The upper lacustrine episode (Vallesian–Turolian) reflects the maximum expansion of shallow carbonate lakes (Fig. 3D).

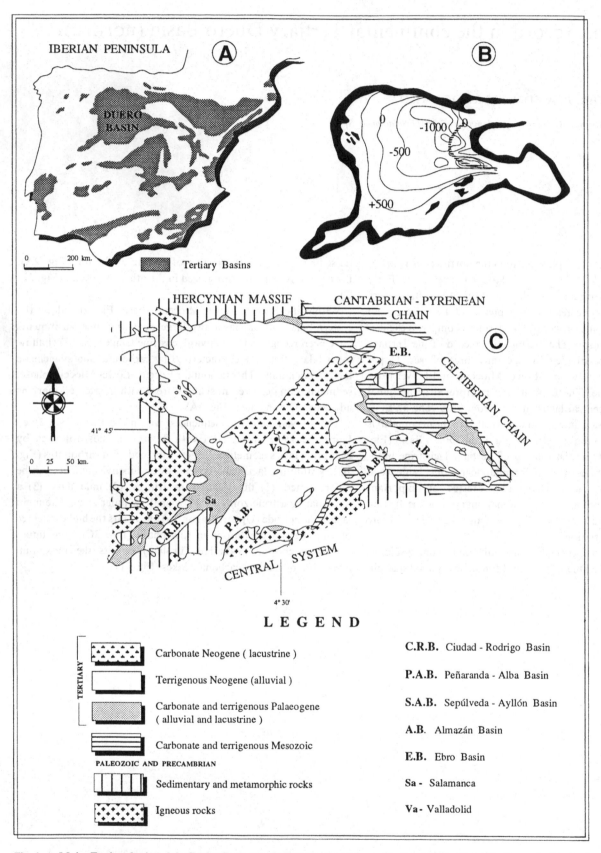

Fig. 1. A: Major Tertiary basins of the Iberian Peninsula. B: Isobath map of Lower Tertiary limit in Duero Basin; maximal axis of subsidence illustrated by dashed line. C: Generalized geologic map of the Tertiary Duero Basin and surrounding basement.

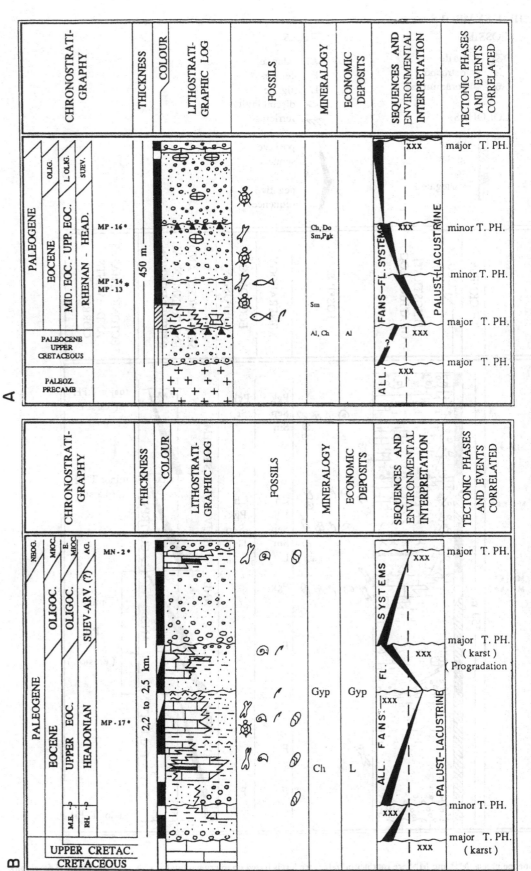

Fig. 2. Stratigraphic record of the Tertiary Duero Basin. Upper Cretaceous–Paleocene and Eocene–Lower Miocene phases in western part (A) and eastern part (B) of basin.

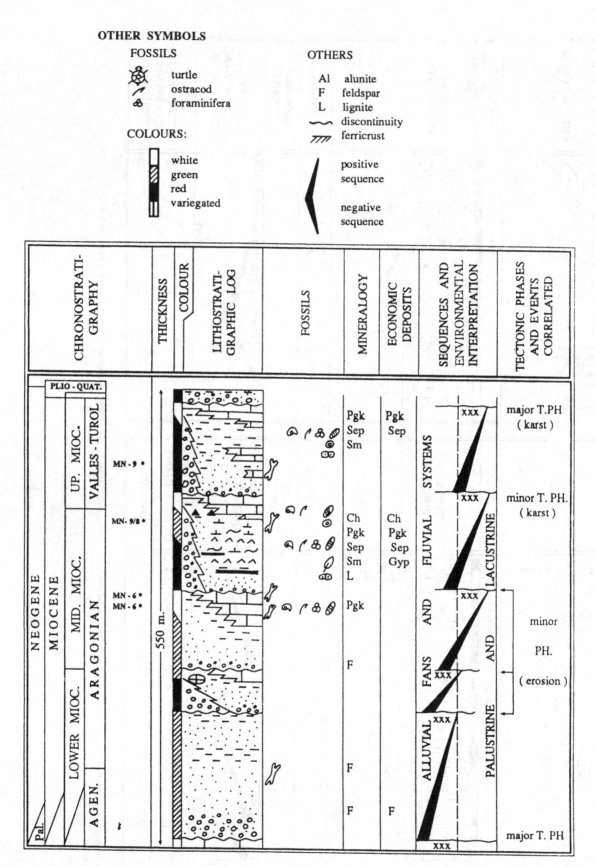

Fig. 2 (cont.)
(C) the Lower Miocene–Upper Miocene phase. MP and MN are continental reference levels based on micromammals. XXX refers to paleosol development.

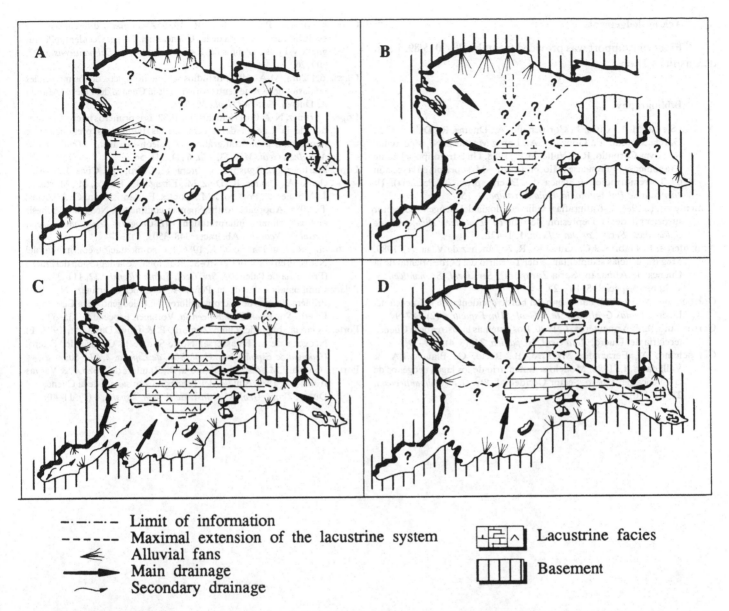

Fig. 3. Paleogeographic sketches of the main lacustrine events in the Tertiary Duero Basin. A: Eocene–Lower Miocene phase. B: Middle Miocene (Aragonian: at level of MN6). C: Middle Miocene (Upper Aragonian: at level of MN 8). D: Upper Miocene (Upper Vallesian: at level of MN9/10). For exact location of landmarks, see Fig. 1C.

Acknowledgements

Financial support was provided from the PB-9015/89, Junta de Castilla y León.

Bibliography

Alvarez Sierra, M.A., Civis, J., Corrochano, A., Daams, R., Dabrio, C.J., García Moreno, E., Gonzalez, A., Lopez Martínez, N., Mediavilla, R., Rivas Carballo, R. & Valle, M.F., 1988. Un estratotipo del límite Aragoniense-Vallesiense (Mioceno medio–M. superior) en la sección de Torremormojon (Cuenca del Duero, prov. de Palencia). IV *Jornadas de Paleontología, Guía de campo*, 2–15.

Armenteros, I., 1986. Estratigrafía y sedimentología del Neógeno del sector suroriental de la Depresión del Duero. *Ediciones Diputación de Salamanca, Serie Castilla y León*, 1, 1–471.

Armenteros, I., Dabrio, J.C., Guisado, R. & Sánchez de Vega, A., 1989. Megasecuencias sedimentarias del Terciario del borde oriental de la Cuenca de Almazán (Soria-Zaragoza). *Stv. Geol. Salmanticensia*, Volumen Especial, 5, 107–27.

Corrochano, A., 1982. El Paleógeno del borde occidental de la Cuenca del Duero. *Temas Geol. Min. Inst. Geol. Min. España*, 6 (2), 687–97.

Corrochano, A. & Armenteros, I., 1990. Los sistemas lacustres de la Cuenca terciaria del Duero. *Acta Geol. Hispanica*, 24, (3–4), 259–79.

Corrochano, A., Fernández Macarro, B., Recio, C., Blanco, J.A. & Valladares, I., 1986. Modelo sedimentario de los lagos neógenos de la Cuenca del Duero. Sector Occidental. *Stv. Geol. Salmanticensia*, 22, 93–110.

Corrochano, A. & Pena dos Reis, R., 1986. Analogías y diferencias en la evolución sedimentaria de las Cuencas del Duero, Occidental Portuguesa y Lousã (Península Ibérica). *Stv. Geol. Salmanticensia*, 22, 309–26.

García del Cura, M.A., 1974. Estudios sedimentológicos de los materiales terciarios de la zona centro-oriental de la Cuenca del Duero (Aranda de Duero). *Estudios Geol.*, 30, 579–97.

López Martínez, N. & Borja Sanchiz, F., 1982. Los primeros microvertebrados de la Cuenca del Duero: Listas faunísticas preliminares e implicaciones biostratigráficas y paleofisiográficas. *Temas Geol. Min. Inst. Geol. Min. España*, 6 (1), 341–56.

López Martínez, N., Agusti, J., Cabrera, L., Calvo, J.P., Civis, J., Corrochano, A., Daams, R., Diaz, M., Elizaga, E., Hoyos, M., Martinez, J., Morales, J., Portero, J.M., Robles, F., Santisteban, C. & Torres, T., 1985. Approach to the Spanish continental Neogene synthesis and paleoclimatic interpretation. *VIII Congress Reg. Comm. Mediterranean Neogene*, Abstracts, 348–350.

Mediavilla, R.M. & Dabrio, C.J., 1986. La sedimentación continental del Neógeno en el sector Centro-Septentrional de la depresión del Duero (Provincia de Palencia). *Stv. Geol. Salmanticensia*, 22, 111–32.

Peláez-Campomanes, P., de la Peña, A. & López Martínez, N., 1989. Primeras faunas de micromamíferos del Paleógeno de la Cuenca del Duero. *Stv. Geol. Salmanticensia*, Volumen Especial 5, 135–57.

Portero García, J.M., del Olmo Zamora, P. & Olive Davo, A., 1983. El Neógeno de la Transversal Norte-Sur de la Cuenca del Duero. *Geología de España, Inst. Geol. Min. de España*, 2, 494–502.

Portero García, J.M., del Olmo Zamora, P., Ramírez del Pozo, J. & Vargas Alonso, I., 1982. Síntesis del Terciario Continental de la Cuenca del Duero. *Temas Geol. Min. Inst. Geol. Min. España*, 6 (1), 11–40.

Cenozoic lacustrine deposits in the Duero Basin (Spain)

R. MEDIAVILLA,[1] A. MARTÍN-SERRANO,[2] C.J. DABRIO,[3] AND J.I. SANTISTEBAN[1]

[1]*Instituto Tecnológico Geominero de España, Cristobal Bordiú 35, E-28003 Madrid, Spain*
[2]*Instituto Tecnológico Geominero de España, Rios Rosas 23, E-28003, Madrid, Spain*
[3]*Dpto. Estratigrafía, Facultad de Ciencias Geológicas, Universidad Complutense, E-28040, Madrid, Spain*

Introduction

The Duero Basin lies in the northwestern Iberian Peninsula with a surface extent of 55,000 km². It rests upon Pre-Mesozoic metasedimentary and granitic rocks of the Hesperic Massif (along the western, southern and northwestern borders) and Mesozoic sedimentary rocks (along the eastern and northeastern borders). Uplift of the proto-Central System during the Paleocene started the differentiation of the Duero Basin (Portero & Aznar, 1984). During the Eocene, a northwest–southeast compression increased the uplift and generated the Central System (Béjar, Gredos & Guadarrama Ranges, Fig. 1) that separated Duero Basin from the Tajo (Madrid) Basin. Most of the sedimentary fill of the Duero Basin consists of terrestrial deposits, however, a connection with the open sea existed to the north and northeast. During the Late Eocene and Oligocene, rapid uplift of the northern and northeastern margins of the Duero Basin and the complete establishment of the Central System (along the southern border) occurred. These major changes of paleogeography triggered rapid erosion of uplifted source areas and a large input of conglomerates to the basin. The largest rates of subsidence and accumulation of sediment are along the northern and eastern margins of the basin; the western and southern borders record much smaller volumes (Fig. 1A).

Paleogene sedimentary rocks crop out only along the margins of the Duero Basin (Fig. 1A), which are interpreted as alluvial-fan and proximal fluvial facies. Distal fluvial and lacustrine facies occur in the central parts of the basin as shown by drill cores (Compañía General de Sondeos, 1985). On the other hand, large outcrops of Neogene deposits in the basin represent various types of facies: alluvial-fan (northern and eastern borders), fluviatile (western border and toward basin center), and fluvio-lacustrine and lacustrine (central sector). The Neogene sedimentary record is limited in the south of the basin (Fig. 1A) because of large-scale erosion.

The Tertiary lacustrine deposits are arranged into two groups based on paleogeography: marginal lacustrine and central lacustrine systems. Marginal lacustrine systems include ephemeral marches or very shallow lakes and ponds related to alluvial-fan or low-sinuosity fluvial systems (Fig. 2A and B). Central lacustrine systems contain more stable, perennial lakes located in the basin center, preferentially aligned along the axis of maximum subsidence (Fig. 1A). These lakes developed at the distal part of fluvial systems, but in areas of high subsidence rates they may have been associated with alluvial fans as well.

Marginal lacustrine deposits

These deposits are lenticular units (0.2 to 10 m thick, several km long) exhibiting many features indicative of subaerial exposure, such as pedogenic textures, brecciation, and carbonate crusts (Fig. 2A and B). They interbed and are associated laterally with conglomerates, sandstones and mudstones, interpreted as alluvial sediments. Three facies associations based upon lithology are recognized: (1) Interbedded carbonate mudstones or marlstones (0.2 to 0.5 cm thick) and ripple cross-laminated sandstones (10 to 20 cm thick) (Fig. 2C, left) with *Chelonia*, crocodylia and fish remains (Jiménez, 1977), interpreted as floodplains of low-sinuosity river systems (Corrochano, 1977; Martín-Serrano, 1988) mostly along the western edge of the basin (Fig. 2A and B) and fed by granitic source areas. (2) Carbonate layers (marlstones or calcareous marlstones) passing upward into brecciated and edaphized limestones (Fig. 2C, right) with ostracods, charophytes, and gastropods, occurring at the front, or between lobes, of alluvial fans (Armenteros, 1986; Alonso Gavilán et al., 1987), mostly along the southern and southeastern edges of the basin and fed by Paleozoic and Mesozoic siliciclastic and carbonate source areas. (3) Alternations of laminated carbonate-siliciclastic mudstones on a cm to dm scale, with abundant remains of palynomorphs or dolostones and gypsum (Fig. 2D), interpreted as lacustrine deposits (marshes/ponds), occupying a position distal to the fan systems (García Talegón & Alonso Gavilán, 1989) along the northern edge of the basin with source areas containing Mesozoic carbonate and gypsiferous rocks.

Central lacustrine deposits

Occuring in areas of maximum subsidence (Fig. 1A), five sedimentary cycles (N-1 to N-5) can be distinguished bounded by

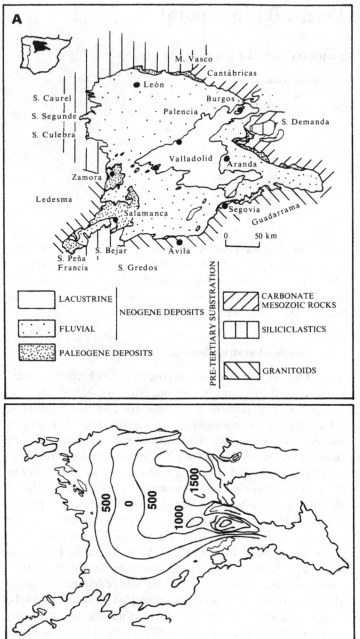

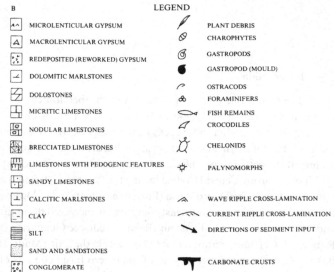

Fig. 1. (A) The Duero Basin and isobath map of the base of the Tertiary (from Compañía General de Sondeos, 1985). Datum, present sea level. (B) Legend for Figs 1–6.

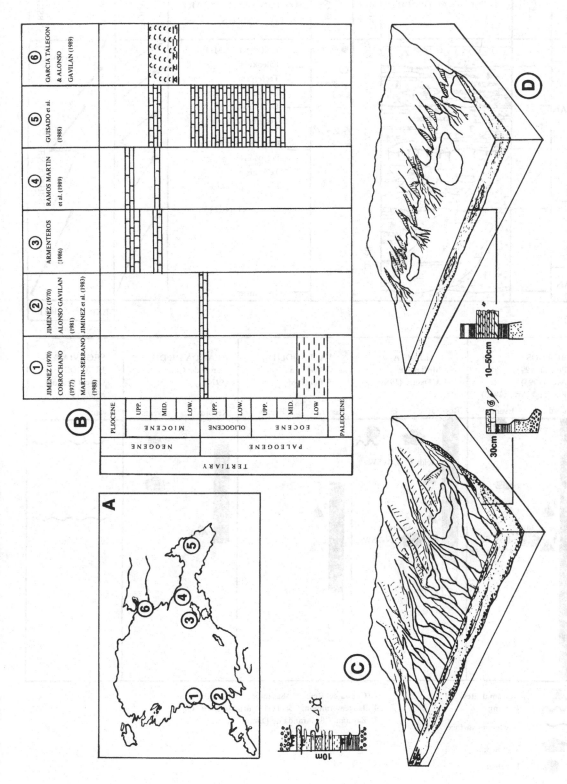

Fig. 2. (A) Location map of the marginal lacustrine deposits. (B) Temporal distribution of the marginal lacustrine deposits. (C) Conceptual models of carbonate ephemeral (modified from Dabrio *et al.*, 1989) and siliciclastic (Martín-Serrano, 1988) ephemeral lakes. (D) Conceptual model of saline (gypsiferous) ephemeral lakes (modified from García Talegón and Alonso Gavilán, 1989).

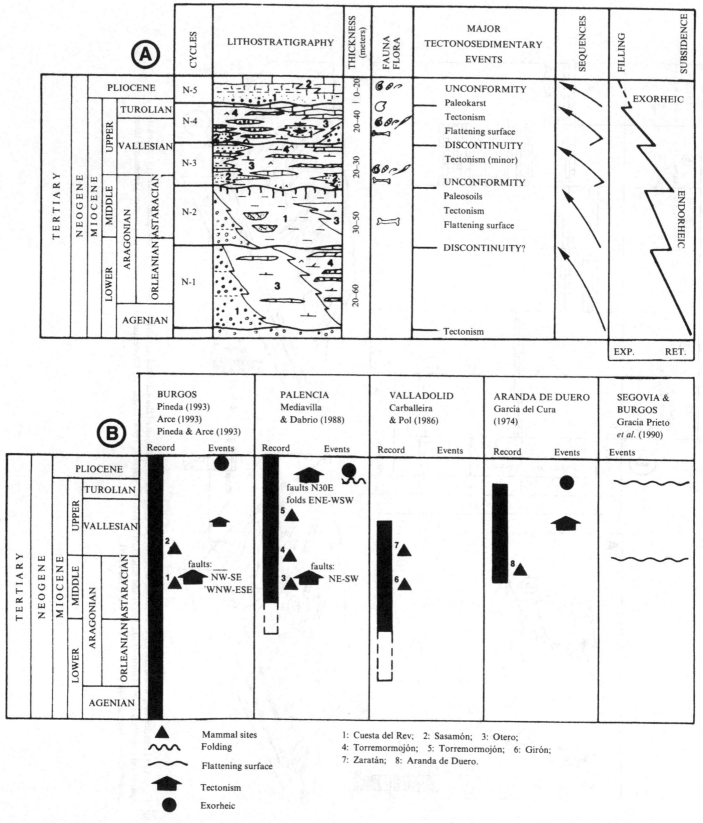

Fig. 3. (A) Generalized stratigraphic section of the Neogene central lake deposits of the Duero Basin. (B) Neogene stratigraphic successions and recorded events in various areas of the Duero Basin. Key: (1) alluvial; (2) fluvio-lacustrine; (3) carbonate lacustrine; (4) saline lacustrine. (Exp.) expansion; (Ret.) regression. Flattening surface = unconformity.

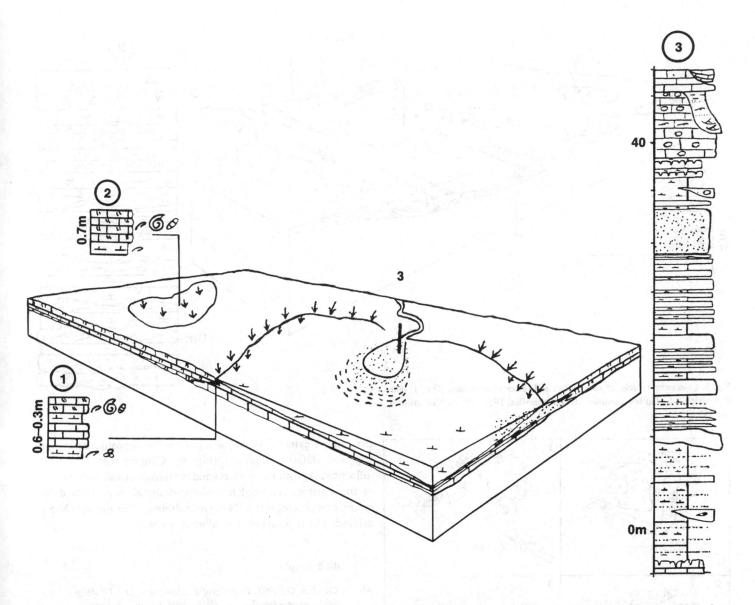

Fig. 4. Conceptual model of carbonate lakes and resulting facies associations. Key: (1) lake deposits; (2) paludal deposits; (3) deltaic deposits (modified after Mediavilla and Dabrio, 1986 and Sanchez Benavides et al., 1988).

unconformities or sedimentary discontinuities (Fig. 3A, B). Each cycle contains alluvial conglomerates, sandstones and mudstones that interfinger with lacustrine deposits. Lacustrine deposits of cycles N-1, N-3 and N-4 are carbonates and evaporites, whereas those of N-2 and N-5 are mostly of carbonates. These perennial lakes experienced continuous water level changes. Very low topographic gradients accounted for large shifts of the shorelines and large areas affected by alternating subaerial exposure and inundation.

A typical sedimentary sequence for carbonate lakes (Fig. 4.1) shows marlstones or biomicrites containing ostracods, charophytes, and some foraminifera that pass upwards into nodular limestones with gastropods and charophytes, topped by brecciated limestones. Paludal carbonates (Fig. 4.2) and deltaic siliciclastic

green mudstones and sandstones (Sánchez Benavides et al., 1988) are associated with these carbonate lakes (Fig. 4.3).

A typical sequence for gypsiferous lakes (Fig. 5) contains alternating centimeter-thick layers of dolostones, or dolomitic marlstones, and lenses of gypsum ('microlenticular') passing upwards into layers of ripple cross-laminated gypsarenites, dolomicrites with gypsum lenses, topped by recrystallized, edaphized dolostones (Fig. 5A). Mediavilla (1986–87) observed layers of turbiditic gypsum in these deposits (Fig. 5B).

The areal distribution of these two types of lakes was controlled by tectonics, sediment input to the lacustrine areas and climate (see also Mediavilla et al., 1991). Tectonic events interrupted the progressive filling of the basin and expansion of the lakes (Figs. 3A & 6), generating the sedimentary cycles. Furthermore, changes in

EVAPORATION

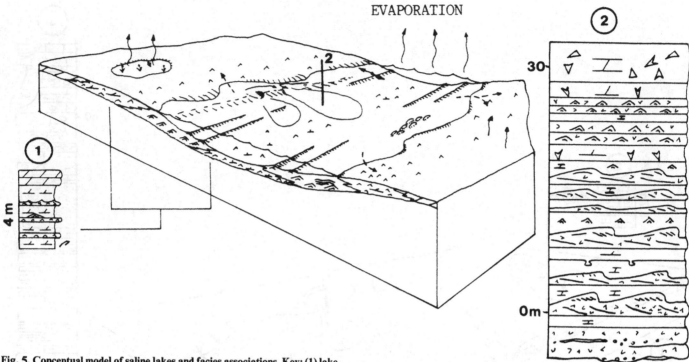

Fig. 5. Conceptual model of saline lakes and facies associations. Key: (1) lake deposits; (2) turbidites (modified after Mediavilla, 1986–87 and Mediavilla *et al.*, 1991).

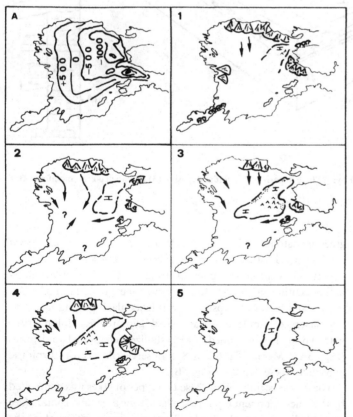

Fig. 6. Paleogeography of central lacustrine environments. (1) Agenien–Orleanian; (2) Orleanian–Astaracian; (3) Astaracian–Vallesian; (4) Valle-sian–Turolian; (5) Turolian–Pliocene. (A) Isobath map.

the rate of external sediment input to the depositional system triggered shifts in lithology (Fig. 6). Climate may also have influenced depositional patterns and the transport and preservation of the gypsum. A closed hydrologic drainage is postulated for sedimentary cycles N-1 to N-4. Open drainage during cycle N-5 is evidenced by deposits of large alluvial systems.

Bibliography

Alonso Gavilán, G., 1981. *Estratigrafía sedimentología del Paleógeno en el borde suroccidental de la Cuenca del Duero (Provincia de Sala-manca)*. PhD dissertation, Univ. Salamanca.

Alonso Gavilán, G., Armenteros, I., Dabrio, C.J. & Mediavilla, R., 1987. Depósitos lacustres terciarios de la Cuenca del Duero (España). *Stv. Geol. Salmanticensia*, **24** (suppl. 1), 1–47.

Arce, M., 1993. *Geological Map of Spain, E. 1:50.000, MAGNA*. Sheet 199 (Sasamón). Inst. Tecno. Geominero España. (In press).

Armenteros, I., 1986. *Estratigrafía y sedimentología del Neógeno del sector suroriental de la Depresión del Duero*. Serie Castilla y León 1, Diputación de Salamanca.

Carballeira, J. & Pol, C., 1986. Características y evolución de los sedimentos lacustres miocenos de la region de Tordesillas ('Facies de las Cuestas') en el Sector Central de la Cuenca del Duero. *Stv. Geol. Salmanticensia*, **24**, 213–46.

Colmenero, J.R., García-Ramos, J.C., Manjón, M. & Vargas, I., 1982. Evolución de la sedimentación terciaria en el borde N. de la Cuenca del Duero entre los valles del Torio y Pisuerga (León-Palencia). *Instituto Geológico y Minero de España, Temas Geol. Min.*, **6**, 171–81.

Compañía General de Sondeos, 1985. *Actualización de la síntesis del Terciario continental de la Cuenca del Duero*. Inst. Tecno. Geominero España Internal Report.

Corrochano, A., 1977. *Estratigrafía y sedimentología del Paleógeno de la Provincia de Zamora*. PhD dissertation, Univ. Salamanca.

Dabrio, C.J., Alonso Gavilán, G., Armenteros, I. & Mediavilla, R., 1989. Tertiary fluvial and fluvio-lacustrine deposits in the Duero Basin. *4th Inter. Conf. Fluvial Sedim. Excursion Guidebook, Servei Geologic de Catalunya*, p. 141.

García del Cura, M.A., 1974. Estudio sedimentológico de los materiales terciarios de la zona centra-oriental de la Cuenca del Duero (Aranda de Duero). *Estud. Geol.*, **30**, 579–97.

García Talegón, J. & Alonso Gavilán, G., 1989. Estudio estratigráfico y paleogeográfico de la Unidad Informal de San Miguel de Pedroso (Neógeno del Pasillo de Los Montes de Oca, Burgos). *Geogaceta*, **6**, 64–6.

Gracia Prieto, F.J., Nozal, F., Pineda, A. & Wouters, P.F., 1990. Superficies de erosión neógenas y neotectónica en el borde NE de la Cuenca del Duero. *Geogaceta*, **7**, 38–40.

Guisado, R., Armenteros, I. & Dabrio, C.J., 1988. Sedimentación continental paleógena entre Almazul y Deza (Cuenca de Almazán oriental, Soria). *Stv. Geol. Salmaticensia*, **25**, 65–83.

Jiménez, E., 1970. *Estratigrafía y paleontología del borde suroccidental de la Cuenca del Duero*. PhD dissertation, Univ. Salamanca.

Jiménez, E., 1977. Sinopsis sobre los yacimientos fosilíferos de la Provincia de Zamora. *Bol. Geol. Min.*, **88** (5), 357–64.

Jiménez, E., Corrochano, A. & Alonso Gavilán, G., 1983. El Paleógeno de la Cuenca del Duero. In *Geología de España* vol. 2, ed. J.A. Comba, pp. 489–94. Libro Jubilar J.M., Rios.

Martín-Serrano, A., 1988. *El relieve de la región occidental zamorana. La evolución geomorfológica de un borde del Macizo Hespérico*. Inst. Estud. Zamor. Florian de Ocampo, Dip. Zamora.

Mediavilla, R., 1985. *Estratigrafía y sedimentología del Neógeno de Palen-cia*. Masters Thesis, Univ. Salamanca, 135 p.

Mediavilla, R., 1986–87. Sedimentología de los yesos del Sector Central de la Depresión del Duero. *Acta Geol. Hispanica*, 21–22, 35–44.

Mediavilla, R. & Dabrio, C.J., 1986. La sedimentación continental del Neógeno en el sector Centro-Septentrional de la depresión del Duero (Provincia de Palencia). *Stv. Geol. Salmanticensia*, **22**, 111–32.

Mediavilla, R. & Dabrio, C.J., 1988. Controles sedimentarios neógenos en la Depresión del Duero (Sector Central). *Rev. Soc. Geol. España*, **1** (1–2), 187–95.

Mediavilla, R., Dabrio, C.J. & Santisteban Navarro, J.I., 1991. Factores alocíclicos que controlan el desarrollo de ciclos evaporíticos en el sector central de la Cuenca del Duero (Provincia de Palencia). *I Congr. Grupo España Terciario, Comunicaciones*, 214–17.

Pineda, A., 1993. *Geological Map of Spain, E. 1:50.000, MAGNA*. Sheet 166 (Villadiego). Inst. Tecno. Geominero España. (In press.)

Pineda, A. & Arce, M., 1993 *Geological Map of Spain, E. 1:50.000, MAGNA*. Sheet 200 (Burgos). Inst. Tecno, Geominero España. (In press.)

Portero, J. & Aznar, J.M., 1984. Evolución morfotectónica y sedimentación terciarias en el Sistema Central y cuencas limítrofes (Duero y Tajo). *I Congr. España Geol.*, **3**, 253–63.

Ramos Martín, M.C., Montes, M.J. & Alonso Gavilán, G., 1989. Caracterización de la sedimentación terciaria en el area area del Burgo de Osma (Soria). *XII Congreso Espanol de Sedimentologia, Comunicaciones*, 51–4.

Sánchez Benavides, F.J., Alonso Gavilán, G. & Dabrio, C.J., 1988. Sedimentología de los depósitos lacustres neógenos de Castrillo del Val (Burgos) España. *Stv. Geol. Salmanticensia*, **25**, 87–108.

Precambrian

Late Proterozoic, northwest Scotland, UK

JOHN PARNELL

Department of Geology, Queen's University, Belfast BT7 1NN, Northern Ireland, UK

In northwest Scotland (Fig. 1), unmetamorphosed/low grade metamorphosed late Precambrian sedimentary rocks, collectively termed the 'Torridonian', lie unconformably upon the Lewisian (Archean-Proterozoic) basement complex. The total thickness of the Torridonian sequence is in excess of 10 km in places, dominated by fluviatile deposits, but also including grey and black shales of probable lacustrine origin (Fig. 2).

The 'Torridonian' is divided into two sequences, divided by a major unconformity. The Stoer Group and overlying Sleat-Torridon Groups (Fig. 3) are dated at about 968 Ma and 777 Ma respectively, by Rb–Sr determinations on shales (Moorbath, 1969).

Distinctive initial Sr-isotopic compositions for the two sequences suggest different source areas. Paleolatitude for the Stoer Group was 15° N (Stewart & Irving, 1974). The 'Torridonian' rocks may have accumulated in a rift setting, bounded by structures approximately coincident with the Outer Hebrides Thrust (west) and the Moine Thrust Zone (east) (Stewart, 1982).

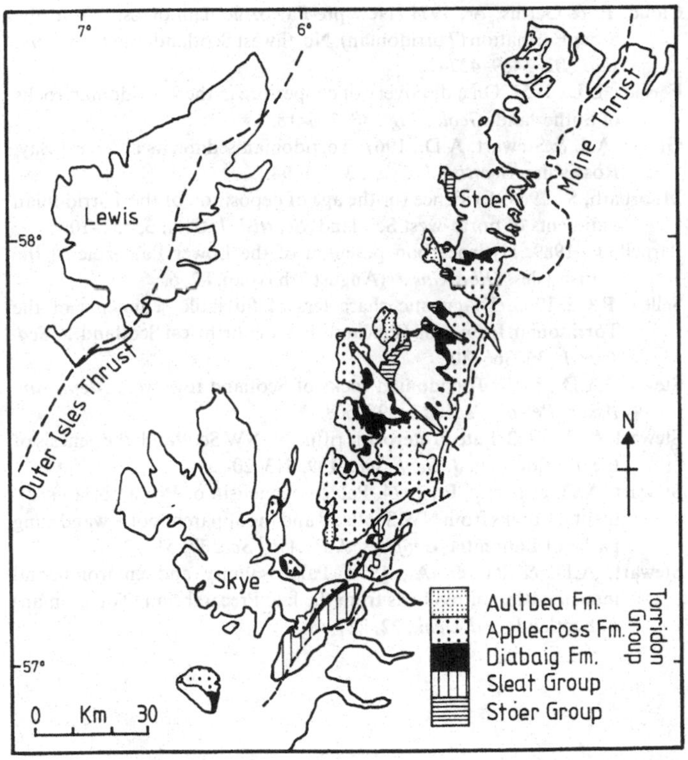

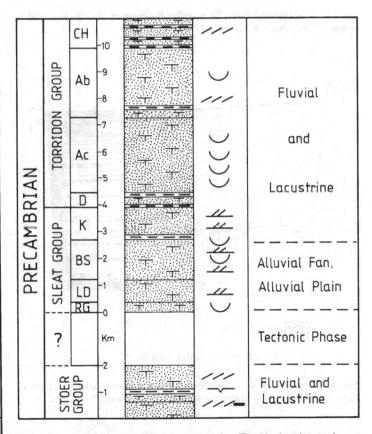

Fig. 2. Summary stratigraphic succession for Torridonian in northwest Scotland. RG: Rubha Guaih Formation; LD: Loch na Dal Formation; BS: Beinn na Seamraig Formation; K: Kinloch Formation; D: Diabaig Formation; Ac: Applecross Formation; Ab: Aultbea Formation; CH: Cailleach Head Formation (adapted from Anderton *et al.*, 1979). Dating method: Isotopic. Reliability index: E.

Fig. 1. Distribution of Stoer and Sleat–Torridon Group sedimentary rocks of the 'Torridonian' in northwest Scotland, after Anderton *et al.*, 1979. Deposit mid-point: 5.5° W, 57.5° N.

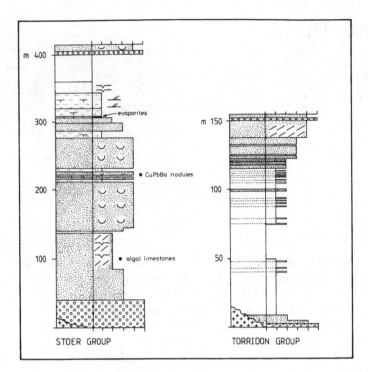

Fig. 3. Sections through Stoer Group (after Barber *et al.*, 1978) and Torridon Group (after Stewart & Parker, 1979).

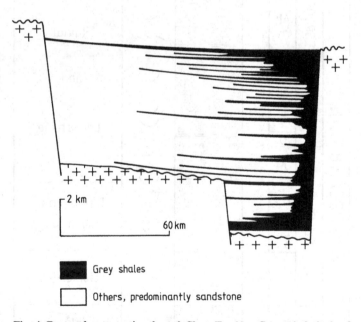

Grey shales

Others, predominantly sandstone

Fig. 4. Restored cross-section through Sleat–Torridon Groups in latitude of Skye, highlighting grey shales (lacustrine) (after Stewart, 1982).

Gray shales which occur at several levels in both the Stoer Group and Sleat-Torridon Groups (Fig. 4) were once thought to be marine (Stewart, 1969), but now non-marine (Stewart, 1982). Evidence from boron-in-illite values supports a non-marine setting (Stewart & Parker, 1979).

The Stoer Group includes a bed of black shale, similar to oil shale (Barber *et al.*, 1978), interpreted as a stratified lake deposit. The bed contains acritarchs (Cloud & Germs, 1971). Pseudomorphs after gypsum occur in associated gray shales. Numerous beds exhibit desiccation cracks. Algal limestones are interbedded with sandstones in the Stoer Group, and oncoid limestone coatings occur on clasts immediately above the Archean basement at one locality. Nodules of copper sulfide-galena-barite occur in siltstone of the Stoer Group (Fermor, 1951; J. Parnell, unpub. data). Phosphate nodules occur in the Aultbea Formation.

During the subsequent Caledonian (Paleozoic) Orogeny, the Sleat and Torridon Groups in Skye, with part of the underlying crystalline basement, experienced westwards thrusting and concomitant deformation and metamorphism. However, much of the original sedimentary fabric is still determinable.

Bibliography

Anderton, R., Bridges, P.H., Leeder, M.R., & Sellwood, B.W., 1979. *A Dynamic Stratigraphy of the British Isles.* Allen & Unwin, London.
Barber, A.J., Beach, A., Park, R.G., Tarney, J. & Stewart, A.D., 1978. The Lewisian and Torridonian rocks of North-West Scotland. *Geol. Assoc. Guide* No. 21.
Cloud, P. & Germs, A., 1971. New pre-Paleozoic nannofossils from the Stoer Formation (Torridonian), Northwest Scotland. *Geol. Soc. Am. Bull.*, **82**, 3469–4374.
Fermor, L.L., 1951. On a discovery of copper-ore in the Torridonian rocks of Sutherland. *Geol. Mag.*, **88**, 215–18.
Gracie, A.J. & Stewart, A.D., 1967. Torridonian sediments at Enard Bay, Ross-shire. *Scottish J. Geol.*, **3**, 181–94.
Moorbath, S., 1969. Evidence for the age of deposition of the Torridonian sediments of north-west Scotland. *Scottish J. Geol.* **5**, 154–70.
Parnell, J., 1989. Hydrocarbon potential of the Lower Paleozoic of the British Isles. *Oil & Gas J.* (August 7th issue), 82–6.
Selley, R.C., 1965. Diagnostic characters of fluviatile sediments of the Torridonian formation (Precambrian) of northwest Scotland. *J. Sed. Petrol.*, **35**, 366–80.
Stewart, A.D., 1969. Torridonian rocks of Scotland reviewed. *Mem. Am. Assoc. Petrol. Geol.*, **12**, 595–608.
Stewart, A.D., 1982. Late Proterozoic rifting in NW Scotland: the genesis of the Torridonian. *J. Geol. Soc.*, **139**, 413–20.
Stewart, A.D. & Irving, E., 1974. Palaeomagnetism of Precambrian sedimentary rocks from NW Scotland and the apparent polar wandering pathe of Laurentia. *Geophys. J. R. Astr. Soc.* **37**, 51–72.
Stewart, A.D. & Parker, A., 1979. Palaeosalinity and environmental interpretation of red beds from the late Precambrian (Torridonian) of Scotland. *Sed. Geol.*, **22**, 229–41.

Cambrian and Devonian

A Cambrian alkaline playa from the Officer Basin, South Australia

P.N. SOUTHGATE

Division of Continental Geology, Bureau of Mineral Resources, P.O. Box 378, Canberra 2601, ACT, Australia

Pseudomorphs after a suite of sodium carbonate evaporite minerals, including trona and shortite, occur in playa sediments in the Observatory Hill Formation in the northeastern parts of the Officer Basin; an intracratonic Late Proterozoic to Early Palaeozoic Basin that occurs in South Australia and Western Australia (Fig. 1). Non-marine lacustrine sediments in the Officer Basin were first recognized in Byilkaoora 1, a fully-cored stratigraphic hole that intersected alkaline evaporite pseudomorphs in laminated siliceous dolostones (Benbow & Pitt, 1979). Poor outcrop and a lack of regional subsurface stratigraphic information precludes a detailed understanding of the Officer Basin sequence so that Cambrian paleoenvironments in this part of Australia are poorly known. In consequence, although detailed sedimentological and geochemical information is available for the Parakeelya Alkali Member, stratigraphic relationships and paleogeography of the associated terrestrial facies and its relationships with them remain poorly resolved. Initial descriptions of the lacustrine carbonates and geochemical characterization of their organic components were based on drill core samples from Byilkaoora 1 (White & Youngs 1980; McKirdy & Kantsler 1980; Pitt, Benbow & Youngs 1980). Benbow (1982) conducted field-based stratigraphic studies in the St Johns Range, where alluvial fan and fluviatile conglomerates and sandstones intertongue with playa facies. Stratigraphic revisions by Brewer *et al.* (1987) summarized the detailed work conducted by Comalco Aluminium Pty Ltd in their search for alkaline evaporite minerals and petroleum. Southgate *et al.* (1989) combined sedimentological observations with geochemical and thermal maturation data to provide a detailed description and discussion of the depositional environments and diagenesis of the siliceous dolostones in the Parakeelya Alkali Member.

Rocks of the Observatory Hill Formation occur in a northeast–southwest trending belt that crops out poorly between Mount Johns Range in the north and Lake Maurice in the south (Fig. 1). Thickness ranges from 178 m to in excess of 494 m in the north and from 178 m to 294 m in the south. Basal clastics of the Observatory Hill Formation unconformably overlie Proterozoic rocks and have conformable and disconformable relationships with Early Cambrian clastics and marine carbonates and evaporites of the Relief Sandstone and Ouldburra Formations. In the Mount Johns Range a lateral intertonguing relationship with alluvial fan conglomerates of the Wallatinna Formation is observed (Fig. 2). The age of the Observatory Hill Formation is poorly constrained. Based on the correlations in Brewer *et al.* (1987) the rocks are younger than Early Cambrian archaeocyathan-bearing limestones of the Ouldburra Formation, but older than shallow marine to fluvial clastics of Late Cambrian to Devonian age (Krieg, 1973). Thus a late Early Cambrian to Middle Cambrian age is the most appropriate for these sediments.

The principal lithologies and facies relationships in the Parakeelya Alkali Member are shown in Fig. 3. The dominance of saline mudflat, dry mudflat and sandflat facies indicates that the lake was rimmed by broad flat areas with negligible relief. The high $\delta^{18}O$ values of primary and penecontemporaneous diagenetic carbonates of $+24‰$ to $+28‰$ (SMOW) indicate strong evaporation of ground and surface waters within the lake system (Southgate *et al.*, 1989). Calcite pseudomorphs of the sodium carbonate minerals trona and shortite have $\delta^{18}O$ values between $+19‰$ and $+22.5‰$, and contain fluid inclusions with variable salinities and homogenization temperatures up to *c.* 110 °C. This suggests that the euhedral alkaline evaporites were dissolved by heated waters; calcite pseudomorphs then precipitated from a mixed solution formed by the interaction of these incoming fluids with the relatively saline interstitial brines. The lacustrine carbonates have $^{87}Sr/^{86}Sr$ ratios of around 0.722. Carbonates and anhydrite/gypsum in the underlying, archaeocyathan-bearing, marine Ouldburra Formation have Sr isotope values close to 0.709 (Donnelly *et al.*, 1990). This suggests that the playa carbonates formed from groundwaters which incorporated radiogenic Sr from older felsic rocks. In the Officer Basin such marked differences in Sr isotope ratios provides a useful method for separating marine and lacustrine carbonate rocks.

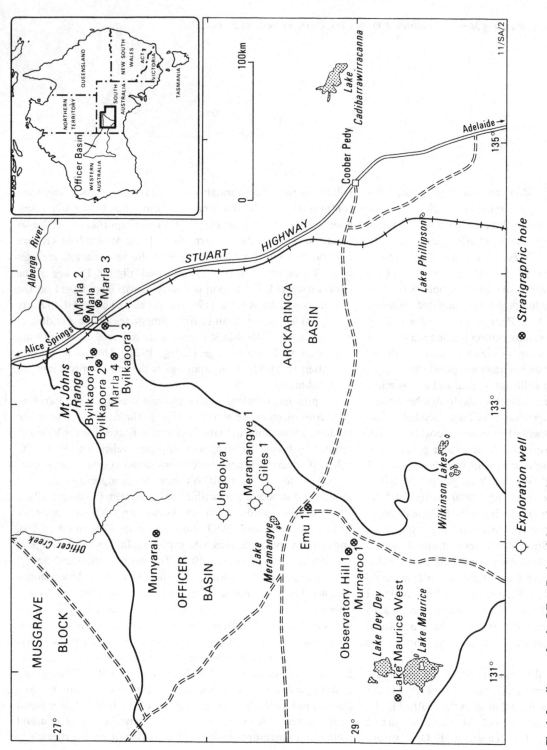

Fig. 1. Locality map for the Officer Basin and the principal areas mentioned in the text.

◇ *Exploration well*   ⊗ *Stratigraphic hole*

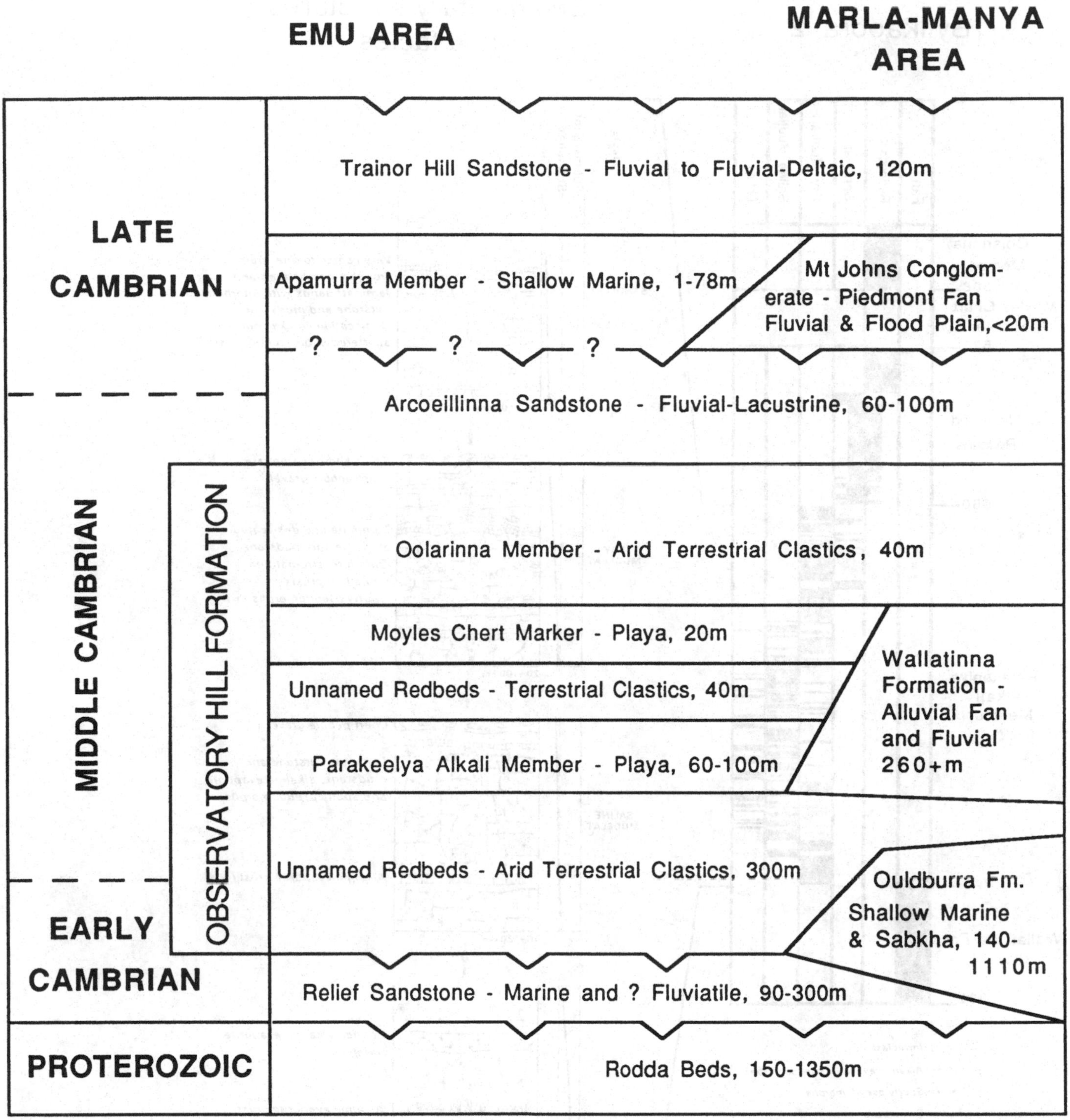

Fig. 2. Cambrian stratigraphy and paleoenvironments for the eastern Officer Basin. Paleoenvironmental interpretations are from Benbow (1982), Brewer *et al.* (1987), and Southgate *et al.* (1989).

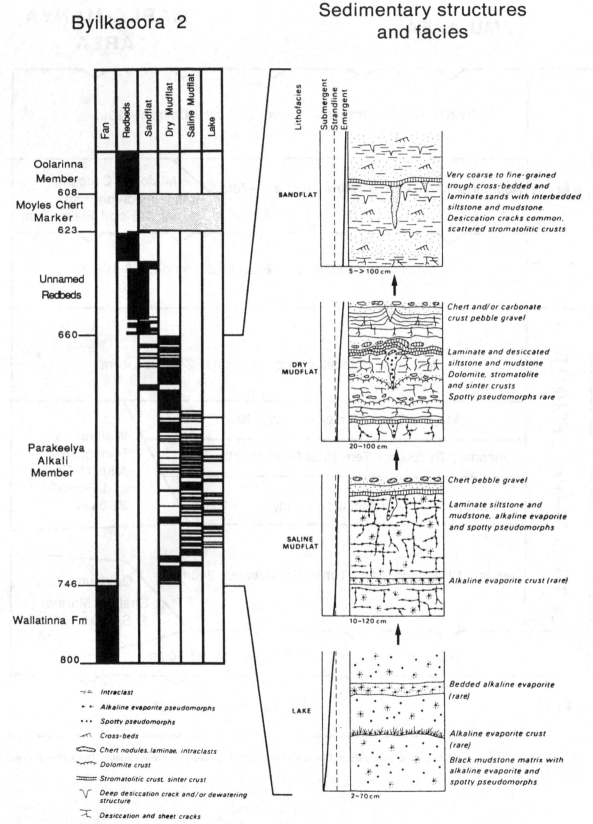

Fig. 3. Cyclic facies relationships in core from drill hole Byilkaoora 2 in the Parakeelya Alkali Member. Note how the facies illustrate the gradual expansion of the playa and its terminal contraction. The principal sedimentary structures and lithologies in each of the facies are illustrated in the generalized diagram on the right-hand side (modified from Southgate *et al.*, 1989).

Bibliography

Brewer, A.M., Dunster, J.N., Gatehouse, C.G., Henry, R.L. & Weste, G., 1987. A revision of the stratigraphy of the eastern Officer Basin. *Quarterly Geological Notes, Geol. Sur. Sth Austr.*, **102**, 2–15.

Benbow, M.C., 1982. Stratigraphy of the Cambrian – ? Early Ordovician, Mount Johns Range, NE Officer Basin, South Australia. *Trans. Roy. Soc Sth Austr.*, **106**, 191–211.

Benbow, M.C. & Pitt, G.M., 1979. *Byilkaoora No. 1 Well Completion Report*. Department of Mines and Energy South Australia Report Book No. 79/115. (Unpublished).

Brewer, A.M., Dunster, J.N., Gatehouse, C.G., Henry, R.L. & Weste, G., 1987. A revision of the stratigraphy of the eastern Officer Basin. *Quarterly Geological Notes, Geol. Sur. Sth Austr.*, **102**, 2–15.

Donnelly, T.H., Shergold, J.H., Southgate, P.N. & Barnes, C.J., 1990. Events leading to global phosphogenesis around the Precambrian/Cambrian boundary. Notholt, A.J.A. & Jarvis, I. (eds) *Geol. Soc. Lond. Spec. Pub.*, **52**, 273–87.

Kreig, G.W., 1973. Everard, South Australia. *Explanatory Notes, 1:250,000 Geological Series, Sheet S/G53–13, Adelaide*, Geological Survey South Australia.

McKirdy, D.M. & Kantsler, A.J., 1980. Oil geochemistry and potential source rocks of the Officer Basin, South Australia. *Austr. Petrol. Expl. Assoc. J.*, **20**, 68–86.

Pitt, G.M., Benbow, M.C. & Youngs, B.C., 1980. A review of recent geological work in the eastern Officer Basin, South Australia. *Austr. Petrol. Expl. Assoc. J.*, **20**, 209–20.

Southgate, P.N., Lambert, I.B., Donnelly, T.H., Henry, R., Etminan, H. & Weste G., 1989. Depositional environments and diagenesis in Lake Parakeelya: a cambrian alkaline playa from the Officer Basin, South Australia. *Sedimentology*, **33**, 1091–112.

White, A.H. & Youngs, B.C., 1980. Cambrian alkali playa-lacustrine sequence in the northeastern Officer Basin, South Australia. *J. Sed. Petrol.*, **50**, 1279–86.

Devonian Orcadian Basin, northern Scotland, UK

JOHN PARNELL,[1] DAVID ROGERS,[2] TIM ASTIN[3] AND JOHN MARSHALL[4]

[1] Dept. of Geology, Queen's University, Belfast BT7 1NN, UK,
[2] Lomond Associates, 48 West Regent St., Glasgow G2 2RA, UK,
[3] Postgraduate Research Institute for Sedimentology, University of Reading, Reading RG6 2AB, UK,
[4] Dept. of Geology, University of Southampton, Southampton SO9 5NH, UK

The Orcadian Basin is a thick development of Devonian (Old Red Sandstone: ORS) facies composed primarily of lacustrine, fluvial and aeolian sediments. They were deposited in a basin extending from the Moray Firth to Shetland, with its eastern margin concealed beneath the North Sea and its western margin exposed in parts of Caithness and the onshore Moray Firth region (Fig. 1). The basin developed as a system of half-grabens, such that the reliability of composite sections from separate areas must be limited.

The ORS sequence in northern Scotland includes Lower (Siegenian-Emsian), Middle (Eifelian-Givetian) and Upper (?Givetian-Famennian) divisions, each separated by an unconformity, and as a whole overlying Precambrian basement (Fig. 2). The three phases of the ORS in northern Scotland represent distinct regimes of sedimentation:

(1) The Lower ORS was deposited in small, fault-bounded basins, where active fault movement ensured a supply of coarse alluvial sediment.

(2) Middle ORS times saw prolonged lacustrine conditions, which were centered upon Orkney and Caithness, and extended at maximum from Shetland to Banffshire.

(3) Upper ORS fluviatile/aeolian sedimentation was the most widespread, probably continuous from Orkney to the Midland Valley of Scotland or even Northern England.

Stratigraphic correlation in the basin is based upon the widely developed Achanarras Fish Bed (Niandt Limestone Member). It can be recognized from Western Shetland in the north to the Gamrie outlier in the southeast. A second younger lacustrine interval of the Eday Group forms a similar though less widespread marker bed. Otherwise long distance correlation is limited to general lithostratigraphic comparisons at the Group level. An established biostratigraphic division is based on fossil fish for which some seven zones are known from the Middle ORS.

The boundary between the Lower ORS and the Middle ORS is unconformable at certain localities. However, these unconformities are not necessarily widespread across the basin and frequently reflect only syn-depositional extensional tectonics. The Lower ORS deposits are dominated by conglomerates with subordinate red siltstones and sandstones, probably reflecting their present exposure around the basin margins. The Middle ORS is the major depositional unit in the Orcadian Basin, overlying and overstepping the Lower ORS onto the metamorphic basement.

Lacustrine deposits occur on a limited scale within Lower ORS successions, but are of major importance within the Middle ORS. In Orkney and Caithness the Middle ORS succession consists of a lower phase dominated by lacustrine deposits up to 3.5 km thick, and an upper phase dominated by fluvial and aeolian deposits with a lesser lacustrine component (Fig. 3).

The Middle ORS lacustrine deposits are particularly well-developed in the basin center (Figs 4 and 5). Basin marginal deposits (Fig. 6) include alluvial facies (alluvial fan and braided river trough and planar cross-bedded sandstones interbedded with wave-rippled and desiccated mudrocks), aeolian facies (large-scale cross-bedded sandstones interbedded with interdune mudrocks), and lacustrine facies which encroached on the margin during periods of high lake stand. The marginal lacustrine facies include limestones deposited on the basement surface (basin-lake coincident margin limestones). The bulk of the lacustrine sediments are thinly bedded laminites, which occur in transgressive-regressive cycles, typically about 10 m thick, with fluviatile incursions at low lake level. Sediment accumulation was generally during the regressive phase, resulting in asymmetric cycles. There are however cyclic sequences which show no evidence of fluviatile sedimentation where the various lithologies in each cycle represent different lake levels in an enclosed basin. Symmetrical cycles (see Donovan, 1980) are composed of four lithological associations which represent the basic lacustrine facies:

(1) grey-black carbonate laminite, composed of 0.5 mm triplets of micritic carbonate, organic carbon and siliciclastic laminae, the deposits of a thermally stratified lake. Some carbonate laminites lack siliciclastic matter and are petrographically limestones;

(2) grey-black alternating carbon-rich shale and grey coarse siltstone (pair up to 7 mm), deposits of a shallow, sediment-starved, permanent lake;

(3) alternating dark grey carbon-rich shale and grey coarse siltstone (pair up to 10 mm); silts are lensoid and often current-rippled, also deposits of a permanent lake;

73

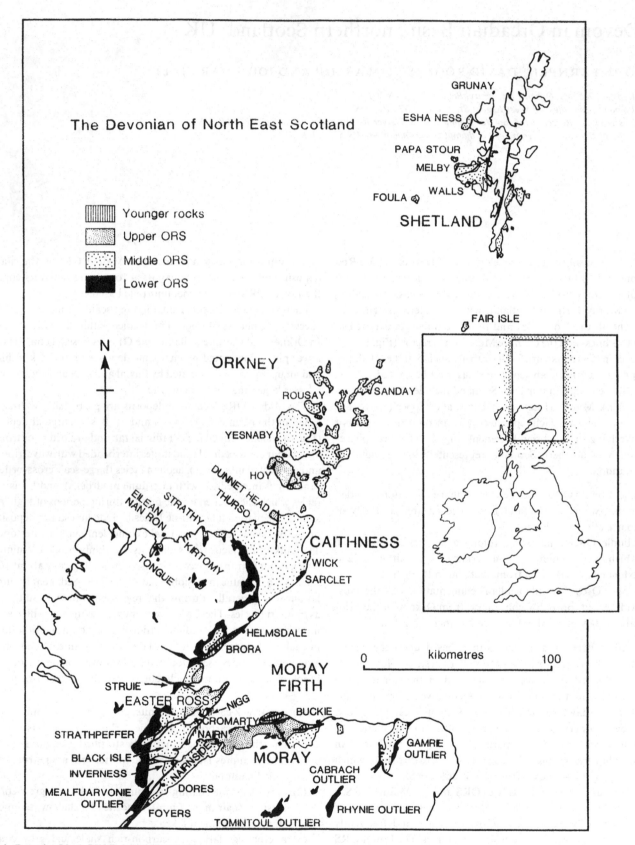

Fig. 1. (Location Map) Mid Point: 4° W, 58½° N.

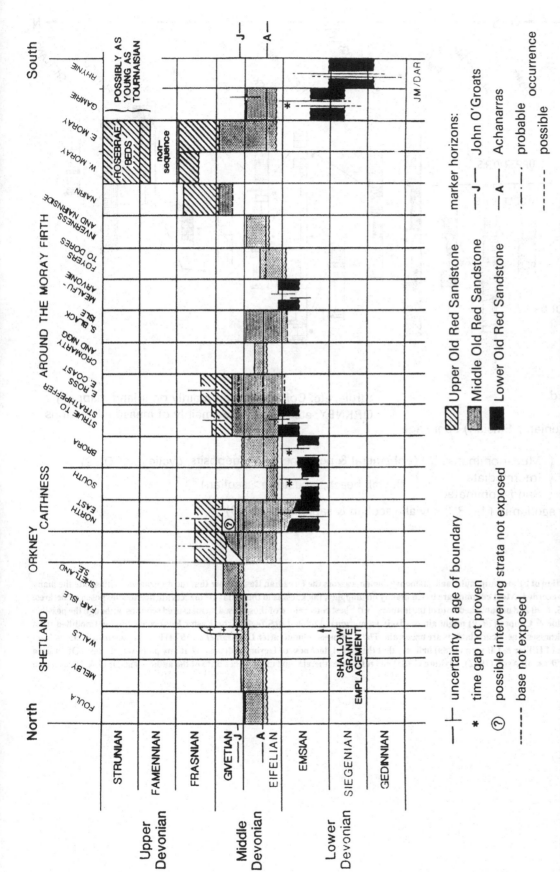

Fig. 2. Chronostratigraphy of the Orcadian Basin. Note the non-equivalence of the Orcadian 'Lower, Middle and Upper ORS' to the threefold subdivision of the Devonian. The Achanarras and John O'Groats 'fish beds' are time markers representing particularly widespread lake transgressions. (Reproduced by permission of the Geological Society from Rogers et al., 1989. J. Geol. Soc. 146).

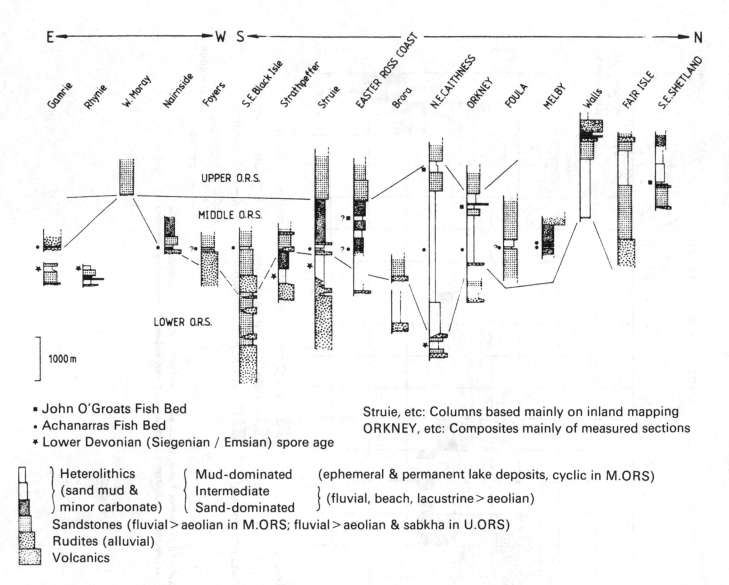

- John O'Groats Fish Bed
- Achanarras Fish Bed
- Lower Devonian (Siegenian / Emsian) spore age

Struie, etc: Columns based mainly on inland mapping
ORKNEY, etc: Composites mainly of measured sections

Heterolithics (sand mud & minor carbonate)
{ Mud-dominated, Intermediate, Sand-dominated } (ephemeral & permanent lake deposits, cyclic in M.ORS)
(fluvial, beach, lacustrine > aeolian)

Sandstones (fluvial > aeolian in M.ORS; fluvial > aeolian & sabkha in U.ORS)
Rudites (alluvial)
Volcanics

Fig. 3. Columns illustrating the variation of lithostratigraphy and paleoenvironments across the Orcadian Basin. Note that the diagram does illustrate the many local facies and thickness variations recorded. Most columns are based mainly on mapping; thicknesses in these may be inaccurate due to unexposed structures. The Orkney column (after Astin 1985, 1990 and unpublished data of the authors) is the best constrained of those based on measured sections, although the possible Lower ORS described there by Michie & Cooper (1979) might alternatively be marginal Middle ORS facies. The Easter Ross coast column (modified from Rogers 1987) assumes minimum thicknesses where correlations are uncertain. The Caithness column (after Donovan *et al.* 1974) is a composite of sections from across a wide area; the maturity data of Hillier & Clayton (1989) indicate that the actual thickness of Devonian deposited at any point was far less. Other data sources used are Allen & Marshall (1981a, b), Astin (1982), Mykura (1983) and references therein, and unpublished data of the authors and Dr G.A. Blackbourn (for Foula).

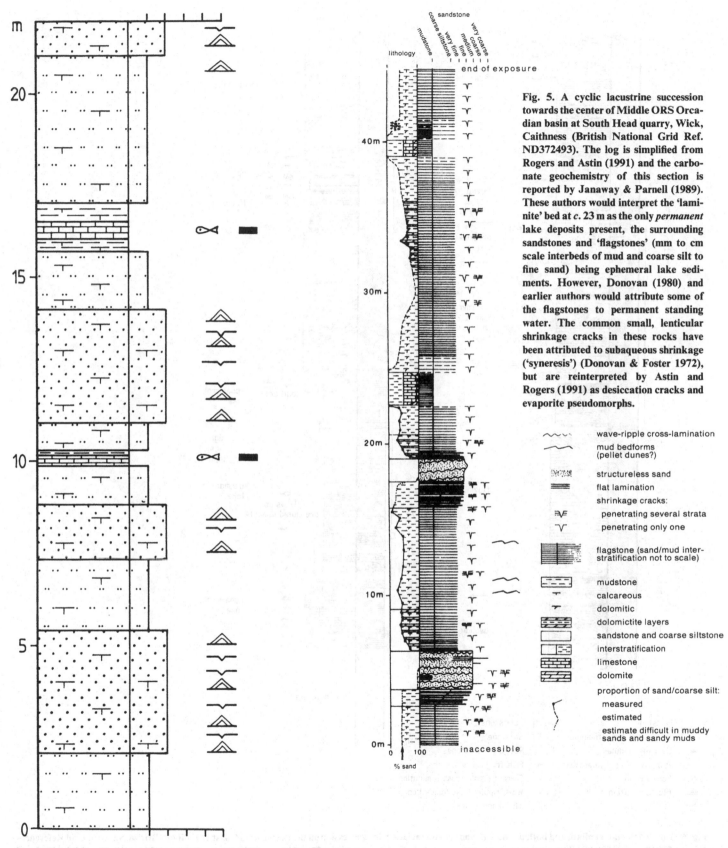

Fig. 5. A cyclic lacustrine succession towards the center of Middle ORS Orcadian basin at South Head quarry, Wick, Caithness (British National Grid Ref. ND372493). The log is simplified from Rogers and Astin (1991) and the carbonate geochemistry of this section is reported by Janaway & Parnell (1989). These authors would interpret the 'laminite' bed at *c.* 23 m as the only *permanent* lake deposits present, the surrounding sandstones and 'flagstones' (mm to cm scale interbeds of mud and coarse silt to fine sand) being ephemeral lake sediments. However, Donovan (1980) and earlier authors would attribute some of the flagstones to permanent standing water. The common small, lenticular shrinkage cracks in these rocks have been attributed to subaqueous shrinkage ('syneresis') (Donovan & Foster 1972), but are reinterpreted by Astin and Rogers (1991) as desiccation cracks and evaporite pseudomorphs.

Fig. 4. Detailed cyclic lacustrine succession, west Caithness (after Donovan 1980).

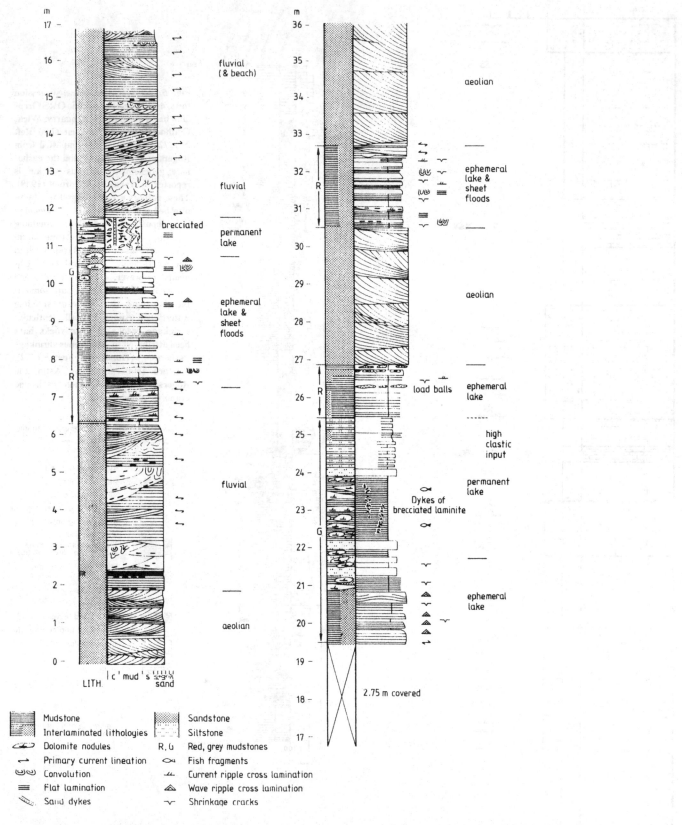

Fig. 6. A mixed fluvial, aeolian, and both permanent and ephemeral lacustrine succession on the Easter Ross Coast at Cadboll (British National Grid References NH89057810 to 89157920). The sequence is attributable to the effects of avulsion of a major alluvial system combined with variations of lake level and ephemerality. It is typical of margins of the Orcadian lake with high clastic inputs.

(4) alternating green-grey shale and siltstone/fine sandstone, deposits of an impermanent lacustrine environment. Current ripples, wave ripples and desiccation cracks are abundant, and pseudomorphs after gypsum may occur.

The lacustrine deposits were for a long time regarded as freshwater. However, gypsum pseudomorphs are now recorded in most facies, and magadi-type cherts with pseudomorphs after trona indicative of an alkaline environment have been identified at several levels.

Bibliography

Allen, P.A., 1981a. Devonian lake margin environments and processes, SE Shetland, Scotland. *J. Geol. Soc. Lond.*, **138**, 1–14.

Allen, P.A., 1981b. Wave generated lacustrine sediments of south-east Shetland and ancient wave conditions. *Sedimentology*, **28**, 369–79.

Allen, P.A. & Marshall, J.E.A., 1981. Depositional environments and palynology of the Devonian south-east Shetland Basin. *Scottish J. Geol.*, **17**, 257–73.

Astin, T.R., 1982. *The Devonian geology of the Walls Peninsula, Shetland.* PhD Thesis, University of Cambridge.

Astin, T.R. (1985). The palaeogeography of the Middle Devonian Lower Eday Sandstone, Orkney. *Scottish J. Geol.*, **21**, 353–75.

Astin, T.R., 1990. The Devonian lacustrine sediments of Orkney, Scotland; implications for climate cyclicity, basin structure and maturation history. *Geol. Soc. Lond.*, **147**, 141–51.

Astin, T.R. & Rogers, D.A., 1991. Subaqueous shrinkage cracks in the Devonian of Scotland reinterpreted. *J. Sed. Petrol.*, **61**, 850–59.

Clarke, P., 1990. *Sedimentological studies in Lower Old Red Sandstone basins, Northern Scotland.* PhD Thesis, Queen's University of Belfast.

Donovan, R.N., 1973. Basin margin deposits of the Middle Old Red Sandstone at Dirlot, Caithness. *Scottish J. Geol.*, **9**, 203–11.

Donovan, R.N., 1975. Devonian lacustrine limestones at the margin of the Orcadian Basin, Scotland. *J. Geol. Soc. Lond.*, **131**, 489–510.

Donovan, R.N., 1980. Lacustrine cycles, fish ecology and stratigraphic zonation in the Middle Devonian of Caithness. *Scottish J. Geol.*, **16**, 35–50.

Donovan, R.N., Archer, R., Turner, P. & Tarling, D.H., 1976. Devonian palaeogeography of the Orcadian Basin and the Great Glen Fault. *Nature*, **259**, 550–1.

Donovan, R.N. & Foster, R.J. 1972. Subaqueous shrinkage cracks from Caithness Flagstone Series (middle Devonian) of northeast Scotland. *J. Sed. Petrol.*, **42**, 309–17.

Donovan, R.N., Foster, R.J. & Westoll, T.S., 1974. A stratigraphical revision of the Old Red Sandstone of North-eastern Caithness. *Trans. Roy. Soc. Edinb.*, **69**, 167–201.

Duncan, A.D. & Hamilton, R.F.M., 1988. Palaeolimnology and organic geochemistry of the Middle Devonian in the Orcadian Basin. *Geol. Soc. Spec. Publ.*, **40**, 173–201.

Fannin, N.G.T. 1969. Stromatolites from the Middle Old Red Sandstone of western Orkney. *Geol. Mag.*, **106**, 77–88.

Fortey, N.J. & Michie, U.McL., 1978. Aegerine of possible authigenic origin in Middle Devonian sediments in Caithness, Scotland. *Mineralog. Mag.*, **42**, 439–42.

Hall, A.J. & Donovan, R.N., 1978. Origin of complex sulphide nodules related to diagenesis of lacustrine sediments of Middle Devonian age from the Shetland Islands. *Scottish J. Geol.*, **14**, 289–99.

Hall, P.B. & Douglas, A.G., 1983. The distribution of cyclic alkanes in two lacustrine deposits. In *Advances in Organic Geochemistry 1981*, ed. M. Bjoroy, pp. 576–587. John Wiley, London.

Hamilton, R.F.M., 1987. *Comparative palaeolimnology of the Middle Devonian Orcadian Basin.* PhD Thesis, University of Aberdeen.

Hamilton, R.F.M. & Trewin, N.H., 1988. Environmental controls on fish faunas of the Middle Devonian Orcadian Basin. *Mem. Can. Soc. Petrol. Geol.*, **3**, 589–600.

Hillier, S. & Clayton, T., 1989. Illite/smectite diagenesis in Devonian lacustrine mudrocks from northern Scotland and its relationship to organic maturity indicators. *Clay Minerals*, **24**, 181–96.

Janaway, T.M., 1987. *Carbonate production in organic-rich playa-lacustrine sequences, Devonian of Northern Scotland.* PhD Thesis, Queen's University of Belfast.

Janaway, T.M., & Parnell, J., 1989. Carbonate production within the Orcadian Basin, northern Scotland: a petrographic and geochemical study. *Palaeogeogr., Palaeoclimat., Palaeoecol.*, **70**, 89–105.

Marshall, J.E.A., Brown, J.F. & Hindmarsh, S., 1985. Hydrocarbon source rock potential of the Devonian of the Orcadian Basin. *Scottish J. Geol.*, **21**, 301–20.

McClay, K.R., Norton, M.G., Coney, P. & Davis, G.H., 1986. Collapse of the Caledonian orogen and the Old Red Sandstone. *Nature*, **323**, 147–9.

Michie, U.McL. & Cooper, D.C., 1979. *Uranium in the Old Red Sandstone of Orkney.* Inst. Geol. Sci. Rpt. No. 78/16.

Muir, R.O. & Ridgway, J.M., 1975. Sulphide mineralisation of the continental Devonian sediments of Orkney (Scotland). *Mineral. Deposita*, **10**, 205–15.

Mykura, W. 1983. Old Red Sandstone. In *Geology of Scotland*, ed. G.Y. Craig, pp. 205–51. Scottish Academic Press, Edinburgh.

Parnell, J., 1985. Hydrocarbon source rocks, reservoir rocks and migration in the Orcadian Basin. *Scottish J. Geol.*, **21**, 321–36.

Parnell, J., 1986. Devonian magadi-type cherts in the Orcadian Basin. *J. Sed. Petrol.*, **56**, 495–500.

Parnell, J., 1988. Significance of lacustrine cherts for the environment of source-rock deposition in the Orcadian Basin, Scotland. *Geol. Soc. Spec. Publ.*, **40**, 205–17.

Parnell, J. & Janaway, T., 1990. Sulphide-mineralized algal breccias in a Devonian evaporitic lake system, Orkney, Scotland. *Ore Geol. Rev.*, **5**, 445–60.

Parnell, J. & Swainbank, I., 1985. Galena mineralization in the Orcadian Basin, Scotland: Geological and isotopic evidence for sources of lead. *Mineral. Deposita*, **20**, 50–6.

Richardson, J.B., 1965. Middle Old Red Sandstone spore assemblages from the Orcadian Basin, north-east Scotland. *Palaeontology*, **7**, 559–605.

Rogers, D.A., 1987. *Devonian correlations, environments and tectonics across the Great Glen Fault.* PhD Thesis, University of Cambridge.

Rogers, D.A. & Astin, T.R. 1991. Ephemeral lakes, mud pellet dunes and wind-blown sand and silt: reinterpretation of Devonian lacustrine cycles in N Scotland. In *Advances in Lacustrine Facies Analysis*, ed. P. Anadon, L. Cabrera & K. Kelts, pp. 199–221. IAS Spec. Publ. 13.

Rogers, D.A., Marshall, J.E.A. & Astin, T.R., 1989. Devonian and later movements on the Great Glen fault system, Scotland. *J. Geol. Soc. Lond.*, **146**, 369–72.

Trewin, N.H., 1986. Palaeoecology and sedimentology of the Achanarras fish bed of the Middle Old Red Sandstone, Scotland. *Trans. Roy. Soc. Edinb.*, **77**, 21–46.

Carboniferous to Permian

Permian lakes in East Africa (Tanzania)

C.Z. KAAYA[1] AND T. KREUSER[2]

[1]Geology Department, University of Dar es Salaam, P.O. Box 35052, Tanzania
[2]Geologisiches Institut, Universität Köln, Zülpicherstr. 49, D-5000 Köln 1, Germany

Introduction

Lake beds of Permian age occur in Tanzania within Karoo basins trending in the north-northeast to south-southwest direction, extending from Kenya in the north to Mozambique in the south and Malawi in the southwest. The basins are fault-controlled, occurring as half-grabens or tilted blocks by post-Karoo fault movements during the Jurassic–Cretaceous and finally in the Mio–Pliocene contemporaneously with the East African rifting (Kreuser, 1990). A polyphase structural history of syn- and post-depositional tectonics, which in concordance with climatic variations, led to the temporary establishment of lakes. The entire thickness of the Permian strata ranges from 600 to 1400 m.

Lacustrine phases in the Permian

Three periods conducive to the formation of lakes were identified in the stratigraphic record of the Ruhuhu Basin (Figs 1 and 2) in southwestern Tanzania:

(1) The initial phase (Idusi Formation) commenced with the deposition of glacio-lacustrine sediments which consisted of tillites at the base and succeeded by rhythmites with dropstones and dark carbonaceous siltstones rich in organic carbon. Extension and duration of these glacial and periglacial lakes are not well known. Palynological results indicate an Asselian age for the upper siltstones in the glacial sequence (Fig. 3) and the tillites are slightly older (Wopfner & Kreuser, 1986; Kreuser, 1987).

The Permo-Carboniferous glaciation extended from a north to south oriented highland, distributing ice lobes to the west and east. Glacially-derived sediments are known from many neighboring African countries and are thought to have originated under a similar climate and depositional environment. The size and depth of these glacial and periglacial lakes were controlled by the pre-Karoo terrain and the glacial scouring. No fossil evidence of organisms is preserved besides pollen and spores. An abundance of pyrite could indicate anoxic conditions well above the water/sediment interface.

(2) The second period (Mchuchuma Formation) was characterized by slight subsidence and faulting, accompanied by a rise in temperature and humidity. In some areas, alluvial plain associations cut into the black siltstones (Kreuser & Semkiwa, 1987). The alluvial plain facies consist of thick channel sands and coals. Braided river to meandering river plains with oxbow lakes and floodplain swamps changed to lacustrine shallow marshes and ponds in the upper third of the sequence (Fig. 3) (Kreuser & Markwort, 1988; Kreuser, 1991). Source rocks for hydrocarbon generation in this sequence indicate the presence of poorly oxygenated, stagnant water bodies over a period of several million years (middle Sakmarian–upper Artinskian, Kreuser et al., 1988).

(3a) Above an unconformity, a new depositional cycle (Ketewaka Formation) was triggered by faulting along the eastern boundary of the Ruhuhu Basin and beyond. Coarse clastics along the margin of the basin are interpreted as deposits of braided streams and flash floods. Toward the basin center, large playa mudflats and playa lakes became established (Fig. 3). Frequent exposure is evidenced by desiccation cracks, hour-glass structures and paleosols. A semi-arid and warm climate favored the deposition of these redbeds.

(3b) The beginning of this phase (Mhukuru, Ruhuhu and Usili Formations) is marked by a color change to grey-green and the occurrence of carbonaceous shales with coaly streaks. Fluvial siliciclastics (Mhurkuru Formation) pass laterally and upward into lacustrine sediments (Ruhuhu Formation) consisting predominantly of parallel-laminated to massive grey and black carbonaceous shales with fossil fish and plant remains. A climatic change to constantly humid conditions was a controlling factor in the development of a large lake system (Fig. 3). This depositional change can be traced all over the basin; the color change acts essentially as a time-line. The lowermost part of this lake sequence is not present in most other Karoo basins in Tanzania (McKinlay, 1965), in Malawi (Yemane & Kelts, 1990), and Kenya or Zambia (Utting, 1978).

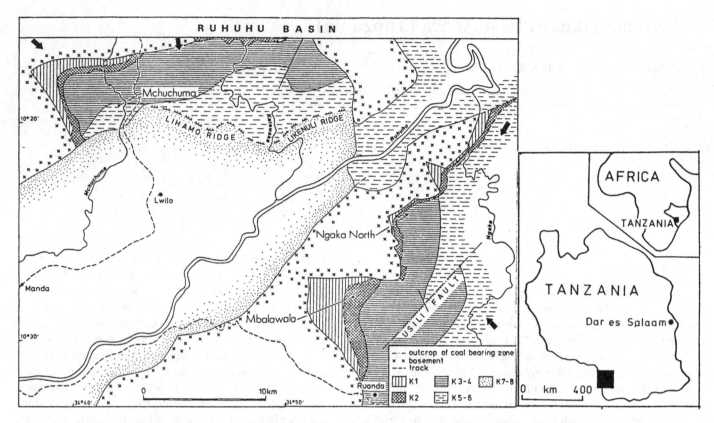

Fig. 1. Locality map of the Ruhuhu Basin with sub-basins and stratigraphic subdivisions. K1: Idusi Formation, K2: Mchuchuma Formation, K3: Ketewaka Formation, K4: Mhukuru Formation, K5: Ruhuhu Formation, K6: Usili Formation, and K7–8: Lower Triassic.

Main phase of lake development

The lake systems of phase 3b increased in size continuously and reached a maximum at the Kazanian–Tatarian boundary (Figs 2 and 3). The original lateral extent is difficult to reconstruct as it reached far beyond the present basin boundaries. Comparable sediments to the west, for example, can be found in Malawi on the opposite shore of Lake Nyasa. Also, similar sequences are found in Kenya (Maji ya Chumvi Beds), in Zambia and Zimbabwe (Madumabisa Shales), and in Mozambique ('Zona Endothiodon') which may represent the southern limit of the lakes of phase 3b. Thus, a north to south extension of roughly 1500 km and a north-northwest to south-southwest extension of 2000 km can be estimated as the areal extent of the lake systems. All the lakes in this lake 'belt' were certainly not all connected and subsidence rates varied from place to place. Lake level fluctuated, as evidenced by alternating shallow water and relatively deep-water facies; changes in rainfall patterns are theorized as the main factor affecting lake levels.

Stromatolites, oolites, and sandstones and mudstones are interpreted as nearshore shoals and deltaic sediments respectively. The morphology of the deltaic rocks include rare lobes and channel forms; a low-energy environment and an absence of large-scale river systems are postulated. The nearshore deposits grade into black shale and mudstones, which are interpreted as the deep-water facies. These shales to mudstones are generally massive with rare fine laminations and organic carbon contents between 1–11% (Kreuser et al., 1988). The presence of burrowing organisms is postulated as the homogenizing agent for these fine-grained sediments. Clay minerals include illite, chlorite and mixed-layer clays. The clays, calcite and the presence of micro-crystalline quartz indicate that the lake waters were alkaline throughout the history of these lake systems. Calcite concentration during periods of higher evaporation rates and/or lower rates of siliciclastic input produced favorable conditions for the formation of carbonate beds, some of which are stromatolitic. The presence of filaments and tube-like structures indicates an algal and/or bacterial origin. Some stromatolites exhibit desiccation cracks, which indicate formation in littoral to sublittoral environments. Laminated microbial mats followed by hemispherical stromatolites in vertical sequence characterize changes from shallower to deeper water conditions. Hemispherical stromatolites topped by oncolites to microbial mats indicate a transgressive lake phase.

A prominent change in lake evolution occured in the Tartarian (Usili Formation) as shown by the presence of a series of 2–6 m thick conglomerates and grits overlying lake mudstones separated by disconformities (Figs 2 and 3). Six such conglomerate/mudstone

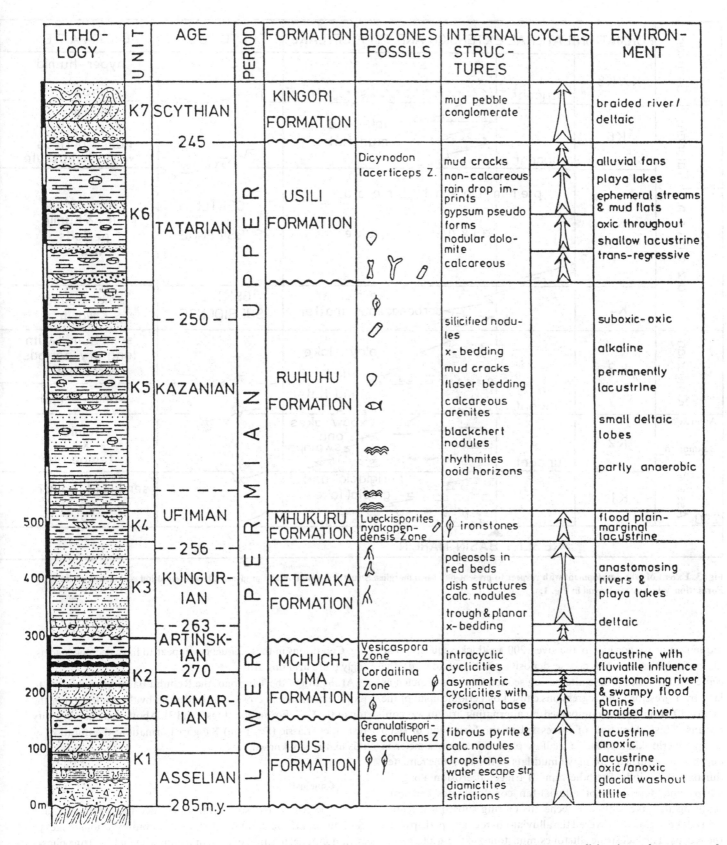

Fig. 2. Stratigraphy of the Permian succession of the Ruhuhu Basin. Additional information includes biozones, fossil content, cyclicites, internal structures and textures, and depositional environments.

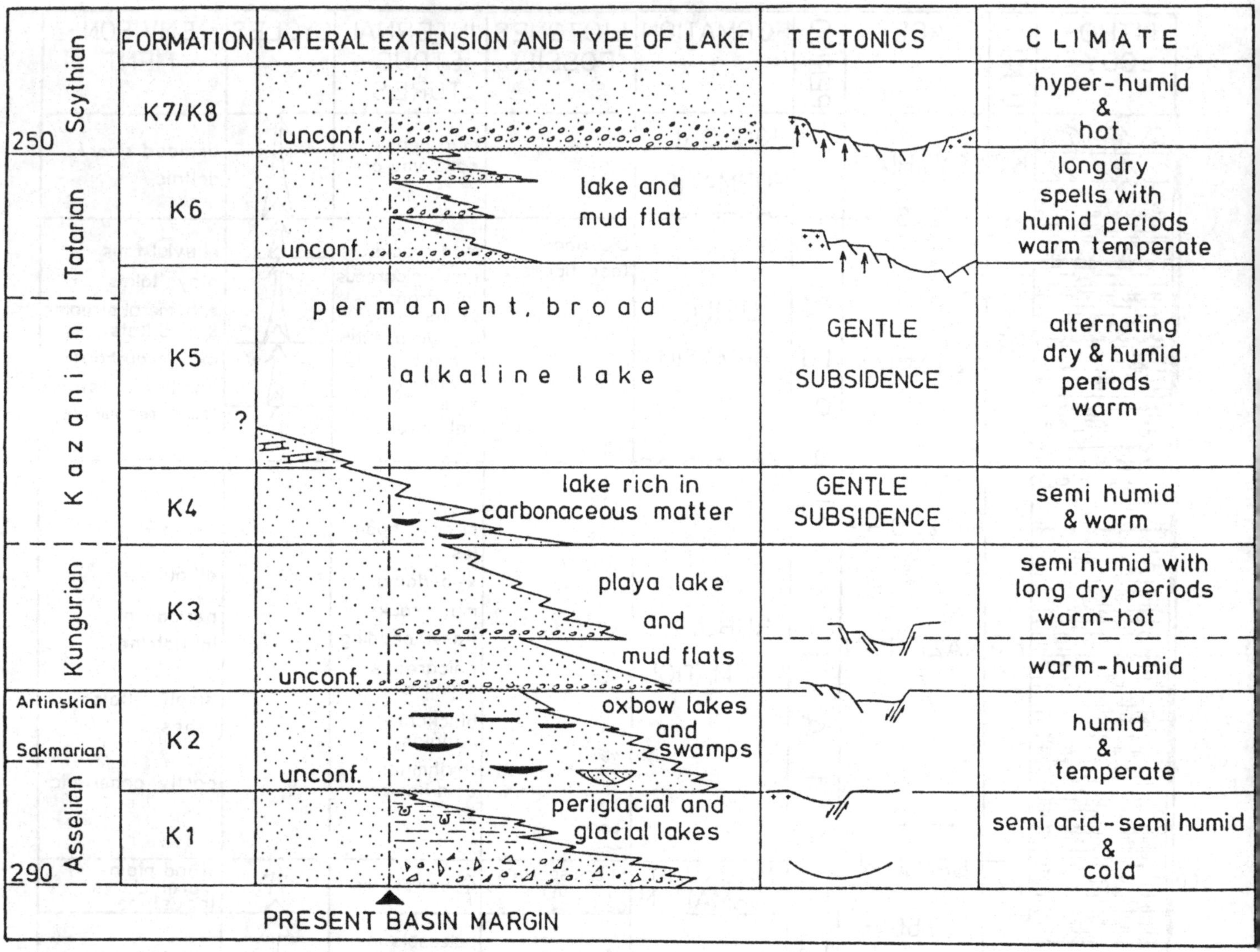

	FORMATION	LATERAL EXTENSION AND TYPE OF LAKE	TECTONICS	CLIMATE
Scythian / 250	K7/K8	unconf.		hyper-humid & hot
Tatarian	K6	lake and mud flat / unconf.		long dry spells with humid periods warm temperate
Kazanian	K5	permanent, broad alkaline lake ?	GENTLE SUBSIDENCE	alternating dry & humid periods warm
Kazanian	K4	lake rich in carbonaceous matter	GENTLE SUBSIDENCE	semi humid & warm
Kungurian	K3	playa lake and mud flats / unconf.		semi humid with long dry periods warm-hot / warm-humid
Artinskian / Sakmarian	K2	oxbow lakes and swamps / unconf.		humid & temperate
Asselian / 290	K1	periglacial and glacial lakes		semi arid-semi humid & cold

PRESENT BASIN MARGIN

Fig. 3. Extent of lake development with respect to present-day basin margins. Structural evolution and proposed climatic evolution is sketched on the right. Formation notation: see legend in Fig. 1.

sequences are recorded in the over 200 m thick sequence. The disconformities and the coarse deposits pinch out towards the center of the basin. These alternating sequences are interpreted as lake transgressions and regressions caused by periodic uplift of the proto-rift shoulders, accompanied by alternating arid to wet conditions. This interaction of forces led to the formation of steep scarps overlooking extensive shallow playa lakes only a meter in depth, which turned into playa mudflats as the shoreline retreated during dry periods. Periodic rainfall triggered erosion along the scarps and deposition of alluvial-fan conglomerates and grits, scouring the lake sediments beneath. During subsequent regressions, lake sediments covered the alluvial-fan toes and perhaps even the scarps. The lacustrine siltstones/mudstones of these alternating sequences contain calcareous nodules, fossil bone horizons (*Dicynodon lacerticeps*) and evidence for subaerial exposure and rework-

ing. Conditions in these sequences appear to have become increasingly more arid upward.

At the top of the Tartarian, the Ruhuhu Basin was subjected to more tectonism and erosion, as evidenced by the truncation of the top of the Usili Formation (Figs 2 and 3). Above this unconformity rests the Triassic (Scythian) Kingori Formation; the duration of this hiatus is unknown.

Conclusions

The Ruhuhu Basin contains evidence for continuous lake development throughout the Permian, exhibiting various stages of evolution. The duration of each of the three main lacustrine phases varies considerably. The glacial to periglacial lakes of the Idusi Formation lasted for approximately 1–2 Ma, depending on the time estimated for the Permo–Carboniferous glaciation (5–10+ Ma). The second phase (Mchuchuma Formation) with its short-lived shallow lakes and ponds which existed only for several ten to hundred thousand years at a time, encompassed a period of approximately 3–5 Ma. The third phase of lake development may have lasted for some 8 Ma, with the establishment of a large lake system (Ruhuhu Formation) existing for 4–5 Ma. This gives an average depositional rate of 7 cm/1000 years.

Bibliography

Bornhardt, W., 1990. *Zur Oberflächengestaltung und Geologie Deutsch-Ostafrikas.* Volume 7. Berlin.

Casshyap, S.M., Kreuser, T. & Wopfner, H., 1987. Analysis of cyclical sedimentation in the Lower Permian Mchuchuma coalfield (SW Tanzania). *Geologisches Rundschau,* 76(3), 869–83.

Cox, L.R., 1935. Karoo lamellibranchs from Tanganyika and Madagascar. *Quart. J. Geol. Soc. London,* 92(1), 32–57.

Dietrich, W.O., 1933. On alleged algal structures from central Africa. *Chronology Mines Colon.* (2nd year), 1, 299–300.

Foster, C.B. & Waterhouse, J.B., 1988. The *Granulatisporites confluens* Oppel zone and early Permian marine faunas from the Grant Formation, Canning Basin, W. Australia. *Austr. J. Earth Soc.,* 35, 135–57.

Haldemann, E.G., 1954. A contribution to the geology of Lumecha–Lukago Karoo area, Songea District. *Records Geol. Surv. Tanganyika,* 1, 9–11.

Hankel, O., 1987. Lithostratigraphic subdivisions of the Karoo rocks of the Luwegu Basin, Tanzania, and their biostratigraphic classification based on micro-, macroflora, fossil wood, and vertebrates. *Geologisches Rundschau,* 76(2), 539–65.

Harkin, D., 1948. The Mbamba Bay coalfield. *Min. Mag.,* 78, 265–72.

Harkin, D., 1953. The geology of the Mhukuru coalfield, Songea District. *Short Paper, Geol. Surv. Tanganyika,* 28, 1–12.

Harkin, D., 1955. Geology of the Songwe-Kiwira coalfield, Rungwe District. *Bull. Geol. Surv. Tanganyika,* 27, 1–41.

Hart, G.F., 1960. Microfloral investigations of lower coal measures in the Ketewaka-Mchuchuma coalfield. *Bull. Geol. Surv. Tanganyika,* 30, 1–18.

Hart, G.F., 1963. Microflora from the Ketewaka-Mchuchuma coalfield, Tanganyika. *Bull. Geol. Surv. Tanganyika,* 36, 1–27.

Hart, G.F., 1965. Miospore zones in Karoo sediments of Tanzania. *Palaeont. Africana,* 9, 139–50.

Haughton, S.H., 1932. One collection of Karoo vertebrates from Tanganyika territory. *Quart. J. Geol. Soc. London,* 88, 634–68.

Huene, F. von, 1939. Die Karoo-Formation im ostafrikanischen Ruhuhu Gebiet. *Zeitblatt Mine. Geol. Paläont.,* 2B, 69–71.

Huene, F. von, 1942. die Anomodontier des Ruhuhu Gebietes in der Tübinger Sammlung. *Paläontographica,* 94A (3/6), 154–84.

Heune, F., von, 1944. Pareiasaurier aus dem Ruhuhu Gebiet. *Paläont. Zeitschrift,* 23 (3/4), 386–410.

Kent, P.E., 1971. The geology and geophysics of coastal Tanzania. *Inst. Geol. Sci. London, Geophysics Papers,* 6, 1–101.

Kreuser, T., 1983. Stratigraphie der Karoo-Becken in Ost-Tansania. *Geol. Inst. Univ. Köln, Sonderveröffnung,* 45, 1–217.

Kreuser, T., 1984. Karoo basins in Tanzania. In *Geologie africaine,* ed. A. Klerkx, & J. Michot, pp. 231–45. Tervuren.

Kreuser, T., 1987. Late Palaeozoic glacial sediments and transition to coal-bearing Lower Permian in Tanzania. *Fazies,* 17, 149–58.

Kreuser, T., 1990. Permo-Trias im Ruhuhu-Becken (Tansania) und anderen Karoo-Becken von SE Afrika. *Geol. Inst. Univ. Köln, Sonderveröffnung,* 75, 1–132.

Kreuser, T., 1991. Facies evolution and cyclicity of alluvial coal deposits in the Lower Permian of East Africa (Tanzania). *Geologisches Rundschau,* 80(1).

Kreuser, T. & Markwort, S., 1988. Facies evolution of a fluvio-lacustrine Permo–Triassic basin in Tanzania. *Zeitblatt Geol. Paläont.,* Part I, 7/8, 821–37.

Kreuser, T., & Semkiwa, P.M., 1987. Geometry and depositional history of a Karoo coal basin in SW-Tanzania. *Neues Jahrbuch Geol. Paläont. Mh.,* 2, 69–98.

Kreuser, T., Schramedei, R. & Rullkötter, J., 1988. Gas prone source rocks from cratogene Karoo basins in Tanzania. *J. Petrol. Geol.,* 11(2), 169–84.

Kreuser, T., Wopfner, H., Kaaya, C.Z., Markwort, S., Semikiwa, P.M. & Aslanidis, P., 1990. Depositional evolution of Permo–Triassic Karoo basins in Tanzania with reference to their economic potential. *J. African Earth Sci.,* 10(1/2), 151–67.

Manum, S.B. & Tien, D., 1973. Palynostratigraphy of the Ketewaka-Mchuchuma coalfield, Tanzania. *Rev. Palaeobot. Palyn.,* 16, 213–27.

Mazurczak, L., 1953. Notes on the Ruhuhu coalfields, southern province, Tanganyika. *Colon. Geol. Min. Resources,* 3(3), 239–41.

McKinlay, A.M.C., 1954. The geology of the Ketewaka-Mchuchuma coalfield, Njombe District. *Bull. Geol. Surv. Tanganyika,* 21, 1–23.

McKinlay, A.M.C., 1965. The coalfields and coal resources of Tanzania. *Bull. Geol. Surv. Tanganyika,* 38, 1–82.

Nowack, E., 1937. Zur Kenntnis der Karoo Formation im Ruhuhu Graben. *Neues Jahrbuch Mineral., Beil.,* 78B, 380–412.

Quennell, A.M., McKinlay, A.M.C. & Spence, J., 1956. Summary of the geology of Tanzania: Part I – introduction and stratigraphy. *Mem. Geol. Surv. Tanganyika,* 1, 1–235.

Spence, J., 1952. Ngaka coalfield, Songea District. *Ann. Rep. Geol. Surv. Tanganyika,* 1950, 18–21.

Spence, J., 1957. The geology of part of the eastern province of Tanganyika. *Bull. Geol. Surv. Tanganyika,* 28, 1–62.

Stockley, G.M., 1932. The geology of the Ruhuhu coalfields, Tanganyika territories. *Quart. J. Geol. Soc. London,* 88(4), 610–22.

Stockley, G.M., 1936. A further contribution on Karoo rocks of Tanganyika territory. *Quart. J. Geol. Soc. London,* 92(1), 1–31.

Stockley, G.M., 1947. The geology and mineral resources of Tanganyika territories. *Bull. Imp. Inst. London,* 45(4), 75–406.

Stockley, G.M. & Oates, F., 1931. The Ruhuhu coalfields, Tanganyika territory. *Min. Mag.,* 45(2), 73–91.

Utting, J., 1978. The Karoo stratigraphy of the northern part of the Luangwa valley. *Mem. Geol. Surv. Tanganyika,* 4, 1–64.

Wopfner, H. & Kreuser, T., 1986. Evidence for Late Paleozoic glaciation in southern Tanzania. *Palaeogeogr., Palaeoclimat. Palaeoecol.,* 56, 269–75.

Wopfner, H., Markwort, S. & Semkiwa, P.M., 1991. Early diagenetic laumontite in the lower Triassic manda beds of the Ruhuhu Basin, southern Tanzania. *J. Sed. Pet.* 61(1), 65–72.

Yemane, K. & Kelts, K., 1990. A short review of palaeoenvironments for Lower Beaufort (U. Permian) Karoo sequences from southern to central Africa: a major Gondwana lacustrine sequence. *J. African Earth Sci.,* 10(1/2), 169–86.

The Stephanian B Lake, Bohemian Basin, Czech Republic

VLADIMÍR SKOČEK

Czech Geological Institute, Malostr. nám. 19, 118 21 Prague 1, Czech Republic

Locality map and facies distribution

Figure 1 shows the remains of one continuous lacustrine basin that originally covered an area exceeding 10,000 km². The lacustrine unit forms a relatively well-defined body within the Late Paleozoic sequence consisting of alluvial red beds, coal-bearing units and other lacustrine deposits of lesser extent. The total thickness of the lacustrine unit varies from 15 to 180 m. The facies distribution is relatively uniform both in vertical and lateral directions.

General stratigraphic column

A summary of the stratigraphic column together with details related to the described lacustrine formation are presented in Fig. 2.

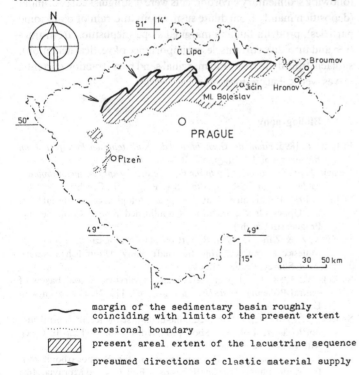

margin of the sedimentary basin roughly coinciding with limits of the present extent

.... erosional boundary

present areal extent of the lacustrine sequence

presumed directions of clastic material supply

Fig. 1. Locality map.

The column shows only the lacustrine unit which was formed over a relatively short period (*c.* 10⁵ years) during the evolution of a tectonically predetermined intermontane basin within the Variscan orogenetic zone. The total thickness of the fill reaches 2500 m and its range is from Westphalian A to Early Triassic. This range is not regionally uniform due to the heterochronous onset and termination of continental sedimentation, and several periods of intraformational and later erosion. The studied lacustrine sequence thus forms only a thin slice of extraordinary high lateral persistence within the Upper Paleozoic beds.

The dating of the lacustrine interval is based upon the identification of a relatively rich association of flora (Němejc, 1953; Šetlík, 1967; Šimůnek & Zajíc, 1984). Various genera and species of pteridosperms, pteridophytes, sphenopsids and algae occur. Remains of *Alethopteris bohemica*, *Pecopteris plumosa* and *Nemejcopteris feminaeformis* are most typical. In addition, the following species occur:

Pecopteris lepidorachis	*Annularia stellata*
Pecopteris polymorpha	*Annularia sphenophyloides*
Pecopteris bredovi	*Sphenophylum oblongifolium*
Asterophyllites equisetiformis	*Sphenophylum longifolium*
Neuropteris neuropteroides	*Alethopteris zeilleri*
Linopteris neuropteroides	*Odontopteris subcrenulata*
Pseudomariopteris riberoni	*Odontopteris intermedia*

Several genera are represented by poorly preserved remains (*Cordaites*, *Poacordaites*, *Sphenopteris*, *Linopteris*, *Sigillaria*, *Walchia*, *Pila*). Various fish species are also present (Zajíc & Štamberg, 1986). The profundal facies is considered an acme zone of *Watsonichthys sphaerosideritarum* (Fritsch, 1895). Rare remains of insects and conchostracas are also reported (Zajíc & Štamberg, 1986).

Important lithological details

The most characteristic feature of the lacustrine sequence is its regular vertical arrangement over the whole area. All sections exhibit a similar succession of sedimentary rocks (Fig. 2). Laminated or massive, dark grey claystones sit with a sharp basal contact on the underlying alluvial sequence. The first laminate unit is

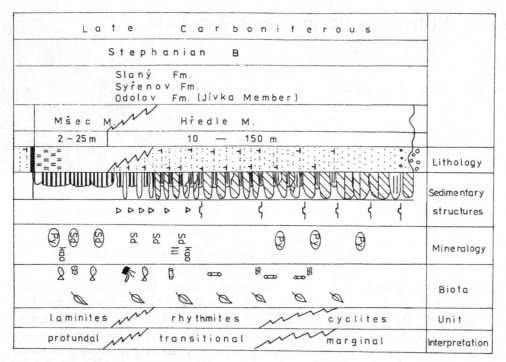

Fig. 2. Summary of stratigraphic column and details of lacustrine formation for Stephanian B Lake.

overlain by a rhythmite unit which consists of alternating thin-bedded graded siltstones, mudstones, claystones, and fine-grained sandstones. A cyclic arrangement of several orders can be recognized. The uppermost cyclite unit involves fine and even medium sandstones intercalated with silty or muddy sediments arranged in cyclothem-like units often with an erosional base. The upper contact of the succession coincides with the remarkable regional unconformity indicating channelling and massive erosion.

The lacustrine sequence forms one large, coarsening upward, sedimentation cycle recording a sudden rise of water level and its successive retreat caused either by climatic fluctuations or epeirogenetic movements. The very regular vertical succession shows progradation of marginal facies over profundal deposits. Cycles or rhythms of several orders were distinguished. Some are considered evidence of seasonal changes (Skoček, 1968); others are related unquestionably to water level fluctuation (Skoček, 1990). Long term climatic oscillations are very probable. The termination of the lake was induced either by climatic or tectonic effects (Skoček, 1990), which transformed the lake into a broad alluvial plain.

Although the studied sequence exhibits distinct variations of thickness, lateral facies changes are less pronounced. Synchronous proximal sediments are practically not known. It is expected (Skoček, 1990) that the former alluvial apron was removed during an erosional period following the lake's existence. Therefore, the sequence does not include any significant amount of coarse material, which is found in the overlying unit.

The synchroneity of the laminite facies is documented by numerous thin layers of argillized and carbonatized tuffs. Their regional persistence supports a probable rapid rise of water level during the initial phase. It is thought that the depth in the central part of the lake was between 100–200 m. Isopach studies and indications of an anoxic bottom regime support this presumption (Skoček, 1990).

Based on sedimentary structures and organic remains, the following sedimentary environments were distinguished: profundal (deposition mainly from dilute suspensions and rain of planktonic particles), prodelta fan and marginal slope (deposition from traction and turbidity currents, lesser importance of vertical accretion), sublacustrine delta front and marginal sand bars (traction currents, waves, vertical accretion).

Bibliography

Fritsch, A., 1895. *Fauna der Gaskohle und der Kalksteine der Permformation Böhmens*, vol. III. Prague: F. Řivnáč. 492 pp.

Němejc, F., 1953. *Introduction to the floristic stratigraphy of the bituminous coal regions in ČSR*. Čs. Akad. Sci., Prague. (In Czech.)

Šetlík, J., 1967. Report on evaluation of paleontological core material from the Upper Grey Formation. Unpublished report. Geol. Survey, Prague. (In Czech.)

Šimůnek, Z. & Zajíc, J., 1984. Report on paleontological findings in the Malesice Formation from the shaft Slaný. Unpublished report. Geol. Survey, Prague. (In Czech.)

Skoček, V., 1968. The Upper Carboniferous varvites in coal basins of central Bohemia. *Věst. Ústř. úst. geol.* **43**, 113–21. (In Czech with English summary.)

Skoček, V. 1990. Stephanian lacustrine-deltaic sequence in central and north-eastern Bohemia. *Sbor. geol. věd.* G. **45**, 91–122. (In Czech with English summary.)

Zajíc, J. & Štamberg, S. 1986. Summary of the Permocarboniferous freshwater fauna of the limnic basins of Bohemia and Moravia. *Acta Mus. Reginaehradec.*, Ser. A, **20** (1985), 61–82.

Lacustrine and palustrine carbonates in the Permian of eastern Provence (France)

NADÈGE TOUTIN-MORIN

Université d'Orléans, U.R.A. au C.N.R.S. no. 1366, BP 6759, F-45067 Orléans cedex 2, France

General

In the south of France, as a result of the Hercynian Orogeny, intermontane basins developed (the Estérel, Bas-Argens, Luc, Solliès Pont-Cuers, and Toulon basins; Fig. 1) during the Carboniferous–Permian transition. The extensional tectonics coupled with intense volcanic activity continued from the Late Carboniferous through the Late Permian. The sedimentary fill of these basins consists of coarse conglomerates and sandstones at basin edges with finer-grained clastics and carbonates at basin centers (Toutin, 1980). Lacustrine and palustrine deposits are associated with the upper portions of sedimentary sequences within these five basins (Toutin-Morin, 1985).

Regional stratigraphy

In eastern Provence (Fig. 2), the Carboniferous is dated as Upper Westphalian and Stephanian through spores, pollen and plant remains. Two of the Permian basins (Bas-Argens and Estérel) contain floral assemblages dated as Thuringian in age, which are conveniently located above a lava flow (A_7) for correlation into the non-fossiliferous Luc basin. Numerous alkaline volcanic flows, with acidic (A) or basic (B, δ) rocks, are intercalated with the sedimentary rocks of the entire Permian sequence for easy correlation. The basal Triassic is represented by sandstones of the Buntsandstein and are Anisian in age (determined from pollen).

Detailed lithologic columns

Only two basin sequences, Bas-Argens (Figs. 3, 4) and Estérel (Fig. 5), are detailed here because they are at present the best documented. The carbonates in the Permian are interpreted as lacustrine to palustrine deposits in the distal portions of the basins. Lacustrine rocks are represented by grey micrite beds some meters in thickness, containing ostracods, algal debris, and other microfauna (e.g. Pradineaux and Mitan Formations). Palustrine rocks are represented by carbonates containing desiccation cracks, rhizoliths, septarian concretions, and authigenic analcime (upper Muy Formation).

Bibliography

Boucarut, M., 1971. *Etude volcanique et géologique de L'Estérel (Var, France)*. Doctoral Dissertation, Sciences, Univ. Nice.

Toutin, N., 1980. *Le Permian continental de la Provence orientale (France)*. Doctoral Dissertation, Sciences, Univ. Nice, 594 p.

Toutin-Morin, N., 1985. Les dépôts carbonatés lacustres et palustres du Permien provençal: différences et significations. *C.R. Acad. Sci Paris*, **301** (20), 1423–8.

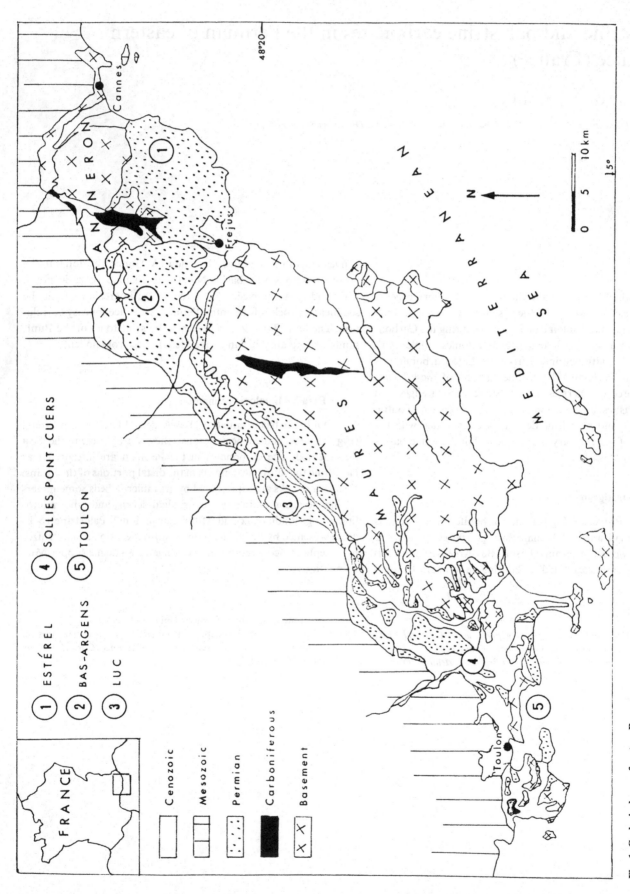

Fig. 1. Geological map of eastern Provence.

Periods		CUERS basin	LUC basin	BAS-ARGENS basin	ESTÉREL basin	Age Ma	Tectonic Events
TRIAS	Anisian	"grès bigarrés" of Buntsandstein facies			no sediments		
	Scythian	EROSION			no sediments		
							Palatine φ
P	T H U R I N G I A N	Gonfaron 50-150m	Pélitique F. 0-500m	Arcs 0-50m Motte F. 0-350m	non- deposition		
E		Gigery Formation 600m	Rouge Supérieure Formation	Serre ⟨ Muy Formation Valette ⟨ 50-300m Paro ⟨ Mitan Formation 200m			
R		Bouisse	200-250m	0-50m ρ A$_{11}$ Pradineaux Formation 0-200m		>247 *	
M	"S A X O N I A A N"	Formation 150m	0-20m ρ A$_7$ Rouge Inférieure Formation 60-350m	300m ρ A$_7$ Bayonne Formation 30-150m	50-150m ρ A$_7$ 0-150m ρ A5 0-30m ρ A2	272 * * *	
I		Bron Formation 200-250m	Tuffique Formation 80-260m		0-30m δ1 0-5m ρ A1	278 * *	
A		Transy F. 100-150m Pellegrin F 20-50m	Claire Formation 100-245m	Ambon Formation	0-70m		
							Saalian
N	AUTU	EROSION or NON-DEPOSITION			10m B1	*	Phase
	NIAN				Avellan 0-200m		
							Asturian φ
LATE	Steph.	Playes F. 20m no sedim.	no sediments Collobrières F. 25m	Upper serie 100m 100m ρ-μγ Lower serie 80m	Auriasque serie 800m	290 *	
CARBO							
	Westph.	EROSION or NON-DEPOSITION		Boson serie 200m			
basement		MAURES and TANNERON massifs (crystalline + metamorphic) 325 - 826 Ma					

~~~~~~ disconformity ; ≈≈≈≈ angular unconformity; φ tectonic phase; F. Formation;

μγ microgranite; volcanic rocks: ρ rhyolite, B-δ basalt; * magmatic events

Fig. 2. Regional stratigraphy encompassing four of the five Permian basins in eastern Provence.

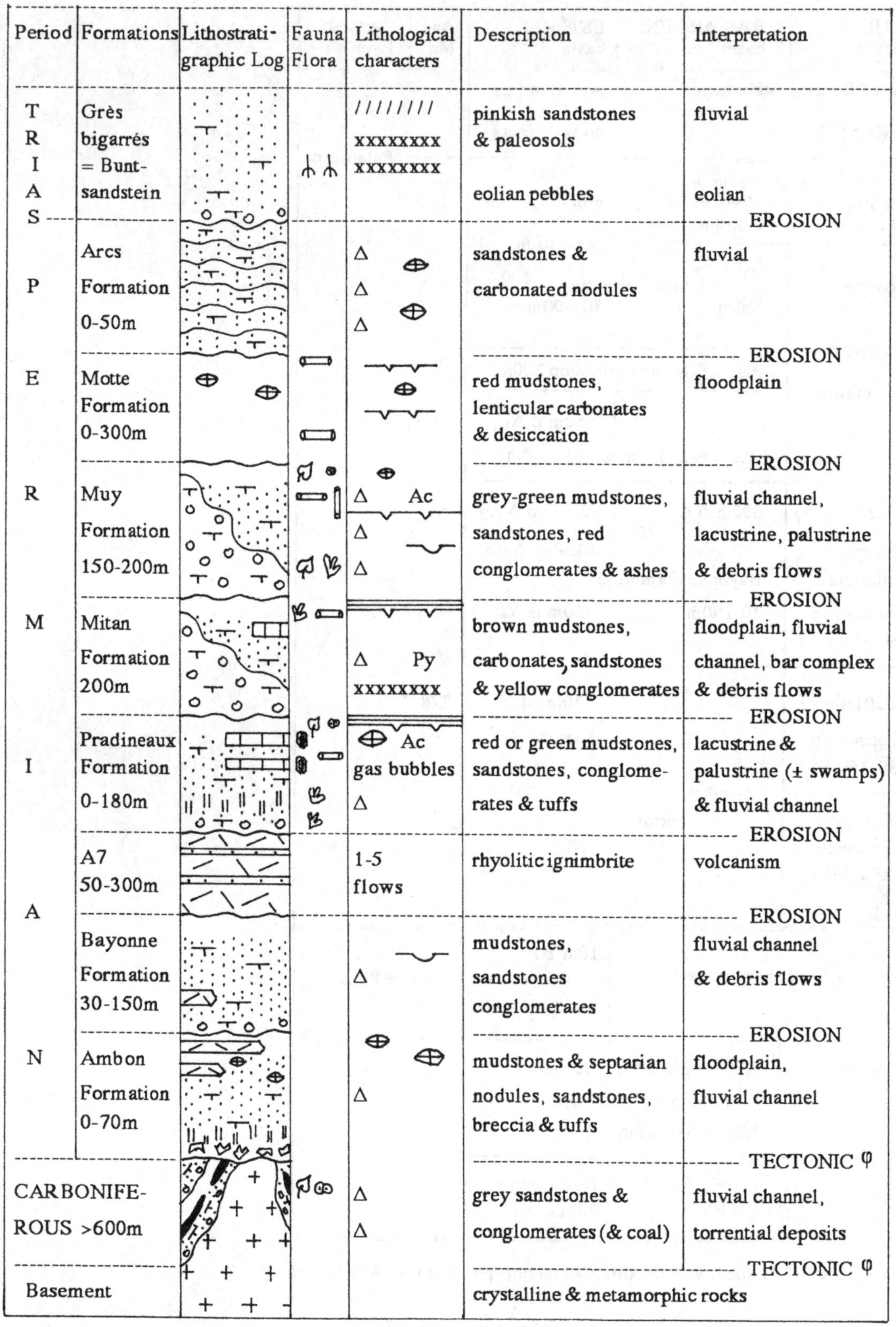

Period	Formations	Lithostratigraphic Log	Fauna Flora	Lithological characters	Description	Interpretation
T R I A S	Grès bigarrés = Bunt-sandstein			///////// xxxxxxxx xxxxxxxx	pinkish sandstones & paleosols / eolian pebbles	fluvial / eolian
					----- EROSION	----- EROSION
P	Arcs Formation 0-50m			Δ / Δ / Δ	sandstones & carbonated nodules	fluvial
					----- EROSION	----- EROSION
E	Motte Formation 0-300m				red mudstones, lenticular carbonates & desiccation	floodplain
					----- EROSION	----- EROSION
R	Muy Formation 150-200m			Δ Ac / Δ / Δ	grey-green mudstones, sandstones, red conglomerates & ashes	fluvial channel, lacustrine, palustrine & debris flows
					----- EROSION	----- EROSION
M	Mitan Formation 200m			Δ Py xxxxxxxx	brown mudstones, carbonates, sandstones & yellow conglomerates	floodplain, fluvial channel, bar complex & debris flows
					----- EROSION	----- EROSION
I	Pradineaux Formation 0-180m			Ac gas bubbles Δ	red or green mudstones, sandstones, conglomerates & tuffs	lacustrine & palustrine (± swamps) & fluvial channel
					----- EROSION	----- EROSION
A	A7 50-300m			1-5 flows	rhyolitic ignimbrite	volcanism
					----- EROSION	----- EROSION
	Bayonne Formation 30-150m			Δ	mudstones, sandstones conglomerates	fluvial channel & debris flows
					----- EROSION	----- EROSION
N	Ambon Formation 0-70m			Δ	mudstones & septarian nodules, sandstones, breccia & tuffs	floodplain, fluvial channel
					----- TECTONIC φ	----- TECTONIC φ
	CARBONIFEROUS >600m			Δ / Δ	grey sandstones & conglomerates (& coal)	fluvial channel, torrential deposits
					----- TECTONIC φ	----- TECTONIC φ
	Basement				crystalline & metamorphic rocks	

Fig. 3. General stratigraphy of the Bas-Argens basin of eastern Provence.

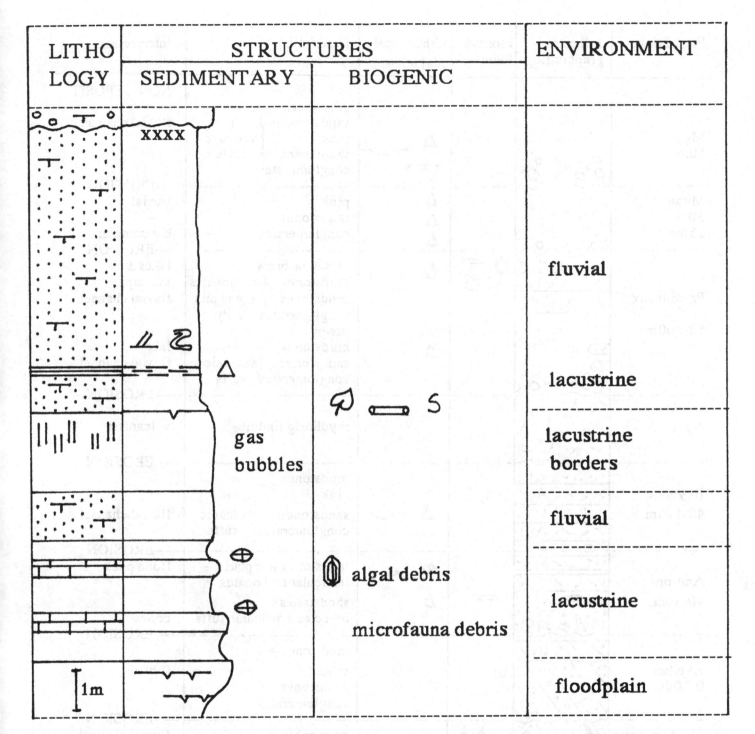

Fig. 4. Detailed lithologic log from the top of the Pradineaux Formation in the Bas-Argens basin of eastern Provence.

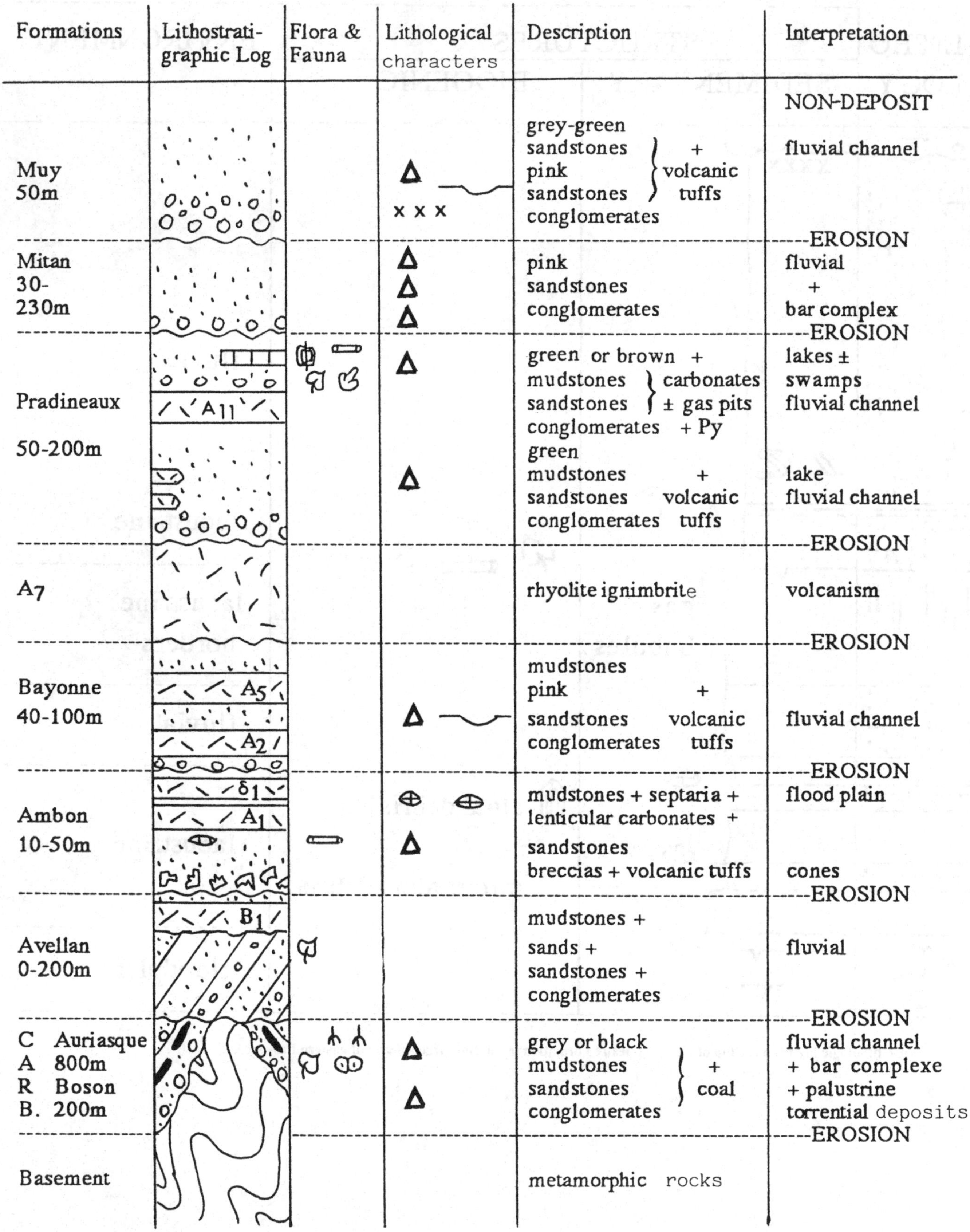

Formations	Lithostrati-graphic Log	Flora & Fauna	Lithological characters	Description	Interpretation
					NON-DEPOSIT
Muy 50m				grey-green sandstones + pink volcanic sandstones tuffs conglomerates	fluvial channel
					--- EROSION
Mitan 30-230m				pink sandstones conglomerates	fluvial + bar complex
					---EROSION
Pradineaux 50-200m	/\\'A11'\\			green or brown + mudstones } carbonates sandstones } ± gas pits conglomerates + Py green mudstones + sandstones volcanic conglomerates tuffs	lakes ± swamps fluvial channel lake fluvial channel
					---EROSION
A7				rhyolite ignimbrite	volcanism
					---EROSION
Bayonne 40-100m	/\\ /\\ A5 /\\ /\\ /\\ A2 /			mudstones pink + sandstones volcanic conglomerates tuffs	fluvial channel
					---EROSION
Ambon 10-50m	/\\ /δ1\\ /\\ /A1\\			mudstones + septaria + lenticular carbonates + sandstones breccias + volcanic tuffs	flood plain cones
					---EROSION
Avellan 0-200m	/\\ /\\ B1 /			mudstones + sands + sandstones + conglomerates	fluvial
					---EROSION
C Auriasque A 800m R Boson B. 200m				grey or black mudstones } + sandstones } coal conglomerates	fluvial channel + bar complexe + palustrine torrential deposits
					---EROSION
Basement				metamorphic rocks	

Fig. 5. General stratigraphy of the Estérel basin of eastern Provence.

Anthracosia shale (Lower Permian), North Sudetic Basin, Poland

KRZYSZTOF MASTALERZ

Institute of Geological Sciences, University of Wrocław, Cybulskiego 30, PL 50-205 Wrocław, Poland

The North Sudetic Basin (Fig. 1) originated during the Stephanian within the Variscan Mid-European basin-and-range province. Its Stephanian–Autunian infill consists of large-scale, fining-upward, alluvial to lacustrine cyclothems related to major events of subsidence and landscape rejuvenation. The Permo-Carboniferous sedimentary succession of the basin (Fig. 2) shows signs of long-lasting climatic control and continual evolution of wet to arid warm climates.

The Anthracosia Shale is a 40 m thick succession of lacustrine deposits terminating the Świerzawa Formation (Stephanian–Lower Permian), the lowermost unit of the North Sudetic Basin (Fig. 3). This transgressive-regressive succession is subdivided into

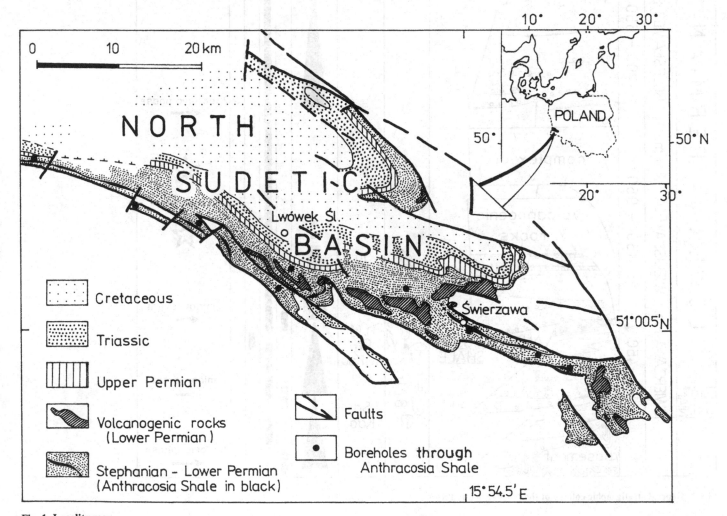

Fig. 1. Locality map.

98   K. Mastalerz

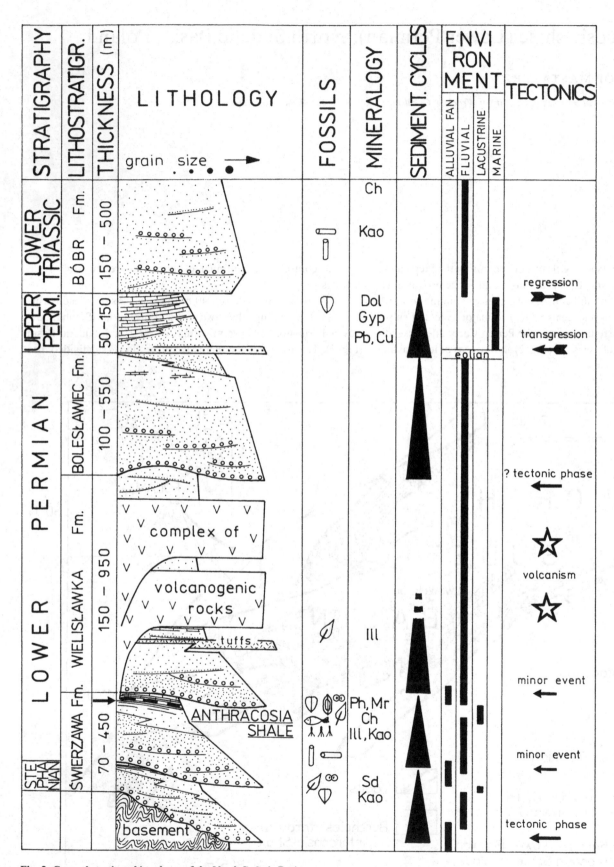

Fig. 2. General stratigraphic column of the North Sudetic Basin.

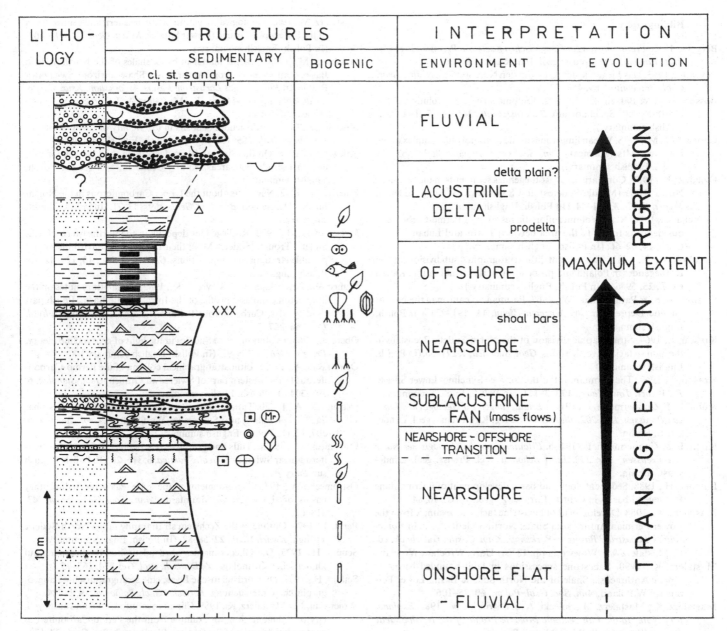

Fig. 3. Stratotype section of Anthracosia Shale, North Sudetic Basin.

four segments related to four major phases of Anthracosian lake development. Onshore facies reflect the initial phase of lake development when the basin floor (previously occupied by a fluvial channel belt) evolved into a fluvially-influenced onshore lake zone resulting in a thick sequence of nearshore deposits. Organic-rich black shales of lacustrine offshore facies correspond to the phase of maximum lake extent. A regressive phase then followed, which resulted in a coarsening-upward sequence of deltaic sediments capped by fluvial channel facies.

Anthracosia Lake developed in an elongated, asymmetrical, intramontane graben. The mobility of the fault-bounded basin margins exerted a strong influence on lacustrine depositional processes. The thick transgressive sequence was a result of tectonic activity coupled with increased subsidence along the northern margin of the basin. This active margin produced a south-southwesterly sediment dispersal pattern and triggered mass resedimentation movements. The regressive phase was related to the fault-induced rejuvenation at the southern basin margin, which led to lake infilling by a prograding fluvial system. Synsedimentary tectonic activity is also indicated by (1) the migration of the basin depocentre, (2) the thick sedimentary sequences of the lake shores, which strongly restricted peat-forming accumulations, (3) local silica precipitation, probably related to hot spring activity and (4) evidence of frequent lake level fluctuations. Anthracosia Lake was probably permanently stratified, with bottom conditions varying from anoxic offshore to oxic/suboxic nearshore.

Bibliography

Berg, G., 1935. *Erläuterungen zur Geologischen Karte von Preussen. Lfg. 333, Blatt Lauban.* Kgl. Preuss. geol. Landesanst., Berlin.

Beyer, K., 1933. Das Liegende der Kreide in den Nordsudeten. *N. Jb. Miner. Geol. Palaeontol.*, **69**, 450–508.

Bossowski, A. & Bałazińska, J., 1982. Tectonic-structural evolution of the North-Sudetic Synclinorium. *Biul. Inst. Geol.*, **341**, 163–7. (In Polish, English summary.)

Chorowska, M., 1978. Visean limestones in the metamorphic complex of the Kaczawa Mts (Sudetes). *Ann. Soc. Geol. Polon.*, **48**, 245–63. (In Polish, English summary.)

Dziedzic, K., 1959. Comparison of Rotliegendes sediments in the region of Nowa Ruda (Middle Sudetes) and Świerzawa (Western Sudetes). *Kwart. Geol.*, **3**, 831–44. (In Polish, English summary.)

Górecka, T., 1970. Results of microfloristic research of Permo-Carboniferous deposits found in the area between Jawor and Lubań. *Kwart. Geol.*, **14**, 52–66. (In Polish, English summary.)

Karnkowski, P., 1981. The current lithostratigraphic subdivision of the Rotliegendes in Poland and proposition of its formalization. *Kwart. Geol.*, **25**, 59–66. (In Polish, English summary.)

Kozłowski, S. & Parachoniak, W., 1967. Permian volcanism in the North-Sudetic Depression. *Pr. Muzeum Ziemi*, **11**, 191–223. (In Polish, English summary.)

Krasoń, J., 1964. Stratigraphic division of the North-Sudetic Zechstein in the light of facial investigations. *Geol. Sudetica*, **1**, 221–56. (In Polish, English summary.)

Krasoń, J., 1967. The Permian of the Bolesławiec Syncline, Lower Silesia. *Pr. Wrocł. Tow. Nauk.*, **133**, 3–151. (In Polish, English summary.)

Kühn, B. & Zimmermann, E., 1918a. *Erlauterungen zur Geologischen Karte von Preussen. Lfg. 202. Blatt Groditzberg.* Kgl. Preuss. geol. Landesanst., Berlin.

Kühn, B. & Zimmermann, E., 1918b. *Erlauterungen zur Geologischen Karte von Preussen, Lfg. 232. Blatt Schonau.* Kgl. Preuss. geol. Landesanst., Berlin.

Lützner, H., 1988. Sedimentology and basin development of intermontane Rotliegend basins in Central Europe. *Z. geol. Wiss.*, **16**, 845–63.

Mastalerz, K., 1988. Development of lacustrine facies: an example from the Lower Permian Anthracosia Shale, North-Sudetic Basin. In *Rotliegendes Lacustrine Basins – Workshop, Książ Castle, Guidebook*, ed. K. Mastalerz & J. Wojewoda pp. 18–31. Univ. Wrocław, Wrocław.

Mastalerz, K., 1990. Lacustrine successions in fault-bounded basins 1. Upper Anthracosia Shale of the North-Sudetic Basin, Lower Permian, SW Poland. *Ann. Soc. Geol. Polon.*, **60**, 75–106.

Mastalerz, K., Mastalerz, M., Solecki, A. & Śliwiński, W., 1983. *Explanatory Notes to the Lithotectonic Molasse Profile of the North Sudetic Basin (SW Poland).* (Unpublished Report.)

Mastalerz, K. & Wojewoda, J., 1988. Rotliegendes sedimentary basins in the Sudetes, Central Europe. In *Rotliegendes Lacustrine Basins – Workshop, Książ Castle, Guidebook*, ed. K. Mastalerz & J. Wojewoda, pp. 1–9. Univ. Wrocław, Wrocław.

Mastalerz, M., 1983. Geological structure of northeastern and central parts of the Świerzawa Trough (Lower Silesia). *Kwart. Geol.*, **27**, 491–502. (In Polish, English summary.)

Mastalerz, M., 1988. Organic matter in black shales of the North Sudetic Basin – an example from Anthracosia Shale horizon, Świerzawa Profile. In *Rotliegendes Lacustrine Basins – Workshop, Książ Castle, Guidebook*, pp. 43–44. ed. K. Mastalerz & J. Wojewoda. Univ. Wrocław, Wrocław.

Milewicz, J., 1965. Rotliegendes deposits in the vicinity of Lwówek Śląski. *Biul. Inst. Geol.*, **185**, 195–217. (In Polish, English summary.)

Milewicz, J., 1968. On the Buntsandstein deposits in the Lwówek Graben and in the adjacent areas. *Kwart. Geol.*, **12**, 547–60. (In Polish, English summary.)

Milewicz, J., 1972. New facts about the Upper Carboniferous in the North-Sudetic Depression. *Biul. Inst. Geol.*, **259**, 153–71. (In Polish, English summary.)

Mroczkowski, J., 1972. Rotliegendes deposits in the northeastern North-Sudetic Trough (Sudetes Mts); their depositional environment and possible stratigraphic implications. *Geol. Sudetica*, **15**, 125–42. (In Polish, English summary.)

Nemec, W., Porębski, S. & Teisseyre, A.K., 1982. Explanatory notes on the lithotectonic molasse profile of the Intra-Sudetic Basin, Polish part (Sudety Mts, Carboniferous-Permian). *Veroff. Zentralinst. Physik Erde*, **66**, 267–72.

Oberc, J., 1966. Evolution of the Sudetes in the light of geosyncline theory. *Pr. Inst. Geol.*, **47**, 3–92. (In Polish, English summary.)

Ostromęcki, A., 1972a. Lithostratigraphical column of the Permo-Carboniferous in the western part of Świerzawa Graben. *Geol. Sudetica*, **6**, 293–304. (In Polish, English summary.)

Ostromęcki, A., 1972b. Tuffaceous sandstones of Marczów Beds (Westphalian?–Stephanian?) near Świerzawa, West Sudetes. *Geol. Sudetica*, **6**, 307–12. (In Polish, English summary.)

Ostromęcki, A., 1972c. Tuffs and eruptive rock pebbles in Upper Carboniferous near Świerzawa. *Geol. Sudetica*, **6**, 315–20. (In Polish, English summary.)

Ostromęcki, A., 1973. Development of the Late Paleozoic sedimentary basins of the Kaczawa Mountains. *Ann. Soc. Geol. Polon*, **43**, 318–62.

Peryt, T., 1978. Outline of the Zechstein stratigraphy in the North-Sudetic Trough. *Kwart. Geol.*, **22**, 58–82. (In Polish, English summary.)

Scupin, H., 1923. Die Gliederung des nordsudetischen Rotliegenden auf klimatischen Grundlage. *Z. Deutsch. geol. Ges.*, **74**, 263–75.

Scupin, H., 1931. Die nordsudetische Dyas, eine stratigraphisch-palaogeographische Untersuchung. *Fortschr. Geol. Paläontol.*, **9**, 1–224.

Wojewoda, J. & Mastalerz, K., 1989. Climate evolution, allo- and autocyclicity of sedimentation: an example from the Permo-Carboniferous continental deposits of the Sudetes, SW Poland. *Prz. Geol.*, **37**, 173–80. (In Polish, English summary.)

The Permian lacustrine Basque Basin (western Pyrenees): an example for high environmental variability in small and shallow carbonate lakes developed in closed systems

BLAS L. VALERO GARCÉS[1]

Departamento de Geología, Universidad de Zaragoza. E-50009 Zaragoza, Spain

Geological setting

In the Pyrenees, Late Hercynian (Stephanian–Permian) continental rocks are exposed as west-northwest to east-southeast trending basins along the southern and northern boundaries of the Axial Paleozoic Zone (Gisbert, 1984). In the Western Pyrenees, Stephanian and Permian rocks are found in several tectonically-bounded Paleozoic Massifs; Les Aldudes-Quinto Real and Cinco Villas being where the most complete series are located (Fig. 1). A detailed paleogeographical reconstruction has not yet been done, nevertheless the available data (Lamare, 1936; Müller, 1969; Lucas, 1985) support the interpretation of all outcrops as a single basin, named the Basque Basin. The Stephanian–Permian succession in this Basque Basin is over 2 km in thickness (Lamare, 1936; Müller, 1969; Müller, 1973; Carbayo *et al.*, 1974; Solé & Villalobos, 1974; Le Pochat, 1978; Lucas, 1985; Lucas, 1989; Valero Garcés 1991). The Upper Stephanian series, representing the 'Unidad de Tránsito' (UT) is composed of quarzitic conglomerates, sandstones, oil shales, coal-bearing siltstones and volcanic tuffs. This UT unit lies unconformably over different Paleozoic rocks and is separated by an angular unconformity from the overlying Permian 'Unidad Roja Superior' (URS). The URS is mainly siliciclastic and very variable in thickness: more than 1000 m in the North (Bidarray region, France) and about 300 m in the South (Spanish outcrops). Some volcanic bodies (basalt facies) intercalate at the top of this Permian unit, which is unconformably overlain by the 'Unidad en Facies Buntsandstein' (UFB, Lower Triassic). Genesis and evolution of these continental basins are related to Late Hercynian strike–slip tectonics (Bixel & Lucas, 1983). The transition from the Stephanian and Autunian compressive conditions to the clearly extensive Triassic setting occurred during the deposition of the URS (Gisbert, 1981; Lucas, 1985; Lucas, 1989).

Lacustrine carbonates in the Basque Basin are restricted to the top of the URS and are only exposed in a small, fault-bounded outcrop located to the south of Les Aldudes-Quinto Real Paleozoic massif, along the Antchignoko Erreka creek which flows into La Nive de Les Aldudes near the village of Banca. Lamare (1931) discovered these facies and Müller (1969) provided the first strati-

[1] Current address: Limnological Research Center, 220 Pillsbury Hall, University of Minnesota, Minneapolis, MN 55455 USA.

graphical data and field descriptions. In this locality, the Permian URS is composed of a fining upward siliciclastic megasequence interpreted as alluvial-fan deposits, followed by about 40 m of grey and pink, massive and nodular limestones interbedded with red siltstones and interpreted as lacustrine sediments (Fig. 2). No paleontological dating is available for the URS. Paleoflora indicate a Middle Stephanian age for the UT Basque series (Lamare, 1936; Lucas, 1985) and Upper Triassic (Carnian-Norian) for the upper part of the UFB in Bidarray, north of the basin (Lucas *et al.*, 1980). Although likely diachronous, the URS is considered as Middle to Upper Permian in the whole Pyrenean domain (Gisbert, 1984; Lucas, 1985; Lucas, 1989).

Facies associations and lacustrine subenvironments

Sedimentological and petrographical features, and interpretation of the five main facies in the Permian Basque lacustrine episode (Valero Garcés, 1991; Valero Garcés, 1992) are summarized in Table 1. A number of shallowing-upward sequences characterize several of the lacustrine subenvironments within this carbonate lacustrine system (Fig. 3):

(1) Low energy, alluvial-influenced littoral area fringed by alluvial plains with rare interbedded siliciclastics. Lake sediments were micrites and intramicrites, reworked by littoral agents (waves, currents) and heavily modified by subaerial exposure (*in situ* brecciation, planar voids, nodulization), pedogenesis (argillans, ferrans, roots, rhizocretions, calcrete cementation) and early diagenesis (vadose silt, mottling and pseudogley textures, calcite cementation). Facies are arranged in shallowing-upwards sequences (Fig. 3, 1).

(2) Low energy, non-alluvial-influenced littoral area, composed exclusively of carbonate. Subaerial exposure and pedogenesis were much more common due to frequent water-level fluctuations and changes in the nature (oxidizing and reducing) of interstitial waters. In some places, hypersaline conditions may have developed and evaporites (gypsum) precipitated within the sediments. The mottled and/or nodulized limestones (palustrine facies) alternate with massive micritic facies (lacustrine) (Fig. 3, 2).

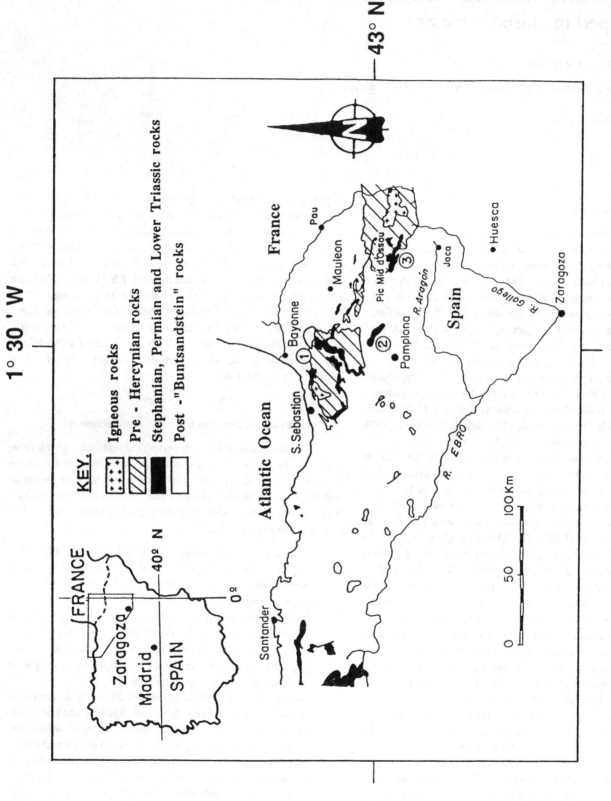

Fig. 1. Geographical and geological setting of the western Pyrenees Late Hercynian Basins: (1) Basque Basin; (2) Pamplona Basin; (3) Aragón–Béarn Basin.

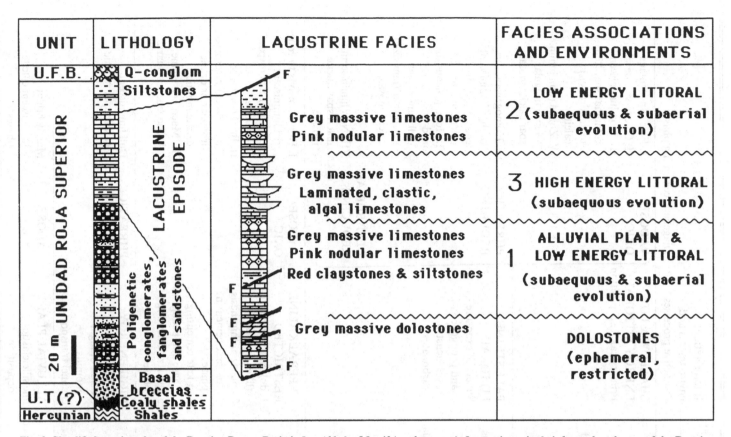

Fig. 2. Simplified stratigraphy of the Permian Basque Basin in Les Aldudes Massif (southern area). Lacustrine episode is located at the top of the Permian 'Unidad Roja Superior' (URS). Facies associations 1, 2, and 3 are described in Fig. 3. 'F' denotes a tectonic fault, UT denotes Unidad de Tránsho; UFB denotes Unidad en Facies Bundsandstein.

(3) High energy littoral area with algae build-ups. Massive, 1 m thick, grey to incipient marmorized micrites associated with decimeter-thick channel-shaped bodies of wackestones with erosional basal surfaces in 2–3 m thick sequences (Fig. 3, 3). Wackestones are interpreted as reworked littoral lithofacies (especially algal mats and mounds) by wave and current action and channeled into more internal (sublittoral) zones. The inferred presence of algal build-ups and the associated clastic limestones are unique of all the Pyrenean Permian basins. The present morphologies are interpreted as the result of calcification (patchy, irregular or around filaments), encrusting, trapping and binding of carbonate by blue-green algae and cyanobacteria. Although algal bioherm position is dependent on many factors, as far as it is known, these bioherms developed in slightly deep water in the littoral and sublittoral zones not colonized by higher plants (to depths of 10 m or more). Micritic limestones represent sedimentation in quieter conditions, likely interchannel or protected zones.

(4) Dolostone. This dolomitic facies is located at the base of the stratigraphical profile and related to palustrine limestones, but outcrop conditions do clearly establish its association with the other facies. The fine-grained micritic texture, the presence of euhedral authigenic quartz, and the very early genesis (prior to pedogenetic calcite nodules, see chronology

in Table 1) point to primary dolomite formation in highly alkaline, Mg-rich lacustrine paleoenvironments where silica may also have precipitated. These conditions would have been reached as a late lacustrine stage after long evaporative processes. Taking into account the importance of ground-waters in the whole lacustrine system, it is also possible that special groundwater recharge played a role in the genesis of these dolostones.

Significance of the Permian Basque lacustrine episode

The Permian Basque lacustrine episode is interpreted as deposition in a closed system of a small and shallow carbonate lake located in the distal areas (mudflat) of an alluvial-fan system. Despite the dominance of low topographic gradients, a high gradient margin would better explain the channelized algal facies. Carbonate deposition was the main sedimentary process in this Permian lake, as we can expect in closed basins with limited alluvial inflows. Hydrology is thought to have been almost exclusively controlled by groundwaters and this would explain the significant variations of water-table levels, responsible for the remarkable suite of vadose, pedogenic and early diagenetic features that characterize these facies.

The presence of this diversified lacustrine system in the Basque

Table 1. *Summarized features of the five main facies defined in the Permian lacustrine episode in the Basque Basin*

Facies	Thickness	Microfacies	Pedogenesis and subaerial processes		Interpretation: environments and distinctive processes	Diagenetic cementing phases	
			Intensity	Features			
Grey massive limestones	dm	micrites intramicrites massive pseudomorphs after gypsum	Low	ferruginous cementation channels planar voids poorly oriented argillans	SHALLOW & LOW ENERGY LITTORAL Primary calcite deposition	VADOSE/OXIC	I. Ferran neocutans II. Calcite vadose silt III. Non luminescent, columnar spar (tens to hundreds μ) calcite IV. Zoned (non lumin.–lum.) irregular spar calcite mosaic
Pink nodular limestones and marly limestones	dm – m	impure micrites and intramicrites wackestones/packstones	Very high	Mottling insepic argillans calcite nodules (cm sized) planar voids, circumgranular and curved cracks cavities vadose silt	PALUSTRINE LITTORAL Subaerial exposure and pedogenesis over lacustrine carbonates	PHREATIC/REDUCING	V. Homogeneous bright orange luminescent equant (hundreds μ) calcite mosaic VI. Homogeneous to zoned dull luminescent macrospar (mm sized) calcite mosaic
						DEEP BURIAL/REDUCING	VII. Non luminescent macrospar calcite mosaic
Grey massive dolostones	dm	dolomicrites with intraclasts euhedral quartz crystals corroded by dolomite	Medium	calcite nodules (cm sized) planar voids	HYPERALKALINE RESTRICTED 1. dolomite formation (likely primary) 2. quartz authigenesis 3. dolomite cementing phase 4. calcite cementing phase and pedogenesis 5. quartz cementing phase	VADOSE/OXIC	I. Homogeneous to zoned red luminescent columnar macrospar dolomite II. Bright or zoned luminescent macrospar irregular or drusic calcite mosaic III. Macrocrystalline homogeneous quartz
Grey laminated algal wackestones and packstones	dm	marly biointramicrites and biointramicrites Normal grading mm to cm lamination	Absent		HIGH ENERGY LITTORAL erosion of algal mounds channel transport	PHREATIC/REDUCING	Dull luminescent calcite spar mosaic
Red massive claystones and siltstones	dm – m	laminated massive	Medium	calcite nodules (mm to cm sized) mottling	ALLUVIAL PLAIN fine siliciclastic deposition	VADOSE/OXIC	Mainly bright luminiscent calcite phases

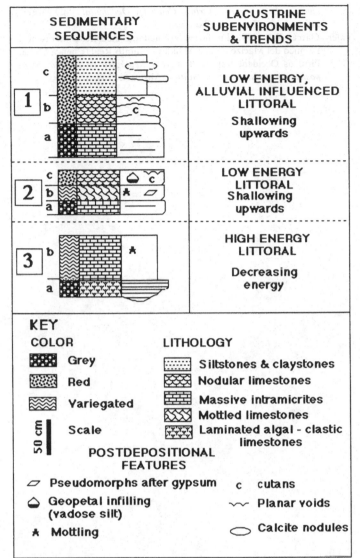

SEDIMENTARY SEQUENCES	LACUSTRINE SUBENVIRONMENTS & TRENDS
1	**LOW ENERGY, ALLUVIAL INFLUENCED LITTORAL** Shallowing upwards
2	**LOW ENERGY LITTORAL** Shallowing upwards
3	**HIGH ENERGY LITTORAL** Decreasing energy

KEY

COLOR

- ▒ Grey
- ▒ Red
- 〰 Variegated
- ▌ Scale (50 cm)

LITHOLOGY

- ⋯ Siltstones & claystones
- ⋈ Nodular limestones
- ▦ Massive intramicrites
- ⋈ Mottled limestones
- 〴 Laminated algal - clastic limestones

POSTDEPOSITIONAL FEATURES

- ▱ Pseudomorphs after gypsum
- ⌂ Geopetal infilling (vadose silt)
- ∗ Mottling
- c cutans
- 〰 Planar voids
- ⬭ Calcite nodules

Fig. 3. Facies associations and lacustrine subenvironments interpreted in the shallow, carbonate lacustrine system in the Basque Basin.

Basin underlines the differences between western (with lake development) and eastern (without lakes) Permian Pyrenean basins. Although a wetter paleoclimate may be invoked, tectonic effects are considered a major agent in lake genesis and evolution. Subsidence in western basins was higher during URS deposition (Middle and Upper Permian) as indicated by the thicker sedimentary series. Besides, the last Permian volcanic phase, which is only recorded in the western Pyrenean basins (Basque and Aragón-Béarn), shows alkaline geochemical affinities (Bixel, 1984; 1989) indicating that extensional conditions may not have been isochronous in the Pyrenean domain. The hypothesis is that extensional tectonics would have controlled the configuration of the basins and favored the origin and evolution of more stable and better developed lakes.

Despite the fact that no *in situ* algal build-ups have been found, the inferred high algal productivity in this lake is a unique feature in Pyrenean Permian lacustrine episodes. On the other hand, it is somehow surprising that in a lacustrine system dominated by extremely shallow and low-energy carbonate facies with conspicuous subaerial exposure and pedogenetic processes, high energy littoral environments were active. Whether the presence of a bench-margin with a small talus or a fluvial influence – or both together – were responsible for this high-energy gradient sequence is still not clear. In any case, the variability of subenvironments and the diversity of sedimentary and post-depositional agents, described in this Permian basin, exemplify how sensitive small and closed carbonate lakes are, and highlight their importance as climatic, tectonic and hydrologic paleorecords.

Acknowledgements

The work in the Permian Basque Basin is part of the author's PhD dissertation submitted to the Departamento de Geología, Universidad de Zaragoza (Spain) and funded by the Spanish Ministry of Science and Education with a 'Formación del Personal Investigador' grant.

Bibliography

Bixel, F. & Lucas, Cl., 1983. Magmatisme tectonique et sedimentation dans les fosses stephano-permiens des Pyrénées occidentals. *Rev. Geogr. Phys. Geol. Dynam.*, **24** (4) 329–42.

Bixel, F., 1984. *Le volcanisme Stephano–Permien des Pyrénées*. PhD Dissertation, Univ. Paul Sabatier, Toulouse.

Bixel, F., 1989. Le volcanisme des Pyrénées. In *Synthese géologique des bassins Permiens français*. Edition Bureau Recherche Geologique et Miniére, Mem. 128, 256–64.

Carbayo, A., Krausse, H.F. & Pilger, A., 1974. *Hoja de Valcarlos*. Mapa Geológico de España, E. 1:50,000. Edition Instituto Geológico y Minero de España.

Gisbert, J., 1981. *Estudio geológico-petrológico del Estefaniense-Pérmico de la Sierra del Cadi. Diagénesis y Sedimentologia*. PhD Dissertation, Univ. Zaragoza.

Gisbert, J., 1984. Las molasas tardihercínicas del Pirineo. In *Libro Jubilar de J.M. Rios*, v. II. Edition Instituto Geológico y Minero de España, 168–86.

Lamare, P., 1931. Sur l'existence du Permien dans les Pyrénées basques entre la vallée du Baztan (Navarre espagnole) et la vallée Baïgorry. *C.R. somm. Soc. Geol. France*, **31** (1), 242–5.

Lamare, P., 1936. *Recherches géologiques dans les Pyrénées basques d'Espagne*. Mem. Societe Geologique de France, n. 27.

Le Pochat, G., 1978. *Feuille St. Jean Pied de Port*. Carte Geologique de France, E. 1:50,000. Edition Bureau Recherche Geologique et Miniere.

Lucas, Cl., 1985. *Le grès rouge du versant nord des Pyrénées. Essai sur la Geodynamique de depots continentaux du Permien et du Trias*. PhD Dissertation, Univ. Toulouse.

Lucas, Cl., 1989. Le Permien des Pyrénées. In *Synthese géologique des bassins permiens francais*. Edition Bureau Recherche Geologique et Miniere, Mem. 128, 139–50.

Lucas, Cl., Doubinger, J. & Broutin, P., 1980. Premieres datations palynologiques dans les grés rouges triasiques des Pyrénées. *C.R. Acad. Sci. Paris*. **129**(D 6), 517–20.

Müller, D., 1969. *Perm und Trias im valle del Baztan (Spanische-Westpyrenäen)*. PhD Dissertation, Fak. Natur.-Geisteswiss. Teknisches Univ. Clausthal, Clausthal-Zellerfeld.

Müller, D., 1973. Perm und Trias im valle del Baztan – ein Beitrag zur stratigraphie und Paläeogeographie der spanische West Pyrenaën. *Neues Jahrbuch Geol. Paäont. Abh.* **142** (1), 30–43.

Solé, J. & Villalobos, L., 1974. *Hoja de Maya de Baztan.* Mapa Geológico de España, E. 1:50,000. Edition Instituto Geológico y Minero de España.

Valero Garcés, B.L., 1991. *Los sistemas lacustres carbonatados del Stephaniense y Pérmico en el Pirineo Central y Occidental.* PhD Dissertation, Univ. Zaragoza, Zaragoza.

Valero Garcés, B.L., 1992. Litofacies carbonatadas lacustres someras en el Pérmico del Macizo de Les Aldudes – Quinto Real (Cuenca Vasca, Pirineos Occidentales). *III Congreso Geológico de España. Symposium on lacustrine sedimentation,* Abstracts, 158–67.

Carbonate lacustrine episodes in the continental Permian Aragón-Béarn Basin (western Pyrenees)

BLAS L. VALERO GARCÉS[1]

Departamento de Geología, Universidad de Zaragoza, E-50009 Zaragoza, Spain

Geological setting

The Aragón-Béarn Basin, located in the Western Pyrenees and dissected by the Spanish–French border, is one of the Late Hercynian Basins exposed along the southern and northern boundaries of the Paleozoic Axial zone (Fig. 1) and related to the Late Paleozoic strike–slip faulting in southern Europe and northern Africa (Arthaud & Matte, 1977). The Aragón-Béarn Basin was filled in the Late Stephanian to Middle Upper Permian with a several thousand meter-thick series comprising continental sediments, volcaniclastics and volcanic materials. A number of formations have been defined in these rocks (Lingen, 1960; Schwarz, 1962; Mirouse, 1966a; Lucas, 1985), but those defined by Gisbert (1981) in the Cadi Basin (Eastern Pyrenees) have proven to be useful for correlation across the Pyrenees (Gisbert, 1983, 1984) and for application to the Aragón-Béarn Basin (Asso *et al.*, 1991; Valero Garcés, 1991). An up to 800 m thick volcanic and volcaniclastic complex, dated as Upper Stephanian (Müller, 1967; Clin *et al.*, 1970; Bixel, 1984; Bixel *et al.*, 1983; Rios *et al.*, 1985), overlies the Hercynian unconformity and represents the 'Unidad Gris' and the lower member of the 'Unidad de Tránsito' (UT). The upper member of the UT is composed of grey alluvial and lacustrine sediments (Upper Stephanian–Autunian) and it is followed by the alluvial, red, clastic sediments of the 'Unidad Roja Inferior' (URI).

A major disconformity separates the lower red unit (URI) from the 'Unidad Roja Superior' (URS). This later unit is mainly composed of alluvial-fan sediments, arranged in several hundred meter-thick fining-upwards megasequences. Some volcanic sills and one lava flow intercalate and a Permian volcanic intrusive complex cuts the URS. No paleontological dating is available for the URS, but this unit is considered of Middle to Upper Permian age in the whole Pyrenean domain (Gisbert, 1984; Lucas, 1985; Lucas, 1989). Nevertheless, some authors (Clin *et al.*, 1970; Bixel *et al.*, 1985) suggest a Lower Triassic age for the most upper part of this unit in the Aragón-Béarn basin.

This contribution summarizes the available data for the lacustrine episodes within the URS in the Aragón-Béarn Basin. During

Permian times, this basin was formed by two sub-basins separated by a high (Candanchú area) where deposition was considerably lower (Fig. 2). In both sub-basins, three southwest on-lapping fining-upwards megasequences have been recognized. URS thickness in the eastern sub-basin is about 700 m and only one lacustrine intercalation, some meters thick, is developed at the top of the second megasequence. Thickness of the URS in the western sub-basin exceeds 1500 m and a total of six lacustrine episodes are intercalated in the first and second megasequences. Although some decimeter-thick lacustrine carbonate layers have been described from the URS of the Cadi Basin, Eastern Pyrenees (Gisbert, 1981), the lacustrine episodes in the Aragón-Bearn Basin constitute the thickest (some tens of meters) and best developed Permian lacustrine series of all the Pyrenean basins. The presence of these lacustrine carbonates have been previously reported (Seunes, 1894; Dalloni, 1950, 1957; Mirouse, 1959, 1966a, b; Bixel *et al.*, 1985), but no detailed work has been done until recent years (Valero Garcés, 1991; Valero Garcés & Gisbert Aguilar, 1991a, b).

Lacustrine episodes

Geographical location of the best exposures of the seven lacustrine series are shown in Fig. 2 and a stratigraphical sketch is provided in Fig. 3. Six of the seven episodes are located in the western sub-basin and only one in the eastern sub-basin. Two of the western episodes appear at the top of the first megasequence in the northern area of the basin, Baralet-Arlet and Acué. Two more are found in the central area of the basin, in the basal part of the second megasequence: Bois de Sansanet, and Candanchú. Finally, in the southern area of the basin, the Aguas Tuertas episode occurs at the top of the second megasequence, and the Escalé episode, in a small and fault-bounded outcrop, tentatively ascribed to the first megasequence. Facies in these URS lacustrine intercalations are described and interpreted in Tables 1 and 2. Most of the lacustrine episodes were associated with the fine alluvial plain deposits extensively developed at the top of the fining-upwards megasequences (Fig. 3). These low gradient alluvial plain areas were suitable locations for small, closed and very shallow carbonate lakes. However, other sedimentary and tectonic settings also

[1] Current address: Limnological Research Center, 220 Pillsbury Hall, University of Minnesota Minneapolis, MN 55455 USA.

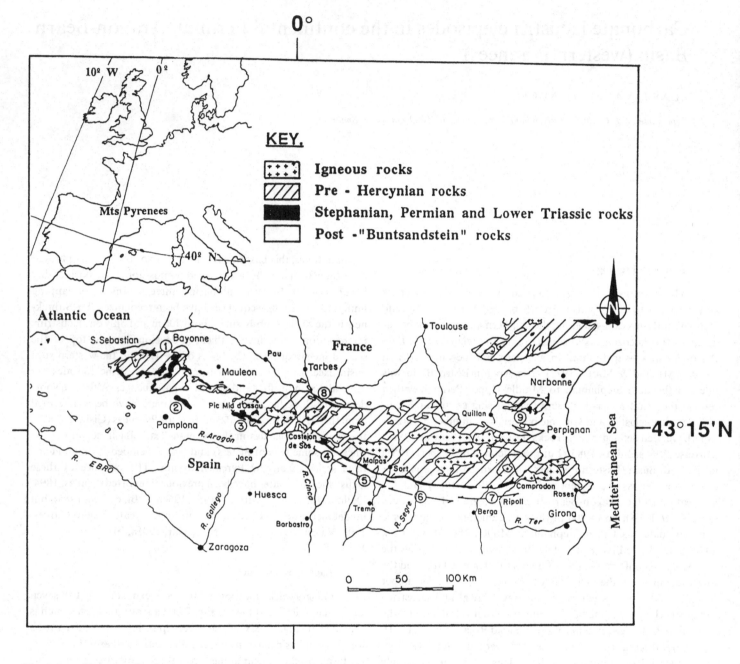

Fig. 1. Late Hercynian (Stephanian, Permian and Lower Triassic) Pyrenean basins. (1) Basque Basin; (2) Pamplona Basin; (3) Aragón-Béarn Basin; (4) Castejón de Sos-Laspaules Basin; (5) Malpas-Sort Basin; (6) Cadi Basin; (7) Castellar–Camprodón Basin; (8) North Pyrenean Basins; (9) Chateaux Segure Basin. (After Gisbert, 1984.)

favored lake development, as the Escalé and the Bois de Sansanet episodes show.

The *Baralet-Arlet lacustrine episode* is an example of a low-energy margin-palustrine facies model with three main subenvironments: alluvial plain, palustrine littoral and evaporitic littoral. The littoral sedimentary sequences are mainly regressive and composed of grey, massive to bedded limestones, nodular limestones, mottled limestones and red siltstones. More alluvial plain-influenced associa-

tions are composed of thin, reworked limestone beds, interpreted as expansion lake events, and yellow dolomitic marls generated in the alluvial plain (Fig. 4A, B).

Although pseudomorphs after gypsum are not a rare feature in nodular, mottled and brecciated limestones in most of the URS lacustrine series, meter-scale gypsum layers are restricted to the *Acué lacustrine episode*. The association of some meter-thick, massive and nodular gypsum layers, travertine limestones and red

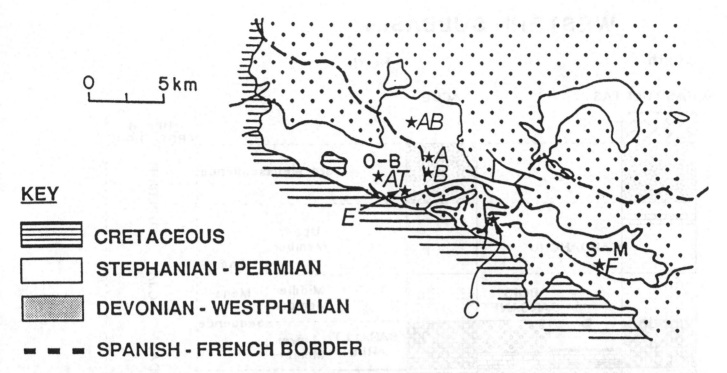

KEY

	CRETACEOUS
	STEPHANIAN - PERMIAN
	DEVONIAN - WESTPHALIAN
- - -	SPANISH - FRENCH BORDER

Fig. 2. Geological sketch of the Aragón–Béarn basin showing the two sub-basins and the location of the studied lacustrine series. Key: O–B = Western sub-basin (Oza-Baralet), S–M = Eastern sub-basin (Sallent de Gállego–Midi d'Ossau), AB-Arlet-Baralet sections, A = Acué section, B = Bois de Sansanet section, AT = Aguas Tuertas sections, C = Candanchú section, E = Escalé, and F = Fuentes de Izas.

siltstones characterize three subenvironments: saline mudflat, saline lake littoral and springs (Fig. 4C, D). Limestones show travertine fabrics (cryptalgal textures, calcite-rafts, spherulites) and a distinctive stable isotope composition: relatively heavy $\delta^{13}C$ values (about -5 per mil PDB) explained as the result of high organic productivity during high lake stands, and lighter values of $\delta^{18}O$ (about -10 per mil PDB) compared to pseudomorphs after gypsum-bearing limestones in the other Permian lakes, indicating less evaporative influence (Valero Garcés & Gisbert Aguilar, 1991b). Travertine features, the absence of intense evaporative isotopic imprint, and the deduced strong variation in water chemistry are congruent with a playa lake system whose paleohydrology was mainly controlled by SO_4^{2-}-rich groundwater recharges in marginal springs. This episode is thought to be synchronous and, likely, physically connected with the Arlet-Baralet one.

The *Bois de Sansanet lacustrine episode* was developed after a tectonic pulse which resulted in the enlargement of the basin and the beginning of the second megasequence. In the Bois de Sansanet area, near the eastern limit of the western subbasin, the second megasequence overlies Westphalian rocks directly. After some meters of conglomeratic breccias (proximal alluvial facies), the lacustrine series is composed of (i) reworked intraclastic limestones (graded & laminated light-colored limestones and massive or laminated, thin-bedded, dark grey limestones), and (ii) nodular and mottled limestones (Fig. 5.IA). As a result of the high-energy wave-

influenced margins, palustrine areas were less developed; higher energy facies appeared in the littoral zone and relatively more open lacustrine areas could be defined (Fig. 5.IB). Palustrine facies associations are progressively more common in the upper part of this series, suggesting more stable conditions and a shallowing-upwards general trend (Fig. 5.IC, D).

Candanchú lacustrine episode, broadly synchronous to the Bois de Sansanet series, represents a carbonate littoral association where intrasediment precipitation of evaporites (gypsum) was very intense. Brecciated limestones are the main facies present in this fault-bounded outcrop located in the high between both sub-basins (Fig. 5.II).

The *Aguas Tuertas lacustrine episode* features three main subenvironments: alluvial plain, palustrine littoral and shallow 'sub-littoral'. Pseudomorphs after gypsum are found in some nodular limestones, indicating that evaporitic conditions were sometimes achieved in littoral subenvironments (Fig. 6A, B). The laminated green-grey siltstones intercalated with light grey limestone lenses are interpreted as deepening-upwards sequences in a shallow 'sub-littoral' setting, relatively oxygen-depleted and far from shoreline influence (Fig. 6C, D).

The *Puerto de Escalé lacustrine episode* is located in the south-eastern margin of the western sub-basin and is exclusively composed of dolomitic facies. Shallowing upward sequences, up to 3 m thick, are composed of grey dolostones, nodular and variegated

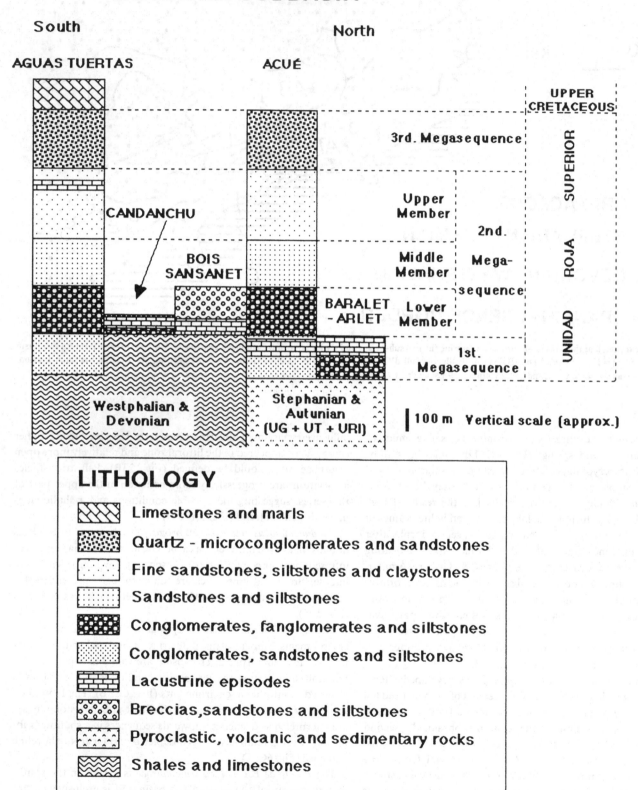

Fig. 3. Simplified sketch showing the stratigraphy of the Permian 'Unidad Roja Superior' in several areas of the western sub-basin of the Aragón–Béarn Basin and the location of the main lacustrine episodes. Escalé lacustrine episode, not shown, is situated in a small, fault-bounded outcrop at the southeastern edge of the sub-basin and it has been tentatively located in the first megasequence. Horizontal distances are not to scale. UG = Unidad Gris, UT = Unidad de Tránsito, URI = Unidad Roja Inferior.

Table 1. *Schematic description and interpretation of the main lithofacies found in the Permian calcitic lacustrine series (Baralet-Arlet, Acué, Candanchú, Bois de Sansanet and Aguas Tuertas episodes)*

Lithofacies	Microfacies	Fossil content	Pedogenic and subaerial exposure (intensity and features)	Early diagenetic cementing phases	Interpretation and subenvironment
Common facies to several lacustrine episodes					
Grey, massive to bedded limestones	Micrites & intramicrites	Scarce (ostracods, bivalves, algae)	MEDIUM: Cavities, planar voids (mainly joint planes). Pseudomicrokarst	VS→NL→Z→BL→DL	Lacustrine littoral, frequently exposed
Nodular limestones	Nodular micrites & intramicrites	Very scarce	HIGH: Glaebulae, crystic fabrics, mottling channels, skew, craze and curve planar voids, pseudomicrokarst pseudomorphs after gysum	VS→NL→BL	Lacustrine littoral, with frequent emersion and soil development
Marmorized or mottled limestones	Mottled micrites & intramicrites	Variable	HIGH: Mottling, clay cutans and ferrans, pseudomorphs after gypsum, cavities and joint planes	NL→BL	Lacustrine littoral with frequent water level changes
Nodule-bearing red and purple siltstones	Massive, rarely laminated siltstones Chlorite <10%	None except root structures	HIGH: Glaebules, mottling, neoferrans & neocalcitans, pedotubules, sepic fabrics, skew, craze and curved planes	NL→BL	Fine grained alluvial plain fringed lacustrine area and with intense pedogenesis
Facies restricted to one lacustrine episode					
Baralet-Arlet					
Yellow dolomitic mottled marls	Marly intradolomicrites with pyroclastic components	None except root structures	HIGH: Glaebulae, mottling, columnar & platty structure, pedotubules, roots & rhizocretions, chlorite sepic fabrics	(NL)→BL Silicification Chlorite authigenesis	Fine grained and pyroclastic influenced alluvial plain with hydromorphism and dolomitization phenomena
Aguas Tuertas					
Light grey, lens-shaped limestones	Biointramicrites and marly intramicrites	Medium (thin bivalves & ostracodes), intense bioturbation	VERY LOW: Scarce planar voids and small cavities	DL	Shallow 'sublittoral' areas
Laminated, green-grey siltstones	Fine grained siltstones chlorite >10% No haematites	Very scarce (TOC = 1%)	VERY LOW: Chlorite cutans, microbrecciation	BL Chlorite authigenesis Pyrite	Shallow, oxygen-depleted hypolimnion
Candanchu					
Brecciated limestones	Micrites & intramicrites with evaporite replacement textures	Very scarce (ostracods)	HIGH: Cavities, planar voids, pseudomorphs after gypsum	NL→BL→(Q & C)	Evaporitic littoral
Bois de Sansanet					
Graded & laminated light-colored limestones	Intrabiomicrites (wackestones), high siliciclastic content	High, up to 15% (ostracods, bivalves, algae)	LOW: Mottling, crystic fabrics, joint & skew planes	BL	High energy and high productivity littoral
Massive or laminated dark grey, thin bedded limestones	Micrites & intramicrites, low siliciclastic content	Medium	VERY LOW: Small cavities	BL	High energy, littoral to sublittoral

Table 1 (*cont.*)

Lithofacies	Microfacies	Fossil content	Pedogenic and subaerial exposure (intensity and features)	Early diagenetic cementing phases	Interpretation and subenvironment
Acué					
Travertine limestones	Micrites, intramicrites, travertine fabrics, very low siliciclastics	High (algal remains & ostracodes)	VERY LOW: joint planes.	BL, Q, C	Springs
Massive and nodular gypsum	Microcrystalline (100 μ to 1 mm sized)	None	NONE	Secondary gypsum BL	Saline lake littoral (subaqueous precipitation)

Note:

Early diagenetic cement phase legend: VS, vadose silt; NL, non-luminescent calcite; Z, zoned calcite; BL, bright luminescent calcite; DL, dull luminescent calcite; Q, quartz; C, celestite-barite. Parentheses indicate rare occurrence and '→' a time-sequence.

Table 2. *Schematic description and interpretation of the main lithofacies found in the Permian dolomite-producing lacustrine series (Escalé and Fuentes de Izas episodes)*

Lithofacies	Microfacies	Fossil content	Pedogenesis and subaerial exposure (intensity and features)	Early diagenetic cementing phases	Interpretation and subenvironment
Escalé					
Grey dolostones	Dolointramicrites & Dolomicrites	Very scarce (algae remains, ostracodes) Bioturbation	MEDIUM: Columnar fabric, planar voids, channels and calcite nodules	BL & DL	Ephemeral littoral frequently exposed
Variegated (mottled) dolostones	Dolointramicrites & Dolomicrites	Very scarce (algae & ostracodes). Intense bioturbation	HIGH: Mottling, pseudogley. Columnar fabric; curved and planar voids; joints, cavities & channels. Argillans & calcitans. Pedotubules	BL & DL	Ephemeral littoral with intense pedogenesis
Red marls	Structureless Clotted, nodular & diffusely laminated zones	None	LOW: Horizontal joint planes	BL	Alluvial plain with pedogenesis & volcanic ashfall influence
Fuentes de Izas					
Grey marly limestones & dolostones	Marly intramicrites & marly laminated dolointramicrites	Scarce	MEDIUM in the limestones (planar voids & cavities) NONE in the dolostones	BL & DL	Shallow, relatively high gradient, rarely exposed littoral
Black dolomitic laminated siltstones	Parallel & convolute lamination. Normal grading	None (TOC = 0.2% (%O.M. = 1.5%)	NONE	BL	Oxygen depleted and saline hypolimnion in a shallow meromictic lake
Greenish-grey calcitic laminated siltstones	Parallel lamination intense hydroplastic deformation	Low: ostracods, algae, bivalves	NONE	BL	Oxygen depleted hypolimnion in a shallow meromictic lake

Note:

See legend for Table 1.

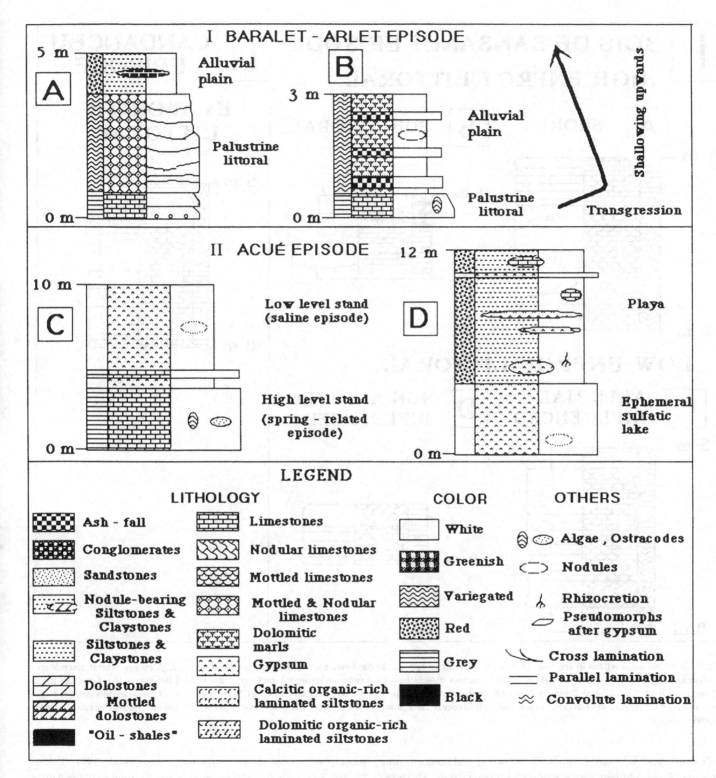

Fig. 4. I: Facies associations in the Baralet–Arlet lacustrine episode. Two subenvironments are defined in this low energy, palustrine littoral: areas where carbonate deposition is more intense and nodulization and mottling processes widely occurred (A) and areas dominated by the alluvial plain depositional processes (fine-grained siliciclastic sedimentation, pedogenesis and dolomitization) (B). II: Facies associations in the Acué lacustrine episode. Massive and nodular gypsum layers associated with travertine limestone in the lacustrine areas (C) and with red siltstones in the saline mudflats (D).

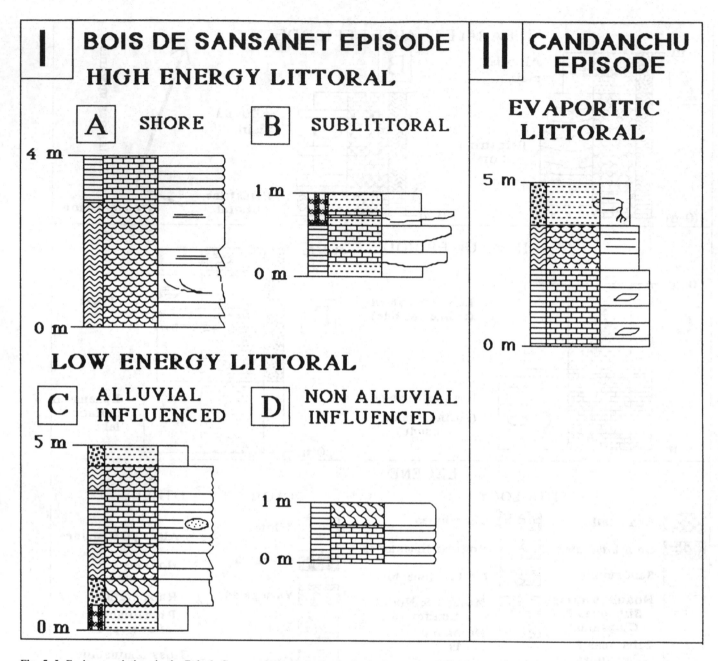

Fig. 5. I: Facies associations in the Bois de Sansanet lacustrine episode. In the lower part of the stratigraphical series, high energy littoral assemblages predominate: graded and laminated light-colored limestones associated with massive or laminated dark-grey, thin bedded limestones (A). Grey and green siltstones with intercalated grey limestone lenses (B) are interpreted as deposition in more internal lake areas. In the upper part of the series, low energy littoral conditions are dominant and associations show alluvial influence (C) and palustrine features (D). II: Facies association in the Candanchú lacustrine episode. Same legend as Fig. 4.

dolostones and red marls (Fig. 7). Fine-grained and homogeneous textures, Sr-enrichment, absence of replacement evaporite textures and a broad suite of pedogenic and vadose features (especially calcite-cementing ones) postdating dolomitization support primary formation of dolomite.

The *Fuentes de Izas lacustrine episode* is the only lake sequence in the eastern subbasin. It was developed at the top of the second megasequence in this subbasin which can be correlated with the second one in the western sub-basin (Asso *et al.*, 1991; Valero Garcés, 1991). So, it might be broadly synchronous with the Aguas Tuertas episode, although sedimentary settings were different. The association of grey marly limestones, dolostones and black organic-rich laminated dolomitic siltstones is thought to represent a chemically-induced water stratification event (ectogenic meromixis) in a small and shallow lake (Fig. 8). Chemically-induced water stratification would be established when less concentrated run-off waters

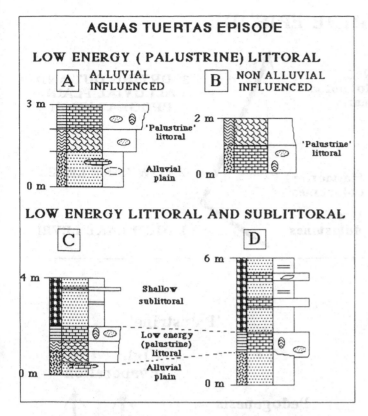

Fig. 6. Facies associations in the Aguas Tuertas lacustrine episode. Low energy, palustrine littoral associations are mainly composed of limestones arranged in deepening upwards (A) or alternating (B) sequences. Deposition of greenish siltstones suggests more internal (sublittoral) lake areas (C & D). Same legend as Fig. 4.

flowed into the lake, leading to the establishment of a saline and anaerobic hypolimnion. There, organic-rich laminated dolomitic and calcitic siltstones could be formed and preserved.

Facies models

The Aragón-Béarn Basin, Western Pyrenees, contains the thickest Permian 'Unidad Roja Superior' in the Late Hercynian Pyrenean basins. High subsidence rates, intense vulcanism with some alkaline-affinities in the last phases (Bixel, 1984, 1987, 1989), presence of disconformities and the general on-lap trend of the three alluvial-fan fining-upward megasequences define a scenario of intermontane basins whose genesis, configuration and sedimentary filling were mainly controlled by contemporaneous strike–slip and extensional tectonics (Soula *et al.*, 1979; Bixel & Lucas, 1983, 1987; Lucas, 1977, 1987a, b, 1989). Conditions for lake genesis and development were achieved during different tectonic, sedimentary and hydrologic settings, leading to deposition of differentiated facies. Main lacustrine episodes were located at the top of the first and second megasequences, associated with fine siliciclastic alluvial plain deposits and related to more stable tectonic conditions. But the changes in the Aragón-Béarn Basin at the beginning of the

second megasequence also led to the development of lacustrine episodes.

A closed, shallow, low gradient margin lake model can be applied to all of these carbonate lacustrine series but characteristic facies and associations allow the definition of four facies models (Valero Garcés, 1991; Valero Garcés & Gisbert Aguilar, 1992):

(1) Low gradient (ramp) margin, holomictic, calcite-producing lakes related to distal zones of large alluvial fans, containing biomicrites, intramicrites, brecciated and nodular limestones, and grey and red siltstones arranged in shallowing-upward sequences. In the regressive phase of the lake, hypersaline conditions were occasionally achieved in marginal littoral areas conducive to intrasediment precipitation of evaporites (gypsum), subsequently replaced by calcite. Mn and Fe enrichment and Sr depletion – caused by pedogenic and vadose modifications – and heavier $\delta^{18}O$ composition (about -7 per mil PDB) due to more evaporative conditions, characterized these littoral 'palustrine' facies. The major lacustrine episodes in the western sub-basin (Baralet-Arlet, Bois de Sansanet-Candanchú and Aguas Tuertas) are interpreted within this model. Two types of littoral facies associations are present:

   (i) the *low-energy littoral (palustrine)* is the most common (Baralet-Arlet, Candanchú and Aguas Tuertas) and it is characterized by intense development of nodulization, mottling and other pedogenic features in the littoral facies; both, transgressive and regressive sequences are present. The sedimentary subenvironments present are alluvial-influenced palustrine littoral (Fig. 4A, 4B, 5.IC and 6A), palustrine littoral (only carbonate deposition, see Fig. 5.ID and 6B), evaporitic littoral intra-sediment gypsum precipitation, (Fig. 5.II), and shallow 'offshore' (fine-grain deposition represented by green-grey siltstones and limestone lenses, Fig. 6C, D). (ii) The *wave influenced littoral* (Bois de Sansanet episode, Fig. 5.IA, B) shows graded and laminated wackestones, with less development of 'subaerial exposure' and 'pedogenic' features, and some 'deeper' facies (grey siltstones).

(2) Ephemeral, holomictic, calcite and gypsum-producing lakes (Fig. 4C, D). This *sulphatic lake model* involves subaqueous precipitation of gypsum; its genesis seems to be constrained by very specific conditions, since gypsum and travertine limestone only appear in the Acué lacustrine episode. A unique spring-fed regime in restricted subenvironments of the alluvial plain-lacustrine system, developed at the end of the first megasequence, is considered responsible for the establishment of the saline lake littoral-playa-springs subenvironments.

(3) Ephemeral, holomictic, dolomite-producing lakes. Dolomitic lakes were smaller than calcitic ones, hydrologically closed, and ephemeral (Fig. 7). Their sedimentary setting was interpreted as marginal or intermediate areas and/or related to smaller alluvial fans than those associated with calcitic lakes. The ephemeral dolomitic facies model is characterized

ESCALE LACUSTRINE EPISODE

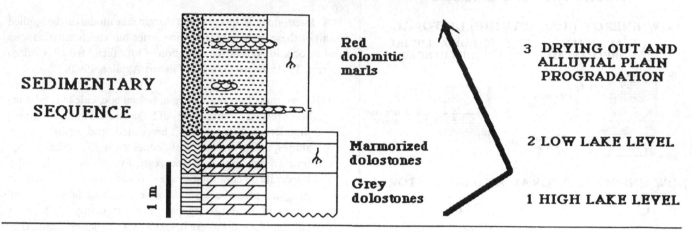

SEDIMENTARY SEQUENCE

Red dolomitic marls

Marmorized dolostones

Grey dolostones

3 DRYING OUT AND ALLUVIAL PLAIN PROGRADATION

2 LOW LAKE LEVEL

1 HIGH LAKE LEVEL

1 m

DEPOSITIONAL MODEL

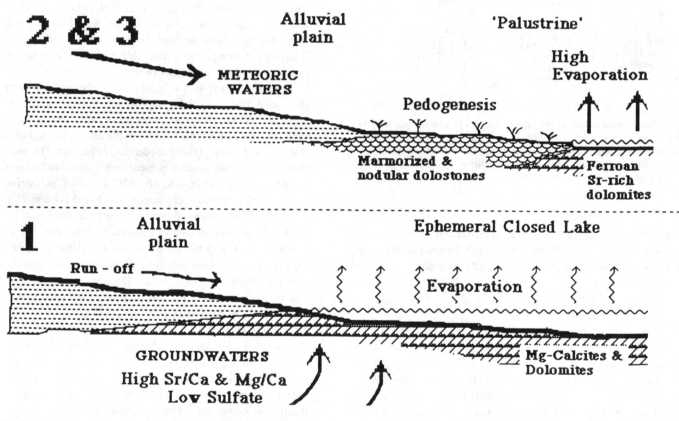

2 & 3

Alluvial plain

'Palustrine'

METEORIC WATERS

High Evaporation

Pedogenesis

Marmorized & nodular dolostones

Ferroan Sr-rich dolomites

1

Alluvial plain

Ephemeral Closed Lake

Run - off

Evaporation

GROUNDWATERS
High Sr/Ca & Mg/Ca
Low Sulfate

Mg-Calcites & Dolomites

Fig. 7. Facies association in the Escalé lacustrine episode and summary diagram to illustrate the evolution of this Permian shallow, ephemeral, holomictic, dolomite-producing lake: 1: progressive evaporation of lake waters led to primary precipitation of Sr-rich, fine-grained dolomite; 2: intense pedogenesis modified depositional facies; 3: drying out and alluvial plain progradation. Same legend as Fig. 4.

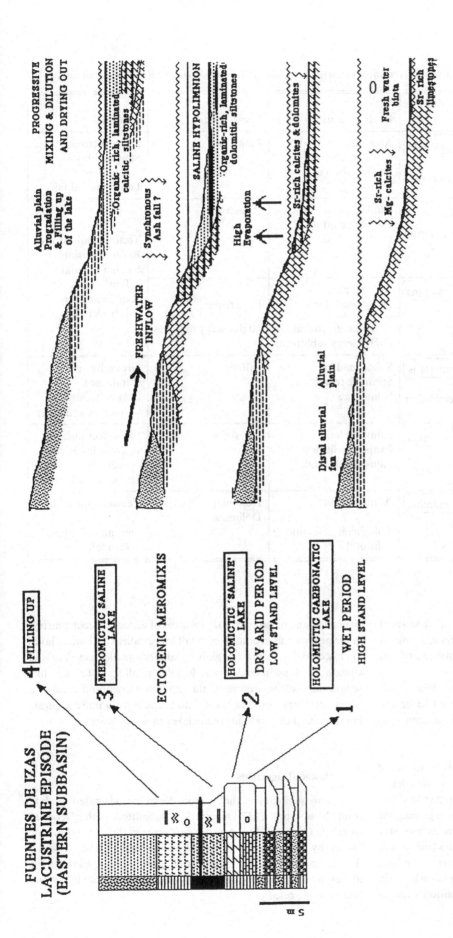

Fig. 8. Simplified sedimentological section of the Fuentes de Izas lacustrine episode (eastern sub-basin) and summary diagram to illustrate the evolution of this shallow, ephemeral, meromictic, carbonate-producing lake: holomictic carbonate-producing, shallow lake facies (grey marly limestones and dolostones) are overlain by meromictic saline dolomite-producing ones (black, dolomitic, organic-rich, laminated siltstones). This transition is interpreted as an ectogenic meromixis event.

Table 3. *Summary of the Permian carbonate lacustrine episodes and interpreted facies models in the Aragón–Béarn Basin (western sub-basin)*

Stratigraphy		Lacustrine episodes	Environments	Facies model	Tectonic
2nd Megasequence Thickness = 900 m	Top	Aguas tuertas — Low energy margins — Large alluvial plains — Incipient meromixis? — Thickness = 30 m	Alluvial plain Palustrine littoral Shallow littoral	Palustrine	Stable tectonics Slow and high subsidence Large alluvial fans
	Base	Candanchu — Low energy margins — Intrasediment gypsum precipitation — Thickness = 40 m	Alluvial plain Evaporitic littoral	Palustrine	Tectonic pulsations Basin extension Smaller alluvial fans Rapid and high subsidence
		Bois sansanet — High energy margins — Alluvial influx — Incipient meromixis? — Thickness = 40 m	Alluvial plain Palustrine littoral ---------------- High energy littoral High energy sublittoral	Palustrine ---------------- High energy margins	
1st Megasequence	North Thickness = 300 m (Top)	Acue — Low energy margins — Travertine & springs — Subaqueous gypsum precipitation — Thickness = 60 m — Groundwaters/springs	Saline mudflat Spring & travertine Saline lake	Saline	Groundwaters influence (related to fault & volcanic activity?)
		Baralet-arlet — Low energy margins — Large alluvial plains — Dolomite formation — Thickness = 60 m	Alluvial plain Evaporitic littoral Palustrine littoral	Palustrine	Stable tectonics volcanic influence (ash fall) Large alluvial fans
	South	Escale — Low energy margins — Primary dolomite formation — Intense pedogenesis — Thickness = 25 m — Groundwaters/springs	Alluvial plain Ephemeral dolomitic littoral	Holomictic Dolomitic	Tectonic pulsation volcanic influence (ash fall)

by primary formation of dolomite in highly alkaline small lakes. Low SO_4^{2-} and high Mg^{2+} groundwaters are thought to have played an important role in the uniqueness of this lake sequence.

(4) Ephemeral, meromictic, carbonate-producing lakes. This *meromictic dolomitic facies model* is considered to be the result of an ectogenic meromixis event in the eastern sub-basin lacustrine episode (Fig. 8).

The presence of different carbonate facies models in the same small continental basin shows how many interrelated factors determine the lithofacies and the sequences to be generated in shallow, low gradient margin lakes (Table 3). High or low energy margins depend on basin configuration, mainly controlled by sedimentary setting (size and type of sedimentary systems), tectonics (subsidence ratio), paleoclimate and paleohydrology (water sources, lake level changes). These differences among lacustrine episodes developed in the same basin stress the fundamental roles of depositional subenvironment and paleohydrology (sources of waters, paleochemistry) in the genesis and evolution of small low-gradient carbonate lakes in closed basins. Although global scale factors like tectonics and climate are fundamental ones, locally, small-scale factors, like sedimentary subenvironment, the groundwater/run-off ratio and water chemistry, must be taken into account for a more accurate interpretation of small carbonate lakes in closed basins.

Acknowledgements

This summary of the Aragón-Béarn basin is based on data from the author's PhD Dissertation, submitted to the Departamento de Geología, Universidad de Zaragoza (Spain) and mainly funded by the Spanish Ministry of Science and Education with a 'Formación del Personal Investigador' grant. Special thanks to the numerous friends and colleagues who collaborated in several phases of the research.

Bibliography

Arthaud, F. & Matte, P., 1977. Late Paleozoic strike–slip faulting in southern Europe and northern Africa: result of a right lateral shear zone between the Appalachians and the Urals. *Geol. Soc. Am. Bull.*, **88**, 1305–20.

Asso, E., Gisbert, J. & Valero Garcés, B., 1991. *El Stephaniense-Pérmico del Alto Aragón*. Instituto de Estudios Aragoneses, Huesca.

Bixel, F., 1984. *Le volcanisme stephano-permien des Pyrénées*. PhD Dissertation, Univ. Paul Sabatier.

Bixel, F., 1987. Le volcanisme stephano-permien des Pyrénées. Petrographie, mineralogie, geochimie. *Cuad. Geol. Ibérica*, **11**, 41–55.

Bixel, F., 1989. Le volcanisme des Pyrénées. In *Synthèse géologique des bassins permiens français*. Edition Bureau de Recherche Geologique et Miniere, Memoir, 128, 258–64.

Bixel, F. & Lucas, Cl., 1983. Magmatisme, tectonique et sedimentation dans les fosses stephano-permiens des Pyrénées occidentales. *Revue Geogr. Phys. Geol. Dyn.*, **24** (4), 329–42.

Bixel, F. & Lucas, Cl., 1987. Approche Geodynamique du Permien et du Trias des Pyrénées dans le cadre du Sudouest europeen. *Cuad. Geol. Ibérica*, **11**, 57–81.

Bixel, F., Kornprobst, J. & Vincent, P., 1983. Le massif du Pic du Midi d'Ossau: un 'cauldron' calco-alcaline stéphano-permien dans la zone axiale des Pyrénées. *Revue Geogr. Phys. Geol. Dyn.*, **24** (4), 315–28.

Bixel, F., Müller, J. & Roger Ph., 1985. *Carte géologique Pic Midi d'Ossau et Haut Bassin du Rio Gállego. E:25.000*. Editión Institut de Geodynamique. Univ. Bordeaux III.

Clin, M., Heddebaut, C., Mirouse, R., Müller, J. & Waterlot, M., 1970. Le cycle hercynien dans les Pyrenees. *Ann. Soc. Géol. Nord*, **90** (4), 253–76.

Dalloni, M., 1950. Revisión de la feuille d'Urdos au 80.000. Terrains primaires. *Bull. Serv. Carte Géol. France*, **48** (231), 209–13.

Dalloni, M., 1957. Sur un horizon marin fossilifere dans le gres rouge permien de la Neste d'Aure (Hautes Pyrenees). *C.R. somm. Soc. Géol. France*, 109–10.

Gisbert, J., 1981. *Estudio geológico-petrológico del Estefaniense-Pérmico de la Sierra del Cadí. Diagénesis y Sedimentología*. PhD Dissertation, Univ. Zaragoza.

Gisbert, J., 1983. El Pérmico de los Pirineos españoles. *X Congreso Intern. de Estr. y Geol. del Carbonífero., Madrid*. Abstracts, 405–20.

Gisbert, J., 1984. Las molasas tardihercínicas del Pirineo. In *Libro Jubilar de J.M. Rios*, v. II, Instituto Geológico y Minero de España, Madrid, 168–86.

Lingen, G.J., 1960. Geology of the Spanish Pyrenees North of Canfranc, Huesca Province. *Est. Geol.*, **16** (4), 205–42.

Lucas, Cl., 1977. Permien et Trias des Pyrénées. Stratigraphie, elements de paléogeographie. *Cuad. Geol. Ibérica*, **4**, 111–22.

Lucas, Cl., 1985. *Le grès rouge du versant nord des Pyrénées. Essai sur la Geodynamique de depots continentaux du Permien et du Trias*. PhD Dissertation, Univ. Toulouse III.

Lucas, Cl., 1987a. Estratigrafía y datos morfo-estructurales sobre el Pérmico y Triásico de fosas nortepirenaicas. *Cuad. Geol. Ibérica*, **11**, 25–40.

Lucas, Cl. 1987b. Fossés stéphano-pérmiens des Pyrénées: dynamique sédimentaire. *Ann. Soc. Géol. Nord*, **106**, 163–72.

Lucas, Cl., 1989. Le Permien des Pyrénées. In *Synthèse géologique des bassins permiens français*. Bureau Recherche Géologique et Miniere, Memoir 128, 139–50.

Mirouse, R., 1959. Extension et relation des series permiennes sur la feuille d'Urdos au 80.000. *Bull. Serv. Carte Géol. France*, **56** (257), 209–18.

Mirouse, R., 1966a. *Recherches géologiques dans la partie occidentales de la zone primaire axiale des Pyrénées*. Mem. Carte Géol. France.

Mirouse, R., 1966b. *Feuille géologique d'URDOS à 1:80.000*. 2nd edition Bureau Recherche Géologique et Miniere.

Müller, J., 1967. Sur la superposition des deformations dans les Pyrénées Occidentales. *C.R. Acad. Sci. Paris*, **265** (D), 400–2.

Rios Aragües, L.M., Lanaja, J.M. & Fernandez, C., 1985. Contribución a la geología del Paleozoico del Valle de Tena, Alto Gállego, Provincia de Huesca. In *Libro Jubilar de J.M. Rios*, Instituto Geológico y Minero de España v. III, 45–61.

Schwarz, E.J., 1962. Geology and paleomagnetism of the valley of the Rio Aragon Subordan, north and east of Oza (Spanish Pyrenees, province of Huesca). *Est. Geol.*, **18**, 193–239.

Seunes, J., 1894. Vallée d'Ossau et vallée d'Aspe. *Bull. Serv. Carte Geol. France*, **6** (38), 97–102.

Soula, J.C., Lucas, Cl. & Bessiere, G., 1979. Genesis and evolution of Permian and Triassic basins in the Pyrenees by regional simple shear acting on older Variscan structures: field evidence and experimental models. *Tectonophysics*, **58** (3–4), T1–T9.

Valero Garcés, B.L., 1991. *Los sistemas lacustres carbonatados del Stephaniense y Pérmico en el Pireneo Central y Occidental*. PhD Dissertation, Univ. Zaragoza, Zaragoza.

Valero Garcés, B.L. & Gisbert Aguilar, J., 1991a. Calcitic-sulphatic lakes in the Western Pyrenees. *VI Reunion Association des Géologes du Permien Paris*. Abstracts, 13.

Valero Garcés, B.L. & Gisbert Aguilar, J., 1991b. Permian saline lakes in the Aragón-Bearn basin (Western Pyrenees). *Sedimentary and Paleolimnological records of saline lakes Conference, Univ. Saskatchewan, Saskatoon*, Abstracts, 15.

Valero Garcés, B.L. & Gisbert Aguilar, J., 1992. Shallow carbonate facies models in the Permian of the Aragón-Béarn basin (western Spanish–French Pyrenees). *Carbonates and Evaporates*, **7** (2), 94–107

Lake Bude (Upper Carboniferous), southwest England

ROGER HIGGS

Sedimentological Consultant, Moreton Cottage, Beer Road, Seaton, Devon EX12 2PR UK

The Bude Formation (Lower Westphalian) consists of 1300 m of turbidite-like sandstone event-beds interbedded with mudstones (Figs 1 and 2). Event beds are individually up to 40 cm thick, and up to 10 m when amalgamated. Deposition took place offshore, as shown by a lack of evidence for emergence. The water body was a fresh to brackish lake, based on: (1) scarcity of body and trace fossils, (2) presence of an endemic fish genus, and (3) confinement of marine fossils (goniatites and pelagic bivalves) to a few thick (2–10 m) shale units, reflecting marine incursions into the lake. There is a dm–m scale cyclicity (Fig. 3), whereby amalgamated event beds

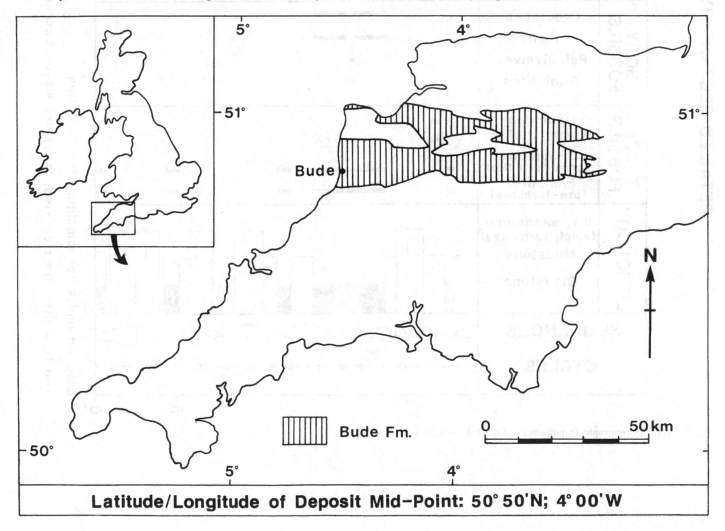

Latitude/Longitude of Deposit Mid–Point: 50° 50'N; 4° 00'W

Fig. 1. Locality map.

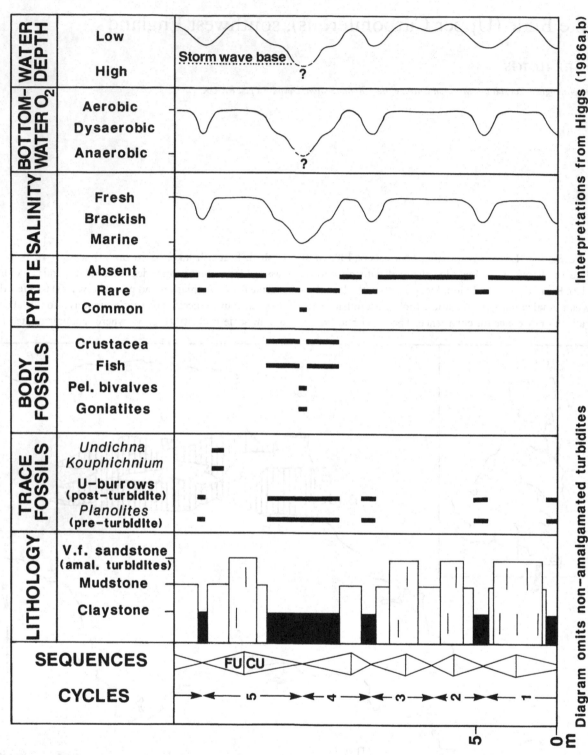

Fig. 2. Detailed stratigraphy (hypothetical) of the Bude formation.

GENERAL SECTION

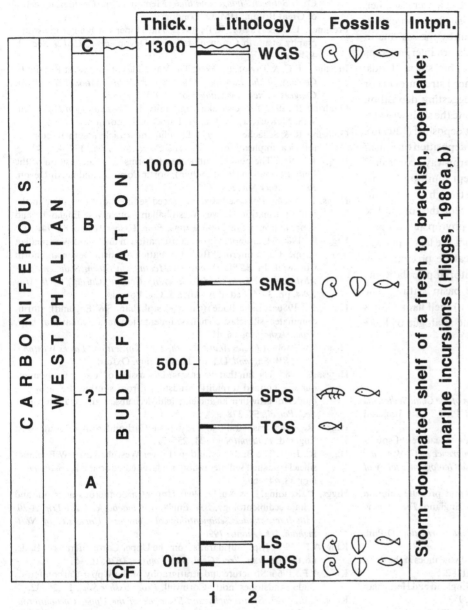

1 Varved claystone; 2–10m units

2 River-fed turbidites; varved mudstones/claystones;
  muddy seismites; in dm–m cycles

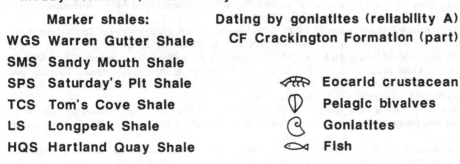

Marker shales:	Dating by goniatites (reliability A)
WGS Warren Gutter Shale	CF Crackington Formation (part)
SMS Sandy Mouth Shale	
SPS Saturday's Pit Shale	Eocarid crustacean
TCS Tom's Cove Shale	Pelagic bivalves
LS Longpeak Shale	Goniatites
HQS Hartland Quay Shale	Fish

Fig. 3. General stratigraphy.

alternate with non-amalgamated beds. Each of these inferred regressive–transgressive cycles is thought to reflect a decrease, then an increase, in water depth and salinity (brackish–fresh–brackish), caused by periodic overtopping of the lake sill by the sea due to glacioeustatic sea-level oscillations. Event beds are interpreted as river-fed turbidites deposited during catastrophic storm floods. Combined-flow ripples and other wave-influenced structures occur in event beds throughout the succession, suggesting deposition entirely above storm wave base. However, despite the shallow water (< 100 m?) and the great thickness of sediment deposited (1300 m), emergence never occurred: this is attributed to deposition on a shelf which could aggrade no higher than the storm-wave-controlled 'equilibrium profile'; surplus sediment was swept southward into a hypothetical (eroded?) deep-water trough. The trough also explains the paleoflow pattern, from all quadrants except the south. The absence of freshwater limestone suggests low phytoplankton productivity (i.e. oligotrophy) in Lake Bude, so that the lake's pH was insufficiently raised (by photosynthesis) for calcite precipitation. Regional paleomagnetic data suggest an equatorial paleolatitude, while regional geology suggests a foreland basin setting. The modern Black Sea, with its broad (northwestern) shelf flanked by a deep trough, is probably a good physiographic analogue of Lake Bude.

Bibliography

Ashwin, D.P., 1957. *The Structure and Sedimentation of the Culm Sediments between Boscastle and Bideford, North Devon.* PhD Thesis, Imperial College, Univ. London.

Ashwin, D.P., 1958. The coastal outcrop of the Culm Measures of south-west England. Abstract. *Proceedings of a Conference on Geology and Geomorphology in South West England, Royal Geological Society of Cornwall,* **2**, 2–3.

Beach, A., 1977. Vein arrays, hydraulic fractures and pressure-solution structures in a deformed flysch sequence, SW England. *Tectonophysics,* **40**, 201–25.

Burne, R.V., 1969. *Sedimentological Studies of the Bude Formation.* D.Phil. Thesis, Univ. Oxford.

Burne, R.V., 1970. The origin and significance of sand volcanoes in the Bude Formation (Cornwall). *Sedimentology,* **15**, 211–28.

Burne, R.V., 1973. Palaeogeography of South West England and Hercynian continental collision. *Nat. Phys. Sci.,* **241**, 129–31.

Cornford, C., Yarnell, L. & Murchison, D.G., 1987. Initial vitrinite reflectance results from the Carboniferous of north Devon and north Cornwall. *Proc. Ussher Soc.,* **6**, 461–7.

Crookall, R., 1930. The plant horizons represented in the Barren Coal Measures of Devon, Cornwall and Somerset. *Proc. Cotteswold Natural. Field Club,* **24**, 27–34.

Edmonds, E.A., McKeown, M.C. & Williams, M., 1975. *British Regional Geology: South-West England.* HMSO, London.

Edmonds, E.A., Williams, B.J. & Taylor, R.T., 1979. *Geology of Bideford & Lundy Island.* Memoirs of the Geological Survey of Great Britain. HMSO, London.

Enfield, M.A., Gillcrist, J.R., Palmer, S.N. & Whalley, J.S., 1985. Structural and sedimentary evidence for the early tectonic history of the Bude and Crackington Formations, north Cornwall and Devon. *Proc. Ussher Soc.,* **6**, 165–72.

Freshney, E.C., Edmonds, E.A., Taylor, R.T. & Williams, B.J., 1979. *Geology of the Country Around Bude and Bradworthy.* Memoirs of the Geological Survey of Great Britain. HMSO, London.

Freshney, E.C., McKeown, M.C. & Williams, M., 1972. *Geology of the Coast Between Tintagel and Bude.* Memoirs of the Geological Survey of Great Britain. HMSO, London.

Freshney, E.C. & Taylor, R.T., 1972. The Upper Carboniferous stratigraphy of north Cornwall and west Devon. *Proc. Ussher Soc.,* **2**, 464–71.

Freshney, E.C. & Taylor, R., 1980. The Variscides of southwest Britain. In *Geology of the European Countries. 26th International Geological Congress, Excursion Guidebook,* No. 1, 379–87.

Goldring, R., 1978. Sea level lake community. In *The Ecology of Fossils.* ed. W.S. McKerrow, pp. 178–81. Duckworth, London.

Goldring, R. & Seilacher, A., 1971. Limulid undertracks and their sedimentological implications. *N. J. Geol. Paläontol., Abh.,* **137**, 422–42.

Higgs, R., 1983. The possible influence of storms in the deposition of the Bude Formation (Westphalian), north Cornwall and north Devon. *Proc. Ussher Soc,* **5**, 477–8.

Higgs, R., 1984a. Possible wave-influenced sedimentary structures in the Bude Formation (Lower Westphalian, south-west England), and their environmental implications. *Proc. Ussher Soc.,* **6**, 88–94.

Higgs, R., 1984b. Hummocky cross-stratification in the Lower Westphalian (Upper Carboniferous) Bude Formation of north Devon and north Cornwall. In *BSRG Discussion Meeting on Storm Sedimentation, Abstracts and Field Guide, University College, Cardiff, 5–7th May 1984,* pp. 34–50, ed. P.A. Allen & C.J. Pound.

Higgs, R., 1986a. 'Lake Bude' (early Westphalian, SW England): storm-dominated siliciclastic shelf sedimentation in an equatorial lake. *Proc. Ussher Soc.,* **6**, 417–18.

Higgs, R., 1986b. *A Facies Analysis of the Bude Formation (Lower Westphalian), SW England.* D.Phil. Thesis, Univ. Oxford.

Higgs, R., 1987. The fan that never was? Discussion of 'Upper Carboniferous fine-grained turbiditic sandstones from southwest England: a model for growth in an ancient, delta-fed subsea fan', by J. Melvin. *J. Sed. Petrol.,* **57**, 378–9.

Higgs, R., 1988. Fish trails in the Upper Carboniferous of south-west England. *Palaeontology,* **31**, 255–72.

Higgs, R., 1991. The Bude Formation (Lower Westphalian), SW England: siliciclastic shelf sedimentation in a large equatorial lake. *Sedimentology* **38**, 445–69.

Higgs, R., Reading, H. & Xu, L., 1990. Upper Carboniferous lacustrine and deltaic sedimentology, SW England. *Guidebook, Field Trip A-10, 13th International Sedimentological Congress, University of Nottingham, UK, August 1990.*

King, A.F., 1965. Xiphosurid trails from the Upper Carboniferous of Bude, north Cornwall. *Proc. Geol. Soc. London,* **1626**, 162–5.

King, A.F., 1966. Structure and stratigraphy of the Upper Carboniferous Bude Sandstones, north Cornwall. *Proc. Ussher Soc.,* **1**, 229–32.

King, A.F., 1967. *Stratigraphy and Structure of the Upper Carboniferous Bude Formation, North Cornwall.* Ph.D. Thesis, University of Reading, UK.

King, A.F., 1971. Correlation in the Upper Carboniferous Bude Formation, north Cornwall. *Proc. Ussher Soc.,* **2**, 285–8.

Lovell, J.P.B., 1965. The Bude Sandstones from Bude to Widemouth. *Proc. Ussher Soc.,* **1**, 172–4.

Lovell, J.P.B., 1975. Correlation and sedimentology of Tom's Cove Shale, Bude Formation. *Proc. Ussher Soc.* 3, 264–70.

McKeown, M.C., Edmonds, E.A., Williams, M., Freshney, E.C. & Masson Smith, D.J., 1973. *Geology of the Country Around Boscastle and Holsworthy.* Memoirs of the Geological Survey of Great Britain. HMSO, London.

Melvin, J., 1976. *Sedimentological Studies in Upper Palaeozoic Sandstones near Bude, Cornwall and Walls, Shetland.* PhD Thesis, University of Edinburgh.

Melvin, J., 1986. Upper Carboniferous fine-grained turbiditic sandstones from southwest England: a model for growth in an ancient, delta-fed subsea fan. *J. Sed. Petrol.,* **56**, 19–34.

Melvin, J., 1987. Upper Carboniferous fine-grained turbiditic sandstones from southwest England: a model for growth in an ancient, delta-fed subsea fan – Reply to discussion by R. Higgs. *J. Sed. Petrol.*, **57**, 380–2.

Owen, D.E., 1934. The Carboniferous rocks of the north Cornish coast and their structures. *Proc. Geol. Assoc.*, **45**, 451–71.

Owen, D.E., 1950. Carboniferous deposits in Cornubia. *Trans. Roy. Geol. Soc. Cornwall*, **18**, 65–104.

Ramsbottom, W.H.C., 1970. Carboniferous faunas and palaeogeography of the south west England region. *Proc. Ussher Soc.*, **2**, 144–57.

Reading, H.G., 1963. A sedimentological comparison of the Bude Sandstones with the Northam and Abbotsham Beds of Westward Ho! *Proc. Ussher Soc.*, **1**, 67–9.

Reading, H.G., 1965a. Recent finds in the Upper Carboniferous of southwest England and their significance. *Nature*, **208**, 745–7.

Reading, H.G., 1965b. Xiphisurid trails from the Upper Carboniferous of Bude, north Cornwall. Discussion. *Proc. Geol. Soc. London*, **1626**, 164–5.

Sedgwick, A. & Murchison, R.I., 1840. On the physical structure of Devonshire, and on the subdivisions and geological relations of its older stratified deposits. *Trans. Geol. Soc. London*, **5**, 633–705.

Thomas, J.M., 1982. The Carboniferous rocks. In *The Geology of Devon.* ed. E.M. Durrance & D.J.C. Laming, pp. 42–65. Exeter: University of Exeter.

Thomas, J.M., 1988. Basin history of the Culm Trough of southwest England. In *Sedimentation in a Synorogenic Basin Complex: the Upper Carboniferous of Northwest Europe.* ed. B.M. Besly & G. Kelling, pp. 24–37. Blackie. Glasgow.

Tyler, D.J., 1988. Evidence and significance of limulid instars from trackways in the Bude Formation (Westphalian), south-west England. *Proc. Ussher Soc.*, **7**, 77–80.

Walker, R.G., 1963. Distinctive types of ripple-drift cross-lamination. *Sedimentology*, **2**, 173–88.

Whalley, J.S. & Lloyd, G.E., 1986. Tectonics of the Bude Formation – the recognition of northerly directed décollement. *J. Geol. Soc. London*, **143**, 83–8.

White, E.I., 1939. A new type of palaeoniscoid fish, with remarks on the evolution of the actinopterygian pectoral fins. *Proc. Zool. Soc. London*, **B109**, 41–61.

Upper Allegheny Group (Middle Pennsylvanian) lacustrine limestones of the Appalachian Basin, USA

SUZANNE D. WEEDMAN

Geosciences Department, Penn State University, University Park, PA 16802, USA

Introduction

The bulk of the Pennsylvanian System of the Appalachian Basin is preserved in a doubly-plunging synclinorium axially aligned from Pittsburgh, Pennsylvania, to Huntington, West Virginia, USA (Fig. 1). The four major stratigraphic units are the Pottsville, Allegheny, Conemaugh and Monongahela Groups, not easily subdivided into formal formations (Edmunds *et al.*, 1979). Overlying these rocks is the Dunkard Group, generally assigned, in part, to the Permian System (Barlow, 1975). The Pennsylvanian and Permian sediments are cyclic and predominantly non-marine, containing many commercial-quality bituminous coals. Lacustrine limestones make their first appearance in the coal cyclothems of the Pennsylvanian in middle Pennsylvanian time, under the Upper Kittanning coal of the Allegheny Group (Fig. 2). Above that horizon, freshwater limestones are a common constituent of the coal cyclothems, both in the relatively non-productive cyclothems of the Conemaugh Group, and in the productive coal cycles of the upper Allegheny and Monongahela Groups (Berryhill *et al.*, 1971; Edmunds *et al.*, 1979). However, very little work has been done on these unusual rocks (Adams, 1954; Williams *et al.*, 1968; Marrs, 1981; Weedman, 1988b). Freshwater limestones, when present, typically occur under coal beds, while marine limestones, if present, are found over the coal beds of the cyclothems.

Thin freshwater limestones interbedded with fluvial sediments, that are similar to those in the Pennsylvanian of the Appalachian Basin, have been reported from the Devonian of New York (Demicco, Bridge & Cloyd, 1987) and of Spitsbergen (Friend & Moody-Stuart, 1970), from the Pennsylvanian of Nova Scotia (Masson & Rust, 1983; Vasey & Zodrow, 1983), the Jurassic of Portugal (Wright, 1985), the Cretaceous of Spain (Gierlowski-Kordesch & Janofske, 1989; Platt, 1989), Cretaceous and Early Tertiary of France (Freytet & Plaziat, 1982) and Morocco (Herbig, 1988), and from the Tertiary of Montana and Wyoming of the United States (Flores, 1981). They are commonly associated with coals.

Geologic setting

The upper Allegheny Group of western Pennsylvania was deposited in late Middle Pennsylvanian time on an upper delta/alluvial plain (Williams *et al.*, 1968; Ferm, 1970) of the eastern North American foreland basin. Sediments infilling the basin were derived primarily from the mountains rising to the south (the Alleghanian orogeny) and are considered to be synorogenic (Donaldson & Schumaker, 1981; Quinlan & Beaumont, 1984). Pennsylvanian sediments have been zoned and correlated by blattoid insects (Durden, 1969), by flora (Read & Mamay, 1964; Darrah, 1969), and fish (Lund, 1970). The Pottsville and Allegheny Groups invertebrates have been classified by environment of deposition, that is, of marine, restricted or freshwater origin (Williams, 1960).

The upper Allegheny Group lacustrine limestones – the Johnstown, the Lower Freeport and the Upper Freeport – are virtually indistinguishable in the field or in core, and so, until further studied, are assumed to have a similar origin: they appear to have been deposited in broad shallow carbonate lakes and marshes, and are now completely encased by terrigenous fluvial and swamp sediments.

Case-study – western Pennsylvania

A study of the thickness distribution of the Upper Freeport sandstone and limestone in a small area in western Pennsylvania (see box labelled 'site 12' in Fig. 1), shows that the distribution and thickness of the limestone there appears to be controlled by the distribution and thickness of the underlying channel sandstone (Weedman, 1988b). Isopach maps made from core-derived data over an area of about 900 km² show that the thickest limestones overlie the sand-deficient areas between the thick channel sandstones, that is, the area interpreted as the floodplain. This offset relationship suggests that the limestone beds were floodplain lakes, as opposed to abandoned channel meanders – oxbow lakes. Over 90% of the five-quadrangle (approximately 160 km² per quadrangle) study area appears to be covered with the limestone, suggesting a composite lake area at its maximum of at least 720 km²

Thin silty, shale beds, interpreted as flood deposits, separate thicker limestone beds. Sandstone beds, however, are absent in the Upper Freeport limestone interval, which suggests that the lakes formed on abandoned or distal parts of the floodplain, far from sand deposition. The abundance of oncoids and 'algal' lamination

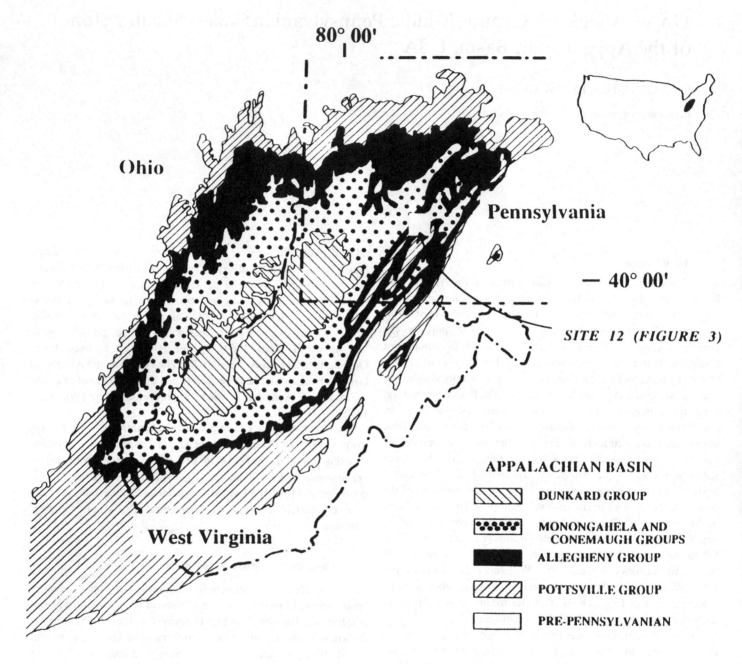

Fig. 1. Map of the Appalachian Basin, western Pennsylvania, USA, showing the distribution of Pennsylvanian and Permian rocks. Basin location shown in upper right on map of USA; location of study area and core site 12, from Fig. 3, shown by arrow and white box.

in the limestone suggests that the lakes were colonized by cyanobacteria; however, charophytes, commonly associated with such deposits, have not been reported. Additional fossils include numerous kinds of ostracodes (not yet studied), *Spirorbis* worm tubes, fish bones and plants. In eastern Ohio, the cannel coal that immediately overlies the Upper Freeport limestone has yielded an abundance of vertebrate fossils of reptiles, amphibians, fish and sharks (Hook, 1986; Hook & Ferm, 1988).

Facies sequence

A Markov chain analysis of lithofacies sequences of the upper Allegheny Group in Indiana and Armstrong Counties, Pennsylvania, has identified a statistically significant vertical sequence of: coal→black shale→dark sandy shale→light sandy shale→sandstone→silty claystone→claystone→limestone→claystone (underclay)→coal (Weedman, 1988b). No marine shales or limestones are reported in the Upper Allegheny Group cycles in the

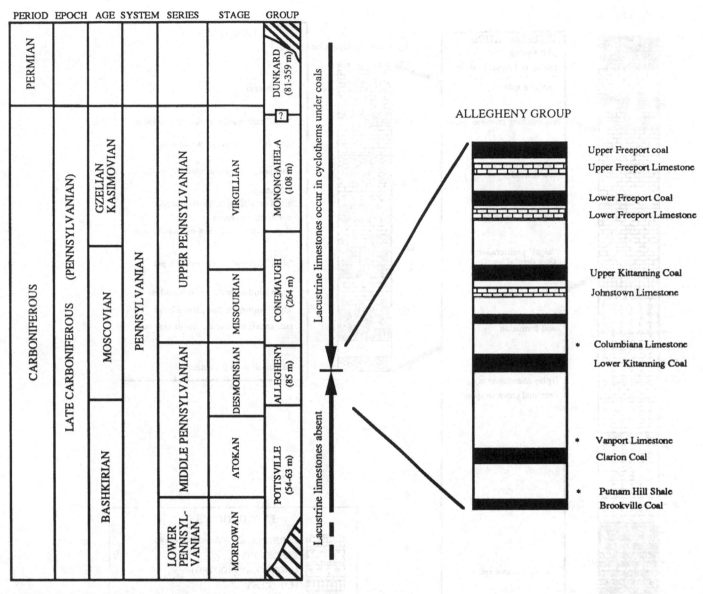

* Marine bed. All other strata are non-marine.

Fig. 2. Stratigraphic column of the Pennsylvanian System in the Appalachian Basin of western Pennsylvania, USA. The distribution of lacustrine limestones is shown with large arrows, and the Allegheny Group is expanded on the right. Thicknesses are given in meters under group names, and are for southwestern Pennsylvania (taken from COSUNA project data, Patchen, *et al.,* **1985). (Compiled from Berg** *et al.,* **1983; Busch & Rollins, 1984; Edmunds** *et al.,* **1979; Patchen** *et al.,* **1985; Phillips** *et al.,* **1985.)**

study area. This coal-to-coal cycle varies in thickness from 15–25 m (Fig. 3). The first part of the cycle, from the coal to the sandstone, is a coarsening-upward sequence. The sandstones are interpreted as the progradational deposits of overbank flows and crevasse splays. The next part of the cycle, from the sandstone to the claystone, is a fining-upward sequence. These sediments are interpreted as the floodplain deposits that accumulated after the abandonment of the main channel. Thick channel sandstones occur in some areas. However, for the Markov analysis, sites with thick sandstones (>6

m) were excluded, so that the analysis would be restricted, primarily, to floodplain processes.

Microfacies description

The Upper Freeport limestone, from examination in core, has been divided into seven calcareous microfacies: a coated grain microfacies (packstone), a massive micrite (wackestone), a laminated micrite, a clastic limestone (reworked limestone clasts in a

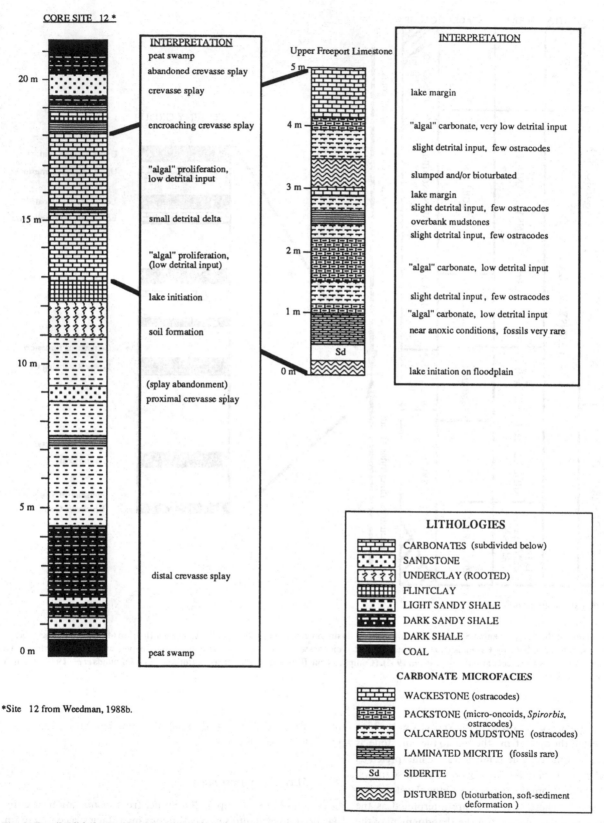

Fig. 3. Detailed lithologic column and interpretation of one core site (site 12, from Weedman, 1988b) of the coal cyclothem that contains the Upper Freeport Limestone, shown expanded on the right. The lithofacies and calcareous microfacies distribution in outcrop and core sites of this sequence is highly variable. This site is one of the thickest, and contains all of the calcareous microfacies with the exception of the clastic limestone and the matrix-supported breccia. See text for further description of microfacies and their sedimentary structures.

micrite matrix), a claystone–limestone disturbed microfacies, a matrix-supported breccia, and a calcareous claystone (Weedman, 1988b). The carbonate minerals present are calcite, dolomite, siderite, and ankerite.

The coated grain microfacies is a packstone of spherical to oblate sparry micro-oncoids (1.0–4.0 mm in diameter) of articulated ostracodes surrounded by radiating filaments (averaging 0.025 mm diameter and 0.5 mm length) in a dark brownish grey micrite matrix. Samples of the coated grain microfacies range from 80–94% calcite. The fauna preserved in this microfacies are ostracodes, *Spirorbis* (worm tubes), very small gastropods (~1 mm), and cyanobacterial filaments (oncoids). A dark (organic-rich?) micrite encases the cyanobacterial filaments and apparently helped protect and preserve their shape. Preservation of the filaments, as well as absence of reworking, suggest that this microfacies was perennially subaqueous.

Sediment injection structures, rare in limestones in general, are common in the coated grain microfacies. Their presence in the 'algal' microfacies is attributed to the probable presence of an algal mucilagious substance there (Schöttle & Müller, 1968; Weedman, 1988a), that may have reduced permeability in that facies to the water escaping from compacting sediments below. Dewatering, then, would be episodic as pore pressures in the underlying compacting shales increased until the confining pressure of the overlying mucilagenous beds was overcome (Swarbrick, 1968).

The micrite microfacies is predominantly massive with scattered peloidal pockets, and ranges in color from light grey to dark brownish grey, depending on calcite content. The highest calcite content (94–97% calcite) samples are very dark brown-grey, and the samples with a higher clay content (89–94% calcite) are light grey. The voids in the peloidal microfacies are filled with either sparry calcite or saddle dolomite. Both articulated and disarticulated ostracodes are abundant but *Spirorbis* is absent. Quartz silt or sand, siderite, pyrite and bone fragments are rare or absent in both the coated grain and micrite microfacies. It is not clear whether all of the micrite precipitated as a primary sediment in the lake, or if it formed as a result of exposure, oxidation and breakdown of the more organic-rich coated grain facies.

The laminated micrite microfacies contains laminae (1–10 mm thick) of light and dark brownish grey micrite, has a calcite content of 82–87%, is delicately microfaulted, and contains virtually no fauna with the exception of rare disarticulated ostracodes. This microfacies is commonly interbedded with a black shale. The absence of fauna or bioturbation, and the association with black shale suggests that this microfacies was deposited either during periods of oxygen depletion in the lake as a whole, or in areas of the lake or marsh where the sediments were deficient in oxygen. The distinctive style of deformation, that is, the delicate microfaulting, is probably due to uneven burial compaction after dewatering, and indicates early cementation of at least some of the laminae. The absence of sediment injection structures of the type seen in the coated grain microfacies, may be due to either slower deposition rates, to a greater permeability, or to the presence of other fluid escape paths along the microfault surfaces.

The clastic limestone microfacies is a poorly sorted but faintly bedded, clast-supported rock comprising clasts of limestone and claystone in a micrite matrix with highly variable clay content. This is the only calcareous microfacies with quartz silt and sand grains. Fossils present in this microfacies are oncoids (0.5–1.0 cm) and fragments of ostracodes, fish bones, *Spirorbis* and micro-oncoids (1.0–4.0 mm). Sediment injection (sudden water escape) structures are observed, but are rare in this microfacies. The variety of clasts, the presence of quartz sand and silt, and the lack of sorting suggests a shoreline environment where bottom sediments may have been periodically exposed and reworked.

The matrix-supported breccia microfacies, very common over the study area, consists of angular clasts of limestone or claystone in a non-bedded calcareous clay and/or micrite matrix. It differs from the clastic limestone in that it is not bedded, and that the clasts are usually of the same rock type, that is, it appears to have formed in place (Freytet & Plaziat, 1982). The fossil assemblage is similar to the coated grain and micrite microfacies. This microfacies is typically found between a micrite and an overlying silty shale, as a transition from limestone deposition to detrital deposition. The brecciated texture is gradational with the underlying micrite, but not with the overlying silt. Brecciation begins with curved fissuring within the micrite followed upward by complete matrix-supported 'clast' separation. This texture is similar to that resulting from autochthonous 'ooid' formation described by Freytet & Plaziat (1982), which is interpreted as a pedogenic process.

Additional evidence for a pedogenic origin of brecciation comes from the location in the depositional system of this microfacies. If these breccias are soils then they must occur in the appropriate locations for soils. As stated earlier, the breccia is nearly always found under a silty shale bed. The problem, then, is placing its formation before or after the deposition of the silty shale. If the breccia formed from subaereal exposure during periods of low lake level, then the overlying shale would be interpreted as the deposits made by re-flooding of the lake basin and the subsequent deepening of the lake.

However, an argument can also be made for breccia formation after the deposition of the silty shale, from the growth of vegetation on small detrital delta lobes built out into the lake. Because the silty shale intervals are too thick (0.5–1.0 m) to represent single flood events (Kesel *et al.*, 1974), they must be attributed to numerous flood deposits; each shale horizon then would be interpreted as a small lacustrine delta, formed by the periodic reactivation of an abandoned crevasse. In modern lacustrine deltas of similar setting and scale, for example, in the freshwater lakes of the Atchafalaya Basin of the Mississippi River Delta, the abandoned parts of the lacustrine deltas quickly become vegetated, and rooted (Tye, 1986). Therefore, the delta lobe is the expected site of rooting. The formation of brecciation is favored here because of the well-preserved state of the organic matter in the breccia and underlying micrite. However, this problem would be better attacked in outcrop than in core.

The disturbed claystone–limestone microfacies is a non-bedded swirled mixture of micrite and claystone. Rarely at the base, and commonly at the top of many of the limestone intervals there is a zone up to 30 cm thick of this texture. This microfacies is probably the result of several processes that cannot be distinguished in a 5.4 cm wide core. The disruptions of primary depositional structures of the sediments can be attributed to bioturbation of organisms or plants, to soft sediment deformation processes such as liquifaction and sudden pore water escape, or to slumping.

There is no significant vertical sequence of microfacies observed in the cores studied, such as a shallowing upward sequence; however, nearly all limestone sequences ended in a claystone overlain by a coal. The microfacies interpreted as shoreline or very shallow water origin (coated grain microfacies, matrix-supported breccia and clastic limestone) are common in most of the cores examined, at all levels in the core. This evidence suggests deposition in a broad shallow lake with fluctuating lake levels.

In general, the thinnest limestone intervals (1–2 m thick) studied in core consist almost entirely of the matrix-supported breccia or a clastic limestone. These poorly bedded rocks are interpreted as very shallow and/or shoreline facies, and the textures due in part to pedogenic processes, i.e., they would be palustrine limestones. However, sites where the limestone interval in the coal-to-coal cycle are thicker (3–5 m thick) comprise a variety of all of the microfacies described above, and show well-preserved bedding. The matrix-supported breccia and clastic limestone are absent in the very thickest limestone core examined (5.3 m), which suggests that this core was taken from near the middle of the lake, not intersecting the shoreline deposits.

Temporal distribution of limestone

Two hypotheses have been proposed for the sudden appearance of lacustrine limestones in the Pennsylvanian of the Appalachian Basin: (1) their occurrence is attributed, by some, to a change from a relatively wetter, to a relatively drier climate during middle Pennsylvanian time (Donaldson et al., 1985; Cecil et al., 1985) and (2) their occurrence is attributed to a decline in the amount of detritus reaching the drainage basin relative to the subsidence rate (Weedman, 1988b). The first argument is supported by the recognition of assumed climatically-controlled changes in the composition of coals and associated sediments. The underlying assumption is that as evaporation increased relative to precipitation, surface waters became more enriched in calcium, and calcite precipitation occurred. However, the climate-control argument does not recognize the abundance of oncoids and coated grains which suggest that much of the calcite was of biogenic origin; excessive supersaturation of calcium was not necessary for precipitation. Neither does this interpretation address the role of and controls on a fluctuating clastic supply to the basin, which must have been reduced to favor carbonate precipitation.

An alternative view is that lacustrine deposition was made possible by a changing fluvial style that favored the creation of lake basins in abandoned areas of the floodplains. The Markov chain analysis has suggested that the limestone was deposited during channel abandonment, and the isopach maps showed that the lakes formed between channel sandbodies, perhaps surrounded by

levees. A change from meandering to anastomosing fluvial system could be a mechanism for creating cyclic, sediment starved floodplain lakes. Such a change in fluvial style can be attributed to the sedimentologic response of the river to an increase in subsidence which exceeded the sedimentation rate (Weedman, 1988b). Studies of modern anastomosed rivers, show that an increase in basin subsidence leads to a reduction in gradient which favors the formation of prominent levees along river channels (vertical accretion), the deposition of shoe-string shaped sandbodies, the occurrence of frequent avulsion events, and ultimately to the development of an extensive wetland complex which blankets nearly all of the abandoned areas (Smith & Putnam, 1980). If there are carbonate rocks in the drainage basin to provide the necessary ions, the broad abandoned areas that receive little clastic supply after avulsion are ideal locations for the deposition of calcareous algal (cyanobacterial) carbonate in marsh and lake environments.

While the classification of rivers into meandering and anastomosing types is fairly straight forward in modern environments it is far more difficult in ancient environments, especially in areas like the Appalachian Basin with so few outcrops. However, Flores (1981) has shown, in well-exposed Paleocene lacustrine limestones in Wyoming and Montana, that the appearance of freshwater limestone in the upper portions of the Tongue River Member coincides roughly with a change in fluvial style from meandering to anastomosed. This change in fluvial style, inferred by sandbody shape, is interpreted by Flores as the response of the fluvial system to an increase in basin subsidence.

Modern analogue

An ideal modern analogue for the upper Allegheny Group of sediment would be an equatorial foreland basin, with calcareous rocks in the source area, and a broad plain of wetlands extending from the thrust belt all the way to distal parts of the basin where deltas build out into marine waters. The Mississippi River Delta has often been cited for a modern analogue for the clastic sediments of the Appalachian Basin (Coleman et al., 1969; Ferm, 1970) but in Great Britain (Fielding, 1984), however, it is not appropriate for the coals or the freshwater or marine limestones. Several modern analogues have been suggested for the lacustrine limestones of the Pennsylvanian of the Appalachian Basin in recent years, such as Sucker Lake, Michigan (Treese & Wilkinson, 1982) and the Florida Everglades (Gleason, 1972). However, neither the Everglades, Michigan, nor the Mississippi River Delta is in a foreland basin, nor are they equatorial. Perhaps a better candidate for a modern analogue for the Upper Allegheny Group depositional setting is the Magdalena Basin of Colombia: it is an equatorial foreland basin, and is also a peat forming environment with anastomosed rivers and extensive wetlands (Smith, 1986). However, very little is known of the chemistry of the lakes and marshes there.

Conclusion

In summary, the thin lacustrine limestones of the Appalachian Basin are part of the coal cyclothems, occurring below many of the coals of the upper half of the Pennsylvanian system. The deposition of calcite in the floodplain lakes occurred during periods of extremely low detrital influx, perhaps after avulsion of the channel draining the plain. The dark, organic rich nature of most of the limestones, as well as the excellent preserved state of the cyanobacterial filaments, suggests that these limestones are predominantly of biogenic rather than inorganic origin, and were perennially subaqueous.

The temporal distribution of lacustrine limestones in the Appalachian Basin has been interpreted in two ways: one hypothesis is based on the need for an adequate calcium supply in lake water (nearly always theoretically sufficient for biogenic carbonate production), and the other is based on the need for a cyclic mechanism for lake basin formation. Unfortunately, due to poor exposures of the appropriate stratigraphic intervals, as well as the rarity of modern analogues for the Pennsylvanian coal swamps, truly conclusive evidence for either hypothesis is not yet available.

Acknowledgement

Support for this research was provided by a Pennsylvania Mining and Mineral Resources Research Institute Fellowship granted by the Department of the Interior's Mineral Institutes program, administered by the Bureau of Mines grant number G1164142. Additional funds were provided by the Krynine Fund of the Geosciences Department of Penn State University and from a Sigma Xi Grant-in-Aid. Cores and stratigraphic data were provided by R & P Coal Company, Indiana, Pennsylvania.

Bibliography

Adams, H.J., 1954. A study of freshwater limestones in the Conemaugh and Monongahela series, Pittsburgh area. MS thesis, Univ. Pittsburgh, Pittsburgh.

Barlow, J.A. (ed.), 1975. *The Age of the Dunkard – Proceedings of the first I.C. White Memorial Symposium, Morgantown, West Virginia.* West Virginia, Geological and Economic Survey, Morgantown.

Berg, T.M., McInerney, M.K., Way, J.H. & MacLachlan, D.B., 1983. *Stratigraphic Correlation Chart of Pennsylvania.* General Geology Report 75. Pennsylvania Topographic and Geologic Survey, Harrisburg.

Berryhill, H.L., Jr., Schweinfurth, S.F. & Kent, B.H., 1971. *Coal-Bearing Upper Pennsylvanian and Lower Permian rocks, Washington area, Pennsylvania.* US Geol. Surv. Prof. Paper 621.

Busch, R.M. & Rollins, H.B., 1984. Correlation of Carboniferous strata using a heirarchy of transgressive-regressive units. *Geology,* 12, 471–4.

Cecil, B.C., Stanton, R.W., Neuzil, S.G., Dulong, F.T., Ruppert, L.F. & Pierce, B.S., 1985. Paleoclimate controls on late Paleozoic sedimentation and peat formation in the central Appalachian Basin (USA). *Int. J. Coal Geology,* 5, 195–230.

Coleman, J.M., Ferm, J.C., Gagliano, S.M., McIntire, W.G., Morgan, J.P. & Van Lopik, J.R., 1969. *Recent and Ancient Deltaic Sediments, a Comparison.* Coastal Studies Institute, Louisiana State Univ., Baton Rouge.

Demicco, R.V., Bridge, J.S. & Cloyd, K.C., 1987. A unique freshwater carbonate from the Upper Devonian Catskill magnafacies of New York State. *J. Sed. Petrol.,* 57, 327–34.

Donaldson, A.C., Renton, J.J. & Presley, M.W., 1985. Pennsylvanian depositems and paleoclimates of the Appalachians. *Int. J. Coal Geol.,* 5, 167–93.

Donaldson, A.C. & Schumaker, R.C., 1981. Late Paleozoic molasse of central Appalachians. In *Sedimentation and Tectonics in Alluvial Basins,* ed. A.D. Miall, pp. 99–124. Geological Assoc. of Canada, Spec. Paper 23.

Darrah, W.C. 1969. *A Critical Review of the Upper Pennsylvania Floras of the Eastern United States with Notes on the Mazon Creek Flora of Illinois.* (Privately published.)

Durden, C.J., 1969. Pennsylvanian correlation using blattiod insects. *Can. J. Earth Sci.,* 6, 1159–77.

Edmunds, W.E., Berg, T.M., Sevon, W.D., Piotrowski, R.C., Heyman, L. & Richard, L.V., 1979. *The Mississippian and Pennsylvanian (Carboniferous) Systems in the United States – Pennsylvania and New York.* US Geological Survey Prof. Paper 1110-B, B1-33.

Ferm, J.C., 1970. Allegheny deltaic deposits. In *Deltaic Sedimentation, Modern and Ancient,* ed. J.P. Morgan, pp. 246–55. Soc. Econ. Mineralogists and Paleontologists Spec. Pub. 15.

Fielding, C.R., 1984. Upper delta plain lacustrine and fluviolacustrine facies from the Westphalian of the Durham coalfield, NE England. *Sedimentology,* 31, 547–67.

Flores, R.M., 1981. Coal deposition in fluvial paleoenvironments of the Paleocene Tongue River Member of the Fort Union Formation, Powder River Area, Powder River Basin, Wyoming and Montana. In *Recent and Ancient Non-marine Depositional Environments.* ed. R.M. Flores & F.G. Etheridge. pp. 169–90. Int. Assoc, Sedimentol. Spec. Pub. 6.

Freytet, P. & Plaziat, J.-C., 1982. *Continental carbonate sedimentation and pedogenesis – Late Cretaceous and Early Tertiary of southern France.* Contributions to Sedimentology vol. 12. E. Schweizerbart'sche Verlagsbuchhandlung, Stuttgart.

Friend, P.F. & Moody-Stuart, M., 1970. Carbonate deposition on the river floodplains of the Wood Bay Formation (Devonian) of Spitsbergen. *Geol. Mag.,* 107, 181–95.

Gierlowski-Kordesch, E. & Janofske, D., 1989. Paleoenvironmental reconstruction of the Weald around Uña (Serranía de Cuenca, Cuenca Province, Spain). In *Cretaceous of the Western Tethys, Proceedings 3rd International Cretaceous Symposium, Tübigen 1987,* ed. J. Wiedmann, pp. 239–64. E. Schweizerbart'sche Verlagsbuchhandlung.

Gleason, P.J., 1972. *The Origin, Sedimentation, and Stratigraphy of a Calcitic Mud Located in the Southern Freshwater Everglades.* PhD Dissertation, The Pennsylvania State University, Pennsylvania.

Herbig, H.-G., 1988. The Upper Cretaceous to Tertiary Hammada West of Errachidia (SE Morocco): A continental sequence involving paleosol development: *N. Jb. Geol. Paläont. Abh.,* 176 (2), 187–212.

Hook, R.W., 1986. *A Paleoenvironmental Model for the Occurence of Vertebrate Fossils in Carboniferous Coal-bearing Stata.* PhD Dissertation, Univ. Kentucky, Lexington.

Hook, R.W. & Ferm, J.C., 1988. Paleoenvironmental controls on vertebrate-bearing abandoned channels in the Upper Carboniferous. *Palaeogeogr., Palaeoclimat., Palaeoecol.,* 63, 159–82.

Kesel, R.H., Dunne, K.C., McDonald, R.C., Allison, K.R. & Spicer, B.E., 1974. Lateral erosion and overbank deposition on the Mississippi River in Louisiana caused by 1973 flooding. *Geology,* 2, 461–4.

Lund, R., 1970. Fossil fishes from southwestern Pennsylvania: Part 1: Fishes of the Duquesne Limestone (Conemaugh Pennsylvania). *Ann. Carneg. Mus.,* 41, 231–61.

Marrs, T.O., 1981. *Lithologic Characteristics and Depositional Environments of the Non-marine Benwood Limestone (Upper Pennsylvanian) in the Dunkard Basin, Ohio, Pennsylvania, and West Virginia.* MS Thesis, Univ. Pittsburgh, Pittsburgh.

Masson, A.G. & Rust, B.R., 1983. Lacustrine stromatolites and algal laminates in a Pennsylvanian coal-bearing succession near Sydney, Nova Scotia. *Can. J. Earth Sci.,* 20, 1111–18.

Patchen, D.G., Avary, K.L. & Erwin, R.B., 1985. *Correlation of Stratigraphic Units of North America (COSUNA) Project. Northern*

Appalachian Region. American Association of Petroleum Geologists Correlation Chart.

Phillips, T.L., Peppers, R.A. & Dimichele, W.A., 1985. Stratigraphic and interregional changes in Pennsylvanian coal-swamp vegetation. Environmental inferences. *Int. J. Coal Geol.*, **5**, 43–109.

Platt, N.H., 1989. Lacustrine carbonates and pedogenesis: sedimentology and origin of palustrine deposits from the Early Cretaceous Rupelo Formation, W. Cameros Basin, N. Spain. *Sedimentology*, **36**, 665–84.

Quinlan, G.M. & Beaumont, C., 1984. Appalachian thrusting, lithospheric flexure, and the Paleozoic stratigraphy of the Eastern Interior of North America. *Can. J. Earth Sci.*, **21**, 973–96.

Read, C.B. & Mamay, S.H., 1964. *Upper Paleozoic Floral Zones and Floral Provinces of the United States.* US Geol. Surv. Prof. Paper 454-K.

Schöttle, M. & Müller, G., 1968. Recent carbonate sedimentation in the Gnadensee (Lake Constance), Germany. In *Recent Developments in Carbonate Sedimentology in Central Europe*, ed. G. Müller & G.M. Friedman. Springer-Verlag, New York.

Smith, D.G., 1986. Anastomosing river deposits, sedimentation rates and basin subsidence, Magdalena River, Northwestern Colombia, South America. *Sed. Geol.*, **46**, 177–96.

Smith, D.G. & Putnam, P.E., 1980. Anastomosed river deposits: modern and ancient examples in Alberta, Canada. *Can. J. Earth Sci.*, **17**, 1396–406.

Swarbrick, E.E., 1968. Physical Diagenesis: intrusive sediment and connate water. *Sed. Geol.* **2**, 161–75.

Treese, K.L. & Wilkinson, B.H., 1982. Peat-marl deposition in a Holocene paludal-lacustrine basin – Sucker Lake, Michigan. *Sedimentology*, **29**, 375–90.

Tye, R.S., 1986. Non-marine Atchafalaya Deltas: processes and products of interdistributary basin alluviation, south-central Louisiana. PhD Dissertation, Louisiana State University, Baton Rouge.

Vasey, G.M. & Zodrow, E.I., 1983. Environmental and correlative significance of a non-marine algal limestone, (Westphalian D), Sydney Coalfield, Cape Breton, Nova Scotia. *Marit. Sediments and Atlantic Geol.*, **19**, 1–10.

Weedman, S.D., 1988a. Physical diagenesis in Pennsylvania lacustrine limestones, Appalachian Basin. *Geol. Soc. Am. Abstracts with Programs*, **20**(7), A52.

Weedman, S.D., 1988b. *The Depositional Environment and Petrography of the Upper Freeport Limestone in Armstrong and Indiana Counties, Pennsylvania.* PhD Dissertation, Pennsylvania State University, Pennsylvania.

Williams, E.G., 1960. Marine and freshwater fossiliferous beds in the Pottsville and Allegheny Groups of western Pennsylvania. *J. Paleontol.*, **34**, 908–22.

Williams, E.G., Bergenback, R.E. & Weber, J.N., 1968. Relationship between paleotopography and the thickness and geochemistry of a Pennsylvanian freshwater limestone. *J. Sed. Petrol.*, **38**, 501–9.

Wright, V.P., 1985. Algal marsh deposits from the upper Jurassic of Portugal. In *Paleoalgology: Contemporary Research and Applications*; ed. D.F. Toomey & M.H. Nitecki, pp. 330–341. Springer-Verlag, New York.

Carboniferous lacustrine deposits from the Paganzo Basin, Argentina

LUIS ALBERTO BUATOIS,[1] CARLOS O. LIMARINO[2] AND SILVIA CÉSARI[2]

[1] Facultad de Ciencias Naturales e Instituto Miguel Lillo, Universidad Nacional de Tucumán, Casilla de correo 1, 4000 San Miguel de Tucumán, Argentina
[2] Facultad de Ciencias Exactas y Naturales, Universidad de Buenos Aires, Dpto. Geología, Pabellón II, Ciudad Universitaria, 1428 Buenos Aires, Argentina

Late Carboniferous lacustrine deposits cover about 50,000 km² of Catamarca, San Juan and La Rioja Provinces in western and northwestern Argentina. They are part of the Paganzo Basin (Azcuy & Morelli, 1970) which comprises 3000 m of non-marine strata of Carboniferous to Permian age. The main lacustrine interval is towards the base of the column (Namurian–Westphalian), although local younger deposits (latest Carboniferous–Early Permian) of similar origin have been recognized towards the eastern side of the basin. The basal lacustrine strata are described here.

The basin is limited to the west by a positive relief called the 'Protoprecordillera' (Fig. 1). Several internal arches divide the basin into different sub-basins. The Paganzo Basin is thought to be a foreland basin related to the subduction of the oceanic Pacific Plate beneath the western continental margin of Gondwana (Ramos, 1988). The lacustrine deposits show features of glacial sedimentation, probably related to the final episode of Gondwana glaciation (Limarino & Césari, 1988).

Five major depositional areas are shown in Fig. 1: the Narváez, Famatina, Malanzán, Guandacol and Jejenes areas. Lacustrine rocks are located at or near the base of the Lower Section (Paganzo I) of the Paganzo Group (Fig. 2).

The maximum thickness of lacustrine sequences is reported from the Guandacol depositional area (926 m thick sequence is estimated at the Agua de la Peña locality; Bossi & Andreis, 1985) (Fig. 3). Sedimentation began with massive sands and diamictites in the inner sublacustrine fan with muds and turbidites of deep-lacustrine origin. Slumps and dropstones are typical features. Later, a delta prograded over the fan-slope complex.

In the Narváez depositional area, a sublacustrine fan related to a prograding delta developed, including a coarsening and thickening-upward sequence from the lake center to sand flat, shallow channels and shoreline bar deposits (Fig. 4). Underflowites (underflow current deposits), low density turbidites, Bouma sequence turbidites and turbidites affected by fluidization and/or liquefaction are very common (Buatois & Mángano, 1990a). Shallow turbidites probably record storm reworking. Underflowites are highly bioturbated with a high-diversity trace fossil assemblage (Buatois & Mángano, 1990b).

The Famatina depositional area is similar in many aspects to the Guandacol and Narváez area (see Fig. 3). Underflow and turbidity currents, and ice-rafting, seem to be the dominant depositional processes (Limarino, Gutierrez, & Césari, 1984; Limarino, 1985; Limarino & Césari, 1988). Marginal deposits are composed of shoreline bars and small deltas.

The Malanzán sequences comprise the smallest depositional area, interpreted as a narrow east to west paleovalley (Andreis et al., 1986). Lacustrine deposits are transgressive over distal alluvial fan deposits (Andreis & Bossi, 1981; Andreis et al., 1986). Turbidites are common deposits and ice-related structures have been observed (Andreis & Bossi, 1981).

In the Jejenes depositional area, interbedded sandstones and siltstones are predominant, and conglomerates are locally abundant. The presence of dropstones indicates glacial processes (Cuerda & Furque, 1983). Sole structures are common features in the turbidite beds (Bercowski, 1985).

Lake deposits commonly occur under fluvial coal beds (Fig. 3) containing plant remains of Late Carboniferous age (Archangelsky et al., 1987). Palynological assemblages have been collected from lacustrine deposits (Azcuy, 1975; Césari & Limarino, 1987; Césari & Vázquez Nístico, 1988; Limarino & Gutierrez, 1990). Cristatisporites-like spores with monosaccate pollen grains dominate the palynological flora in the lacustrine deposits (Césari & Limarino, 1987) with rare small lycopods, Bumbudendron versiforme Gutierrez, Césari, & Limarino and Malanzania nana Archangelsky, Azcuy & Wagner (Gutierrez, 1984; Andreis et al., 1986; Césari et al., 1989).

Lacustrine conditions dominated the western margin of Argentina during early Late Carboniferous times; the deposits are clearly of glaciolacustrine origin. Glacial features include: (1) close association of lake deposits with diamictites interpreted as tillites (Limarino & Gutierrez, 1990), (2) very common occurrences of dropstones in most outcrops, (3) similar stratigraphic position of the lacustrine deposits throughout the Paganzo Basin, (4) existence of short sequences which contain lamination resembling varves and (5) the common presence of lake turbidites.

Lakes probably developed in response to a base-level rise related to a deglaciation event. During these times, the lake margin was coincident with the basin margin. At a later stage, as a result of erosion of mountain chains, fluvial systems were formed. These

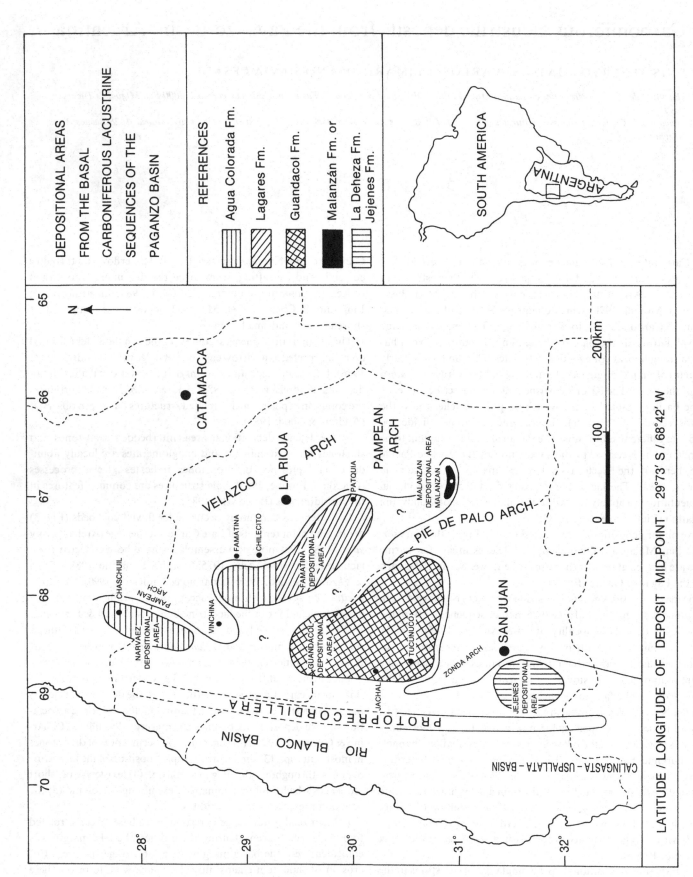

Fig. 1. Locality map and depositional areas of the Carboniferous basal lacustrine deposits of the Paganzo Basin, Argentina.

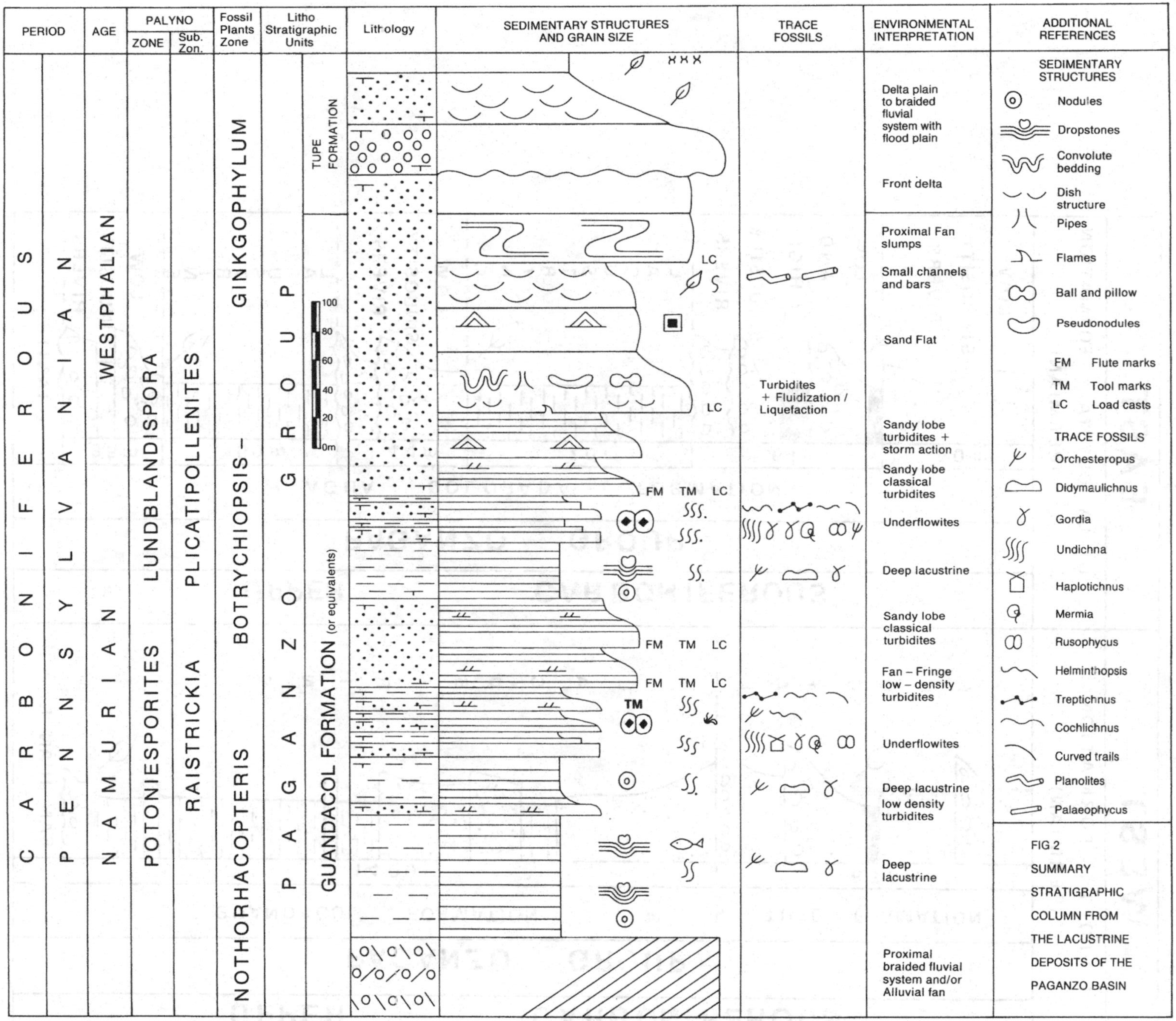

Fig. 2. Summary stratigraphic column of the Carboniferous lacustrine deposits. Palynozones and fossil plant zones from Limarino & Césari (1988). Reliability Index: B.

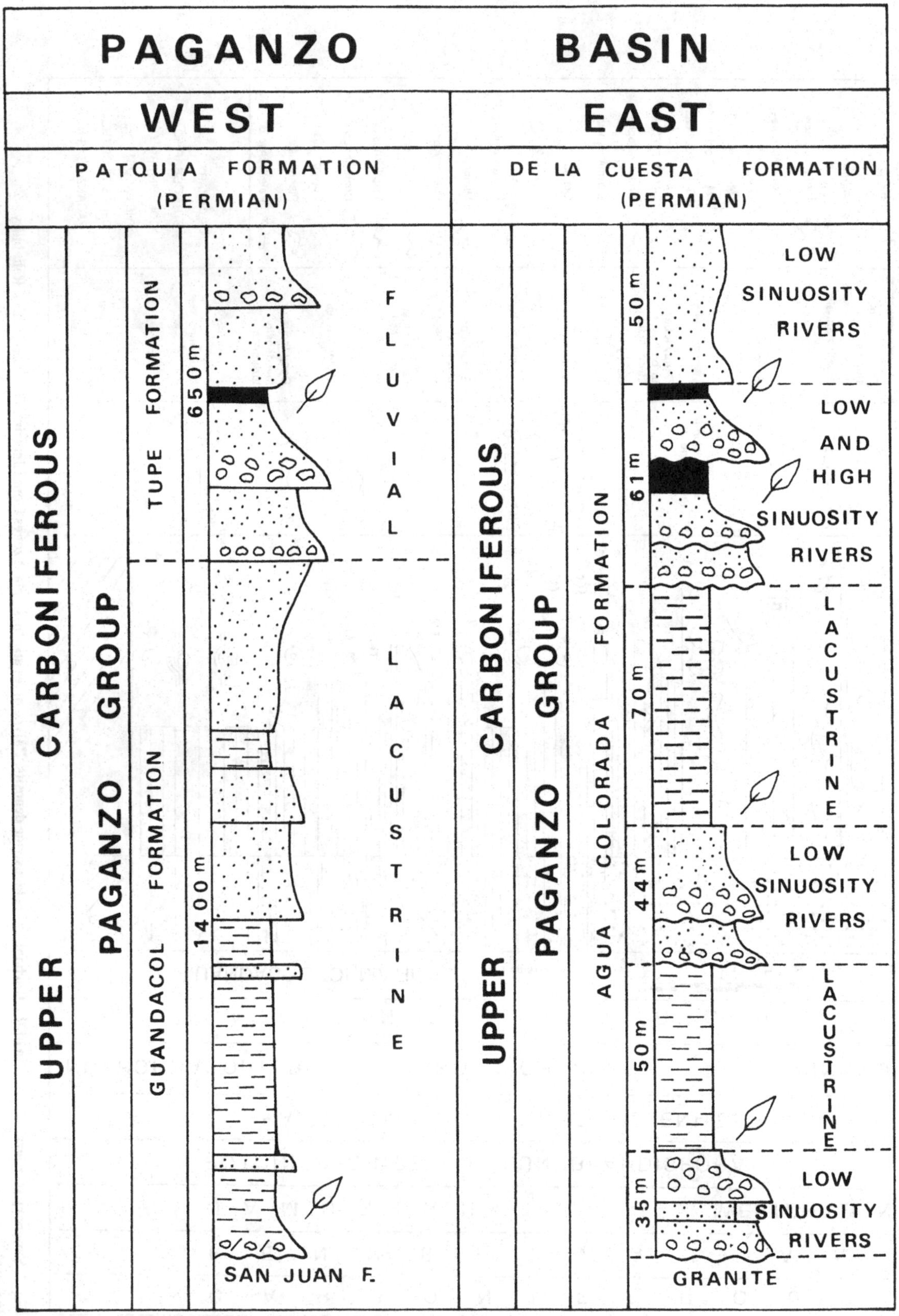

Fig. 3. Schematic vertical profile showing lacustrine deposits from the western Paganzo Basin (Guandacol Formation) and the eastern Paganzo Basin (Agua Colorada Formation).

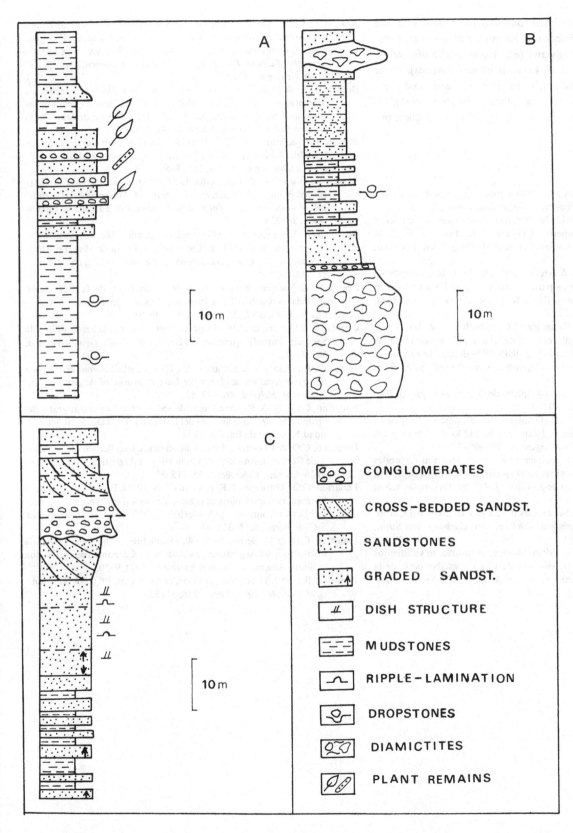

	CONGLOMERATES
	CROSS-BEDDED SANDST.
	SANDSTONES
	GRADED SANDST.
	DISH STRUCTURE
	MUDSTONES
	RIPPLE-LAMINATION
	DROPSTONES
	DIAMICTITES
	PLANT REMAINS

Fig. 4. Schematic profiles showing lacustrine lithofacies indentified in the Agua Colorada Formation: (A) laminated siltstones and claystones containing dropstones associated with conglomerates and fine sandstones, (B) thin beds of intercalated mudstones and sandstones associated with sandstones and diamictites, and (C) interlayered mudstones and sandstones (with sole marks).

rivers fed deltas of variable dimensions that prograded over the lakes. Comparable Upper Palaeozoic lacustrine deposits are present in different parts of Gondwana (e.g. latest Carboniferous or Early Permian of Brazil and Early Permian of South Africa). These sequences show a striking similarity in terms of facies and trace fossils with the Paganzo Basin (Aceñolaza & Buatois, 1991), although the Argentinian glaciolacustrine deposits are unquestionably more ancient.

Bibliography

Aceñolaza, F.G. & Buatois, L.A., 1993. Non-marine perigondwanic trace fossils from the Late Paleozoic of Argentina. *Ichnos*, **2**, 183–201.

Andreis, R.R. & Bossi, G.E., 1981. Algunos ciclos lacustres en la Formación Malanzán (Carbónico superior) en la región de Malanzán, Sierra de los Llanos, Provincia de La Rioja. *Actas VIII Cong. Geol. Argentina, San Luis*, **4**, 639–55.

Andreis, R.R., Leguizamón, R. & Archangelsky, S., 1986. El paleovalle de Malanzán: nuevos criterios para la estratigrafía del Neopaleozoico de la Sierra de Los Llanos, la Rioja, República Argentina. *Bol. Acad. Nac. Cienc.*, **27**, 3–119.

Archangelsky, S., Azcuy, C.L., Gonzalez, C.R., Sabattini, N. & Aceñolaza, F.G., 1987. Paleontología, bioestratigrafía y paleoecología de las Cuencas Paganzo, Calingasta-Uspallata y Río Blanca. In *El Sistema Carbonífero en la República Argentina, Córdoba*, ed. S. Archangelsky, pp. 133–52.

Azcuy, C.L., 1975. Palinología estratigráphica de la Cuenca Paganzo. *Rec. Asoc. Geol. Argentina*, **30**, 104–9.

Azcuy, C.L. & Morelli, J.R., 1970. Geologíca de la comarca Panganzo-Amaná. El Grupo Paganzo. Formaciones que lo componen y sus relaciones. *Rev. Asoc. Geol. Argentina*, **25**, 405–29.

Bercowski, F., 1985. Estructuras sedimentarias y características texturales de las sedimentitas carbónicas de la Quebrada de Las Lajas, Sierra Chica de Zonda, Provincia de San Juán. *I. J. Geol. Precordillera, San Juán*, **I**, 253–8.

Bossi, G. & Andreis, R., 1985. Secuencias deltaicas y lacustres del Carbónico del centro-oeste argentina. *Xth Cong. Int. Carboniferous Strat., Madrid, 1983*, **3**, 285–309.

Buatois, L.A. & Mángano, M.G., 1990a. Flujos gravitatorios de sedimentos en el Carbonífero lacustre del área de Los Jumes, Noroeste de la Sierra de Narváez, Catamarca. *II Reun. Argentinas Sedim., San Juán, Actas*, 66–71.

Buatois, L.A. & Mángano, M.G., 1990b. Una asociación de trazas fósiles del Carbónico lacustre del área de Los Jumes, Catamarca, Argentina: su comparación con la icnofacies de *Scoyenia*. *V Congress Argentino Paleont. Bioestrat., San Miguel de Tucumán, Serie Correlación Geológica*, **7**, 77–81.

Buatois, L.A. & Mángano, M.G. 1992. Abanicos sublacustres, abanicos submarinos o plataformas glacimarinas? Evidencias icnologícas para una interpretación paleoambiental del Carbonífero de la Cuenca Paganzo. *Ameghiniana*, **29** (4), 323–35.

Buatois, L.A. & Mángáno, M.G. 1993. Trace fossils from a Carboniferous turbiditic lake: implications for the recognition of additional non-marine ichnofacies. *Ichnos*, **2**, 237–58.

Césari, S., Brussa, E. & Benedetto, L., 1989. *Malanzania nana* Archangelsky, Azcuy, & Wagner, en sedimentitas de la Formación Guandacol, al oeste del Cerro del Fuerte, Provincia de San Juán. *Ameghiniana*, **26**, 225–8.

Césari, S. & Limarino, C., 1987. Análisis estratigráfico del perfil de la Quebrada de la Cortadera (Carbonífero), Sierra de Maz, La Rioja, Argentina. *IV Cong. Latinoamer. Paleont. (Bolivia), Actas I*, 217–33.

Césari, S. & Vázquez Nístico, B., 1988. Palinología de la Formación Guandacol (Carbonífero), Provincia de San Juán, Repúblico Argentina. *Rev. Española Micropal.*, **20**(1), 39–58.

Cuerda, A. & Furque, G., 1983. Depósitos carbónicos de la Precordillera de San Juán. Parte II. Quebrada La Deheza. *Rev. Asoc. Geol. Argentina*, **38**, 381–91.

Gutierrez, R., Césari, S. & Limarino, C., 1984. *Bumbudendron versiforme* a new lycophyte species from the Late Paleozoic of Argentina. *Rev. Palaeobot. Palynol.*, **46**, 377–86.

Limarino, C.O., 1985. Paleoambiente de sedimentación y estratigrafia del Grupo Paganzo en el Sistema del Famatina. Tesis Doctoral, Universidad Nacional de Buenos Aires.

Limarino, C.O. & Césari, S., 1988. Paleoclimatic significance of the lacustrine Carboniferous deposits in northwest Argentina. *Palaeogeogr., Palaeoclimat., Palaeoecol.*, **65**, 115–31.

Limarino, C.O., Gutierrez, P.R. & Césari, S.N., 1984. Facies lacustres de la Formación Agua Colorada (Paleozoico superior): aspectos sedimentológicos y contenido paleoflorístico. *IXth Cong. Geol. Argentina, S.C. de Bariloche*, **5**, 324–41.

Limarino, C.O. & Gutierrez, R., 1990. Diamictites in the Agua Colorada Formation, nw Argentina: new evidence of Carboniferous glaciation in South America. *J. S. Am. Earth Sci.*, **3** (1), 9–20.

Ramos, V.R., 1988. The tectonics of the Central Andes, 30° to 33° S latitude. *Geol. Soc. Am. Spec. Paper*, **218**, 31–54.

Permo–Triassic to Jurassic

Late Triassic lacustrine deposits from the Middle European Keuper Basin: a Lower Middle Keuper sequence in northwest Germany

NORBERT HAUSCHKE

Department of Geology and Palaeontology, University of Münster, Corrensstrasse 24, D-48149 Münster, Germany

During the Triassic, the evolution of western and central Europe with the development of a series of troughs and grabens is closely related to the initial phase of the break-up of Pangaea (Ziegler, 1982). The Middle European Keuper Basin (Fig. 1) was created by crustal extension. Subsidence and sedimentation were generally balanced throughout the evolution of the basin; facies distribution and isopachs evidence these subsidence patterns (Schröder, 1982).

Fig. 1. Paleogeography of the Middle European Keuper Basin for the Ladinian to Carnian (Gipskeuper, km 1), with study area enclosed. (Redrawn from Ziegler, 1982: Fig. 5.)

The Lower Middle Keuper (Gipskeuper, km 1) is the lower unit of the Middle Keuper (Late Triassic). In the study area (Lippisches Bergland, northwest Germany; see Fig. 1), as well as in other areas of the Middle European Keuper Basin, the Lower Middle Keuper is composed of a pelitic-evaporitic redbed sequence. This sequence is stratigraphically enclosed by predominantly psammitic units: the Lower Keuper (ku) and the Schilfsandstein (km 2), both of which are interpreted as deltaic/fluvial deposits (Wurster, 1964b, 1972; Dittrich, 1989). Clastic sources were from northern and eastern directions (Fig. 1).

In the study area (SA; Fig. 1), the Lower Middle Keuper is about 140–150 m in thickness. The lack of index fossils does not allow a biostratigraphical zonation. Palynological data indicate a Ladinian to Carnian age (Hauschke & Heunisch, 1989, 1990; Schröder, 1982). Correlations are mainly based on lithostratigraphy (e.g. Duchrow 1968a, b, 1984). The Lower Middle Keuper in the study area is interpreted as predominantly lacustrine in origin; in the uppermost part, a change to fluvial conditions occurs (Hauschke, 1988). Evidence of marine ingressions is present in the southern part of the Keuper Basin (e.g. Richter, 1985; Aigner & Bachmann, 1989), but it is lacking in the study area.

Characteristic for the Lower Middle Keuper in the study area are the occurrence of two distinct cycles within the Ladinian and Carnian sequences respectively (Fig. 2). The Ladinian cycle is composed of three facies types (from bottom to top) (Fig. 2; column 3a): A. Interbeds of siltstone and claystone. Claystone is commonly laminated and greenish-grey or red in color. Siltstone contains horizontal lamination, ripple cross-lamination (current and wave) and deformation structures (convolute lamination, fill structures above and below halite crystal marks, and textural alterations caused by early diagenetic growth of evaporitic minerals). B. Dolomitic mudstone. This mudstone is mostly red-colored with some mottling. It is massive and contains mudcracks and desiccation breccias. C. Dolomitic mudstone with intercalations of nodular residual rocks (primary Ca-sulfates). In the upper portions, intergrowth of nodules build vertical columns. This cycle, A–C, is interpreted as playa–playa lake deposits (Hauschke, 1987, 1989). The interbeds of siltstone and claystone represent distal sheetflood sediments, whereas the dolomitic claystone resulted from suspension settling in the ponded flood waters of a playa lake. As the playa

143

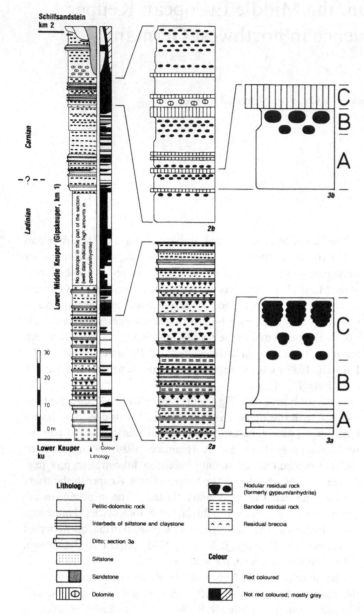

Fig. 2. Vertical succession of the Lower Middle Keuper in the study area (1). Close-ups show cyclically developed sections (2a, 2b), and individual cycles (3a, 3b). See also text.

lake dried up, early diagenetic processes caused alterations within the sediment. The intrasedimentary growth of hopper-shaped halite cubes and nodular Ca-sulfates occured from brines diffusing upward. A hot, dry climate with repeated flooding and desiccation of the playa basin is postulated for the Ladinian.

The Carnian cycle (Fig. 2, column 3b) is distinctly different; it is restricted to a 10–11 m-thick sequence completely grey in color. The three facies types from bottom to top include: A. Marlstone. This facies is mostly finely laminated with rare fish remains. Some intraformational pebbles are present at its base. B. Marlstone. This facies is massive with intercalations of nodular residual rocks (primary Ca-sulfates). C. Dolomite. This facies is mostly finely laminated. This cycle, A–C, is interpreted as a perennial lake

(Hauschke & Heunisch, 1989) in which a more regular water input prevented complete desiccation. The cyclicity probably reflects lake level changes and a general shallowing upward tendency is observed in each cycle and in the entire Carnian sequence as a whole.

Bibliography

These references include more comprehensive information about the Keuper and Late Triassic evolution in Europe.

Aigner, T. & Bachmann, G.H., 1989. Dynamic stratigraphy of an evaporite to redbed sequence, Gipskeuper (Triassic), southwest German Basin. *Sediment. Geol.*, **62**, 5–25.

Aigner, T., Bachmann, G.H. & Hagdorn, H., 1990. Zyklische Stratigraphie und Ablagerungsbedingungen von Hauptmuschelkalk, Letten-keuper und Gipskeuper in Nordost-Württemberg (Exkursion E). *Jber. Mitt. oberrh. geol. Ver.*, **72**, 125–43.

Bachmann, G.H. & Wild, H., 1976. Die Grenze Gipskeuper/Schilfsandstein (Mittlerer Keuper) bei Heibronn/Neckar. *Jber. Mitt. oberrh. geol. Ver.*, **58**, 137–52.

Bergh, J. van den, 1987. Aspects of middle and late Triassic palynology. Palynological investigations in the Keuper (Upper Ladinian, Carnian, Norian, and Rhaetian) and Lower Jurassic (Lias α1 + α2) of Franken, SE West Germany. *Stuifmail*, **5**, 26–33.

Bertelsen, F., 1980. Lithostratigraphy and depositional history of the Danish Triassic. *Danm. geol. Undersol.*, **B4**, 1–59.

Beutler, G., 1980. Beitrag zur Stratigraphie des Unteren und Mittleren Keupers. *Z. geol. Wiss.*, **8**, 1001–18.

Beutler, G. & Schüler, F., 1979. Über Vorkommen salinarer Bildungen in der Trias im Norden der DDR. *Z. geol. Wiss.*, **7**, 903–12.

Beutler, G. & Haüsser, I., 1982. Über den Schilfsandstein der DDR. *Z. geol. Wiss.*, **10**, 511–25.

Brandner, R., 1984. Meeresspiegelschwankungen und Tektonik in der Trias der NW-Tethys. *Jb. geol. Bundesanst., Wien*, **126**, 435–75.

Brunner, H., 1973. Stratigraphische und sedimentpetrographische Untersuchungen am Unteren Keuper (Lettenkeuper, Trias) im nördlichen Baden-Württemberg. *Arb. geol.-paläont. Inst. Univ. Stuttgart*, **70**, 1–85.

Brunner, H. & Bruder, J., 1981. Standardprofile des Unteren Keupers (Lettenkeuper, Trias) im nördlichen Baden-Württemberg. *Jber. Mitt. oberrh. geol. Ver.*, **63**, 253–69.

Dittrich, D., 1989. Der Schilfsandstein als syn sedimentär-tektonisch geprägtes Sediment–eine Umdeutung bisheriger Befunde. *Z. dtsch. geol. Ges.*, **140**, 295–310.

Dockter, J., Langbein, R., Seidel, G. & Unger, K.P., 1970. Die Ausbildung des Unteren und Mittleren Keupers in Thüringen. *Jb. Geol.*, **3**, 145–94.

Duchrow, H., 1968a. Stratigraphie und Lithologie des Keupers im Lippischen Berglande. *Z. dtsch. geol. Ges.*, **117**, 371–87.

Duchrow, H., 1968b. Zur Keuperstratigraphie in Südostlippe (Trias, Nordwestdeutschland). *Z. dtsch. geol. Ges.*, **117**, 620–62.

Duchrow, H., 1984. Keuper. In *Geologie des Osnabrücker Berglandes*, ed. H. Klassen, pp. 221–334.

Emmert, U., 1965. Ist der Schilfsandstein des Mittleren Keupers eine Flußablagerung? *Geol. bavarica*, **55**, 146–68.

Emmert, U., 1977. Faziestypen der Schilfsandstein-Schichten (Mittlerer Keuper) und ihre Genese. *Jber. Mitt. oberrh. geol. Ver.*, **59**, 195–203.

Frank, M., 1930. Stratigraphie und Bildungsgeschichte des süddeutschen Gipskeupers. *Jber. Mitt. oberrh. geol. Ver.*, **19**, 25–77.

Frey, M., 1968. Quartenschiefer, Equisetenschiefer und germanischer Keuper – ein lithostratigraphischer Vergleich. *Eclogae geol. Helv.*, **61**, 141–56.

Freyberg, B. von, 1965. Cyklen und stratigraphische Einheiten im Mittleren Keuper Nordbayerns. *Geol. bavarica*, **55**, 130–45.

Geyer, G., 1987. Die Fossilien der Modiola-Bank Frankens (Karn, Gipskeuper, km 1 γ). *N. Jb. Geol. Paläont. Abh.*, **173**, 271–302.

Gravesen, P., Rolle, R. & Surlyk, F., 1982. Lithostratigraphy and sedimentary evolution of the Triassic, Jurassic and Lower Cretaceous of Bornholm, Denmark. *Danm. Geol. Undersol.*, B7, 1–51.

Gwinner, M.P., 1970a. Gipskrusten im Gipskeuper bei Obersontheim? (Baden-Württemberg). *N. Jb. Geol. Paläont. Mh.*, 88–90.

Gwinner, M.P., 1970b. Über Resedimentation im Schilfsandstein (Mittlerer Keuper). *N. Jb. Geol. Paläont. Mh.*, 141–8.

Hahn, G.G., 1984. Paläomagnetische Untersuchungen im Schilfsandstein (Trias, km 2) Westeuropas. *Geol. Rundschau*, **73**, 499–516.

Hahn, G.G., 1986. Die Umpolung des Erdmagnetfeldes zur Zeit der Schilfsandsteinsedimentation. *Jber. Mitt. oberrh. geol. Ver.*, **68**, 197–215.

Haunschild, H., 1981. Ein Beitrag zur Lithologie und zur Stratigraphie des Unteren Gipskeupers im südlichen Franken sowie zur Mächtigkeitsentwicklung des gesamten Gipskeupers. *Jber. Mitt. oberrh. geol. Ver.*, **63**, 293–313.

Haunschild, H., 1983. Der Gipskeuper im Bereich der nördlichen Frankenhöhe. *Geol. Bl. Nordost-Bayern*, **32**, 146–65.

Hauschke, N., 1982. Untersuchungen zur Stratigraphie und Fazies im Unteren Gipskeuper (km 1) des Lippischen Berglandes. *Münster Forsch. Geol. Paläont.*, **55**, 113–47.

Hauschke, N., 1987. Knollige und tepeeartige Strukturen – Indikatoren für die frühdiagenetische Bildung von Ca-Sulfaten unter Playa-Bedingungen im Unteren Gipskeuper (km 1) des Lippischen Berglandes. *N. Jb. Geol. Paläont. Abh.*, **175**, 147–79.

Hauschke, N., 1988. Unterer Gipskeuper, km 1. In *Zyklen im oberen Muschelkalk und Keuper Ostwestfalens*, 3. *Treffen deustchspr.* ed. N. Hauschke & U. Röhl, pp. 9–19. Sedimentologen, Bochum, Excursion Guide D.

Hauschke, N., 1989. Steinsalzkristallmarken – Begriff, Deutung und Bedeutung für das Playa-Playasee-Faziesmodell. *Z. dtsch. geol. Ges.*, **140**, 355–69.

Hauschke, N. & Heunisch, C., 1989. Sedimentologische und palynologische Aspekte einer zyklisch entwickelten lakustrischen Sequenz im höheren Teil des Unteren Gipskeupers (km 1, Obere Trias) Nordwestdeutschlands. *Lippische Mitt. Ges. Landeskd.*, **58**, 233–56.

Hauschke, N. & Heunisch, C., 1990. Lithologie und Palynologie der Bohrung USB 3 (Horn-Bad Meinberg, Ostwestfalen): ein Beitrag zur Faziesentwicklung im Keuper. *N. Jb. Geol. Paläont. Abh.*, **181**, 79–105.

Häusser, I. & Kurze, M., 1975. Sedimentationsbedingungen und Schwermineralführung im Mesozoikum des Nordteils der DDR. *Z. Geol. Wiss.*, **3**, 1317–32.

Heling, D., 1965. Zur Petrographie des Schilfsandsteins. *Beitr. Mineral. Petrol.*, **11**, 272–96.

Heling, D., 1967. Die Salinitätsfazies von Keupersedimenten aufgrund von Borgehaltsbestimmungen. *Sedimentology*, **8**, 63–72.

Heling, D., 1979. Zur Faziesanalyse des Schilfsandsteins. *Jber. Mitt. oberrh. geol. Ver.*, **61**, 153–6.

Heunisch, C., 1986a. Palynologie des Unteren Keupers in Franken, Süddeutschland. *Palaeontographica*, **200B**, 33–110.

Heunisch, C., 1986b. Gliederung und Milieuinterpretation des fränkischen Unteren Keupers aus palynologischer Sicht. *Geol. bavarica*, **89**, 151–9.

Heunisch, C., 1990. Palynologie der Bohrung 'Natzungen 1979', Blatt 4321 Borgholz (Trias; Oberer Muschelkalk 2, 3, Unterer Keuper). *N. Jb. Geol. Paläont. Mh.*, 17–42.

Jenkner, B., 1983. Eine Erosionsdiskordanz an der Grenze Gipskeuper-Schilfsandstein (Mittlerer Keuper) bei Diefenbach/Kraichgau. *Jh. geol. Landesamt Baden-Württemberg*, **25**, 91–106.

Kozur, H., Ökologisch-fazielle Probleme bei der stratigraphischen Gliederung und Korrelation der germanischen Trias und faziell ähnlicher Triasablagerungen. *Jb. Geol.*, 7/8, 87–108.

Krimmel, V., 1980. Epirogene Paläotektonik zur Zeit des Keupers (Trias) in Südwest-Deutschland. *Arb. geol-paläont. Inst. Univ. Stuttgart*, **76**, 1–74.

Kruck, W. & Wolff, F., 1975. Ergebnisse einer Fazieskartierung im Schilfsandstein des Weserberglandes. *Mitt. geol. Staatsinst. Hamburg*, **44**, 417–21.

Kühl, K.W., 1957. *Stratigraphisch-fazielle Untersuchungen im Mittleren Keuper zwischen Weser und Osning.* PhD Dissertation, Tech. Univ. Braunschweig.

Larsen, G. & Friis, H., 1975. Triassic heavy-mineral associations in Denmark. *Danm. geol. Undersol.* (1974), 33–47.

Lewandowski, J., 1988. Tonminerale im Keuper zwischen Osnabrück und Helmstedt. *Jb. Geol. Paläont. Abh.*, **176**, 157–85.

Linck, O., 1943. Fossile Wurzelböden aus dem Mittleren Keuper. *Natur. u. Volk*, **73**, 226–34.

Linck, O., 1970. Eine neue Deutung der Schilfsandstein-Stufe (Trias, Karn, Mittlerer Keuper 2). *Jh. geol. Landesamt Baden-Württemberg*, **12**, 63–99.

Lippmann, F. & Steiner, K., 1983. Der Mineralbestand des Gipskeuper von Pfäffingen (Tübingen) und Schwenningen, Württemberg. *Oberrh. geol. Abh.*, **32**, 15–43.

Michel, G., 1977. Die Schichtenfolge der Bohrung Alexander-von-Humboldt-Sprudel in Bad Oeynhausen. *Fortschr. Geol. Rheinland Westfalen*, **26**, 63–80.

Richter, D.K., 1985. Die Dolomite der Evaporit- und der Dolcrete-Playasequenz im mittleren Keuper bei Coburg (NE-Bayern). *N. Jb. Geol. Paläont. Abh.*, **170**, 87–128.

Rosenfeld, U., 1968. Beobachtungen zur Stratigraphie des Schilfsandsteins im ostlippischen Bergland (Lithostratigraphische Untersuchungen in Sandstein-Folgen III). *Geol. Rundschau*, **57**, 402–24.

Rosenfeld, U., 1978. Beitrag zur Paläogeographie des Mesozoikums in Westfalen. *N. Jb. Geol. Paläont. Abh.*, **156**, 132–55.

Salger, M., 1965. Zur Petrographie des Schilfsandsteins. *Geol. bavarica*, **55**, 169–78.

Salger, M., 1973. Untersuchungen zur Tonmineralogie des Gipskeupers im Raum Frankenhöhe-Steigerwald. *Geol. bavarica*, **67**, 145–54.

Sander, A., 1977. Rote Wand und Steinmergelkeuper (Mittlerer Keuper) in der Bohrung Alexander-von-Humboldt-Sprudel im Vergleich mit Übertageaufschlüssen des Weserberglandes. *Fortsch. Geol. Rheinland Westfalen*, **26**, 89–108.

Schlenker, B., 1971. Petrographische Untersuchungen am Gipskeuper und Lettenkeuper von Stuttgart. *Oberrh. geol. Abh.*, **20**, 69–102.

Schmitz, H.H., 1959. Über die mineralogische Zusammensetzung des unteren Gipskeupers, nach einem Bohrprofil. *Geol. Jb.*, **77**, 59–94.

Schröder, B., 1972, Zur Stratigraphie des höheren Gipskeupers im östlichen Süddeutschland. *Z. dtsch. geol. Ges.*, **123**, 215–28.

Schröder, B., 1977a. Bemerkungen zum Schilfsandstein von Lichtenau bei Ansbach/Mittelfranken. *Jber. Mitt. oberrh. geol. Ver.*, **59**, 205–14.

Schröder, B., 1977b. Unterer Keuper und Schilfsandstein im germanischen Trias-Randbecken. *Zbl. Geol. Paläont.*, Part I, 1976, 1030–56.

Schröder, B., 1982. Entwicklung des Sedimentbeckens und Stratigraphie der klassischen Germanischen Trias. *Geol. Rundschau*, **71**, 783–94.

Starke, R., 1970. Verteilung und Faziesabhängigkeit der Tonminerale in den geologischen Systemen. *Freiberger Forsch.*, C254, 1–185.

Trümpy, R., 1982. Das Phänomen Trias. *Geol. Rundschau*, **71**, 711–23.

Urlichs, M., 1982. Zur Stratigraphie und Fossilführung des Lettenkeupers (Obere Trias) bei Schwäbisch Hall (Baden-Württemberg). *Jber. Mitt. oberrh. geol. Ver.*, **64**, 213–24.

Warth, M., 1969. Conchostraken (Crustacea, Phyllopoda) aus dem Keuper (Obere Trias) Zentral-Württembergs. *Jh. Ges. Naturk. Württemberg*, **124**, 123–45.

Warth, M., 1988. Lebten die Muscheln des Schilfsandsteins (Trias, Karn, km 2) im Meer? *Jber. Mitt. oberrh. geol. Ver*, **70**, 245–66.

Wild, R., 1982. Die Evolution der Reptilien in der Triaszeit. *Geol. Rundschau*, **71**, 725–39.

Wild, R., 1987. Die Tierwelt der Keuperzeit (unter besonderer Berücksichtigung der Wirbeltiere). *Natur an Rems u. Murr*, **6**, 17–43.

Wolburg, J., 1956. Das Profil der Trias im Raum zwischen Ems und Niederrhein. *N. Jb. Geol. Paläont. Mh.* 305–30.

Wolburg, J., 1969a. Die epirogenetischen Phasen der Muschelkalk-und Keuper-Entwicklung Nordwest-Deutschlands, mit einem Rückblick auf den Buntsandstein. *Geotekt. Forsch.*, **32**, 1–65.

Wolburg, J., 1969b. Zum Wesen der altkimmerischen Hebung mit einem Überblick über die Muschelkalk- und Keuper-Entwicklung in NW-Deutschland. *Z. dtsch. geol. Ges.*, **119**, 516–23.

Wurster, P., 1964a. Geologie des Schilfsandsteins. *Mitt. geol. Staatinst. Hamburg*, **33**, 1–140.

Wurster, P., 1964b. Delta sedimentation in the German Keuper Basin. In *Deltaic and Shallow Marine Deposits, Developments in Sedimentology* 1, ed. L.M.J.U. Van Straaten, pp. 436–46.

Wurster, P., 1964c. Krustenbewegungen, Meeresspiegelschwankungen und Klimaänderungen der deutschen Trias. *Geol. Rundschau*, **54**, 224–40.

Wurster, P., 1968. Paläogeographie der deutschen Trias und die paläogeographische Orientierung der Lettenkohle in Süddeutschland. *Eclogae geol. Helv.*, **61**, 157–66.

Wurster, P., 1972. Entgegnung auf LINCK's neue Deutung der Geologie des Schilfsandsteins (GdSch). *Jh. Geol. Landesamt Baden-Württemberg*, **14**, 53–67.

Ziegler, P.A., 1978. Northwestern Europe: tectonics and basin development. *Geol. en Mijnb.*, **57**, 589–626.

Ziegler, P.A., 1982. Triassic rifts and facies patterns in western and central Europe. *Geol. Rundsch.*, **71**, 747–72.

Semi-permanent lacustrine conditions in the Upper Triassic Mercia Mudstone Group, South Wales, UK

ALICK B. LESLIE[1] & MAURICE E. TUCKER[2]

[1] Marine Geology, British Geological Survey, Murchison House, West Mains Road, Edinburgh EH9 3LA, UK
[2] Department of Geological Sciences, University of Durham, South Road, Durham DH1 3LE, UK

Introduction

The Upper Triassic Mercia Mudstone Group consists of up to 1000 m of red, dolomitic mudstones and siltstones (Fig. 1), with local coarser clastics, laid down in fault-controlled graben and half-graben which developed during the Permian and Lower Triassic (Holloway, 1985). As opposed to the Lower Triassic in which sandstones and conglomerates of the Sherwood Sandstone Group were laid down, in the Upper Triassic there was little significant fault activity and sediments passively onlapped basin margins as a result of thermal subsidence (Chadwick, 1985).

The mudstones and siltstones contain few fossils of any biostratigraphic value (Warrington, 1976) and sedimentary structures are rare. Horizons of sulfate are common and in certain sections of the deeper basins halite is developed (Fig. 1). Other than the Blue Anchor Formation, which comprises the topmost 10–40 m of the Mercia Mudstones and is composed of grey and green mudstones and dolomites, the Group has not been formally subdivided (Warrington et al., 1980). The lack of both sedimentary structures and diagnostic faunas has led to some discussion of the exact depositional environment in which the mudstones were formed. Interpretations have ranged from hypersaline sea (Evans et al., 1968; Warrington, 1970; Jeans, 1978) to playa–alluvial basin (Tucker, 1977, 1978). At present a modified playa–alluvial model, in which there were periodic incursions of marine-derived waters into the basins, is favored (Arthurton, 1980; Taylor, 1983; Leslie, 1989). This interpretation is given credence by marine signatures derived from bromine analyses of halite horizons within Upper Triassic mudstones in Somerset (Figs 1 and 2) and Devon. The marginal South Wales deposits contain abundant sulfate and celestite (Figs 2 and 3). There are no massive halite horizons, although these have been identified in cores from the Bristol Channel Basin to the west (Kamerling, 1979).

In most of the exposures of Mercia Mudstones in southwest Britain, sedimentary structures indicate that the mudstones were laid down in an arid distal floodplain environment subject to periodic sabkha conditions (Leslie, 1989). In the South Wales area, however, there is evidence to suggest the existence of more permanent lacustrine conditions (Tucker, 1978; Leslie, 1989). At several

coastal outcrops south of Cardiff, wave-cut platforms and beach deposits which are the lateral equivalents of the Mercia Mudstone Group are exposed (Figs 2 and 3). These sedimentary structures set the South Wales exposures apart from those in other parts of southwest Britain. The mudstones in South Wales are also geochemically different from equivalent exposures at St Audrie's Bay in Somerset (Fig. 2), 25 km to the south, indicating a different depositional environment.

The Mercia Mudstones in South Wales which contain evidence of more permanent lacustrine conditions are late Norian or early Rhaetian in age (Warrington et al., 1980). Exact dating of the mudstones is not usually possible and in South Wales the proximity of overlying Rhaetian beds containing a marine fauna is the most reliable dating method (reliability index: C). Within the red mudstones, which contain evidence of lacustrine conditions, the only fauna present are rare ostracods within the freshwater limestones and some vertebrate tracks (Tucker & Burchette, 1977). In Somerset a sparse fauna has been recovered from sandstones giving a late Carnian age (Fig. 1) but these coarse beds have not been recovered in South Wales.

The Upper Triassic in South Wales

The total thickness of Mercia Mudstone Group sediments in South Wales is 100 m (Waters & Lawrence, 1987), although only some 50 m are exposed (Figs 2 and 3). The sediments are generally flat lying or slightly inclined, there having been little deformation in South Wales during the Mesozoic and Cenozoic. At least the topmost 30 m of red mudstones underlying the Blue Anchor Formation were deposited in semi-permanent lacustrine conditions. At Penarth Cliffs (Fig. 3) 30 m of red mudstones with nodular sulfate horizons and beds of massive celestite were laid down in continental-derived waters (Taylor, 1983). At Barry Island, 7 km to the west, the Mercia Mudstone Group directly overlies Carboniferous limestones, and scree deposits are associated with several Triassic wave-cut platforms (Tucker, 1978). The presence of sulfate horizons which are laterally equivalent to the wave-cut platforms indicates that there was periodic desiccation of the lake and a retreat of the paleoshoreline. At Sully Island

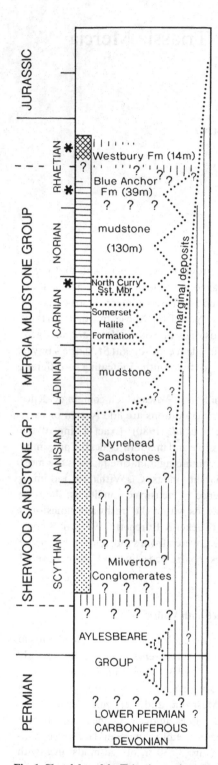

Fig. 1. Sketch log of the Triassic stratigraphic succession in South Wales and Somerset. The South Wales succession is a maximum of 100 m in thickness and does not include an equivalent of the North Curry Sandstone Member in Somerset. An asterix indicates paleontological dating of the succession in Somerset. Adapted from Warrington _et al._, 1980.

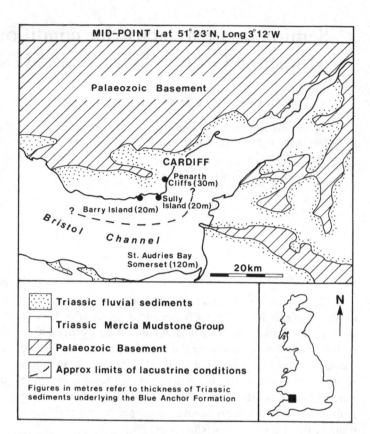

Fig. 2. Location map for the Bristol Channel Basin and the outcrops in South Wales and Somerset covered in the text.

Carboniferous limestones are overlain by a wedge of beach-type clastic sediments (Tucker, 1978). These are themselves overlain by a succession of evaporitic mudflat dolomites and freshwater limestones which formed in an embayment of the paleoshoreline associated with resurgent groundwaters (Leslie, 1989). The limestones represent a lacustrine environment and travertines associated with stromatolites and other microbial structures, developed around spring sites. This marginal lacustrine environment is very different from that of the laterally equivalent hypersaline mudstones. Although the exact stratigraphic position of the Sully Island succession is not known, it lies no more than 50 m below the base of the Blue Anchor Formation.

Lacustrine shoreline deposits are only exposed at Sully Island and Barry Island (Fig. 3). At Penarth Cliffs the mudstones are not visibly different from those in Somerset and Devon, and have a similar mineralogy. Geochemically the mudstones can be distinguished both in the composition of the carbonate, which is interpreted to be syndepositional or early diagenetic in origin, and by differences in sulfur stable isotopic composition.

Taylor (1983) described an enrichment in ^{34}S by several per mil from gypsum nodules in Somerset relative to samples taken from Penarth Cliffs. These data were interpreted to indicate a greater marine influence in the Somerset mudstones. The South Wales sulfates are, in fact, depleted in ^{34}S relative to all other laterally

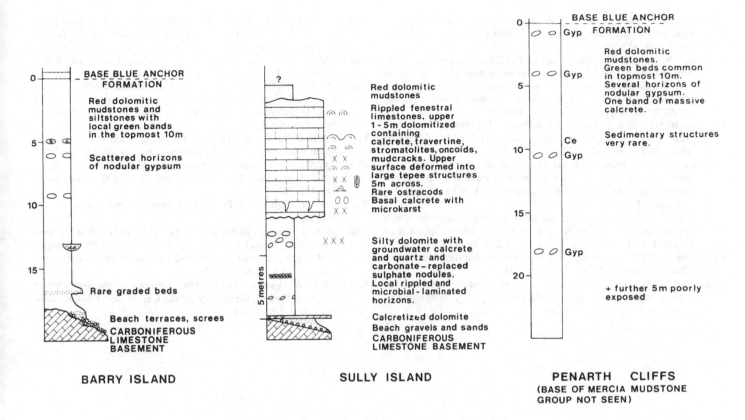

Fig. 3. Sedimentological logs of the semi-permanent lacustrine beach clastics and carbonates exposed at Sully Island and Barry Island, which pass up into Mercia Mudstones, and of the Mercia Mudstone at Penarth Cliffs south of Cardiff, South Wales. Figures on the left of the logs are in meters.

equivalent samples examined by Taylor (1983). This would tend to indicate that the South Wales area was more influenced by continental-derived waters during the Late Triassic than other basins.

The carbonate from the mudstones in South Wales is enriched in ^{18}O relative to laterally equivalent deposits in Somerset and Devon (Leslie, 1989), but still lies within the field of evaporative continental waters of Taylor (1983). The greater enrichment in ^{18}O may be the result of prolonged exposure of waters in the more permanent lake to evaporative effects. The geochemical differences in South Wales relative to Somerset and Devon are best explained by a different hydrogeochemical environment since the mudstones are texturally indistinguishable. The gradually increasing influence of marine-derived waters during deposition of the Blue Anchor Formation, leading to more enriched ^{18}O values, can be seen in both South Wales and Somerset. The reasons for the presence of a relatively permanent lake in the South Wales area are not certain, although it is possible that high land in central Wales provided a relatively high volume of runoff which was channeled through canyons into the area. Brooks (1987) and Brooks & Al-Saadi (1977) have described faulting in the Inner Bristol Channel which, during the Late Triassic, may have formed an area of high ground, separating the South Wales sub-basin from the Somerset and Devon basins.

Conclusions

The Triassic Mercia Mudstone Group deposits in South Wales were laid down in a semi-permanent lake. Several wave-cut platforms indicate significant periods of erosion during which time the lake level was stable, although episodes of desiccation also took place.

Oxygen stable isotopic values indicate that the lake waters were derived from continental runoff and then evaporatively enriched in ^{18}O before carbonate was precipitated.

The semipermanent lacustrine conditions did not extend to the Somerset area. There is some evidence for the presence of a Permo-Triassic graben in the Bristol Channel. This structure may have separated the South Wales and Somerset areas into separate sub-basins with different depositional environments.

Bibliography

Arthurton, R.S., 1980. Rhythmic sedimentary sequences in the Triassic Keuper Marl (Mercia Mudstone Group) of Cheshire, northwest England. *Geol. J.*, **15**, 43–58.

Brooks, M., 1987. Geophysical investigations of deep structure in the Bristol Channel area. *Proc. Geol. Assoc.*, **98**, 397–8.

Brooks, M. & Al-Saadi, R.H., 1977. Seismic refraction studies of geologic structures in the inner part of the Bristol Channel. *J. Geol. Soc. Lond.*, **133**, 433–45.

Chadwick, R.A., 1985. Permian, Mesozoic and Cenozoic structural evolution of England and Wales in relation to the principles of extension and inversion tectonics. In *Atlas of Onshore Sedimentary Basins in England and Wales: Post-Carboniferous Tectonics and Stratigraphy*, ed. A. Whittaker, pp. 9–25. Blackie, London.

Evans, W.B., Wilson, A.A., Taylor, B.J. & Price, R.H., 1968. Geology of the country around Macclesfield, Congleton, Crewe and Middlewitch. *Mem. Geol. Surv. UK.*, 1–328

Holloway, S., 1985. Triassic: Mercia Mudstone and Penarth Groups. In *Atlas of Onshore Sedimentary Basins in England and Wales: Post-Carboniferous Tectonics and Stratigraphy*, ed. A. Whittaker, pp. 34–6. Blackie, London.

Jeans, C.V., 1978. The origin of the Triassic clay assemblages of Europe with special reference to the Keuper Marl and Rhaetic of parts of England. *Phil. Trans. Roy. Soc. Lond.* A.**289**, 549–639.

Kamerling, P., 1979. The geology and hydrocarbon habitat of the Bristol Channel Basin. *J. Petrol. Geol.*, **2**, 75–93.

Leslie, A.B., 1989. Sedimentology and Geochemistry of the Upper Triassic Mercia Mudstone Group and Marginal Deposits, Southwest Britain. *PhD Thesis, Univ. Durham.*

Taylor, S.R., 1983. A stable isotopic study of the Mercia Mudstones (Keuper Marl) and associated sulphate horizons in the English Midlands. *Sedimentology*, **30**, 11–31.

Tucker, M.E., 1977. The Marginal Triassic deposits of South Wales: Continental Facies and palaeogeography. *Geol. J.* **12**, 169–89.

Tucker, M.E., 1978. The marginal clastic deposits from South Wales: shore Zone clastics, evaporites and carbonates. In *Modern and Ancient Lake Sediments*, ed. A. Matter & M.E. Tucker, *Int. Assoc. Sediment. Spec. Pub. No.* 2, 205–24.

Tucker, M.E. & Burchette, T.P., 1977. Triassic dinosaur footprints from South Wales. *Palaeogeog. Palaeoclimat. Palaeoecol.*, **22**, 195–208.

Warringon, G., 1970. The stratigraphy and palaeontology of the 'Keuper' Series of the central Midlands of England. *Quart. J. Geol. Soc. Lond.*, **126**, 183–223.

Warrington, G., 1976. British Triassic Palaeontology. *Proc. Ussher Soc.*, **3**, 341–553.

Warrington, G., Audley-Charles, M.G., Elliot, R.E., Evans, W.B., Ivimey-Cook, H.C., Kent, P.E., Robinson, P.L., Shotton, F.W. & Taylor, F.M., 1980. A correlation of Triassic rocks in the British Isles. *Spec. Rep. Geol. Soc. Lond.* **13**, 1–78.

Waters, R.A. & Lawrence, D.J.D., 1987. Geology of the South Wales Coalfield, Part III, the country around Cardiff. 3rd. Edition. *Mem. Geol. Surv. UK.*, 1–114.

Lacustrine deposits of the Upper Triassic Chinle Formation, Colorado Plateau, USA

RUSSELL F. DUBIEL

U.S. Geological Survey MS 919 Box 25046 DFC Denver, CO 80225, USA

The Upper Triassic Chinle Formation is exposed throughout the Colorado Plateau of the western United States; the most extensive exposures occur in the Four Corners region of Arizona, Utah, New Mexico, and Colorado (Fig. 1). Because of the wide geographic distribution and rapid lateral facies changes within the continental strata of the Chinle, stratigraphic nomenclature is complex, and a variety of members have been defined in different regions. This report summarizes aspects of the Chinle Formation in the Four Corners region, although many of the lacustrine features of the Chinle are present throughout the Colorado Plateau. The stratigraphic sections (Figs 2 and 3) are composites constructed from three localities in southeastern Utah and northeastern Arizona and represent salient features of the major components of the Chinle (Stewart et al., 1972; Dubiel, 1989b, c) (Fig. 2).

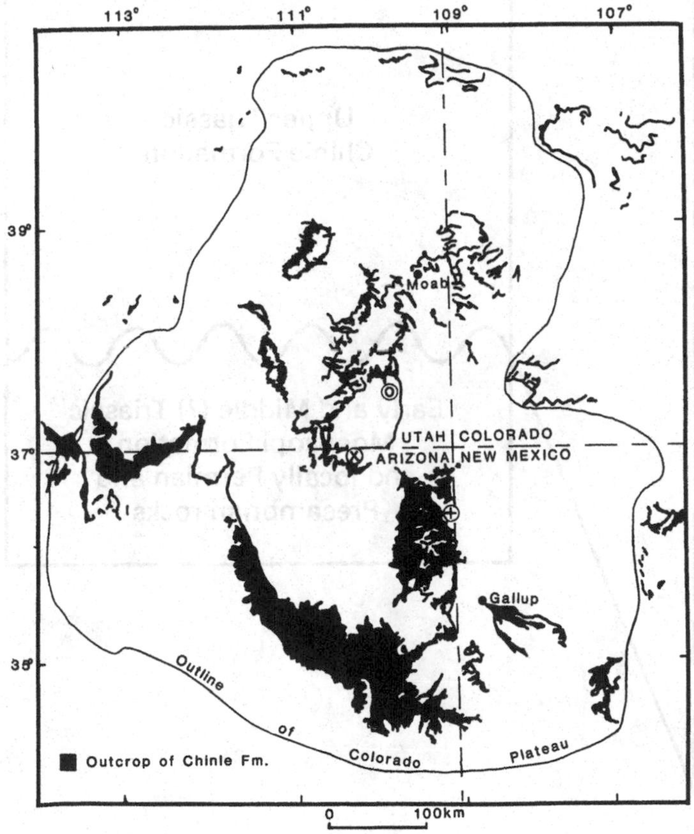

The Chinle Formation unconformably overlies mainly the Lower and Middle(?) Triassic Moenkopi Formation over much of the Colorado Plateau and locally overlies Permian strata in New Mexico and in the Defiance uplift of northern Arizona, and Precambrian rocks in west-central Colorado. The Chinle is unconformably overlain by the Lower Jurassic Wingate Sandstone and locally by the Middle Jurassic Entrada Sandstone. The Chinle Formation is equivalent to the Upper Triassic Dolores Formation, which crops out in a small area of southwestern Colorado.

The Chinle Formation was deposited in a continental back-arc basin (Dickinson, 1981; Blakey & Gubitosa, 1983; Dubiel, 1987a, b, c, 1989d). The Chinle records sedimentation in a geographically widespread continental depsystem containing complexly interfingered fluvial, floodplain, lacustrine-deltaic, lacustrine and eolian facies. Lacustrine facies predominate in the Monitor Butte and Owl Rock Members and are minor components of the Petrified Forest and Rock Point Members (Fig. 3). The Chinle is as much as 500 m thick in the Four Corners region.

The diverse flora and fauna in the Chinle Formation, including abundant plants, non-marine aquatic invertebrates, and aquatic and terrestrial vertebrates, have permitted paleoecological and paleoclimatological reconstructions of the Chinle (Good et al., 1987; Dubiel et al., 1989; Parrish et al., 1989). Paleontologic studies of the Chinle megaflora (Ash, 1972, 1978, 1980), pollen (Gottesfeld, 1972; Litwin, 1986a, b), invertebrates (Good et al., 1987; Parrish & Good, 1987; Good, 1989) and vertebrates (Schaeffer, 1967; Parrish, 1989) have established a Late Triassic age for the Chinle; the Carnian–Norian boundary occurs within the Petrified Forest Member (Fig. 2).

The Chinle Formation was deposited in large paleovalleys eroded into underlying rocks. The Shinarump Member consists of fluvial and minor floodplain strata deposited in the bottoms of the

Fig. 1. Distribution of Upper Triassic Chinle Formation (black) on the Colorado Plateau in the Four Corners region. Symbols (+, x, o) indicate locations of components of stratigraphic sections used in composite sections depicted in Figs 2 and 3.
Basin Mid-point: 36° 00′ 109° 00′

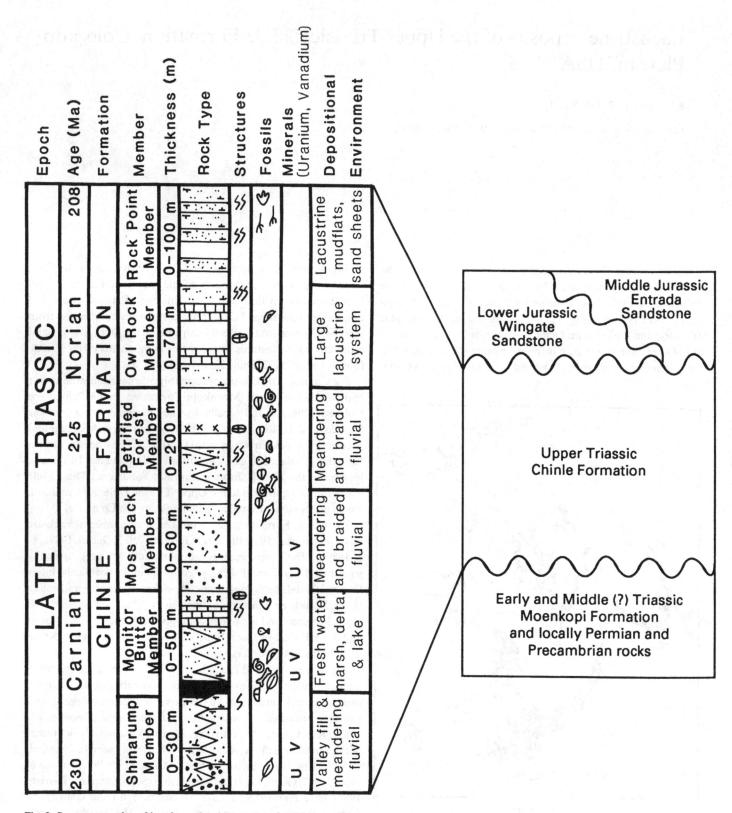

Fig. 2. Summary stratigraphic column showing stratigraphy, lithology, paleontology and depositional environments.

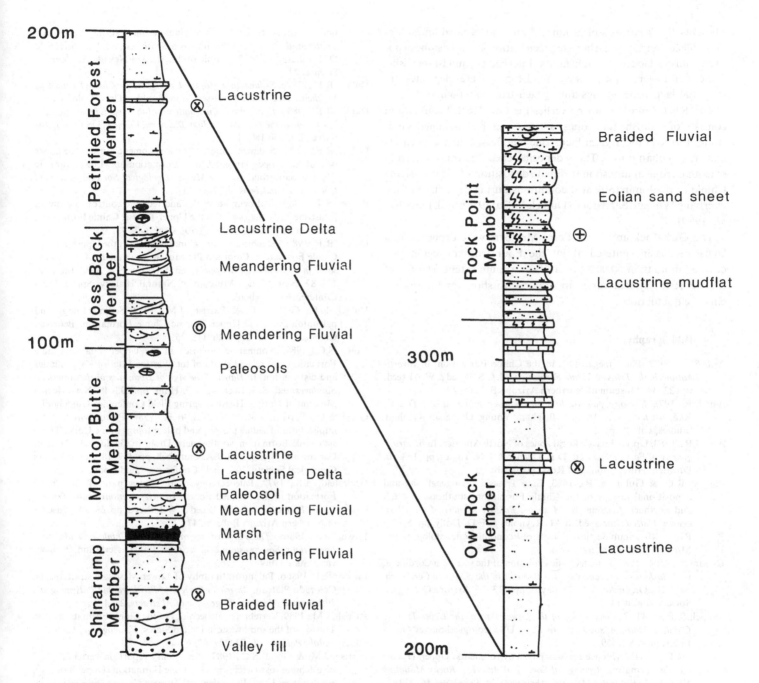

Fig. 3. Detailed lithologic column of the Chinle Formation from near the depocenter of the Chinle basin in northern Arizona (see Fig. 1 for location of component measured stratigraphic sections used to construct this composite section).

paleovalleys. The overlying Monitor Butte Member consists of a variety of sandstones, siltstones, mudstones and limestones formed in lacustrine, lacustrine-deltaic, lacustrine-marsh and fluvial settings within a northwest-trending depositional system (Stewart *et al.*, 1972; Blakey & Gubitosa, 1983; Dubiel, 1983, 1984, 1985, 1986, 1987a, b, c, 1989a, b, c). These fluvial and lacustrine deposits filled the pre-existing paleovalleys. Green facies within the Monitor Butte Member represent lacustrine, lacustrine-delta, marginal-lacustrine, and freshwater marsh deposits that contain abundant organic

material and plant, invertebrate, and vertebrate fossils (Fig. 2). Red facies within the Monitor Butte Member represent fluvial and floodplain strata. Lacustrine and marsh deposits of the Monitor Butte Member and equivalent strata cover approximately 20,000 km² on the Colorado Plateau.

Lacustrine strata predominate in the Owl Rock Member of the Chinle Formation. The underlying Moss Back and Petrified Forest Members consist of primarily fluvial strata, although the Petrified Forest Member locally contains marsh mudstones (Dubiel, 1987a).

The Owl Rock consists of laminated and bioturbated limestones that reflect alternating carbonate precipitation and mudstone deposition during lacustrine highstands. Interbedded mudstones, siltstones and very fine-grained sandstones were deposited in marginal-lacustrine settings during lacustrine lowstands.

The Rock Point Member overlies the Owl Rock Member and consists of reddish mudstones, siltstones and fine-grained sandstones formed on marginal-lacustrine mudflats, eolian sand sheets and small eolian dunes. These deposits reflect the transition from seasonal, tropical monsoonal climate deposition of the underlying Chinle to predominantly arid conditions during deposition of the overlying Wingate Sandstone (Dubiel, 1987a; 1989b, c, d; Parrish et al., 1989).

The Owl Rock and Rock Point Members were deposited in a lacustrine basin centered about the Four Corners region that covered more than 60,000 km² at its maximum extent. The size of the lacustrine system varied in response to short- and long-term climatic fluctuations.

Bibliography

Ash, S.R., 1972. Plant megafossils in the Chinle Formation. In *Investigations in the Triassic Chinle Formation*, ed. C.S. Breed & W.J. Breed, pp. 23–44. Museum of Northern Arizona Bulletin 47.

Ash, S.R., 1978. *Geology, paleontology, and paleoecology of a Late Triassic lake, western New Mexico*. Brigham Young University Geology Studies, vol. 25, pt. 2.

Ash, S.R., 1980. Upper Triassic floral zones of North America. In *Biostratigraphy of Fossil Plants*, ed. D.L. Dilcher & T.N. Taylor, pp. 153–70. Dowden, Hutchinson, and Ross, Stroudsburg.

Blakey, R.C. & Gubitosa, R., 1983. Late Triassic paleogeography and depositional history of the Chinle Formation, southeastern Utah and northern Arizona. In *Mesozoic Paleogeography of the West-central United States*, ed. R.M. Reynolds & E.D. Dolly, pp. 57–76. Rocky Mountain Section, Society of Economic Paleontologists and Mineralogists, Denver, Colorado.

Dickinson, W.R., 1981. Plate tectonic evolution of the southern Cordillera. In *Relations of Tectonics to Ore Deposits in the Southern Cordillera*. ed. W.R. Dickinson & W.D. Payne, pp. 113–35. Arizona Geological Society Digest 14.

Dubiel, R.F., 1983. *Sedimentology of the Lower Part of the Upper Triassic Chinle Formation, southeastern Utah*. US Geological Survey Open-File Report 83–459.

Dubiel, R.F., 1984. Evidence for wet paleoenvironments, Upper Triassic Chinle Formation. *Geological Society of America, Rocky Mountain Section, 37th Annual Meeting, Abstracts with Programs*, **16**, 220.

Dubiel, R.F., 1985. *Preliminary Report on Mudlumps in Lacustrine Deltas of the Monitor Butte Member of the Upper Triassic Chinle Formation, southeastern Utah*. US Geological Survey Open-File Report 85–27.

Dubiel, R.F., 1986. Evolution of a fluvial-lacustrine system – Tectonic and climatic controls on sedimentation in the Upper Triassic Chinle Formation, Colorado Plateau. *Geological Society of America, Rocky Mountain Section, 39th Annual Meeting, Abstracts with Programs*, 18, 352.

Dubiel, R.F., 1987a. Sedimentology and new fossil occurrences in the Upper Triassic Chinle Formation, southeastern Utah. In *Geology of Cataract Canyon and Vicinity*, ed. J.A. Campbell, pp. 99–107. Four Corners Geological Society, 10th Field Conference Guidebook.

Dubiel, R.F., 1987b. Sedimentology of the Upper Triassic Chinle Formation, southeastern Utah – Paleoclimatic implications, In *Triassic Continental Deposits of the American Southwest*, ed. M. Morales & D.K. Elliot, pp. 35–45. Journal of the Arizona-Nevada Academy of Science, **22**.

Dubiel, R.F., 1987c. *Sedimentology of the Upper Triassic Chinle Formation, Southeastern Utah*. PhD Dissertation, Boulder, Univ. Colorado.

Dubiel, R.F., 1989a. *Depositional Environments of the Upper Triassic Chinle Formation in the Eastern San Juan Basin and Vicinity*. US Geological Survey Bulletin 1808-B.

Dubiel, R.F., 1989b. Sedimentology and revised nomenclature of the upper part of the Upper Triassic Chinle Formation and Lower Jurassic Wingate Sandstone. In *New Mexico Geological Society, 40th Field Conference Guidebook*, ed. S.G. Lucas.

Dubiel, R.F., 1989c. Paleoclimate cycles and tectonic controls on fluvial, lacustrine and eolian strata in the Upper Triassic Chinle Formation, San Juan Basin. *Am. Assoc. Petrol. Geol Bull.*, **73**, 351.

Dubiel, R.F., 1989d. Depositional and climatic setting of the Upper Triassic Chinle Formation, Colorado Plateau. In *Dawn of the Age of Dinosaurs in the American Southwest*, ed. S.G. Lucas & A.P. Hunt, pp. 171–87. New Mexico Museum of Natural History, Spring Field Conference Guidebook.

Dubiel, R.F., Good, S.C. & Parrish, J.M., 1989. Sedimentology and paleontology of the Upper Triassic Chinle Formation, Bedrock, Colorado. *Mount. Geol.*, **26**, 113–26.

Good, S.C., 1989. Nonmarine mollusca in the Upper Triassic Chinle Formation and related strata of the Western Interior – systematics and distribution. In *Dawn of the Age of Dinosaurs in the American Southwest*. ed. S.G. Lucas & A.P. Hunt, pp. 233–48. New Mexico Museum of Natural History, Spring Field Conference Guidebook.

Good, S.C., Parrish, J.M. & Dubiel, R.F., 1987. Paleoenvironmental implications of sedimentology and paleontology of the Upper Triassic Chinle Formation, southeastern Utah. In Geology of Cataract Canyon and Vicinity, ed. J.A. Campbell, pp. 117–18. Four Corners Geological Society, 10th Field Conference Guidebook.

Gottesfeld, A.S., 1972. Paleoecology of the lower part of the Chinle Formation in the Petrified Forest. In *Investigations in the Triassic Chinle Formation*. ed. C.S. Breed & W.J. Breed, pp. 59–74. Museum of Northern Arizona Bulletin, **47**.

Litwin, R.J., 1986a. *The Palynostratigraphy of the Chinle and Moenave Formations, Southwestern United States*. PhD Dissertation, Pennsylvania State Univ.

Litwin, R.J., 1986b. Palynostratigraphy of Lower Mesozoic strata on the Colorado Plateau. *American Society of Stratigraphic Palynologists, 19th Annual Meeting, Program with Abstracts*, 23–4.

Parrish, J.M., 1989. Vertebrate paleoecology of the Chinle Formation (Late Triassic) of the southwestern United States. *Palaeogeogr, Palaeoclimatol., Palaeoecol.*, **72**, 227–47.

Parrish, J.M. & Good, S.C., 1987. Preliminary report on vertebrate and invertebrate fossil occurrences, Chinle Formation (Upper Triassic), southeastern Utah. In *Geology of Cataract Canyon and Vicinity*. ed. J.A. Campbell, pp. 108–16. Four Corners Geological Society, 10th Field Conference Guidebook.

Parrish, J.M., Dubiel, R.F. & Parrish, J.T., 1989. Triassic tropical monsoonal climate in Pangaea – Evidence from the Upper Triassic Chinle Formation, Colorado Plateau, USA. *Washington, D.C., International Geologic Congress, Symposium on Global Aspects of the Triassic*.

Schaeffer, B., 1967. Late Triassic fishes from the Western United States. *Bull. Am. Mus. Nat. Hist.*, **135**, 287–342.

Stewart, J.H., Poole, F.G. & Wilson, R.F., 1972. *Stratigraphy and origin of the Upper Triassic Chinle Formation and related strata in the Colorado Plateau region*. US Geological Survey Professional Paper 693.

Triassic–Jurassic lacustrine deposits in the Culpeper Basin, Virginia, USA

PAMELA J.W. GORE

Department of Geology, DeKalb College, 555 North Indian Creek Drive, Clarkston, Georgia 30021, USA

Overview

The Culpeper basin is a half-graben containing Late Triassic and Early Jurassic non-marine sediments belonging to the Culpeper Group and Newark Supergroup of eastern North America. The sedimentary record of the basin spans 25–30 million years, from the Carnian (Late Triassic) to the late Hettangian or early Sinemurian, and possibly the Pliensbachian (Early Jurassic). The Culpeper basin is the southernmost basin of the Newark Supergroup known to contain rocks of Early Jurassic age. The basin is a complex mosaic of lacustrine, fluvial and alluvial-fan deposits, along with a series of Early Jurassic basaltic lava flows and diabase intrusives. Several types of lacustrine deposits are present in the basin, representing deposition in lakes both shallow and deep, freshwater and saline. The lacustrine sequences are commonly interbedded with mudflat and fluvial sheetflood deposits, as well as massive mudstones interpreted as paleosols. Numerous episodes of lake expansion and contraction are recorded. Cumulative thickness of the stratigraphic units in the basin has been calculated at about 7900–9000 m, but the complete thickness is not preserved in any one place.

The overall stratigraphic pattern is similar to that in the Gettysburg and Newark basins to the north, and it is fairly certain that all three basins were interconnected at one time. Stratigraphic nomenclature has had several complete revisions. The more recent stratigraphic nomenclature (Lee & Froelich, 1989) is shown on the geologic map (Fig. 1). Equivalency of stratigraphic terms and summary of geologic data are shown in Fig. 2.

Lacustrine sequences

Lacustrine sequences are present in the Balls Bluff, Midland and Waterfall Formations (see Gore, 1988a for more information.) The lacustrine deposits range from less than 1 m thick to more than 35 m thick, and are dominated by red, grey, or black shales and coarser terrigenous clastic units, many of which are calcareous. Thin beds of limestone (including stromatolitic and oolitic limestone) are also present locally.

The lacustrine sequences in the Balls Bluff Siltstone range from less than 1 m thick ('very thin' lacustrine sequences) to approximately 7 m thick ('thin' lacustrine sequences). The lacustrine units are interbedded with red beds interpreted as mudflat, paleosol, sheetflood and fluvial deposits, along with conglomeratic alluvial fan deposits. The 'very thin' lacustrine sequences consist of beds of limestone and/or red to grey laminated shale with invertebrate fossils (primarily conchostracans, ostracodes, and pelecypods) and sparse, isolated fish scales (Fig. 3a, A–F). Evaporite crystal molds are present locally. The 'very thin' lacustrine sequences are interpreted as shallow lacustrine deposits.

The 'thin' lacustrine sequences (1–7 m thick) in the Balls Bluff Siltstone (Fig. 3b, G–J) contain black shale (interpreted as deeper, anoxic water deposits), grey shale and sandstone (interpreted as shallower water deposits) and red and grey siltstone (interpreted as mudflat, paleosol and sheetflood deposits). The facies are arranged in transgressive–regressive sequences, representing the expansion and contraction of perennial lakes.

The Midland Formation contains lacustrine sequences ranging from less than 1 m thick to more than 5 m thick, as well as fluvial deposits and paleosols. One of the lacustrine sequences (Fig. 3b, K) is known as the Midland fish-bed because of the presence of well-preserved fish fossils. The fish-bed sequence is transgressive–regressive in character, and is topped by a thick unit of fluvial trough-cross-bedded sandstone. (For more information on this section, see Gore, 1986, 1987b, 1987c, 1988a; Gore & Talbot, 1988; Roberts & Gore, 1988a, 1988b; Gore & Roberts, 1989; Roberts, 1989).

The Waterfall Formation contains at least ten 'thick' lacustrine sequences (averaging about 35 m thick), separated by mudflat and fluvial deposits. The lacustrine units are primarily grey to black mudshale and siltshale, with interbedded sandstone layers interpreted as lacustrine turbidites (Fig. 3c, L). Invertebrate remains (conchostracans, ostracodes and gastropods) and fish fossils are present in many of the grey to black units. (For more information, see Hentz, 1981, 1985.)

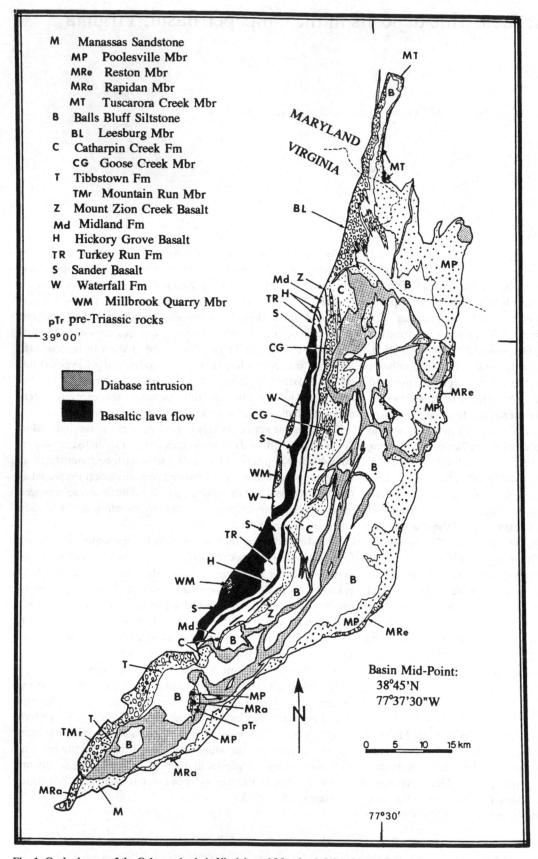

M   Manassas Sandstone
   MP   Poolesville Mbr
   MRe   Reston Mbr
   MRa   Rapidan Mbr
   MT   Tuscarora Creek Mbr
B   Balls Bluff Siltstone
   BL   Leesburg Mbr
C   Catharpin Creek Fm
   CG   Goose Creek Mbr
T   Tibbstown Fm
   TMr   Mountain Run Mbr
Z   Mount Zion Creek Basalt
Md   Midland Fm
H   Hickory Grove Basalt
TR   Turkey Run Fm
S   Sander Basalt
W   Waterfall Fm
   WM   Millbrook Quarry Mbr
pTr   pre-Triassic rocks

Diabase intrusion

Basaltic lava flow

Basin Mid-Point:
38°45'N
77°37'30"W

0   5   10   15 km

Fig. 1. Geologic map of the Culpeper basin in Virginia and Maryland, USA. Modified from Lee & Froelich (1989) and Olsen *et al.* (1989).

Age	Lindholm (1979)	Lee and Froelich (1989)	Thickness (m)	Description
JURASSIC Sinemurian - Pliensbachian?	Waterfall Fm	Waterfall Fm	1150-1719	Sandstone, siltstone, mudstone, shale, and conglomerate. Fish, conchostracans, ostracodes, gastropods, dinosaur tracks. Lacustrine cycles and fluvial.
		Millbrook Quarry Mbr	450	Conglomerate and sandstone. Alluvial fan.
JURASSIC Hettangian - Sinemurian	Buckland Fm	Sander Basalt	140-690	Basalt with interbedded sandstone & siltstone
		Turkey Run Fm	<150-330	Sandstone, siltstone, silty shale. Dinosaur tracks, fish scales, plant fragments. Mostly fluvial, some deltaic to lacustrine.
		Hickory Grove Basalt	50-380	Basalt with interbedded sandstone & siltstone
		Midland Formation	150-300	Red sandstone & siltstone interbedded with dark gray-black calcareous, microlaminated, fossiliferous shale & argillaceous limestone. Fish, conchostracans, ostracodes, plants. Fluvial and lacustrine.
		Mt. Zion Church Basalt	3-180	Discontinuous layer of basalt. Lenses of red sandstone and siltstone.
TRIASSIC - JURASSIC Norian - Hettangian	Bull Run Fm	Catharpin Creek Fm	ca. 500	Red sandstone & clayey siltstone. Fluvial; braided streams & debris flows on fans.
		Goose Creek Mbr	>900	Conglomerate. Clasts from Blue Ridge. Grades into arkosic sandstones. Fan.
TRIASSIC Norian		Tibbstown Fm	avg 300	Arkosic sandstone and conglomerate. Fluvial.
		Mountain Run Mbr	0-640	Conglomerate. Fluvial fan with clasts from Blue Ridge.
TRIASSIC Carnian - Norian		Balls Bluff Siltstone	80-1690	Red siltstone with gray fossiliferous shale & siltstone, thin limestone beds. Ostracodes, conchostracans, notostracans, clams, rare stromatolites, fish teeth and scales, plants, dinosaur tracks, trails, and a phytosaur. Evaporite crystal molds. Lacustrine (closed basin), paleosols, fluvial.
		Leesburg Mbr	40-1070	Conglomerate. Clasts derived from west and northwest. Debris flows and fluvial fans.
TRIASSIC Carnian	Manassas Sandstone	Manassas Sandstone		
		Poolesville Mbr	200-1000	Pink to red arkosic sandstone, with minor siltstone & mudstone. Plants & footprints. Fluvial; braided streams.
		Tuscarora Creek Mbr	21-67	Conglomerate with limestone and dolomite clasts from Frederick Valley. Fan deposits.
		Rapidan Mbr	70-140	Indurated conglomerate. Fluvial fan deposit with clasts from Blue Ridge.
TRIASSIC Carnian	Reston Fm	Reston Member	<3-100	Conglomerate with clay-silt matrix and associated siltstone and sandstone. Fluvial fan deposit with clasts from Piedmont.

Fig. 2. Stratigraphy of the Culpeper Basin showing equivalency of stratigraphic terms, ages, and thicknesses of units. Ages and thicknesses are from Lee and Froelich (1989). Ages are based on pollen and spore data (Cornet, 1977).

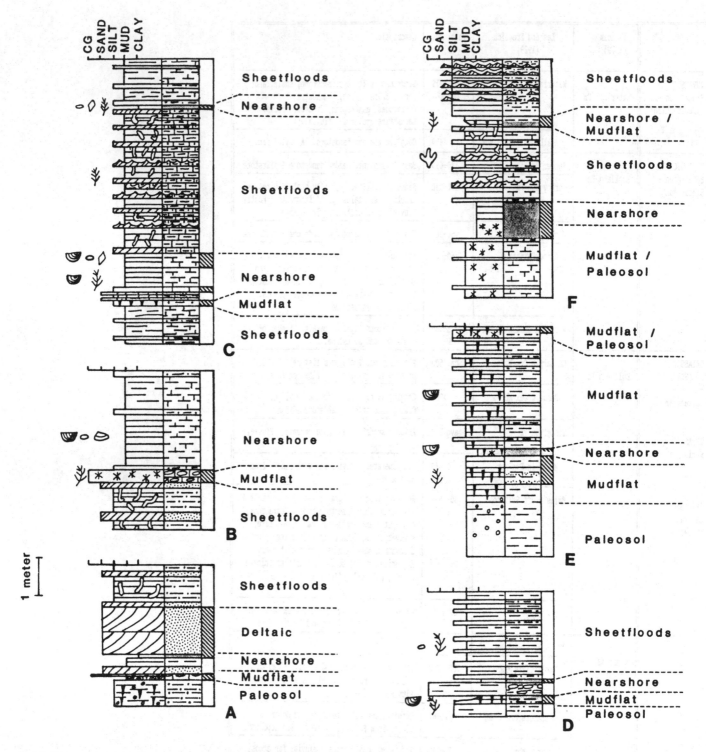

Fig. 3. Detailed lithologic sections of lacustrine sequences in the Culpeper Basin (from Gore, 1988a). These are primarily short sections (a few meters thick) presented in stratigraphic order (oldest to youngest), to illustrate changes in lacustrine style in the basin through time. In general, the lacustrine units thicken up-section (or toward the western border fault). These sections are not correlative with one another. See Gore (1983) for detailed section locations.
(A–F) 'Very thin' lacustrine sequences in the Balls Bluff Siltstone.

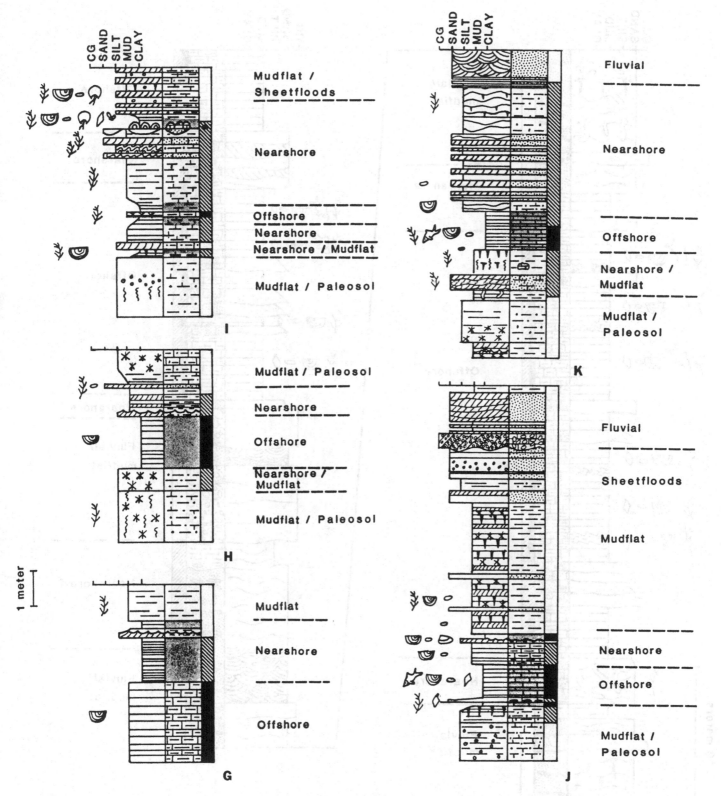

Fig. 3 (cont.) (G–J) 'Thin' lacustrine sequences in the Balls Bluff Siltstone. (K) Fish-bearing lacustrine sequence in the Midland Formation.

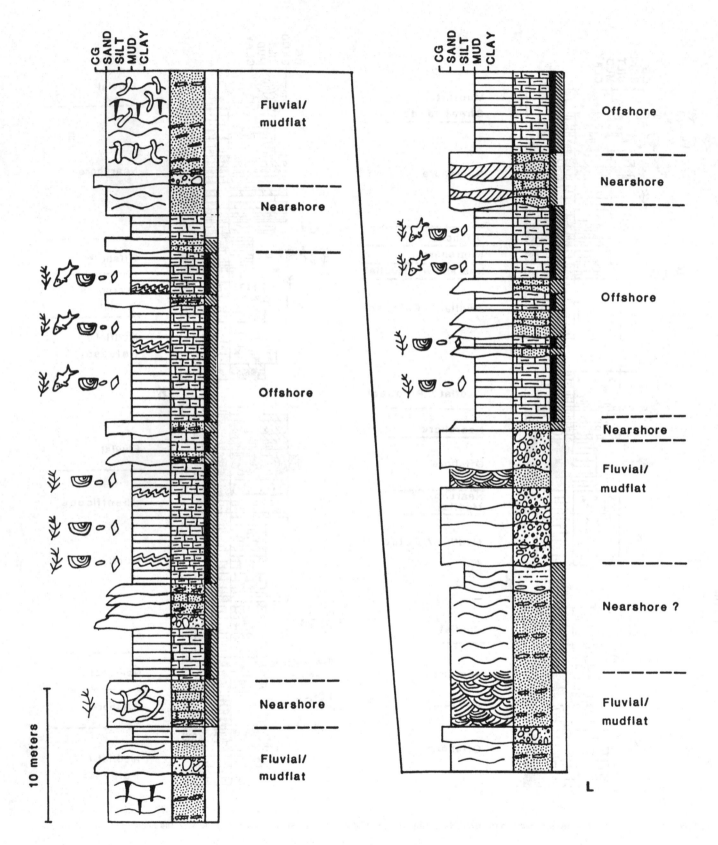

Fig. 3 (*cont.*) (L) 'Thick' lacustrine sequences in the Waterfall Formation (see Hentz 1981, 1985).

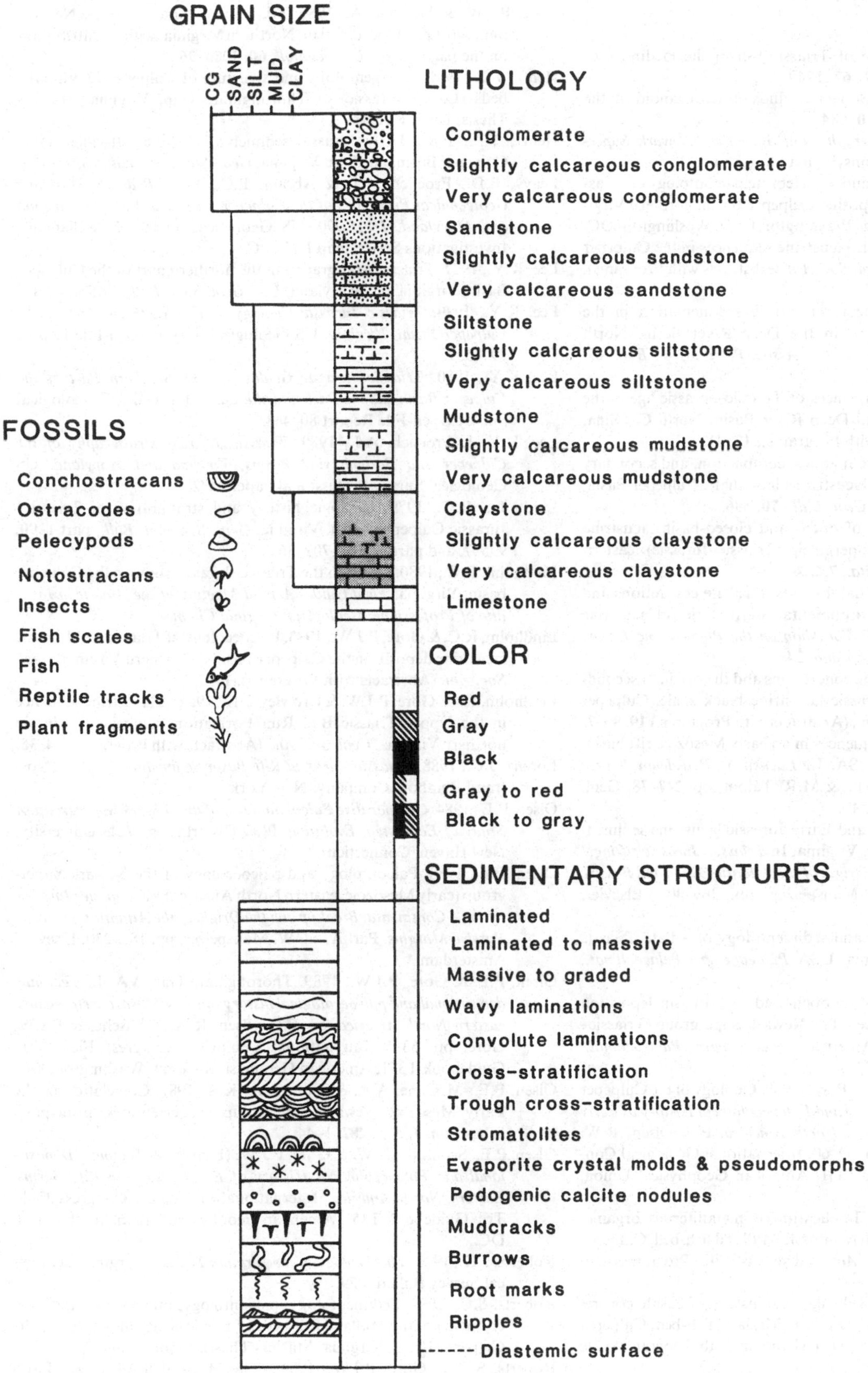

GRAIN SIZE

CG
SAND
SILT
MUD
CLAY

LITHOLOGY

Conglomerate
Slightly calcareous conglomerate
Very calcareous conglomerate
Sandstone
Slightly calcareous sandstone
Very calcareous sandstone
Siltstone
Slightly calcareous siltstone
Very calcareous siltstone
Mudstone
Slightly calcareous mudstone
Very calcareous mudstone
Claystone
Slightly calcareous claystone
Very calcareous claystone
Limestone

FOSSILS

Conchostracans
Ostracodes
Pelecypods
Notostracans
Insects
Fish scales
Fish
Reptile tracks
Plant fragments

COLOR

Red
Gray
Black
Gray to red
Black to gray

SEDIMENTARY STRUCTURES

Laminated
Laminated to massive
Massive to graded
Wavy laminations
Convolute laminations
Cross-stratification
Trough cross-stratification
Stromatolites
Evaporite crystal molds & pseudomorphs
Pedogenic calcite nodules
Mudcracks
Burrows
Root marks
Ripples
------ Diastemic surface

Fig. 3 (*cont.*) Key.

Bibliography

Applegate, S.P., 1956. Distribution of Triassic fish in the Piedmont of Virginia. *Geol. Soc. Am. Bull.*, **67**, 1749.

Baer, F.M. & Martin, W.H., 1949. Some new finds of fossil ganoids in the Virginia Triassic. *Science*, **110**, 684–6.

Cornet, B., 1977. *The Palynostratigraphy and Age of the Newark Supergroup.* PhD Dissertation, Pennsylvania State Univ.

Gore, P.J.W., 1983. Sedimentology and invertebrate paleontology of Triassic and Jurassic lacustrine deposits, Culpeper Basin, northern Virginia. PhD Dissertation, George Washington Univ. Washington, DC.

Gore, P.J.W., 1984. Triassic–Jurassic lacustrine sequences in the Culpeper Basin, northern Virginia. *Geol. Soc. Am.* (Abstracts with Programs), **16**, 141.

Gore, P.J.W., 1985a. Early Mesozoic lacustrine sedimentation in the Culpeper Basin, Virginia and in the Deep River Basin, North Carolina. A comparative study. *Am. Assoc. Petrol. Geol. Bull.*, **69**, 1438.

Gore, P.J.W., 1985b. Lacustrine sequences of Triassic–Jurassic age in the Culpeper Basin, Virginia and Deep River Basin, North Carolina. *Geol. Soc. Am.*, (Abstracts with Programs), **17**, 21.

Gore, P.J.W., 1986. Early diagenetic nodules, compaction, and secondary lamination in Early Jurassic lacustrine black shale, Culpeper Basin, Virginia. *Am. Assoc. Petrol. Geol. Bull.*, **70**, 596.

Gore, P.J.W., 1987a. Comparison of open- and closed-basin lacustrine deposition in the Newark Supergroup (Triassic–Jurassic), eastern North America. *Terra Cognita.*, **7**, 224.

Gore, P.J.W., 1987b. Precompactional textures in calcite concretions and their significance for paleoenvironmental interpretation of lacustrine shales. *Abstracts, IGCP-219 Workshop on the Phanerozoic Lacustrine Record, Kehrsiten, Switzerland*, 24.

Gore, P.J.W., 1987c. Early diagenetic concretions and the origin of secondary lamination in Early Jurassic lacustrine black shale, Culpeper Basin, Virginia. *Geol Soc. Am.*, (Abstracts with Programs), **19**, 86–7.

Gore, P.J.W., 1988a. Lacustrine sequences in an early Mesozoic rift basin: Culpeper Basin, Virginia, USA. In *Lacustrine Petroleum Source Rocks*, ed. A.J. Fleet, K. Kelts & M.R. Talbot, pp. 247–78. Geol. Soc. London Spec. Pub. No. 40.

Gore, P.J.W., 1988b. Late Triassic and Early Jurassic lacustrine sedimentation in the Culpeper Basin, Virginia. In *Triassic–Jurassic rifting – Continental Breakup and the Origin of the Atlantic Ocean and Passive Margins*, part A, ed. W. Manspeizer, pp. 369–400. Elsevier, Amsterdam.

Gore, P.J.W., 1988c. Paleoecology and sedimentology of a Late Triassic lake, Culpeper Basin, Virginia, USA *Palaeogeogr., Palaeoclimat., Palaeoecol.*, **62**, 593–608.

Gore, P.J.W., 1989. Toward a model for open- and closed-basin deposition in ancient lacustrine sequences: The Newark Supergroup (Triassic–Jurassic), eastern North America. *Palaeogeogr., Palaeoclimat., Palaeoecol.*, **70**, 29–51.

Gore, P.J.W., Olsen, P.E. & Schlische, R.W., 1989. Geology of the Culpeper Basin. In *Tectonic, Depositional and Paleoecological History of Early Mesozoic Rift Basins, Eastern North America*. P.E. Olsen, R.W. Schlische & P.J.W. Gore, pp. 59–60. International Geological Congress Field Trip Guidebook T351, American Geophysical Union, Washington, DC.

Gore, P.J.W. & Roberts, S.C., 1989. Implications of metalliferous, organic-rich lacustrine shale in the Early Jurassic Midland fish-bed, Culpeper Basin, Virginia. *Geol. Soc. Am.* (Abstracts with Programs), **21**, 17–18.

Gore, P.J.W. & Talbot, M.R., 1988. Diagenetic history of calcite concretions in lacustrine shale: Early Jurassic Midland fish-bed, Culpeper Basin, Virginia. *Geol. Soc. Am.* (Abstracts with Programs), **20**, A376–7.

Gore, P.J.W. & Traverse, A., 1986. Triassic notostracans in the Newark Supergroup, Culpeper Basin, Northern Virginia, with a contribution on the palynology. *J. Paleontol.*, **60**, 1086–96.

Hentz, T.F., 1981. Sedimentology and structure of Culpeper Group lake beds (Lower Jurassic) at Thoroughfare Gap, Virginia. Masters Thesis, Univ. Kansas.

Hentz, T.F., 1985. Early Jurassic sedimentation of a rift-valley lake: Culpeper Basin, northern Virginia. *Geol. Soc. Am. Bull.*, **96**, 92–107.

Leavy, B.D., Froelich, A.J. & Abram, E.C., 1983. *Bedrock Map and Geotechnical Properties of the Culpeper Basin and Vicinity, Virginia and Maryland, 1:125,000.* US Geological Survey, Miscellaneous Investigations Series, map I-1313-C.

Lee, K.Y., 1977. Triassic stratigraphy in the northern part of the Culpeper Basin, Virginia and Maryland. *U.S. Geol. Surv. Bull.*, 1422-C, 1–17.

Lee, K.Y., 1979. *Triassic–Jurassic Geology of the Northern Part of the Culpeper Basin, Virginia.* US Geological Survey Open-File Report 79–1557.

Lee, K.Y., 1980. *Triassic–Jurassic Geology of the Southern Part of the Culpeper Basin and the Barboursville Basin, Virginia.* US Geological Survey Open-File Report 80–468.

Lee, K.Y. & Froelich, A.J., 1989. *Triassic–Jurassic stratigraphy of the Culpeper and Barboursville Basins, Virginia and Maryland.* US Geological Survey Professional Paper 1472.

Lindholm, R.C., 1979. Geologic history and stratigraphy of Triassic–Jurassic Culpeper Basin, Virginia. *Geol. Soc. Am. Bull.*, part I, **90**, 995–7, and part II, **90**, 1702–36.

Lindholm, R.C., 1980. Guide to the Triassic–Jurassic rocks in the Culpeper basin, Virginia. *Field Guide, Annual Meeting of the American Institute of Professional Geologists, Virginia Chapter*.

Lindholm, R.C. & Gore, P.J.W., 1983. Paleoecology of Triassic and Jurassic lacustrine deposits in the Culpeper Basin of northern Virginia. *Geol. Soc. Am.* (Abstracts with Programs), **15**, 192.

Lindholm, R.C., Gore, P.J.W. & Crowley, J.K., 1982. A lacustrine sequence in the Upper Triassic Bull Run Formation (Culpeper Basin) in northern Virginia. *Geol. Soc. Am.* (Abstracts with Programs), **14**, 35.

Lorenz, J.C., 1988. *Triassic–Jurassic Rift-Basin Sedimentology.* Van Nostrand Reinhold Company, New York.

Olsen, P.E., 1984. *Comparative Paleolimnology of the Newark Supergroup. A Study of Ecosystem Evolution.* PhD Dissertation, Yale University, New Haven, Connecticut.

Olsen, P.E., 1988. Paleontology and paleoecology of the Newark Supergroup (early Mesozoic, eastern North America). In *Triassic–Jurassic rifting – Continental Breakup and the Origin of the Atlantic Ocean and Passive Margins*. Part A, ed. W. Manspeizer, pp. 185–230. Elsevier, Amsterdam.

Olsen, P.E. & Gore, P.J.W., 1989. Thoroughfare Gap, VA. In *Tectonic, depositional and paleoecological history of early Mesozoic rift basins, eastern North America*. ed. P.E. Olsen, R.W. Schlische, & P.J.W. Gore, pp. 63–5. International Geological Congress Field Trip Guidebook T351, American Geophysical Union, Washington, DC.

Olsen, P.E., McCune, A.R. & Thompson, K.S., 1982. Correlation of the Early Mesozoic Newark Supergroup by vertebrates, principally fishes. *Am. J. Sci.*, **282**, 1–44.

Olsen, P.E., Schlische, R.W. & Gore, P.J.W. (Eds) 1989. *Tectonic, Depositional and Paleoecological History of Early Mesozoic Rift Basins, Eastern North America.* International Geological Congress Field Trip Guidebook T351, American Geophysical Union, Washington, DC.

Roberts, J.K., 1928. *The Geology of the Virginia Triassic.* Virginia Geological Survey Bulletin, 29.

Roberts, S.C., 1989. Sedimentology, paleontology, mineralogy, and geochemistry of the Midland fish-bed, a Jurassic lacustrine deposit in the Culpeper Basin, Virginia. Masters Thesis, Emory Univ.

Roberts, S.C. & Gore, P.J.W., 1988a. The Midland fish-bed: An Early

Jurassic transgressive-regressive lacustrine sequence in the Culpeper Basin, Virginia (USA). (Program with Abstracts), *8th International Conference on Basement Tectonics, Butte, Montana*, p. 30.

Roberts, S.C. & Gore, P.J.W., 1988b. Sedimentary features and fossil distribution in an Early Jurassic transgressive-regressive lacustrine sequence in Virginia. *Geol. Soc. Am.* (Abstracts with Programs), 20, 311.

Schaeffer, B., Dunkle, D.H. & McDonald, N.G., 1975. *Ptycholepis marshi* Newberry, a chondrostean fish from the Newark Group of eastern North America. *Fieldiana Geol.*, **33**, 205–33.

Smith, M.A. & Robison, C.A., 1986. Lacustrine source rock potential, Culpeper Basin, Virginia. *Am. Assoc. Petrol. Geol. Bull.*, **70**, 650.

Smith, M.A. & Robison, C.R., 1988. Early Mesozoic lacustrine petroleum source rocks in the Culpeper basin, Virginia. In *Triassic–Jurassic Rifting – Continental Breakup and the Origin of the Atlantic Ocean and Passive Margins*, part B, ed. W. Manspeizer, pp. 697–709. Elsevier, Amsterdam.

Smoot, J.P. & Olsen, P.E., 1988. Massive mudstones in basin analysis and paleoclimatic interpretation of the Newark Supergroup. In *Triassic–Jurassic Rifting – Continental Breakup and the Origin of the Atlantic Ocean and Passive Margins*, part A, ed. W. Manspeizer, pp. 249–74.

Elsevier, Amsterdam.

Smoot, J.P. & Olsen, P.E., 1989. Culpeper Crushed Stone Quarry, Culpeper (Stevensburg), VA. In *Tectonic, Depositional and Paleoecological History of Early Mesozoic Rift Basins, Eastern North America*, ed. P.E. Olsen, R.W. Schlische, & P.J.W. Gore, pp. 60–3. International Geological Congress Field Trip Guidebook T351, American Geophysical Union, Washington, DC.

Smoot, J.P. & Robinson, G.R., Jr., 1988. Base- and precious-metal occurrences in the Culpeper Basin, northern Virginia. In *Studies of the Early Mesozoic Basins of the Eastern United States*, ed. A.J. Froelich & G.R. Robinson, pp. 403–23. US Geological Survey Bulletin 1776.

Sobhan, A.N., 1985. Petrology and depositional history of Triassic red beds, Bull Run Formation, eastern Culpeper Basin, Virginia. Masters Thesis, George Washington Univ., Washington, DC.

Sobhan, A.N., 1987. Fluvial-lacustrine ichnofossils from Triassic red beds, Virginia, USA. *Bangladesh J. Geol.*, **6**, 61–7.

Sobhan, A.N. & Lindholm, R.C., 1986. Alluvial paleosol of the Bull Run Formation, Culpeper Basin, Virginia, USA. *Bangladesh J. Geol.*, **5**, 49–57.

Toewe, E.C., 1966. *Geology of the Leesburg Quadrangle, Virginia*. Virginia Division of Mineral Resources Report of Investigations, 11.

Late Triassic rift valley lacustrine sequence in the Dan River (North Carolina)–Danville (Virginia) Basin, USA

PAUL A. THAYER[1] AND ELEANORA I. ROBBINS[2]

[1] *Department of Earth Sciences, University of North Carolina at Wilmington, Wilmington, NC 28403, USA*
[2] *US Geological Survey, Reston, VA 22092 USA*

Introduction

The Dan River-Danville basin (Fig. 1) is one of a series of tilted fault-block basins of the Late Triassic and Early Jurassic Newark rift system of eastern North America; it contains a thick sequence of lacustrine strata. The basin formed during the same tectonic event responsible for opening the Atlantic Ocean (Robbins, 1981; Gore & Olsen, 1989). The basin is a half-graben with present-day dimensions of 165 km by 2 to 10 km. It is bounded on the northwest by a complex system of steeply-dipping normal faults called the Dan River fault zone in North Carolina (NC) and the Chatham fault zone in Virginia (VA); faults within this zone dip 45°–65° SE (Meyertons, 1959; Thayer, 1970; Lineberger, 1983). The fault zone trends N30° E–N40° E and locally is cut by northwest- to northeast-trending cross faults. Rocks adjacent to the fault zone are brecciated and silicified and commonly display shear foliation. Movement along the Dan River–Chatham fault zone occurred during and after basin filling. Total vertical displacement along the fault zone may be as much as 4500 m (Meyertons, 1963; Thayer, 1970).

The contact between Triassic and adjacent crystalline rocks on the southeastern basin margin is an unconformity except near the North Carolina–Virginia state line, where discontinuous, northwest-dipping normal faults occur (Thayer, 1970; Henika & Thayer, 1977, 1983) (Figs 1 and 2). These faults are shorter and exhibit less displacement than those along the northwest border.

Upper Triassic non-marine sedimentary rocks that fill the basin are collectively termed the Dan River Group (Thayer, 1970) (Fig. 3). These rocks dip steeply toward the northwest fault margin at 20°–50° and are deformed into broad, gentle folds whose axes are transverse to the longitudinal basin axis. Total stratigraphic thickness is estimated at 1300 m in the narrowest part of the basin and 4500 m in the widest part (Thayer, 1970; Henika & Thayer, 1977, 1983) (Fig. 3). Analysis of gravity profiles suggests a maximum depth to basement of about 1.5 to 2 km just south of the North Carolina–Virginia state line near the Dan River fault zone (Geddes & Thayer, 1971). Post-depositional normal faults occur throughout the basin; near its northern terminus, faults uplift a small block of pre-Triassic granitic basement (Meyertons, 1963) (Fig. 1).

The basin is surrounded by late Precambrian and early Paleozoic medium- to high-rank metamorphic rocks that have been intruded by small granitoid bodies (Thayer, 1970; Henika & Thayer, 1977, 1983; Price *et al.*, 1980a, b, c; Marr, 1984). Near-vertical northwest- to northeast-trending diabase dike swarms of probable Jurassic age cut Triassic and adjacent country rocks (Meyertons, 1963; Thayer, 1970; Johnson *et al.*, 1985) and produce a characteristic magnetic signature (US Geological Survey, 1971).

Stratigraphy

Upper Triassic sedimentary rocks of the Dan River Group are red, reddish-grey, and grey to black terrigenous clastic rocks consisting of interbedded lithic conglomerate, arkose, lithic arkose, mudrock and thin coal beds. Individual units are characterized by abrupt lateral and vertical changes in texture, color, composition and thickness. Typically, conglomerate and coarse sandstone dominate the basin margins, whereas finer sandstone and mudrock occur in the axial part. The age of Dan River Group is middle Carnian to middle early Norian (Fig. 3), on the basis of pollen, spores and vertebrates (Olsen *et al.*, 1978, 1982; Robbins & Traverse, 1980; Robbins, 1982a, 1985; Olsen, 1984; R.E. Litwin, pers. comm., 1988; Smoot *et al.*, 1988).

In the wider northeastern and southwestern parts of the basin, the Dan River Group is divided into three units; from the base upward, these are the Pine Hall, Cow Branch and Stoneville Formations. The three formations intertongue, but, in most sections across strike, the main bodies of the three units lie one above the other (Figs 1–3) (Thayer, 1970; Henika & Thayer, 1977, 1983). In the narrow central part of the basin these units intertongue with the Dry Fork Formation (Meyertons, 1963), which fills the entire central and south-central part of Danville basin (Henika & Thayer, 1977, 1983; Price *et al.*, 1980b). The Dry Fork is synchronous with the Pine Hall, Cow Branch and Stoneville Formations, and is separated from them by inferred lithofacies boundaries drawn at the termination of the Cow Branch Formation, beyond which distinctive dark beds of the Cow Branch do not extend. The Pine Hall, Stoneville and Dry Fork Formations are divided into conglomerate, sandstone and red mudrock facies (Fig. 1).

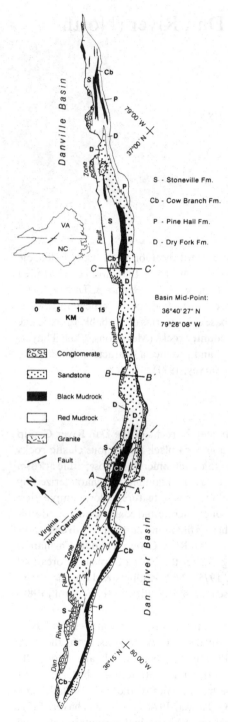

Fig. 1. Location and geologic map of the Dan River-Danville basin, North Carolina and Virginia. Black indicates predominantly lacustrine deposits. Most intrabasinal faults and numerous north- to northwest-trending diabase dikes are not shown. Geologic cross-sections, A–A′, B–B′, and C–C′ are given in Fig. 2. Abbreviations are: P, Pine Hall Formation; Cb, Cow Branch Formation; S, Stoneville Formation; and D, Dry Fork Formation. Numbers identify location of measured sections shown in Figures 4–6. Approximate basin center is at 36° 40′ 27″ N and 79° 28′ 08″ W. Compiled from Meyertons (1963), Thayer (1970, 1975a,b), Henika & Thayer (1977, 1983), Marr (1984), and Price et al. (1980a,b).

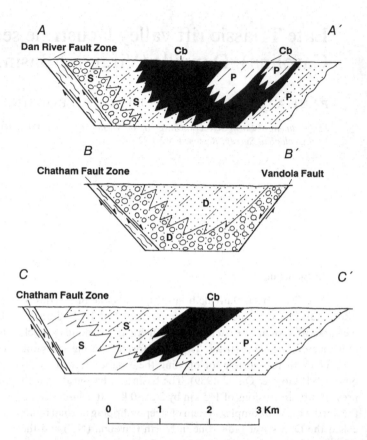

Fig. 2. Geologic cross-sections of the Dan River-Danville basin. Black indicates predominantly lacustrine deposits. Abbreviations are: P, Pine Hall Formation; Cb, Cow Branch Formation; S, Stoneville Formation; and D, Dry Fork Formation. Lithologic symbols are the same as those in Fig. 1. Intensely sheared cataclastic rocks along the faulted northwest basin margin (Dan River-Chatham fault zone) are indicated by wavy lines.

A comparison of sedimentary features in the Dan River Group with those reported from modern environments indicates deposition in alluvial fan, fluvial, floodplain, playa, lacustrine and swamp environments (Thayer, 1967; Olsen et al., 1978; Robbins 1982a, 1983; Olsen, 1984). Coarse, crudely-stratified conglomerate in the Pine Hall, Stoneville and Dry Fork Formations represents alluvial fan deposition adjacent to fault scarps that formed the basin margins. Cross-bedded, lenticular sandstone bodies in these units accumulated in high- to low-sinuosity streams. Finer-grained, reddish-brown to grey, uniformly thin- and medium-bedded mudrocks were deposited on floodplains, mudflats and playas and in shallow oxygenated lakes (Thayer, 1967; Smoot & Olsen, 1988). Fine-grained, dark, well bedded strata of the Cow Branch Formation were deposited in lakes and swamps that formed by damming axial and transverse drainage in the rift basin.

Cow Branch lacustrine sequence

In outcrop, the Cow Branch Formation occurs as elongate, discontinuous, lens-shaped bodies that are likely remnants of post-Triassic erosion (Fig. 1). The outcrop pattern suggests the existence

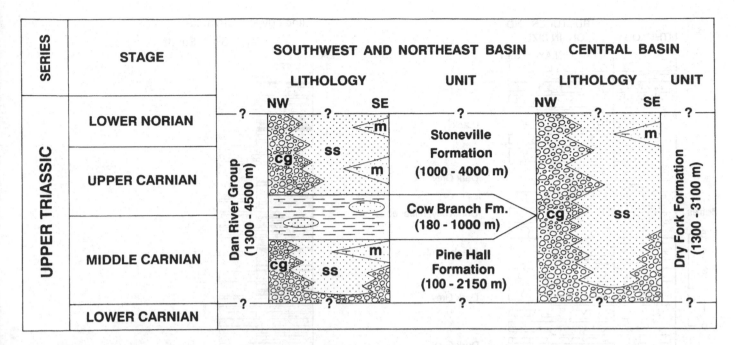

Fig. 3. Correlation chart of lithostratigraphic units in the Dan River-Danville basin. The usage shown is that of Smoot *et al.* (1988). The relationship between lithology and age is schematic. Abbreviations are: NW, northwest basin margin; SE, southeast basin margin; cg, conglomerate facies; ss, sandstone facies; and m, brown and reddish-brown mudrock facies. Based on Thayer (1970), Henika & Thayer (1977, 1983), Robbins (1982a, 1985), Olsen *et al.* (1982), and Smoot *et al.* (1988).

of multiple lakes and swamps of varying size throughout the history of basin filling. The lakes and swamps formed in local topographic lows that were probably associated with synsedimentary grabens within the basin. Palynomorphs recovered from these lens-shaped bodies show that they contain two discrete floral suites, the lower of Julian substage and the upper of Tuvalian substage (Robbins & Traverse, 1980; Robbins, 1982a; Brugman, 1983). The flora is not facies-controlled because the sampled lithologies were the same. On the basis of these floral differences, Robbins (1982a) informally designated lower and upper members of the Cow Branch. The lower member is late middle Carnian whereas the upper Cow Branch is early late Carnian (Fig. 3).

The lower Cow Branch crops out in a 50-km-long belt near the southeastern basin margin in North Carolina, where it averages about 180 m in thickness. The upper member is more widespread; it crops out or has been encountered in the subsurface throughout the northeastern two-thirds of the basin. Extensive quarrying by Solite Corporation at the state line in Eden, North Carolina, exposes a spectacular, nearly continuous 1000-m-thick section in the upper member (Locality 2, Fig. 1). Robbins (1982a) referred to this 17-km-long lens-shaped body of upper Cow Branch as 'Lake Danville'. Detailed studies of the sedimentology, paleontology and paleoecology of 'Lake Danville' have been conducted by Thayer (1967), Robbins (1982a), Olsen *et al.* (1978, 1982) and Olsen (1984).

Northeast of this locality, poor exposures of upper Cow Branch range from 1 to 20 km in length and 50 to 750 m in thickness (Fig. 1).

The lower Cow Branch is characterized by dark-grey, massive and laminated claystone, siltstone, shale, coal and tan to grey sandstone (Fig. 4). Finer-grained clastics display very thin to medium bedding; interbedded sandstones are lenticular and thick bedded. Ripple marks, ripple cross-lamination, desiccation cracks, burrow casts, graded bedding, convolute lamination, slump folds and load casts occur locally. Carbonate concretions up to 30 cm long, parallel to layering, are common in the basal 50 m, and are best developed in organic-rich shales. Thin lenticular layers of coal with autochthonous plant fragments, root structures, vivianite and abundant sulfide minerals are sporadically distributed throughout the lower Cow Branch.

These features suggest that the lower Cow Branch accumulated in a transitional fluvial-lacustrine environment. Organic-rich, concretion-bearing shales and coal beds formed in swamps and shallow stagnant lakes. Coarser grained, lenticular sandstones are probably delta plain deposits that formed along the shoaling margin, where small transverse rivers entered the basin (Thayer, 1967).

The upper Cow Branch is characterized by well-bedded, dark-grey to black shale and mudstone with subordinate reddish-brown, reddish-grey and green mudstone, and dark-grey, very fine to medium sandstone (Figs 5 and 6). Table 1 summarizes salient

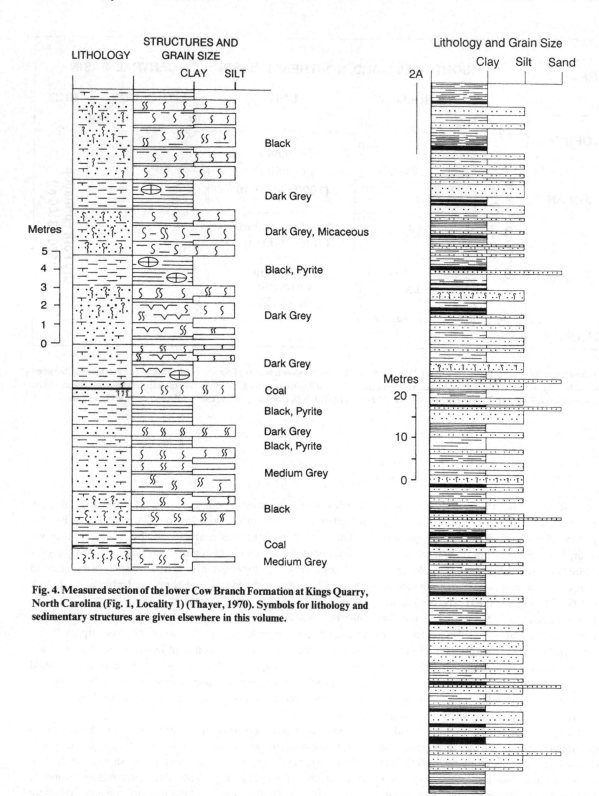

Fig. 4. Measured section of the lower Cow Branch Formation at Kings Quarry, North Carolina (Fig. 1, Locality 1) (Thayer, 1970). Symbols for lithology and sedimentary structures are given elsewhere in this volume.

Fig. 5. Measured section of the upper Cow Branch Formation at Solite Quarry, Virginia (Fig. 1, Locality 2). Black represents grey to black, finely-laminated mudstone. Other lithologic symbols are standard for this volume. Upper lacustrine units labeled 2A are shown in Fig. 6. Modified from Olsen (1984).

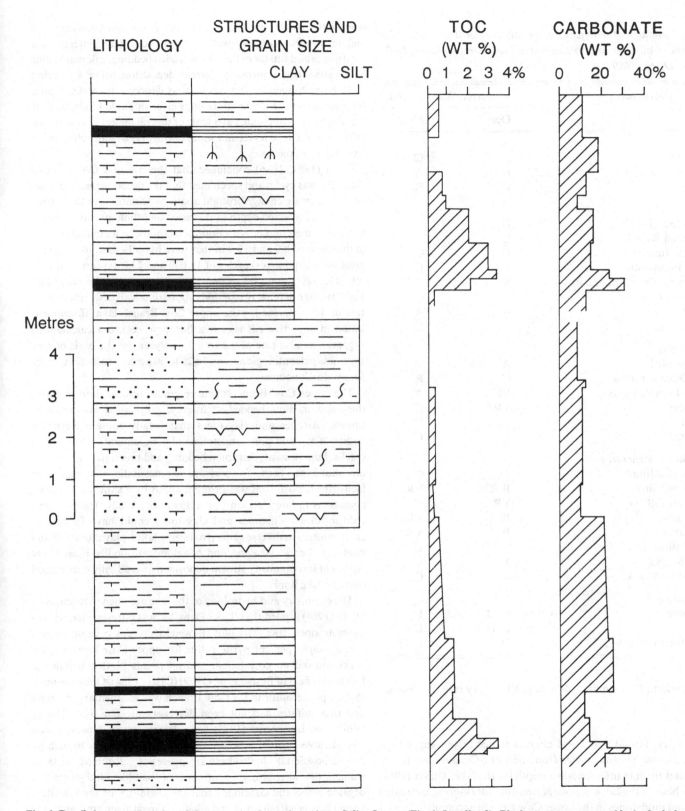

Fig. 6. Detailed measured section of the uppermost lacustrine units at Solite Quarry (Fig. 1, Locality 2). Black represents grey to black, finely-laminated mudstone. Symbols for lithology and sedimentary structures are given elsewhere in this volume. Total organic carbon (TOC) and carbonate are expressed as weight percent. Modified from Olsen (1984).

Table 1. *Summary of sedimentary, faunal, and floral characteristics of the upper Cow Branch Formation at Solite Quarry (Thayer, 1967; Gore and Olsen, 1989)*

Feature or Structure	Inferred Water Depth	
	Deep	Shallow
Lithology		
Claystone	A	C
Siltstone	R	A
Sandstone	R	R
Fossils		
Articulated Fish	C	—
Articulated Reptiles	C	—
Complete Insects	C	C
Reptile Footprints	—	C
Carbonized Plant Fragments	R	C
Plant Stem Casts	—	C
Root Structures	—	A
Burrows	—	C
Laminations		
Even, Parallel	A	—
Even, Discontinuous	C	R
Wavy, Discontinuous	VR	A
Lenticular	VR	C
Rippled	—	C
Brecciated	—	C
Sedimentary Structures		
Massive Bedding	—	C
Graded Bedding	R–C	VR
Scour and Fill	VR	C
Load Casts	R	VR
Sole Marks	R	VR
Bioturbation	—	C
Ripple Marks	VR	C
Desiccation Cracks	—	C
Composition		
Carbonate	L–H	L
Pyrite	L–H	L
Total Organic Carbon	H	L–H

Notes:
A – abundant; C – common; R – rare; VR – very rare; H – High; L – Low.

sedimentary, faunal and floral characteristics of the upper Cow Branch that have been reported from the Solite Corporation quarry and relates them to inferred water depth for the lake. Thayer (1967) and Robbins (1982a) have suggested that dark-grey, laminated shales and mudstones in the upper Cow Branch accumulated under anoxic conditions in deep water (>100 m) below lacustrine wave base. Variation in oxygenation of the water column produced lithologic and color changes in muds accumulating on the lake bottom (Robbins, 1982a). Dark-grey and black colors resulted from bottom-water anoxia, green from suboxic conditions and red and brown from bottom-water oxygenation. Thin, interbedded sandstones and siltstones that show graded bedding, sole marks and load casts suggest turbidity current deposition below lacustrine wave base. Numerous features such as desiccation cracks, reptile footprints, root structures and others given in Table 1 indicate that the water depth fluctuated at a scale of several meters. The cause of fluctuations in lake levels and water chemistry, whether climatic or tectonic, is unknown.

Olsen (1984, 1986) concluded that the rise and fall of 'Lake Danville' was cyclic and speculated that the cyclicity resulted from periodic climate change brought about by variations in the earth's orbit. Using inferred depth ranks based on lithology and sedimentary structures coupled with Fourier methods, he identified periods in thickness of 8.9 and 12.0 m that were hypothesized to represent precession (equinoxes) cycles of 19,000 and 23,000 years, respectively (Gore & Olsen, 1989). Olsen (1984) pointed out that the 21,000 years average of these two periods approximates to a sedimentation rate of 0.3 mm/year for the upper Cow Branch lake. If sedimentation in an active rift lake is a linear process, our calculations suggest a similar rate of 0.2 to 0.5 mm/year, on the basis of unit thickness paleontologic age, and a 20 to 50 percent reduction factor for mudrock volume.

The upper Cow Branch member grades upward into massive, thin- and medium-bedded, light-grey, greyish-red and reddish-brown siltstones and claystones that locally contain lenticular bodies of tan and grey sandstone and conglomerate. Desiccation cracks, burrow casts, disrupted bedding and brecciated mudstones are abundant. These features are characteristic of a regressive lacustrine deposit that was brought about by a shift toward drier conditions in late Carnian time (Hay *et al.*, 1982). The grey and reddish-brown siltstones and claystones could have formed in environments as diverse as deep oxidizing lakes, playas or oxidizing mudflats. Lenticular sand and conglomerate bodies most likely represent low-sinuosity stream deposits that gradually encroached over the lake floor.

The chemistry and hydrology of the lake are open to discussion. Olsen (1984) argued that 'Lake Danville' was a freshwater, hydrologically open lake. His plot of weight percent organic carbon versus weight percent pyrite sulfur for upper Cow Branch mudrocks showed no correlation, and all points plotted within the freshwater field of Berner *et al.* (1979) (Olsen, 1984). Olsen assumed that all pyrite sulfur in the Cow Branch was sedimentary in origin and that sulfates had not been diagenetically removed. Thayer (1967) and Lineberger (1983) have demonstrated, however, that considerable pyrite has been introduced into the Cow Branch by post-depositional hydrothermal processes. Robbins (1982a) studied the alternating occurrence of ball-like and sheet-like organic tissue and concluded that the chemistry of the lake fluctuated between fresh water and alkaline. Green algae are commonly ball-shaped and indicators of fresh water; cyanobacteria, which often proliferate in high-nutrient, alkaline waters, disintegrate into sheet-like masses (Robbins *et al.*, 1979; Robbins, 1983, 1984; Robbins *et al.*, 1990).

The original extent of Cow Branch lakes is difficult to estimate because of sparse surface exposures and subsurface borings, post-depositional faulting, and extensive post-Triassic erosion that may have removed as much as several kilometers of Triassic strata (Thayer, 1967; Lineberger, 1983). The lower Cow Branch lake was at least 60 km long and 5–10 km wide. 'Lake Danville' attained minimum width of 10–15 km and had a minimum length of 20 km. Upper Cow Branch lakes in the northeastern part of the basin probably were from 2–30 km long and 5–10 km wide.

Economic potential

Cow Branch Formation shales and mudstones are used for the manufacture of lightweight aggregate, building brick and structural wares (Meyertons, 1963; Thayer et al., 1970). The Dan River-Danville basin was targeted for uranium exploration (Dribus, 1978; Thayer & Cook, 1982) because of proximity to granitic source terrane, the likely existence of roll-front traps near the contact between reduced Cow Branch mudrocks and sandstone units, and indication of uranium from ground and airborne surveys. Although commercial uranium has not been found in the Cow Branch, Marline Uranium Corporation disclosed discovery of significant uranium mineralization along the Chatham fault zone in the Spring Garden and Java Quadrangles, Virginia (Henika & Thayer, 1983). Reserves are estimated at 30 million pounds of uranium oxide (U_3O_8) with an average grade in excess of four pounds per ton of ore. Thin, discontinuous seams of non-commercial high-volatile bituminous to anthracite coal occur in the basal part of the lower Cow Branch and sporadically in the upper Cow Branch; reserve estimates are given by Stone (1910), Robbins (1982b) and Robbins et al. (1988). The hydrocarbon potential of the basin is considered to be poor owing to deep burial and high heat flow associated with initial rifting and post-depositional diabase intrusions (de Boer & Snider, 1979; Thayer et al., 1982).

Acknowledgements

J.P. Smoot and L.C. Wnuk of the US Geological Survey reviewed an earlier draft of this contribution.

Bibliography

Berner, R.A., Baldwin, T. & Holdren, G.R., Jr., 1979. Authigenic iron sulfides as paleosalinity indicators. *J. Sed. Petrol.*, **49**, 1345–50.

Brugman, W.A., 1983. *Permian–Triassic Palynology*. Written Communication of the Laboratory of Palaeobotany and Palynology, State University, Utrecht.

de Boer, J. & Snider, F.G., 1979. Magnetic and chemical variations of Mesozoic diabase dikes from eastern North America: evidence for a hotspot in the Carolinas. *Geol. Soc. America Bull.*, **90**, 185–98.

Dribus, J.R., 1978. *Preliminary Study of the Uranium Potential of the Dan River Triassic Basin System, North Carolina and Virginia*. US Dept. Energy Rept. GJBX-131(78).

Geddes, W.H. & Thayer, P.A., 1971. Gravity investigation of the Dan River Triassic basin of North Carolina. *Geol. Soc. Am.* (Abstracts with Programs)., **3**, 312–13.

Gore, P.J.W. & Olsen, P.E., 1989. Geology of the Dan River-Danville basin. In *Tectonic, Depositional, and Paleoecological History of Early Mesozoic Rift Basins, Eastern North America.* ed. P.E. Olsen & P.J.W. Gore, pp. 35–45. 28th Int. Geol. Cong. Field Trip Guidebook T351.

Hay, W.W., Behensky, J.F., Jr., Barron, E.J. & Sloan, J.L., II, 1982. Late Triassic–Liassic paleoclimatology of the proto-central North Atlantic rift system. *Palaeogeogr., Palaeoclimatol., Palaeoecol.*, **40**, 13–30.

Henika, W.S. & Thayer, P.A., 1977. *Geology of the Blairs, Mount Hermon, Danville, and Ringgold Quadrangles*, Virginia. Va. Div. Min. Res. Pub. 2.

Henika, W.S. & Thayer, P.A., 1983. *Geological Map of the Spring Garden Quadrangle, Virginia*. Va. Div. Min. Res. Pub. 48, map scale 1:24,000.

Johnson, S.S., Wiener, L.S. & Conley, J.F., 1985. *Simple Bouguer Gravity Anomaly Map of the Danville-Dan River Basin and Vicinity, Virginia–North Carolina and the Scottsville Basin and Vicinity, Virginia.* Va. Div. Min. Res. Pub. 58, map scale 1:125,000.

Lineberger, D.H., Jr., 1983. *Geology of the Chatham fault zone, Pittsylvania County, Virginia*. MS thesis, Univ. North Carolina.

Marr, J.D., Jr., 1984. *Geological Map of the Pittsville and Chatham Quadrangles, Virginia, Triassic System by P.A. Thayer*. Va. Div. Min. Res. Pub. 49, map scale 1:24,000.

Meyertons, C.T., 1959. *The Geology of the Danville Triassic Basin of Virginia*. PhD Dissertation, Virginia Polytechnic Institute, Blacksburg, Virginia.

Meyertons, C.T., 1963. *Triassic Formations of the Danville Basin*. Va. Div. Min. Res. Rept. Inv. 6.

Olsen, P.E., 1984. *Comparative Paleolimnology of the Newark Supergroup: A Study of Ecosystem Evolution*. PhD Dissertation, Yale Univ.

Olsen, P.E., 1986. A 40-million year lake record of early Mesozoic orbital climatic forcing. *Science*, **234**, 842–8.

Olsen, P.E., McCune, A.R. & Thomson, K.S., 1982. Correlation of the early Mesozoic Newark Supergroup by vertebrates, principally fishes. *Am. J. Sci.*, **282**, 1–44.

Olsen, P.E., Remington, C.L., Cornet, B. & Thomson, K.S., 1978. Cyclic change in Late Triassic lacustrine communities. *Science*, **201**, 729–33.

Price, V., Conley, J.F., Piepul, R.G., Robinson, G.R. & Thayer, P.A., 1980a. *Geology of the Axton and Northeast Eden Quadrangles, Virginia.* Va. Div. Min. Res. Pub. 22, map scale 1:24,000.

Price, V., Conley, J.F., Piepul, R.G., Robinson, G.R., Thayer, P.A. & Henika, W.S., 1980b. *Geology of the Whitmell and Brosville Quadrangles, Virginia.* Va. Div. Min. Res. Pub. 21, map scale 1:24,000.

Price, V., Thayer, P.A. & Ranson, W.A., (Eds) 1980c. *Geologic Investigations of Piedmont and Triassic Rocks, Central North Carolina and Virginia.* Carolina Geological Society Field Trip Guidebook, Raleigh, N.C.

Robbins, E., Creamer, A. & Rose, S., 1990. Preliminary analysis of biostratigraphic changes and mineralogy of short cores from Oneida Lake, New York. In *The Biogeochemistry of Metal Cycling.* ed. K.H. Nealson, M. Nealson, and F.R. Dutcher, pp. 154–68. NASA Contractor Rept. 4295, Washington, D.C.

Robbins, E.I., 1981. A preliminary account of the Newark rift system. Lunar and Planetary Inst. Contr. 465, pp. 107–9. *Conference on the Processes of Planetary Rifting, St. Helena, CA.*

Robbins, E.I., 1982a. *'Fossil Lake Danville': The Paleoecology of a Late Triassic Ecosystem on the North Carolina-Virginia border*. PhD Dissertation, Pennsylvania State Univ.

Robbins, E.I., 1982b. Economic potential of ancient lakebeds in the Newark rift system (eastern North America). *Geol. Soc. Am.* (Abstracts with Programs), **14**, 77.

Robbins, E.I., 1983. Accumulation of fossil fuels and metallic minerals in active and ancient rifts. *Tectonophysics*, **94**, 633–58.

Robbins, E.I., 1984. The degradation of metal-bearing organic tissues (kerogen): Experiments, measurements, observations, and interpretations. *Palynology*, **8**, 250.

Robbins, E.I., 1985. Palynostratigraphy of coal-bearing sequences in Early Mesozoic basins of the eastern United States. In *Proceedings of the Second US Geological Survey Workshop on the Early Mesozoic Basins of the Eastern United States*. eds, G.R. Robinson, Jr. & A.J. Froelich, 27–9. US Geol. Survey Circ, 946.

Robbins, E.I., Niklas, K.J. & Sanders, J.E., 1979. Algal kerogens in the Newark Group lakebeds – their bearing on the Early Mesozoic history of the Atlantic continental margin. *Palynology*, **3**, 291.

Robbins, E.I. & Traverse, A., 1980. Degraded palynomorphs from the Dan River (North Carolina)-Danville (Virginia) basin. In *Geological Investigations of Piedmont and Triassic Rocks, Central North Carolina and Virginia*. ed. V. Price, Jr., P.A. Thayer & W.A. Ranson. Carolina Geological Society Field Trip Guidebook, Raleigh, N.C., B X1-X11.

Robbins, E.I., Wilkes, G.P. & Textoris, D.A., 1988. Coal deposits of the Newark rift system. In *Triassic-Jurassic Rifting, Continental Breakup, and the Origin of the Atlantic Ocean and Passive Margins*, ed. W. Manspeizer, pp. 649–82. Elsevier, New York.

Smoot, J.P. & Olsen, P.E., 1988. Massive mudstones in basin analysis and paleoclimatic interpretation of the Newark Supergroup. In *Triassic–Jurassic Rifting, Continental Breakup, and the Origin of the Atlantic Ocean and Passive Margins*. ed. W. Manspeizer, pp. 249–74. Elsevier, New York.

Smoot, J.P., Froelich, A.J. & Luttrell, G.W., 1988. Uniform symbols for the Newark Supergroup. In *Studies of the Early Mesozoic Basins of the Eastern United States*. ed. A.J. Froelich & G.R. Robinson, Jr. pp. 1–6, US Geol. Survey Bull, 1776.

Stone, R.W., 1910. *Coal on Dan River, North Carolina*. pp. 137–69. US Geol. Survey Bull. 471.

Thayer, P.A., 1967. *Geology of the Dan River and Davie County Triassic basins, North Carolina*. PhD Dissertation, Univ. North Carolina.

Thayer, P.A., 1970. Stratigraphy and geology of Dan River Triassic basin, North Carolina. *Southeastern Geology*, **12**, 1–31.

Thayer, P.A., 1975a. *Triassic Geology of the Gretna Quadrangle, Virginia*. Va. Div. Min. Res. Open-File Map, scale 1:24,000.

Thayer, P.A., 1975b. *Triassic Geology of the Java Quadrangle, Virginia*. Va. Div. Min. Res. Open-File Map, scale 1:24,000.

Thayer, P.A. & Cook, J.R., 1982. *Detailed Geochemical Study of the Dan River–Danville Triassic basin, North Carolina*. US Dept. of Energy Rept. GJBX-148(82).

Thayer, P.A., Kirstein, D.S. & Ingram, R.L., 1970. *Stratigraphy, Sedimentology and Economic Geology of Dan River basin, North Carolina*. Carolina Geological Survey Field Trip Guidebook, Raleigh, North Carolina.

Thayer, P.A., Robbins, E.I. & Ziegler, D.G., 1982. Hydrocarbon potential of the Dan River-Danville Triassic basin, North Carolina and Virginia. *Geol. Soc. Am.* (Abstracts with Programs), **14**, 89.

US Geological Survey, 1971. *Aeromagnetic map of the Danville Quadrangle, Pittsylvania County, Virginia, and Caswell County, North Carolina*, US Geol. Survey Map GP-745, map scale 1:62,500.

Upper Triassic organic-rich lake, Sanford sub-basin, North Carolina, USA

DANIEL A. TEXTORIS[1] AND PAMELA J.W. GORE[2]

[1] Department of Geology, University of North Carolina, Chapel Hill, North Carolina 27599, USA
[2] Department of Geology, DeKalb College, Clarkston, Georgia 30021, USA

Location and general geology

The Sanford sub-basin is located in central North Carolina, USA and constitutes the middle sub-basin of the Deep River basin (Fig. 1). The Durham sub-basin to the northeast is reported in the next contribution in this volume (see Textoris). The Sanford sub-basin is a half-graben bordered on the east by the Jonesboro fault zone, which was active primarily during filling of the sub-basin. The zone is several kilometers wide, and appears to consist of a series of steep normal step faults (Bain & Harvey, 1977; Bain & Brown, 1981; Lai et al., 1985; Brown, 1987; Ziegler, 1990), although recent seismic evidence indicates a low angle main fault (Davis et al., 1991). The western side of the half-graben is the passive portion consisting of nonconformities, and both syn- and post-depositional faults. The depocenter lies several kilometers northwest of Sanford, and seismic evidence indicates a total basin thickness of up to 2800 m. Drainage was fluvial to the southwest from the Durham sub-basin (Fig. 1). Overall, the beds dip to the southeast at 20°, but numerous major faults within the sub-basin affect the general pattern (Fig. 1). The sub-basin is riddled with Lower Jurassic diabase sills and dikes. The country rock surrounding the sub-basin consists of Upper Precambrian to Paleozoic metamorphics and intrusives.

The Sanford sub-basin is part of the extensive Newark rift system of eastern North America, which heralded the opening of the Atlantic Ocean basin during the Late Triassic–Early Jurassic (Manspeizer, 1988), and like other exposed basins, it is filled almost entirely by non-marine siliciclastics and intrusive diabase (Lorenz, 1988; Olsen et al., 1991).

Stratigraphy

The rocks in the Sanford sub-basin are divided into three rock units (defined by Campbell & Kimball, 1923; Reinemund, 1955), the Pekin, Cumnock and Sanford Formations (Fig. 2). They constitute the Chatham Group of the Newark Supergroup, and are Upper Ladinian? through Carnian, and possibly Lower Norian in age (Smoot et al., 1988; Luttrell, 1989; Gore et al., 1989). The Cumnock Formation, representing the major lake sequence nearly 300 m thick, is Middle Carnian.

The Pekin and Sanford Formations consist primarily of red and tan alluvial and fluvial conglomerate and sandstone; and siltstone and claystone of floodplain origin (Reinemund, 1955; Gore, 1986). Where the Cumnock Formation is not present, the Pekin and Sanford Formations merge and are indistinguishable from one another. Rare shale and limestone in the Pekin and Sanford constitute local floodplain lakes, and thick gypsum represents deposition in playas.

The Cumnock Formation consists of organic-rich shale, siltstone, sandstone, and coal. Dating and correlation have been done on paleontologic evidence; amphibians, reptiles, fish and invertebrates were reviewed by Olsen (1988); palynomorphs were reviewed by Robbins (1985) and Traverse (1986, 1987); and plants in the upper Pekin were reviewed by Gensel (1986). The Middle Carnian date is the official United States Geological Survey acceptance (Smoot et al., 1988) for the Cumnock Formation, although Luttrell (1989) indicates an Upper Carnian date based on an unpublished palynomorph study. The dikes and sills are considered to be Lower Jurassic (Brown et al., 1985). No extrusive volcanics or volcaniclastics have been identified.

This overall stratigraphic sequence represents a nonmarine, primarily siliciclastic basin filling. Lithologic, sedimentologic, and paleontologic evidence indicates evolution from a humid, warm climate to a dry, warm climate (Traverse, 1986; Textoris et al., 1989). On a regional scale, this climatic change occurred as the North American plate moved northward, with the Sanford sub-basin moving from about 10° N with wet easterly winds, northward to a dry zone (Hay et al., 1982; Manspeizer, 1982). Local variations in climate could have taken place due to tectonic activity along the Jonesboro fault zone. With uplift of the highlands southeast of the fault, the wet summer easterly winds coming off the new Atlantic Ocean and Tethys Sea would have been cut off, causing temporary dryness on the leeward side of the highlands.

Cumnock lake sequence

The Cumnock Formation represents the best-developed hydrologically-open lake sequence in the Deep River basin (Reinemund, 1955; Gore, 1989; Textoris et al., 1989), and the thickest continuous lake record in the entire Newark Supergroup. The

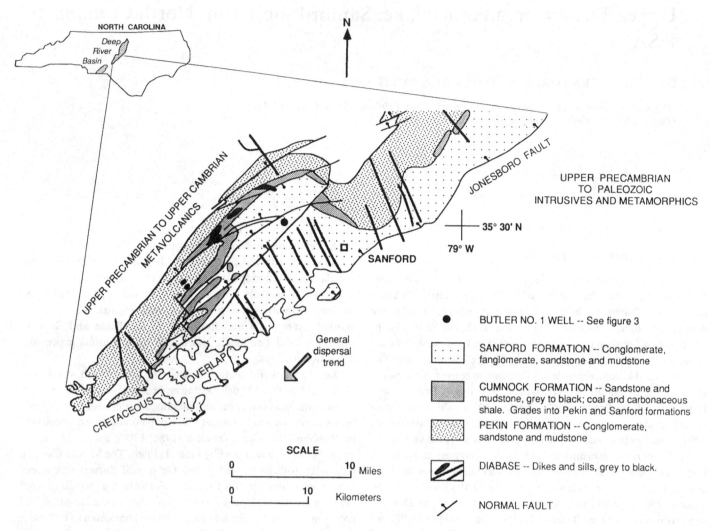

Fig. 1. Location and geologic map of the Sanford sub-basin, North Carolina, USA. The geology is modified from Reinemund (1955), Bain & Harvey (1977), Bain & Brown (1981), and Brown *et al.* (1985). General dispersal trend of fluvial sediment to the southwest is generalized from Reinemund (1955) and Patterson (1969). The approximate center of the lake sequence is at the Butler No. 1 well, 35° 30′ N and 79° 15′ W. This sub-basin is connected to the Durham sub-basin, to the northeast by the Colon cross-structure (an accommodation zone), and to the southwest to the Wadesboro sub-basin by another cross-structure (accommodation zone), most of which is covered by Cretaceous sedimentary rocks. Together, these three sub-basins constitute the Deep River basin.

sequence is nearly 300 m thick, and is best developed in the northwestern part of the sub-basin, in the vicinity of the Butler No. 1 well (Figs 1 and 3).

The Cumnock Formation consists of coal beds representing forested wetlands, carbonaceous shales representing emergent and shrub wetlands, black and grey shales representing lake deposition (oxygenated, suboxic and anoxic waters), and coarser-grained grey siliciclastics representing deltas and shorelines (Robbins & Textoris, 1988a; Textoris *et al.*, 1989). Authigenic apatite, siderite, pyrite, ammonium illite (Krohn *et al.*, 1988) and possibly chamosite and evaporite nodules, are found in the mudrocks. Illite, illite-smectite, chlorite and kaolinite (Robbins & Textoris, 1988b) vary in abundance depending on environment of deposition and extent of diagenesis.

The Cumnock lake was fed primarily by a southwest-flowing axial river from the Durham sub-basin to the north, and by major

rivers from a river-delta-beach system west and northwest of the depocenter. The lake reached dimensions of at least 20 × 60 km, and existed for one to two million years (Textoris *et al.*, 1989). Sedimentation rates of up to 0.6 mm/year seem reasonable, based on some calculations involving pre-compaction thickness and duration of lake existence (Hu & Textoris, 1991; Hu *et al.*, 1990). With the use of gamma-ray logs from five deep wells, Hu & Textoris (1991) and Hu *et al.* (1990) show the probable existence of cycles caused by astronomical climate forcing (i.e., Milankovitch and Van Houten cycles). Periodicities may represent 21,700, 100,000 and 400,000 year cycles.

The Cumnock Formation contains abundant ostracodes, conchostracans, fish scales, coprolites, plant fragments, pollen and spores, zooplankton fecal pellets, algal balls (Textoris *et al.*, 1989), as well as other fossil groups listed in the preceding section.

One of the economic resources of the Cumnock Formation is a

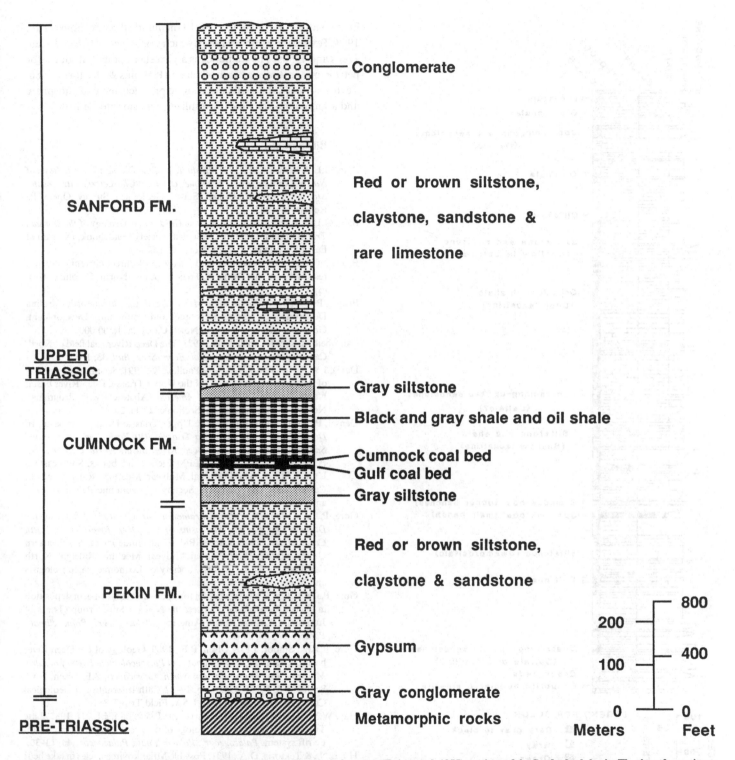

Fig. 2. Generalized stratigraphy for the central (Butler No. 1 well) and northwestern (Reinemund, 1955) portions of the Sanford sub-basin. The three formations belong to the Chatham Group, Newark Supergroup. The lower part of the Pekin Formation may be Upper Ladinian, with the remainder Lower to Middle Carnian. The Cumnock Formation, the major lake sequence, is Middle Carnian. The Sanford Formation is Carnian to Upper Carnian to possibly Lower Norian (Smoot *et al.*, 1988; Luttrell, 1989; Gore *et al.*, 1989). Other lake beds in the sub-basin, not discussed in this report, are (1) the 50 m of gypsum and gypsiferous beds near the base of the Pekin, representing playa conditions, (2) several thin gray to green shales representing local, shallow lakes in the Pekin, and (3) several reddish limestone and shale beds in the Sanford representing local shallow lakes.

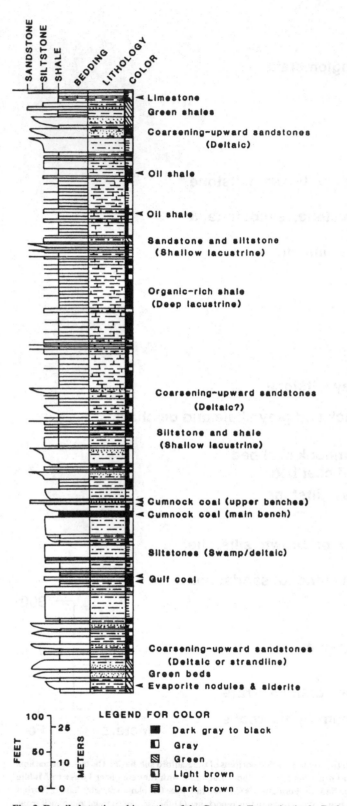

LEGEND FOR COLOR

■	Dark gray to black
▯	Gray
▨	Green
▤	Light brown
▦	Dark brown

Fig. 3. Detailed stratigraphic section of the Cumnock Formation in the Butler No. 1 well, located near the center of the Sanford sub-basin (see Fig. 1). Information was derived from air-drilled cuttings and various logs. Note variety of paleoenvironments including swamp, strandline, delta, and shallow and deep lake.

high volatile A bituminous coal (Reinemund, 1955; Robbins *et al.*, 1988; Textoris & Robbins, 1988), which was mined between the late 1700s and 1950s. The Cumnock is an excellent potential source for petroleum, natural gas and oil shale (Robbins & Textoris, 1986; Textoris *et al.*, 1989). Other economic products include phosphate and ammonium-rich illite for fertilizer, and siderite for iron.

Bibliography

Bain, G.L. & Brown, C.E., 1981. *Evaluation of the Durham Triassic Basin of North Carolina and Technique Used to Characterize its Waste-Storage Potential.* United States Geological Survey Open-File Report 80–1295.

Bain, G.L. & Harvey, B.W., 1977. *Field Guide to the Geology of the Durham Triassic Basin.* Carolina Geological Society Guidebook, Division of Earth Resources, Raleigh, North Carolina.

Brown, C.E., 1987. Modeling and analysis of direct-current electrical resistivity in the Durham Triassic basin, North Carolina. *Geoexploration*, **24**, 429–40.

Brown, P.M. *et al.*, 1985. *Geologic Map of North Carolina.* North Carolina Department of Natural Resources and Community Development, Geological Survey, Raleigh, North Carolina, 1:500,000.

Campbell, M.R. & Kimball, K.W., 1923. The Deep River coal field of North Carolina. *North Carolina Geol. Econ. Surv. Bull.* **33**, 1–95.

Davis, J.W.S., Jr., Textoris, D.A. & Paull, C.K., 1991. Seismic observations of the geometry and origin of the Upper Triassic Deep River basin, North Carolina. *Geol. Soc. America* (Abstracts with Programs), Northeastern–Southeastern Sections., **23**(1), 20.

Gensel, P., 1986. Plant fossils of the Upper Triassic Deep River basin. In *Depositional Framework of a Triassic Rift Basin: the Durham and Sanford Subbasins of the Deep River Basin, North Carolina.* ed. P.J.W. Gore, pp. 82–6. In SEPM Field Guidebooks, Southeastern United States, Third Annual Midyear Meeting, Raleigh, North Carolina ed. D.A. Textoris. Society of Economic Paleontologists and Mineralogists.

Gore, P.J.W., 1986. *Depositional Framework of a Triassic Rift Basin: the Durham and Sanford Subbasins of the Deep River basin, North Carolina,* pp. 53–115. In SEPM Field Guidebooks, Southeastern United States, Third Annual Midyear Meeting, Raleigh, North Carolina. ed. D.A. Textoris, Society of Economic Paleontologists and Mineralogists.

Gore, P.J.W., 1989. Toward a model for open- and closed-basin deposition in ancient lacustrine sequences: the Newark Supergroup (Triassic–Jurassic), Eastern North America. *Palaeogeogr. Palaeoclimat., Palaeoecol.,* **70**, 29–51.

Gore, P.J.W., Smoot, J.P. & Olsen, P.E., 1989. Geology of the Deep River basin. In *Tectonic, Depositional, and Paleoecological History of Early Mesozoic Rift Basins, Eastern North America,* ed. P.E. Olsen, R.W. Schlische & P.J.W. Gore, pp. 19–22. 28th International Geological Congress, Washington, DC, USA, Field Trip T-351.

Hay, W.W., Behensky, J.F., Jr., Barron, E.J. & Sloan, J.L., II, 1982. Late Triassic–Liassic paleoclimatology of the proto-central North Atlantic rift system. *Palaeogeogr., Palaeoclimat., Palaeoecol.,* **40**, 13–30.

Hu, L.N., & Textoris, D.A., 1991. Possible Milankovitch cycles in lake beds of the Triassic Sanford subbasin of North Carolina. *Geol. Soc. America* (Abstracts with Programs), Northeastern–Southeastern Sections, **23** (1), 47.

Hu, L.N., Textoris, D.A. & Filer, J.K., 1990. Cyclostratigraphy from gamma-ray logs, Upper Triassic lake beds (Middle Carnian), North Carolina, USA. *13th International Sedimentological Congress, Nottingham, England, Abstracts,* 548–9.

Krohn, M.D., Evans, J. & Robinson, G.R., Jr., 1988. Mineral-bound ammonium in black shales of the Triassic Cumnock Formation, Deep River basin, North Carolina. In *Studies of the Early Mesozoic Basins of the Eastern United States*, ed. A.J. Froelich & G.R. Robinson, Jr., pp. 86–98. US Geol. Survey Bull., 1776.

Lai, S.F., Ferguson, J.F., Aiken, C.L.V. & Ziegler, D.G., 1985. A test of gravity and magnetic inversion in defining the structure of the Sanford basin, North Carolina, as an example of a Triassic basin. *Society of Exploration Geophysicists, Expanded Abstracts with Biographies, Technical Programs*, 207–10.

Lorenz, J.C., 1988. *Triassic–Jurassic Rift-Basin Sedimentology*. Van Nostrand Reinhold. New York.

Luttrell, G.W., 1989. *Stratigraphic Nomenclature of the Newark Supergroup of Eastern North America*. US Geol. Survey Bull. 1572.

Manspeizer, W., 1982. Triassic-Liassic basins and climate of the Atlantic passive margins. *Geol. Rundschau*, **71**, 895–917.

Manspeizer, W., 1988. Triassic–Jurassic rifting and opening of the Atlantic: an overview. In *Triassic–Jurassic Rifting: Continental Breakup and the Origin of the Atlantic Ocean and Passive Margins*, part A, ed. W. Manspeizer, pp. 41–79. Elsevier, Amsterdam.

Olsen, P.E., 1988. Paleontology and paleoecology of the Newark Supergroup (early Mesozoic, eastern North America). In *Triassic–Jurassic Rifting: Continental Breakup and the Origin of the Atlantic Ocean and Passive Margins*, part A, ed. W. Manspeizer, pp. 185–230. Elsevier, Amsterdam.

Olsen, P.E., Froelich, A.J., Daniels, D.L., Smoot, J.P. & Gore, P.J.W., 1991. Rift basins of Early Mesozoic age. In *The Geology of the Carolinas*, ed. J.W. Horton, Jr. and V.A. Zullo, chapter 9. Carolina Geological Society 50th Anniversary Volume, University of Tennessee Press, Knoxville.

Patterson, O.F., III, 1969. *The Depositional Environment and Paleoecology of the Pekin Formation (Upper Triassic) of the Sanford Triassic Basin, North Carolina*. Masters Thesis, North Carolina State Univ.

Reinemund, J.A., 1955. *Geology of the Deep River Coal Field, North Carolina*. US Geol. Survey Prof. Paper 246.

Robbins, E.I., 1985. Palynostratigraphy of coal-bearing sequences in Early Mesozoic basins of the Eastern United States. In *Proceedings of the Second US Geological Survey Workshop on the Early Mesozoic Basins of the Eastern United States*, ed. G.R. Robinson, Jr. & A.J. Froelich, pp. 27–9. US Geol Survey Circ. 946.

Robbins, E.I. & Textoris, D.A., 1986. Fossil fuel potential of the Deep River basin, North Carolina. In *Depositional framework of a Triassic rift basin: the Durham and Sanford subbasins of the Deep River basin, North Carolina*, ed. P.J.W. Gore, pp. 75–9. SEPM Field Guidebooks, Southeastern United States, ed. D.A. Textoris. Third Annual Midyear Meeting, Raleigh, North Carolina. Society of Economic Paleontologists and Mineralogists.

Robbins, E.I. & Textoris, D.A., 1988a. Origin of Late Triassic coal in the Deep River basin of North Carolina (USA). *International Association of Sedimentologists, International Symposium on Sedimentology Related to Mineral Deposits, Beijing, PRC, Abstracts Volume*, 219–20.

Robbins, E.I. & Textoris, D.A., 1988b. *Analysis of Kerogen and Biostratigraphy of Core from the Dummit-Palmer No. 1 Well, Deep River Basin, North Carolina*. United States Geological Survey Open-File Report 88–670.

Robbins, E.I., Wilkes, G.P. & Textoris, D.A., 1988. Coal Deposits of the Newark rift system. In *Triassic–Jurassic Rifting: Continental Breakup and the Origin of the Atlantic Ocean and Passive Margins*, part B, ed. W. Manspeizer, pp. 649–82. Elsevier, Amsterdam.

Smoot, J.P., Froelich, A.J. & Luttrell, G.W., 1988. Uniform symbols for the Newark Supergroup. In *Studies of the Early Mesozoic Basins of the Eastern United States*, ed. A.J. Froelich & G.R. Robinson, Jr. US Geol. Survey Bull. 1776.

Textoris, D.A. & Robbins, E.I., 1988. *Coal resources of the Triassic Deep River basin, North Carolina*. US Geol. Survey Open-File Report 88–682.

Textoris, D.A., Robbins, E.I. & Gore, P.J.W., 1989. Origin of organic-rich strata in an Upper Triassic rift basin lake, eastern USA. *28th International Geological Congress, Washington, DC, USA, Abstracts*, vol. 3, pp. 229–30.

Traverse, A., 1986. Palynology of the Deep River basin, North Carolina. In *Depositional Framework of a Triassic Rift Basin: The Durham and Sanford subbasins of the Deep River Basin, North Carolina*, ed. P.J.W. Gore, pp. 66–71. SEPM Field Guidebooks, Southeastern United States, ed. D.A. Textoris. Third Annual Midyear Meeting, Raleigh, North Carolina. Society of Economic Paleontologists and Mineralogists.

Traverse, A., 1987. Pollen and spores date origin of rift basins from Texas to Nova Scotia as Early Late Triassic. *Science*, **236**, 1469–72.

Ziegler, D.G., 1990. Seismic stratigraphy in the Triassic Sanford basin, North Carolina. *Geol. Soc. Am. Ann. Meeting, Dallas, Texas* (Abstracts with Programs), **22**, 203.

Upper Triassic playa, Durham sub-basin, North Carolina, USA

DANIEL A. TEXTORIS

Department of Geology, University of North Carolina, Chapel Hill, North Carolina 27599, USA

Location and general geology

The Durham sub-basin, named and first described in detail by Prouty (1931), represents the northern one-third of the Deep River basin of central North Carolina (Fig. 1). The sub-basin is a half-graben bounded on the east by the Jonesboro fault zone which was active primarily during filling of the sub-basin (Bain & Harvey, 1977; Bain & Brown, 1981; Brown, 1987). This fault zone, several kilometers wide, may consist of a series of high angle, normal, step faults. Preliminary seismic evidence, however, indicates a low angle for the major fault in the southern part of the sub-basin (Davis *et al.*, 1991). The west, or passive, side of the sub-basin is bounded by nonconformities, and both syn- and post-depositional normal faults (Harrington, 1951; Cobrain & Textoris, 1991). The country rock consists of Upper Precambrian to Paleozoic intrusives, meta-volcanics and high-grade metamorphics (Fig. 1).

The Durham sub-basin is part of the eastern North American Newark rift zone formed during the Late Triassic–Early Jurassic which resulted in the opening of the Atlantic Ocean basin (Manspeizer, 1988); and like those other basins which are exposed, consists entirely of non-marine sedimentary rocks and diabase intrusives and extrusives (Lorenz, 1988; Olsen *et al.*, 1991).

The sub-basin contains between 1800 m (Bain & Harvey, 1977; Bain & Brown, 1981; Brown, 1987) and 3000 to 4200 m (Hoffman & Gallagher, 1989a; Parker, 1979) of alluvial and fluvial conglomerates, fluvial sandstones and floodplain mudstones. The central portion contains the limestone, chert and mudstone of the playa sequence (Wheeler & Textoris, 1978), the subject of this contribution. Lower Jurassic diabase dikes are common, as are intrabasinal normal faults (Fig. 1). The beds generally dip 10° to 20° east and southeast toward the Jonesboro fault zone. Overall drainage was axial to the southwest (Dittmar, 1979). At least 600 m has been eroded from the sub-basin and the surrounding country rock since faulting (Harrington, 1951).

Stratigraphy

The rocks in the sub-basin have been assigned to the undivided Chatham Group, Newark Supergroup, of Upper Ladinian? through Carnian age (Fig. 2). The playa sequence is Middle to Upper Carnian. Ages are based primarily on physical stratigraphic correlation from the next sub-basin to the south, the Sanford sub-basin. The Sanford sub-basin has been studied in more detail because of coal, gas and oil exploration, and has been the subject of a monograph by Reinemund (1955). The Sanford sub-basin (elsewhere in this volume, Textoris and Gore), has a three-part stratigraphic sequence, in the middle of which are 300 m of fossiliferous organic-rich shales, siltstones and sandstones of lake origin, and several extensive coal beds. The fossils used to establish dates are fish, amphibians, reptiles, plants, pollen and spores, and invertebrates reviewed by Olsen (1988), Traverse (1986, 1987), Robbins *et al.* (1988) and Gensel (1986). Correlation from this sub-basin into the Durham sub-basin by Bain & Harvey (1977) and Bain & Brown (1981) was accomplished by physical stratigraphic inference. The overall compilation by the US Geological Survey of the Newark Supergroup is also used in Fig. 2 (Smoot *et al.*, 1988; Luttrell, 1989). An informal stratigraphic correlation has been introduced (Gore *et al.*, 1989), which basically matches that in Fig. 2. Where rare floodplain lake beds exist (Gore, 1986), fossils may be very abundant and confirm the inferred dates. The dikes and sills are considered Lower Jurassic (Brown *et al.*, 1985). Thus far, no extrusive volcanics have been identified in the sub-basin, either as flows or volcaniclastics.

Nearly all of the basin-fill rocks are siliciclastic and alluvial-fluvial-floodplain in origin. Except for silicified logs and *Scoyenia* burrows, fossils are rare, especially those that would serve as index fossils. The upper part of the sub-basin sequence represents a general evolution to a dry climate experienced by the Deep River basin (Traverse, 1986) as the North American plate moved northward from 10° N during the Carnian (Hay *et al.*, 1982; Manspeizer, 1982; Textoris, Robbins, & Gore, 1989), from a warm humid climatic zone to one which was warm but dry.

Playa sequence

The stratigraphic section of the playa sequence of the Chatham Group is at least 8.8 m thick (Fig. 3), and includes distinctive beds of limestone with caliche fabric overlays and pure

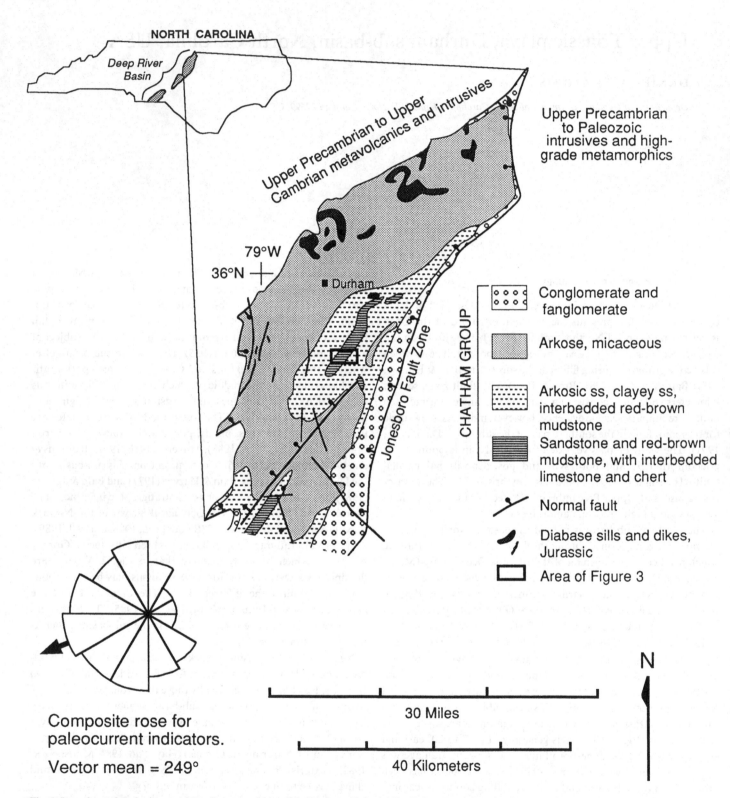

Composite rose for
paleocurrent indicators.

Vector mean = 249°

Fig. 1. Location and geologic map with sedimentary facies of the Durham sub-basin, North Carolina, USA. Modified from Bain & Harvey (1977), Bain & Brown (1981), and Brown *et al.* (1985). Paleocurrent rose diagram is from Dittmar (1979). Playa location is 35° 52′ N and 78° 52′ W, in the area of Fig. 3. Another probable playa is located in the southern part of the sub-basin in an accommodation zone (the Colon cross-structure), as shown by the pattern of parallel lines.

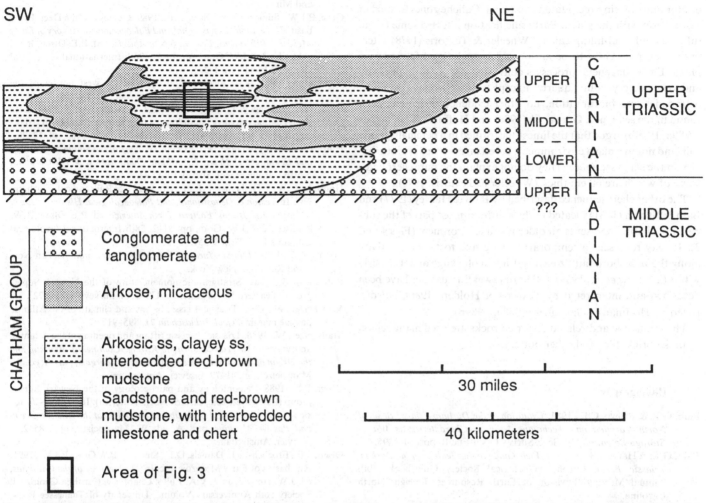

DURHAM SUBBASIN

Fig. 2. Stratigraphic and facies diagram of the Durham sub-basin. Depocenter thickness may reach 4200 m. The line of section is axial from the southwest to the northeast. The information is modified from Bain & Harvey (1977), Bain & Brown (1981), Traverse (1987), Smoot *et al.* (1988) and Luttrell (1989). The Chatham Group is undifferentiated as to formal members and formations. Note the Middle to Upper Carnian stratigraphic position for the playa sequence of this report in the area of Fig. 3. Not included in this report is the probable playa sequence in the southwestern part of the sub-basin, which appears to be Lower Carnian.

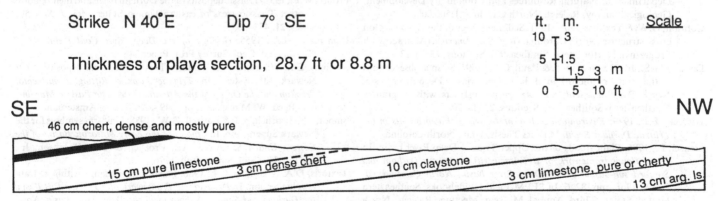

Fig. 3. Outcrop sketch of playa limestone and chert locality of the Durham sub-basin, slightly modified from Wheeler & Textoris (1978). The measured section was located at the corner of North Carolina Highway 54 and County Road 1999, 10 km south of Durham (see Figs 1 and 2 for location). This outcrop was removed by road construction in 1980, but the stratigraphic sequence is now exposed along the southward extension of County Road 1999, which is essentially parallel to strike.

chert, both associated with red-brown mudstone. Wheeler & Textoris (1978) interpreted the limestone to have been a calcareous tufa of inorganic origin precipitated in a playa. Caliche zones formed in association with the playa. Early silicification affected some of the tufas, as did void-filling calcite. Wheeler & Textoris (1978) interpreted the pure chert to be an inorganic opaline gel formed in a playa. Early diagenetic processes converted the gel to chalcedony and very fine crystalline quartz. Except for rare woody remnants, ostracodes and bioturbation, no other evidence of life has been found in the sequence. Gore *et al.* (1989) and Hoffman & Gallagher (1989a, 1989b) argue that the limestone formed in a perenially moist soil, and not in a playa environment, based on evidence of bioturbation in the same sequence. They do not explain the pure chert beds, some of which are 60 cm thick (Wheeler & Textoris, 1978).

The other chert sequence described by Bain & Harvey (1977) and Bain & Brown (1981) located in the southern-most part of the sub-basin probably represents an older playa environment (Figs 1 and 2). It may represent a temporary drying due to tectonic activity along the Jonesboro fault zone resulting in blockage of wet easterly winds (Textoris *et al.*, 1989). Other dry–wet fluctuations have been described and interpreted by Textoris & Holden (1986), but disputed by Hoffman & Gallagher (1989a, 1989b).

The mudstone and related clay-rich rocks are used in the region to make brick, tiles and other pottery.

Bibliography

Bain, G.L. & Brown, C.E., 1981. *Evolution of the Durham Triassic Basin of North Carolina and Technique Used to Characterize its Waste-Storage Potential*. US Geol. Survey Open-File Report 80–1295.

Bain, G.L. & Harvey, B.W., 1977. *Field Guide to the Geology of the Durham Triassic Basin*. Carolina Geological Society Guidebook, 70th Annual Meeting, Division of Earth Resources, Raleigh, North Carolina.

Brown, C.E., 1987. Modeling and analysis of direct-current electrical resistivity in the Durham Triassic basin, North Carolina. *Geoexploration*, **24**, 429–40.

Brown, P.M., *et al.*, 1985. *Geologic map of North Carolina*. North Carolina Department of Natural Resources and Community Development, Geological Survey, Raleigh, North Carolina, 1:500,000.

Cobrain, D.W. & Textoris, D.A., 1991. Sedimentology of the Triassic Colon cross-structure, North Carolina. *Geol. Soc. America*, (Abstracts with Programs), Northeastern–Southeastern Sections, **23** (1), 17.

Davis, J.W.S., Jr., Textoris, D.A. & Paull, C.K., 1991. Seismic observations of the geometry and origin of the Upper Triassic Deep River basin, North Carolina. *Geol. Soc. America* (Abstracts with Programs), Northeastern–Southeastern Sections, **23** (1), 20.

Dittmar, E.I., 1979. *Environmental Interpretations of Paleocurrents in the Triassic Durham Basin*. Masters Thesis, Univ. North Carolina.

Gensel, P., 1986. Plant fossils of the Upper Triassic Deep River basin. In *Depositional framework of a Triassic rift basin: the Durham and Sanford sub-basins of the Deep River basin, North Carolina*, ed. P.J.W. Gore, pp. 82–6. In SEPM Field Guidebooks, Southeastern United States, Third Annual Midyear Meeting, Raleigh, North Carolina, ed. D.A. Textoris. Society of Economic Paleontologists and Mineralogists.

Gore, P.J.W. (Ed.), 1986. *Depositional framework of a Triassic rift basin: the Durham and Sanford sub-basins of the Deep River basin, North Carolina*, pp. 53–115. In *SEPM Field Guidebooks*, Southeastern United States, Third Annual Midyear Meeting, Raleigh, North Carolina, ed. D.A. Textoris, Society of Economic Paleontologists and Mineralogists.

Gore, P.J.W., Smoot, J.P. & Olsen, P.E., 1989. Geology of the Deep River basin. In *Tectonic, Depositional, and Paleoecological History of Early Mesozoic Rift Basins, Eastern North America*, ed. P.E. Olsen, R.W. Schlische & P.J.W. Gore, pp. 19–22. 28th International Geological Congress Field Trip T-351.

Harrington, J.W., 1951. Structural analysis of the west border of the Durham Triassic basin. *Geol. Soc. Am. Bull.*, **62**, 149–58.

Hay, W.W., Behensky, J.F., Jr., Barron, E.J. & Sloan, J.L., II, 1982. Late Triassic–Liassic paleoclimatology of the proto-central North Atlantic rift system. *Palaeogeogr., Palaeoclimat., Palaeoecol.*, **40**, 13–30.

Hoffman, C.W. & Gallagher, P.E., 1989a. *Geology of the Southeast Durham and Southwest Durham, 7.5 Minute Quadrangles, North Carolina*. North Carolina Geol. Survey Bull., 92.

Hoffman, C.W. & Gallagher, P.E., 1989b. Stop 1.5: Railroad cut, Durham, NC. In *Tectonic, Depositional, and Paleoecological History of Early Mesozoic Rift Basins, Eastern North America*. ed. P.E. Olsen, R.W. Schlische & P.J.W. Gore, pp. 31–2. 28th International Geological Congress Field Trip T-351.

Lorenz, J.C., 1988. *Triassic–Jurassic Rift-Basin Sedimentology*. Van Nostrand Reinhold, New York.

Luttrell, G.W., 1989. Stratigraphic nomenclature of the Newark Supergroup of eastern North America. US Geol. Survey Bull., 1572.

Manspeizer, W., 1982. Triassic–Liassic basins and climate of the Atlantic passive margins. *Geol. Rundschau*, **71**, 895–917.

Manspeizer, W., 1988. Triassic–Jurassic rifting and opening of the Atlantic: an overview. In *Triassic–Jurassic Rifting: Continental Breakup and the Origin of the Atlantic Ocean and Passive Margins*, part A, ed. W. Manspeizer, pp. 41–79. Elsevier, Amsterdam.

Olsen, P.E., 1988. Paleontology and paleoecology of the Newark Supergroup (Early Mesozoic, eastern North America). In *Triassic–Jurassic Rifting: Continental Breakup and the Origin of the Atlantic Ocean and Passive Margins*, part A, ed. W. Manspeizer, pp. 185–230. Elsevier, Amsterdam.

Olsen, P.E., Froelich, A.J., Daniels, D.L., Smoot, J.P. & Gore, P.J.W., 1991. Rift basins of Early Mesozoic age. In *The Geology of the Carolinas*, ed. J.W. Horton, Jr. & V.A. Zullo, Chapter 9. Carolina Geological Society 50th Anniversary Volume, University of Tennessee Press, Knoxville.

Parker, J.M., III, 1979. *Geology and mineral resources of Wake County*. North Carolina Department of Natural Resources and Community Development. Geol. Survey Sect. Bull. 86.

Prouty, W.F., 1931. Triassic deposits of the Durham basin and their relation to other Triassic areas of eastern United States. *Am. J. Sci.*, 5th Series, **21**, 473–90.

Reinemund, J.A., 1955. *Geology of the Deep River Coal Field, North Carolina*. US Geol. Survey Prof. Paper 246.

Robbins, E.I., Wilkes, G.P. & Textoris, D.A., 1988. Coal deposits of the Newark rift system. In *Triassic–Jurassic Rifting: Continental Breakup and the Origin of the Atlantic Ocean and Passive Margins*, part B, ed. W. Manspeizer, pp. 649–82. Elsevier, Amsterdam.

Smoot, J.P., Froelich, A.J. & Luttrell, G.W., 1988. Uniform symbols for the Newark Supergroup. In *Studies of the Early Mesozoic Basins of the Eastern United States*. ed. A.J. Froelich & G.R. Robinson, Jr., pp. 1–6. US Geol. Survey Bull. 1776.

Textoris, D.A. & Holden, C.J., 1986. Paleoclimate change within a stratigraphic section. In *Depositional Framework of a Triassic Rift Basin: the Durham and Sanford Subbasins of the Deep River Basin, North Carolina*, ed. P.J.W. Gore, pp. 101–2. In SEPM Field Guidebooks, Southeastern United States, Third Annual Midyear Meeting, Raleigh, North Carolina, ed. D.A. Textoris. Society of Economic Paleontologists and Mineralogists.

Textoris, D.A., Robbins, E.I. & Gore, P.J.W., 1989. Origin of organic-rich strata in an Upper Triassic rift basin lake, eastern USA. *28th International Geological Congress, Washington, DC, USA*, Abstracts Volume 3, 229–30.

Traverse, A., 1986. Palynology of the Deep River basin, North Carolina. In *Depositional Framework of a Triassic rift Basin: the Durham and Sanford Subbasins of the Deep River Basin, North Carolina*, ed. P.J.W. Gore, pp. 66–71. In SEPM Field Guidebooks, Southeastern United States, Third Annual Midyear Meeting, Raleigh, North Carolina, ed. D.A. Textoris. Society of Economic Paleontologists and Mineralogists.

Traverse, A., 1987. Pollen and spores date origin of rift basins from Texas to Nova Scotia as Early Late Triassic. *Science*, **236**, 1469–72.

Wheeler, W.H. & Textoris, D.A., 1978. Triassic limestone and chert of playa origin in North Carolina. *J. Sedim. Petrol.*, **48**, 765–76.

Lower Jurassic Kota Limestone of India

DHIRAJ KUMAR RUDRA AND PRADIP KUMAR MAULIK

Geological Studies Unit, Indian Statistical Institute, 203 Barrackpore Trunk Road, Calcutta, 700 035, India

Introduction

In the Godavari Valley of south-central India, a 3600 m-thick sequence of fossiliferous and Permian coal-bearing Gondwana rocks (Early Permian to Early Cretaceous; Table 1) is exposed in a narrow northwest–southeast trending basin flanked on two sides by Precambrian and basement complex rocks (Fig. 1). The basin is currently believed to be a highly asymmetric rift structure (Quereshy *et al.*, 1968, Mishra *et al.*, 1987). The Kota Limestone is a thin

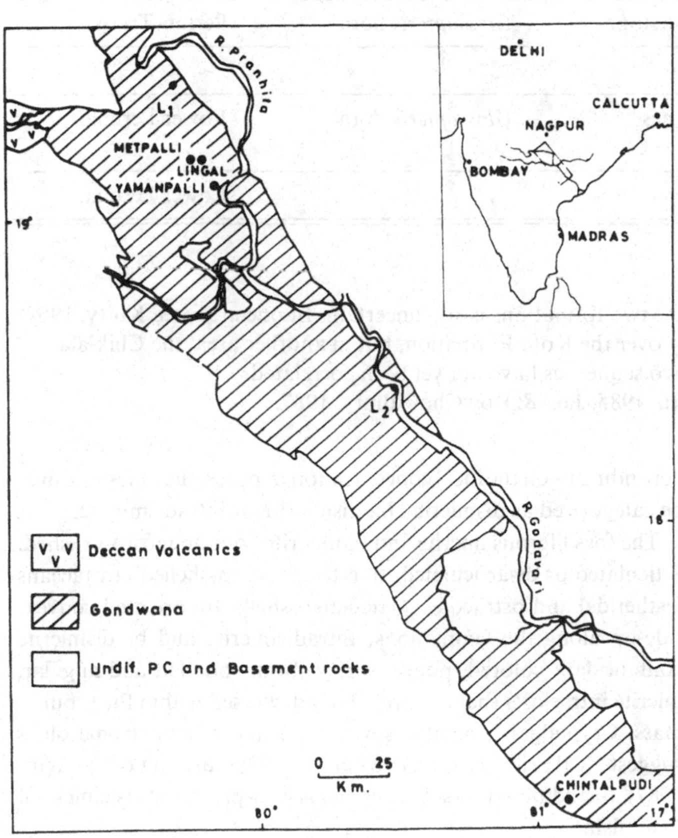

Fig. 1. The Gondwana outcrop in the Godavari Valley. L₁ marks the northern and L₂ the southern extent of the Kota Limestone facies exposure.

facies (maximum 15 m thick), consisting essentially of alternating limestone and calcareous clay beds. It is part of the 590 m-thick Kota Formation (Upper Gondwana) of the Gondwana sequence (Table 1; Figs 2 and 3). The Kota Limestone facies can be traced laterally for 120 km (from L1 to L2 in Fig. 1).

The Yamanpalli-Metpalli region in Adilabad district, Andhra Pradesh, is the only area where the Kota Limestone and associated siliciclastics have been mapped in some detail within the Kota Formation (Figs 2 and 3). The Kota Formation has been divided into a lower member and an upper member (Rudra, 1982). The lower member consists of a siliciclastic facies (Fig. 3) containing a lower coarser subfacies and an upper finer subfacies (unconsolidated red and green clay). The upper member consists of the lower Kota Limestone facies and an upper siliciclastic facies (Fig. 3) which has been divided into a coarser lower subfacies and a finer upper subfacies (unconsolidated red and green clay). Some of the clays of the finer siliciclastic subfacies of the lower member and the limestones of the Kota Limestone facies of the Kota Formation have yielded a rich biota (Table 2), indicating a Lower Jurassic (Liassic) age. Fossils include fish, a flying reptile, crocodiles, insects, crustaceans, ostracods, charophytes, wood fragments, burrows, coccoliths, etc. The biota has been reviewed by Jain (1980) and Jain & Roy Chowdhury (1987).

Kota Limestone facies

The Kota Limestone sharply overlies red clay of the lower member and appears to intertongue with red clay of the upper member of the Kota Formation (Fig. 3). Only at two localities (1 and 2 of Fig. 4) is the entire spectrum of the Kota Limestone facies exposed. Otherwise, while the limestone rock exposures are quite common, a few good, but somewhat incomplete sections (for example, 3 of Fig. 4), are available. The facies attains its maximum thickness near the location of section 1 (NNE of Metpalli) in the northern sector and decreases very gradually to the north and south.

The Kota Limestone facies contains four subfacies (Fig. 4): (1) limestone; (2) calcareous clay; (3) interbedded calcareous clay and

185

Table 1. *Gondwana stratigraphy of the Godavari Valley, south-central India*

Main Division	Subdivision (Formation)	Lithology	Major Fossils	Age
Upper Gondwana	**Chikiala (160 m thick)/Gangapur	Chikiala: Highly ferruginous, calcareous, sandstones, siltstones, breccio-conglomerates.	Chikiala: none	Chikiala: undecided
		Gangapur: Mudstones, ferruginous siltstones and sandstones with pebbly beds and conglomerates.	Gangapur: *Ptilopyllum* flora, ganoid fish	Gangapur: Early Cretaceous
	Kota (590 m thick)	Ferruginous & calcareous sandstones, mudstones, red clays, and *bedded limestones with intercalated calcareous clay. At base, calcareous conglomeratic sandstones.	fish, sauropod dinosaurs, mammals	Early Jurassic
	Dharmaram	Ferruginous sandstones and siltstones, some oolitic limestones (lower part), red clays	prosauropod dinosaurs	Late Triassic (Norian & Rhaetian)
	Maleri	Red clays, oolitic sandstones	metaoposaur, phytosaur, rhynchosaur, aetosaur	Late Triassic (Carnian)
	Bhirmaram	Calcareous sandstones, red clays		
	Yerrapalli	Red clays, oolitic sandstones	capitosaur, dicynodont, erythrosuchid, cynodont	Middle Triassic
Lower Gondwana	Kamthi	Ferruginous sandstones and siltstones	*Glossopteris* flora	Permo-Triassic
	Ironstone Shale	Ferruginous shales		
	Barakar	Feldspathic argilleous sandstones with pebbles; coal beds	*Glossopteris* flora	Lower Permian
	Talchir	tillite and stratified drift		Lower Permian

Notes:
Individual thicknesses of many of the formations are not yet available.
* Kota Limestone facies.   ** Stratigraphic relationship between these two formations is still uncertain. In one area (see Kutty, 1969), the Gangapur Formation occurs with a distinct angular unconformity over the Kota Formation, but in another area, the Chikiala Formation occurs with a possible unconformity over the Kota. The two sequences have not yet been correlated.
After King, 1881; Kutty, 1969; Rudra, 1982; Bandyopadhyay & Rudra, 1985; Jain & Roy Chowdhury, 1987.

limestone; (4) calcarenite–calcareous sandstone. Subfacies identification is based on general sedimentological features at outcrop scale, not on a microfacies level. These four subfacies generally grade into one another gradually, with most beds found to be lenticular in character when followed over a distance of tens of meters (Fig. 4). Individual sedimentary units vary in thickness from 1.5 to 30 cm with subfacies thicknesses varying from 7 to 200 cm.

Limestone subfacies

This limestone is cream-colored, locally pinkish or greyish. The limestone is micrite, containing varying amounts of fossil shells and intraclasts. The rock grades from fossiliferous micrite to intrabiomicrite to intramicrite within sedimentary units. Also, depending upon the local concentration of pores, the limestone may be categorized as dismicrite, biodismicrite or intradismicrite.

The fossiliferous micrite and biomicrite contain mostly crushed, articulated or disarticulated, thin to very thin-shelled crustaceans (estherids) and ostracods. Articulated shells are oriented convex-side-up along the laminations. Intradismicrite and biodismicrite contain dark-colored, poorly sorted, both rounded and angular, micrite intraclasts that are irregularly dispersed within the groundmass. Cryptalgal laminations, oncolites and domal stromatolites suggest algal origin (Rudra & Maulik, 1987; Gururaja & Yadagiri, 1987). In addition, this subfacies yields well-preserved specimens of fossil fish.

It is thin- to medium-bedded with bed thickness ranging from 7–30 cm. Contacts between beds are sharp and planar to wavy, with

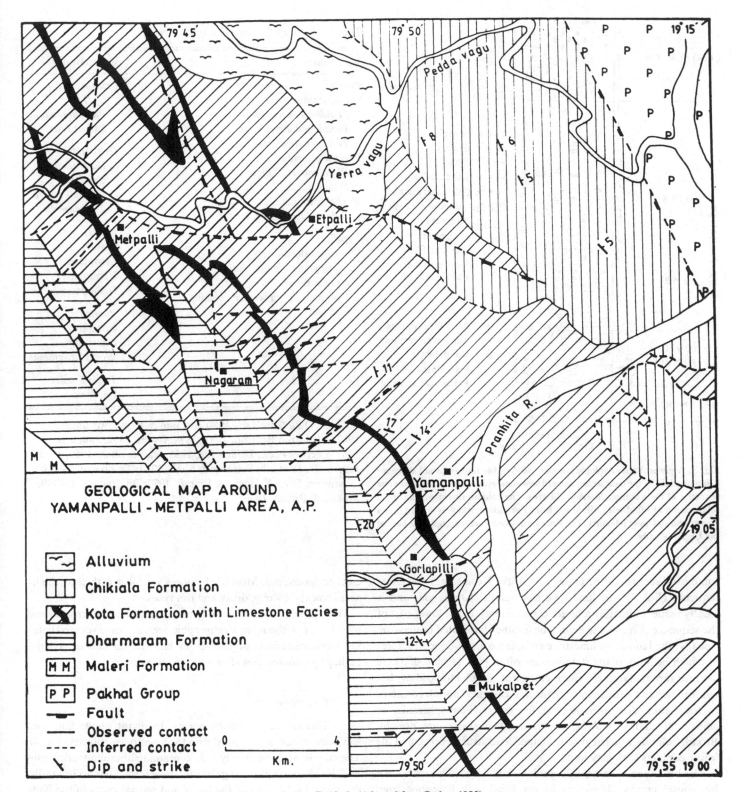

Fig. 2. Geological map of the Yamanpalli-Metpalli area, Andhra Pradesh. (Adapted from Rudra, 1982).

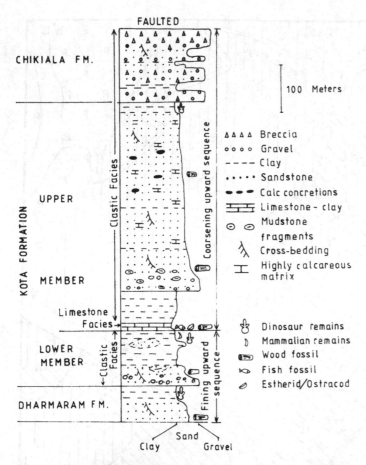

Fig. 3. A generalized lithologic column showing the fining-upward sequence being overlain by the coarsening upward sequence within the Upper Gondwana sequence of the Yamanpalli–Metpalli area. The column also shows the broad relationship of lithofacies within the Kota Formation, which is divided into a Lower and an Upper Member. The Kota Limestone Facies occupies the base of the Upper Member.

Table 2. *Biota of the Kota limestone facies*

Vertebrates	Invertebrates
Fishes	Crustaceans
Semionotidae	*Cyzicus*
Lepidotes deccanensis	*Paleolimnadia*
Paradapedium equertoni	Estherinids
Tetragonolepsis oldhami	
Pholidophoridae	Ostracods
Philidophorus kingii	*Darwinula*
P. indicus	*Candona*
Coelacanthidae	*?Limnocythere*
Indocoelacanthus robustus	*Timiriaseria*
Reptiles	Insects
Dimorphodontidae	Blatids
Campylognathoides indicus	Cleoptera
?Teleosauridae	Hemiptera
Fragmentary crocodile remains	Orthoptera
	Plants
	Coccoliths
	Charophytes
	Wood Fragments
	Trace Fossils
	Burrows

Sources: Data from Jain, 1980 and Jain & Roy Chowdhury, 1987. Also from Egerton, 1851, 1854; Owen, 1852; Jones, 1862, 1863; Egerton *et al.*, 1878; King, 1881; Rao & Shah, 1959, 1963; Jain, 1973, 1974 a,b; Satsangi & Shah, 1973; Tasch *et al.*, 1973; Govindan, 1975; Yadagiri & Prasad, 1977; Bhattacharya, 1980; Maulik & Rudra, 1986.

some local scours. In the upper part of the Kota Limestone facies sequence (Fig. 4), the limestone has parallel laminations and is locally fissile (indicating a larger clay content). In the lower part of the sequence (Fig. 4), the limestone is either laminated or massive. Locally, the laminated limestone contains cone-in-cone structures in which bedding plane stylolites are often well-developed. There are also rare replacement chert, elongated parallel to bedding, but without definite stratigraphic control. Also, chert nodules coalesce to form thin, discontinuous parallel layers. Few spheroidal to ellipsoidal chalky pebbles, containing fish scales, are present. Broad, bulbous, and irregularly-shaped load structures occur at the lower contacts with clay units.

The limestone is commonly mudcracked with both vertical and horizontal cracks. These cracks are filled with limestone from overlying beds. Some of the cracks penetrate the entire thickness of the bed. Angular limestone intraclasts are commonly associated and are lithologically similar to the enclosing limestone. Pore spaces occur as intraparticle, fenestral, vug, channel and fracture porosity. Fenestral, vug and channel porosity often occur together in mud-

cracked limestones. Most of the pores are filled with clear calcite spar cement – blocky, drusy and rim types.

The micrite itself is commonly neomorphosed to microspar and pseudospar, without any stratigraphic control. Some very restricted local concentrations of needle- to lath-shaped calcite crystals, perhaps pseudomorphs after gypsum, also occur.

Calcareous clay

This subfacies, though not as dominant as the limestone subfacies, occurs interbedded with the limestone subfacies (Fig. 4). The clay is white (or rarely red) containing abundant calcareous nodules. Angular to subangular limestone intraclasts occur locally, especially in the units of the lower and central parts of the Kota Limestone facies (Fig. 4). Calcareous clay beds occur as thin- to medium-bedded units of varying thicknesses. Contacts with the overlying and underlying limestone beds are always sharp and undulatory, with local scours and load structures at the base of overlying limestones. Mineralogy of the clay is mostly montmoril-

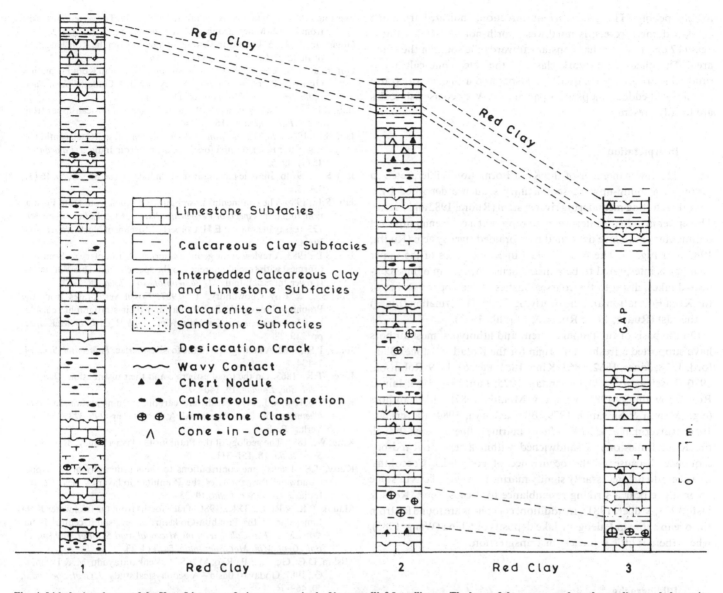

Fig. 4. Lithologic columns of the Kota Limestone facies sequence in the Yamanpalli–Metpalli area. The base of the sequence abruptly overlies a red clay unit. Column 1: 3 km NNE of Metpalli (Fig. 2; NW-sector); Column 2: W of Metpalli; Column 3: E of Mukalpet (Fig. 2; southern sector).

lonite with some kaolinite and minor quantities of degraded illite (Bhattacharya, 1980). Minor amounts of gypsum intraclasts are locally present in the clay beds near locality L_1 (Fig. 1). Coccoliths are mainly found in this subfacies.

Interbedded calcareous clay and limestone

This subfacies is composed of thin- to very thin-bedded alternations of calcareous clay and limestone beds. The regularity of the alternations and the equal thickness of both units, giving this subfacies a shaley appearance, define the uniqueness of this sub-facies. The thickness of a subfacies unit is variable with a maximum of 200 cm. Limestones can be laminated or massive; cone-in-cone

structures are fully to partially developed. Limestone clasts can be incorporated into the clays. This subfacies occurs in the lower and central portions of the Kota Limestone facies (Fig. 4).

Calcarenite–Calcareous sandstone

In the northern part of the Yamanpalli–Metpalli region (Fig. 2), near the top of the Kota Limestone facies sequence, a thin, laminated lens of light-colored calcarenite grades into the limestone subfacies above and below it (section 1 of Fig. 4). In the central part of the region (section 2 of Fig. 4), this calcarenite grades laterally into a grey, planar cross-stratified, moderately-sorted, medium-grained calcareous sandstone (sublitharenite) which is locally pebbly. The paleocurrent direction, indicated from the cross-bedding foresets, is northwest–north-northwest. Bed thickness (17 cm) is somewhat constant towards the south in the study area. The major framework clasts of the calcarenite–calcareous sandstone are limestone, quartz, feldspar and iron oxide ore (rare) and are embedded in a pseudospar matrix. Very coarse fish scales are locally present.

Interpretation

The lower member of the Kota Formation is interpreted to represent a fining-upward sedimentary sequence deposited by a laterally-shifting meandering river system (Rudra, 1982 and Fig. 3). The upper member is interpreted to represent a coarsening-upward sedimentary sequence deposited by a braided river system (Rudra, 1982 and Fig. 3). The basal Kota Limestone facies of the upper member is interpreted to be a distal lacustrine system of interconnected lakes, distal to the coarser clastics of the upper member of the Kota Formation and the overlying Chikiala Formation (Fig. 3) to the east (Rudra, 1982; Rudra & Maulik, 1987).

On the basis of the faunal content and lithotypes, most authors have supported a freshwater origin for the Kota Limestone (Blanford, 1878; Jones, 1862, 1863; King, 1881; Pascoe, 1959; Robinson, 1970; Tasch et al., 1973; Govindan, 1975; Jain, 1980, 1983; Jain & Roy Chowdhury, 1987; Rudra & Maulik, 1987). Other authors (e.g., Mishra & Satsangi, 1979; Bhattacharya, 1980; Raiverman, 1986; Raiverman et al., 1985) favor marine influence, even though the Kota Limestone is sandwiched within a terrestrial red-bed sequence, because of the occurrence of coccoliths. Coccoliths, however, do not necessarily signify marine influence. Tests of some green algae bear a striking resemblance to coccospheres (Kelts & Hsü, 1978; Müller, 1981). In addition, coccoliths are found far from the ocean shore in Paleogene lake deposits of China (Wang, 1985), where there is little evidence for transgression.

Bibliography

Bandyopadhyay, S. & Rudra, D.K., 1985. Upper Gondwana stratigraphy, north of Pranhita–Godavari confluence, southern India. *J. Geol. Soc. India*, **26** (4), 261–6.

Bell, T.L., 1853. Further account of the boring at Kotah, Deccan, and note on an Ichthyolote from the place. *Quart. J. Geol. Soc. London*, **9** (1), 351–2.

Bhattacharya, N., 1980. Depositional patterns in limestones of Kota Formation (Upper Gondwana), Andhra Pradesh, India. In *Gondwana Five, Fifth Int. Gondwana Symp., New Zealand*, ed. M.M. Cresswell & P. Vella, pp. 135–9.

Blanford, W.T., 1878. The palaeontological relation of the Gondwana System. *Rec. Geol. Surv. India*, **11**, 104–50.

Egerton, P.M.G., 1851. Description of the specimens of fossil fish from the Deccan. *Quart. J. Geol. Soc. London*, **7**, 273.

Egerton, P.M.G., 1854. Palichthyological notes no. 7 on two new species of *Lepidotes* from the Deccan. *Quart. J. Geol. Soc. London*, **10** (2), 371–3.

Egerton, P.M.G., 1878. On some remains of ganoid fishes from the Deccan. *Palaeon. Indica*, Ser. 4, pt. 1 (2), 1–8.

Egerton, P.M.G., Miall, L.C. & Blanford, W.T., 1878. The vertebrate fossils from Kotah-Maleri group. *Palaeon. Indica*, Ser. 4, pt. 1(2), 26.

Govindan, A., 1975. Jurassic freshwater ostracods from the Kota limestone of India. *Palaeontology*, **18** (1), 207–16.

Gururaja, M.N. & Yadagiri, P., 1987. Stromatolites from Kota Formation (Jurassic) Pranhita Godavari Valley, Andhra Pradesh. *Geol. Surv. India, Spec. Publ.* No. 11, (vol. 1) 213–15.

Jain, S.L., 1973. New specimens of Lower Jurassic holostean fishes from India. *Palaeontology*, **16** (1), 149–77.

Jain, S.L., 1974a. *Indocoelacanthus robustus* n. gen., n. sp. (Coelacanthidae; Lower Jurassic), the first fossil coelacanth from India. *J. Palaeontol.*, **48** (1), 49–62.

Jain, S.L., 1974b. Jurassic pterosaur from India. *J. Geol. Soc. India*, **15** (3), 330–5.

Jain, S.L., 1980. The continental Lower Jurassic fauna from Kota Formation, India. In *Aspects of Vertebrate History*, ed. L.L. Jacobs, pp. 99–123. (Essay in honor of E.H. Colbert), Museum of Northern Arizona Press.

Jain, S.L., 1983. A review of the genus Lepidotes (Actinopterygii: Semionotiformes) with special reference to the species from Kota Formation (Lower Jurassic), India. *J. Palaeontol. Soc. India*, **28**, 7–42.

Jain, S.L. & Roy Chowdhury, T., 1987. Fossil vertebrates from the Pranhita-Godavari Valley (India) and their stratigraphic correlation. In *Gondwana Six, Geophys. Monograph* 41, ed. G.D. McKenzie, pp. 219–28.

Jones, T.R., 1862. A monograph of the fossil estheriae. *Palaeontol. Soc.*, **14**, 1–134.

Jones, T.R., 1863. On the fossil estheriae and their distribution. *Quart. J. Geol. Soc. London*, **19**, 140–57.

Kelts, K. & Hsü, K.J., 1978. Freshwater carbonate sedimentation, In *Lakes: Chemistry, Geology, Physics*, ed. A. Lerman, pp. 295–323. Springer-Verlag, Berlin.

King, W., 1881. The geology of the Pranhita-Godavari Valley. *Mem. Geol. Surv. India*, **18**, 151–311.

Kutty, T.S., 1969. Some contributions to the stratigraphy of the Upper Gondwana formations of the Pranhita-Godavari Valley, Central India. *J. Geol. Soc. India*, **10**, 33–48.

Maulik, P.K. & Rudra, D.K., 1986. Trace fossils from the freshwater Kota Limestone of the Pranhita-Godavari valley, south central India. *Proc. XI Indian Colloquium on Micropal. and Stratigraphy* (pt. 2), *Bull. Geol. Min. Met. Soc. India*, **54**, 114–23.

Mishra, D.C., Gupta, S.B., Rao, M.B.S.V., Venkatarayudu, M. & Laxman, G., 1987. Godavari Basin – A geophysical study. *J. Geol. Soc. India*, **30**, 469–76.

Mishra, R.S. & Satsangi, P.P., 1979. Ostracodes from Kota Formation. Proc. Colloq. Palaeontological Studies in Southern Region. *Misc. Publ. Geol. Surv. India*, **45**, 81–8.

Müller, G., 1981. The occurrence of calcified planktonic green algae in freshwater carbonates. *Sedimentology*, **28**, 897–902.

Owen, R., 1852. Note on the crocodilian remains accompanying Dr. T.L. Bell's paper on Kotah. *Proc. Geol. Soc. London*, **7**, 233.

Pascoe, E.H., 1959. *A Manual of the Geology of India and Burma*, vol. 2 (3rd edn.), pp. 485–1343.

Quereshy, M.N., Krishna Brahman, N., Gorde, S.C. & Mathur, B.K., 1968. Gravity anomalies and the Godavari rift, India. *Bull. Geol. Soc. Amer.*, **79**, 1221–30.

Raiverman, V., 1986. Depositional model of Gondwana sediments in Pranhita–Godavari graben, south India. *Proc. XI Indian Colloquium on Micropal. and Stratigraphy* (pt. 2). *Bull. Min. Met. Soc. India*, **54**, 69–90.

Raiverman, V., Rao, M.R. & Pal, D., 1985. Stratigraphy and structure of the Pranhita–Godavari graben. *Petrol. Asia J.*, **8** (11), 174–90.

Rao, C.N. & Shah, S.C., 1959. Fossil insects from the Gondwanas of India. *Indian Minerals*, **12** (1), 3.

Rao, C.N. & Shah, S.C., 1963. On the occurrence of pterosaur from the Kota–Maleri beds of Chanda district, Maharastra. *Rec. Geol. Surv. India*, **92**, 315–18.

Robinson, P.L., 1970. The Indian Gondwana Formations – a review. *First Int. Symp. on Gondwana Stratigraphy, IUGS, South America*, Abstracts, p. 201–68.

Rudra, D.K., 1982. Upper Gondwana stratigraphy and sedimentation in the Pranhita-Godavari Valley, India. *Quart. J. Geol. Min. Met. Soc. India*, **54**, 56–79.

Rudra, D.K. & Maulik, P.K., 1987. Stromatolites from Jurassic freshwater limestones, India. *Mesozoic Res.*, **1** (3), 135–46.

Satsangi, P.P. & Shah, S.C., 1973. A new fish from Kota Formation, Pranhita–Godavari basin. *Abstracts, Proc. 60th session, Indian Sci. Congress, II, India*, p. 193.

Sykes, C., 1851. On a fossil fish from the table land of the Deccan, in the Peninsula of India, with a description of specimens by P.M.G. Egerton. *Quart. J. Geol. Soc. London*, **7**, 272–3.

Tasch, P., Sastry, M.V.A., Shah, S.C., Rao, B.R.J., Rao, C.N. & Ghosh, S.C., 1973. Estheriids of the Indian Gondwana: significance for continental drift. *3rd Int. Gondwana Symp. Australia*, pp. 443–52.

Walker, W., 1840. Memoir on the coal found at Kotah. *J. Asiatic Soc. Bombay*, **10**, 341–4.

Wall, P.W., 1857. Report on a reputed coal formation at Kota on the (Upper) Godavari River. *Madras J. Lit. and Sci.*, **18** (4), 256–69 (new series).

Wang, P., 1985. Palaeontological evidence for the formation of seas neighboring China. In *Marine Micropaleontology of China*, ed. P. Wang, pp. 1–14. China Ocean Press, Beijing.

Yadagiri, P., 1979. Observations on Kota Formation of Pranhita-Godavari Valley, south India. *Geol. Surv. India, Misc. Publ.*, **45**, 73–9.

Yadagiri, P. & Prasad, K.N., 1977. On the discovery of *Pholidophorus* fishes from the Kota Formation, Adilabad district, Andhra Pradesh. *J. Geol. Soc. India*, **18** (8), 436–44.

Yadagiri, P. & Rao, B.R.J., 1987. Contribution to the stratigraphy and vertebrate fauna of Lower Jurassic Kota Formation, Pranhita–Godavari Valley, India. *The Palaeobotanist*, **36**, 230–44.

Yadagiri, P., Satsangi, P.P. & Prasad, K.N., 1980. The piscean fauna from the Kota Formation of the Pranhita-Godavari Valley, Andhra Pradesh. *Mem. Geol. Surv. India., Palaeontol. Indica*, **45**, 1–31 (new series).

Early Cretaceous

The western Cameros Basin, northern Spain: Rupelo Formation (Berriasian)

NIGEL H. PLATT

GECO-PRAKLA, Ltd. Schumberger House, Buckingham Gate, Gatwick Airport West Sussex RH6 0NZ, UK

Introduction, stratigraphy and geological setting

The Cameros Basin lies in northern Spain (Fig. 1A, B), approximately 200 km to the north of Madrid. The Cameros was a major Mesozoic rift basin (Salomon, 1983; Platt, 1989a), subsequently inverted during Pyrenean compression (Platt, 1990). A thick sequence of Late Jurassic–Early Cretaceous continental sediments was described by Salomon (1982). The stratigraphy of this succession in the western half of the basin is outlined in Fig. 2 (Platt, 1986, 1989a; Wright *et al.*, 1988). This contribution provides a brief sedimentological summary of the lacustrine Rupelo Formation, which forms the upper part of the Tierra de Lara Group.

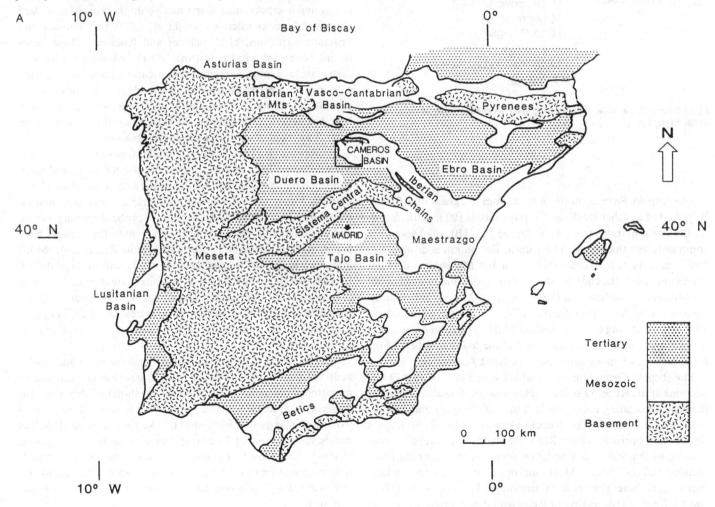

Fig. 1. A: Simplified geological map of Iberia showing location of the Cameros Basin and location of the study area (after Platt, 1990).

195

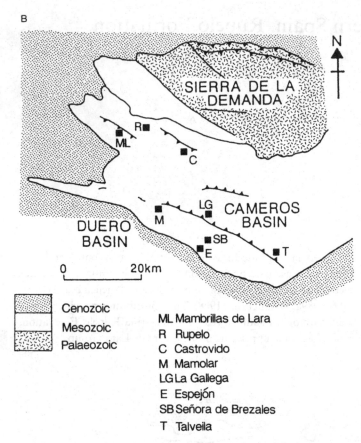

The Rupelo Formation has a maximum thickness of 200 m at Rupelo, but at other localities it rarely exceeds 100 m in thickness. At the basin margins (e.g., Talveila, see Fig. 1B), thicknesses are commonly less than 10 m. The Rupelo Formation is divided into four members (Fig. 3). Samples from the Mambrillas de Lara Member contain the charophytes *Globator maillardi protoincrassatus* Mojon and *Globator maillardi praecursor* Mojon, both indicating the top of the Lower Berriasian (P.-O. Mojon, pers. comm., 1985, 1990). Dating of the remainder of the formation is uncertain; it cannot at this stage be excluded that the lower members of the formation may include some strata of latest Jurassic age.

The Rupelo Formation is traceable over an area of 75 × 50 km. Sections of the Rupelo Formation are now tilted, faulted, folded or thrusted, depending upon location, this reflecting Pyrenean deformation (Platt, 1990). The Rupelo Formation locally includes a detachment horizon (within Rio Cabrera Member evaporites, see below), so that care must be taken when compiling stratigraphic sections and thicknesses. Most outcrops occur on reactivated fault block highs; facies present in the unexposed basin center may show some differences (deeper water facies would be expected to be more abundant in central basin areas).

Flora/fauna

Floral and faunal elements present in the Rupelo Formation include charophytes (stems and gyrogonites), ostracods (Cypridae), gastropods and sauropod bones (Platt 1989b, c; Platt & Meyer, 1991). Sauropod footprints occur at Mambrillas de Lara (see Fig. 1B; Platt & Meyer, 1991). The biota clearly indicates a lacustrine environment.

Sedimentology

The following paragraphs outline the main facies present at outcrop (no borehole data is currently available). For a fuller description of facies, the reader is referred to Platt (1986, 1989a, 1989c). Schematic sedimentological logs are shown in Fig. 4.

The Las Vinas Member consists of terrigenous, red or mottled marls and immature paleosols. This facies assemblage represents a distal alluvial, lake margin environment. The peloidal carbonates, red marls and paleosols of the Ladera Member show a wide variety of sedimentary structures indicating pedogenetic modification, including mottling, pedogenetic and desiccation brecciation, circumgranular cracks, microkarst and solution cavities, abundant rhizoliths, laminar calcrete (Wright *et al.*, 1988), columnar and prismatic structures, black pebbles and fenestrae. These facies record deposition in a 'palustrine' setting, within an ephemeral carbonate lake subject to frequent desiccation, pedogenesis during subaerial exposure and terrigenous progradation. A sandstone bed present at the boundary of the Ladera Member and overlying Mambrillas de Lara Member in the extreme northwest of the basin records an input of clastic material from that direction.

Charophyte-ostracod-gastropod wackestones and marls of the Mambrillas de Lara Member have a freshwater fauna and flora. Sauropod bones and abundant footprints also occur (Platt 1989b; Platt & Meyer, 1991). Lamination is rare, indicating shallow, oxygenated conditions. Lake depth was probably mostly < 5 m, and is unlikely to have exceeded 15 m; only the darkest open lacustrine carbonates show diffuse lamination. Rarely, interbedded grey marls in the northwest of the basin display mm-scale graded silt laminae, which are interpreted as deposits of small-scale lacustrine density currents. The local preservation of lamination indicates periods of reduced bioturbation, due to increased salinity or oxygen deficiency at the lake floor during times of increased organic productivity.

The limestones show rare microkarst, erosive bases and graded beds of intraclastic and black pebble wackestone. Intraclastic grainstones are interpreted as storm or shoreline deposits. The presence of abundant intraclasts suggests significant internal reworking of marginal facies washed in from the area around the lake during storms, and of lacustrine facies along internal drainage channels during low lake stands. This facies assemblage probably reflects deposition in a low-salinity shallow-water 'open lacustrine' environment with occasional storm influence and minor subaerial exposure.

The Rio Cabrera Member consists of marly shallow lacustrine

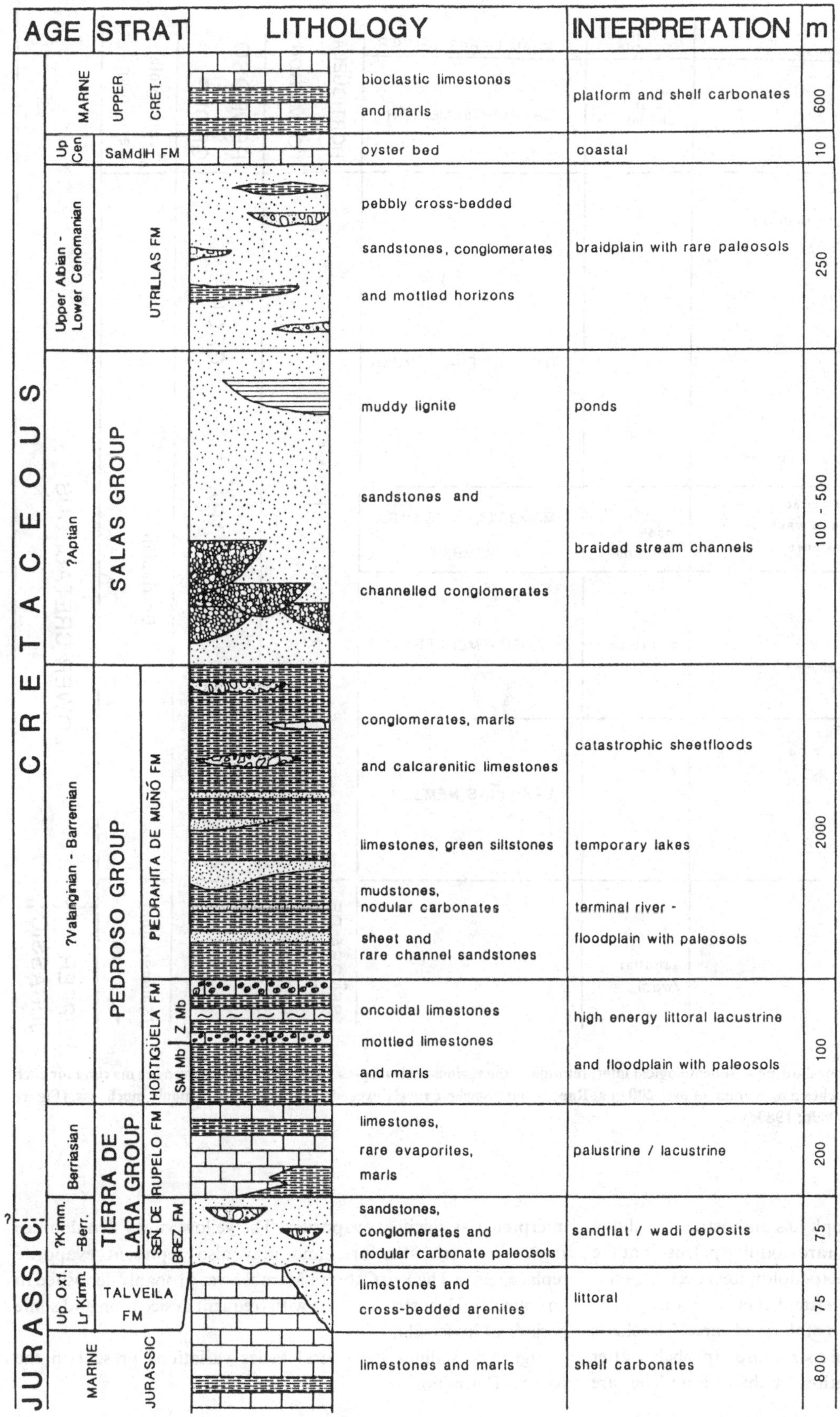

AGE	STRAT	LITHOLOGY	INTERPRETATION	m
UPPER CRET.	MARNE	bioclastic limestones and marls	platform and shelf carbonates	600
Up. Cen	SaMdlH FM	oyster bed	coastal	10
Upper Abian - Lower Cenomanian	UTRILLAS FM	pebbly cross-bedded sandstones, conglomerates and mottled horizons	braidplain with rare paleosols	250
?Aptian	SALAS GROUP	muddy lignite	ponds	100 - 500
?Aptian	SALAS GROUP	sandstones and	braided stream channels	100 - 500
?Aptian	SALAS GROUP	channelled conglomerates	braided stream channels	100 - 500
?Valanginian - Barremian	PEDROSO GROUP / PEDRAHITA DE MUÑÓ FM	conglomerates, marls and calcarenitic limestones	catastrophic sheetfloods	2000
?Valanginian - Barremian	PEDROSO GROUP / PEDRAHITA DE MUÑÓ FM	limestones, green siltstones	temporary lakes	2000
?Valanginian - Barremian	PEDROSO GROUP / PEDRAHITA DE MUÑÓ FM	mudstones, nodular carbonates	terminal river -	2000
?Valanginian - Barremian	PEDROSO GROUP / PEDRAHITA DE MUÑÓ FM	sheet and rare channel sandstones	floodplain with paleosols	2000
	HORTIGÜELA FM / Z Mb	oncoidal limestones	high energy littoral lacustrine	100
	HORTIGÜELA FM / SM Mb	mottled limestones and marls	and floodplain with paleosols	100
Berriasian	TIERRA DE LARA GROUP / RUPELO FM	limestones, rare evaporites, marls	palustrine / lacustrine	200
?Kimm.-Berr.	SEÑ DE BREZ FM	sandstones, conglomerates and nodular carbonate paleosols	sandflat / wadi deposits	75
Up.Oxf. Lr.Kimm	TALVEILA FM	limestones and cross-bedded arenites	littoral	75
MARINE JURASSIC		limestones and marls	shelf carbonates	800

Fig. 2. General stratigraphic column for the Upper Jurassic–Lower Cretaceous of the western Cameros Basin (modified after Platt, 1989a).

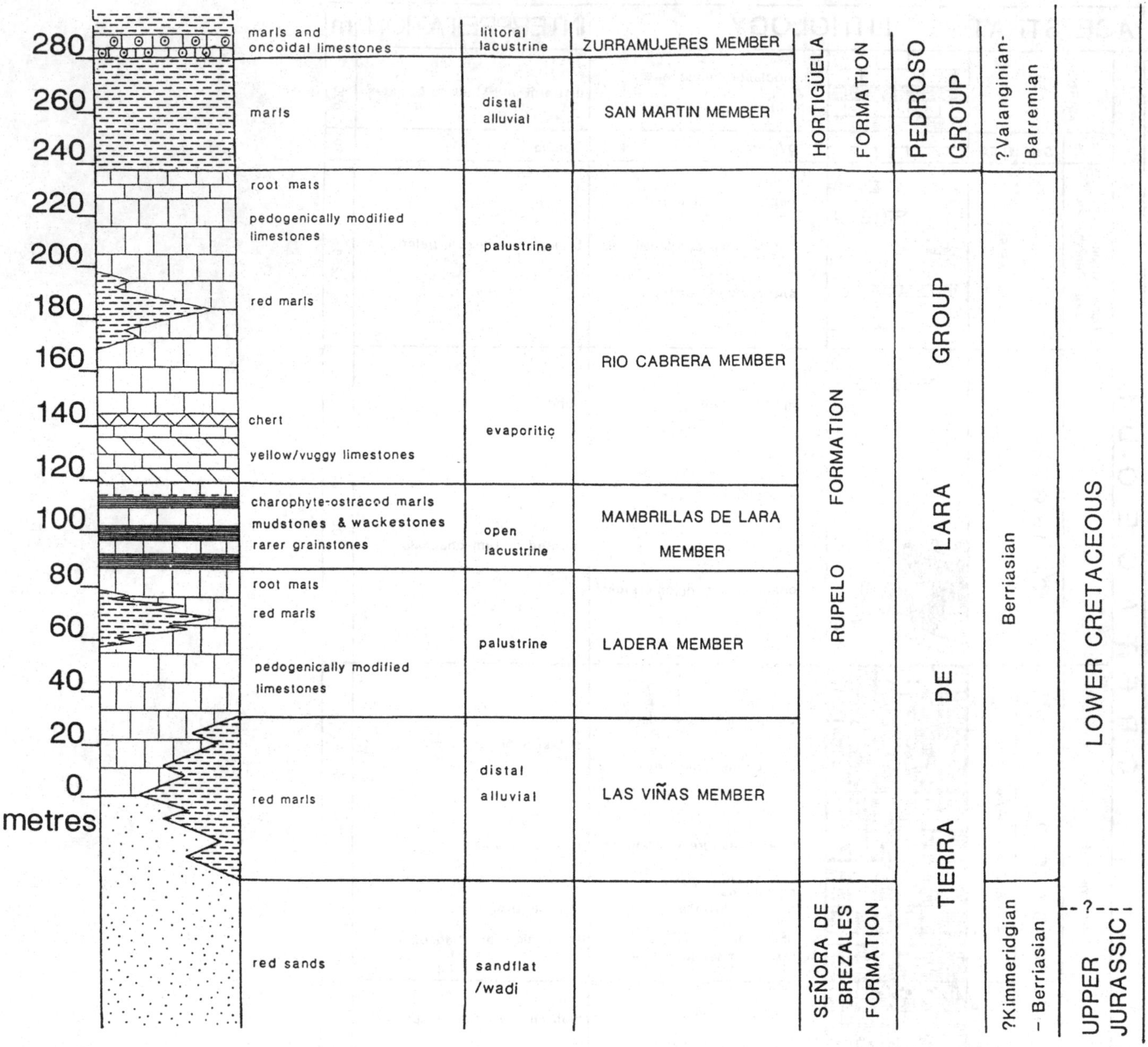

Fig. 3. Stratigraphy of the Rupelo Formation, showing sedimentological interpretation of the various facies represented. Thicknesses given are maxima for each unit; thickness of the Rupelo Formation reaches a maximum of over 200 m at Rupelo, but elsewhere rarely exceeds 100 m, and is commonly much less. (Figure modified after Wright *et al.*, 1988; Platt, 1989b; 1989c).

and palustrine limestones, with charophytes and ostracods, showing polygonal desiccation cracks and rare nodular paleosols at the tops of beds. Yellow vuggy limestones (dedolomites) occur at a few localities, and at least one laterally persistent chert horizon from 2–80 cm in thickness is present over much of the study area. The cherts show enterolithic and pseudomorphic structures (probably after anhydrite and gypsum), and rare length-slow chalcedony. They are

interpreted as silicified evaporites. The yellow vuggy dedolomites show box-work structure, and may also represent evaporite replacements. The Rio Cabrera Member was probably deposited in an ephemeral lacustrine setting with frequent desiccation and some periods of hypersalinity.

Figure 5 outlines the lateral facies variations present in the Rupelo Formation.

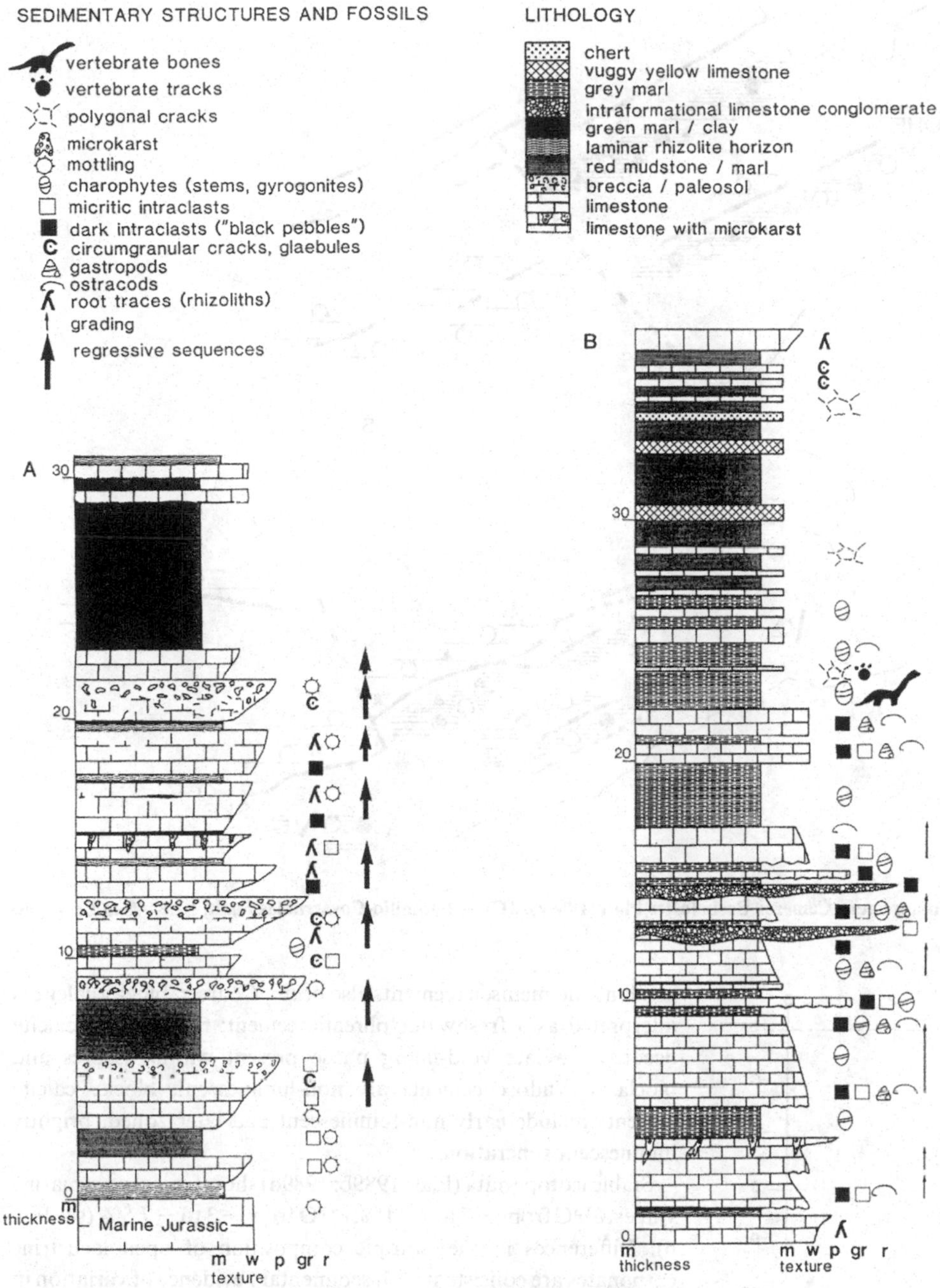

Fig. 4. Detailed sedimentological logs from the Rupelo Formation (after Platt, 1989b). A: Marginal lacustrine facies association (e.g. Ladera Member). Broad arrows indicate regressive cycles, with increasing pedogenetic modification and brecciation of lacustrine limestones towards the top of each cycle. Microkarst or biogenic laminar calcrete beds 10–100 cm thick occur at the top of some cycles, the latter commonly associated with thin green marl horizons, which are probably of pedogenetic origin (see Platt, 1989b for a fuller discussion of these). Red mudstone/marl units represent distal alluvial facies deposited during times of lacustrine regression. Fine clastic input reflects terrigenous progradation at times of low lake stand. Section is schematic, cut is mainly based on the Ladera Member section at Mambrillas de Lara (see Fig. 1B). B: Open lacustrine and evaporitic facies associations (e.g. Mambrillas de Lara and Rio Cabrera Members). Thin arrows indicate sense of grading in wackestone beds. Each graded unit probably records a single storm event. Erosive-based intraformational conglomerates represent low-stage, lake-flat drainage channels (Platt, 1989b). Grey charophytic marls and charophyte-gastropod-ostracod limestones are interpreted as open lacustrine facies, with only minor microkarst modification at the top of some beds. Thin grainstone units were probably deposited on higher-energy beaches or shoals. Chert and vuggy limestone represent evaporitic facies. Section is schematic, but the lower part is based upon the Mambrillas de Lara Member section on the road 1 km north of Mambrillas de Lara (see Fig. 1B), while the upper part is based upon the Rio Cabrera Member sections on the north side of Rupelo, near the village football pitch, and on the road leading south out of Rupelo (see Fig. 1B).

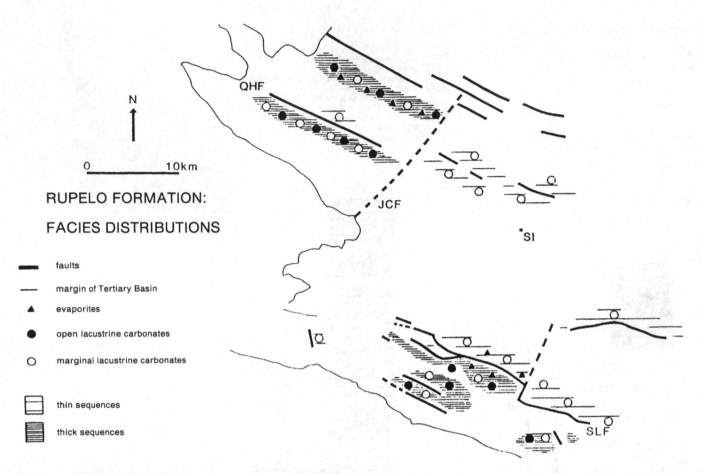

Fig. 5. Facies distributions for the Rupelo Formation, western Cameros Basin (after Platt, 1989c). JCF = Jaramillo-Covarrubias Fault, SLF = San Leonardo Fault, and QHF = Quintanilla-Hortigüela Fault.

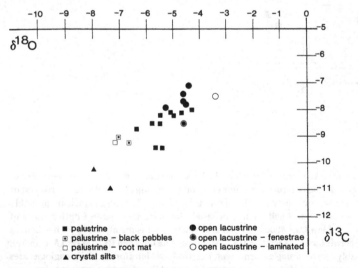

Fig. 6. Stable isotope data from the Rupelo Formation (after Platt, 1989b).

Diagenesis and stable isotopes

Characteristic diagenetic features present in the Rupelo Formation (Platt, 1986, 1989b) include vadose crystal silts and layered internal sediment occurring in microkarstic cavities. Rare pendant and meniscus cements also occur. Coarser, blocky calcite is interpreted as a freshwater phreatic cement; rare ferroan calcite occurs as a late void-filling phase present within fissures and bioclasts. Vadose cements are non-luminescent; blocky calcite cements include early non-luminescent and later zoned, brightly luminescent generations.

Stable isotope data (Platt 1989b; 1989d) show typical non-marine values; $\delta^{13}C$ from -7 to -11%; $\delta^{18}O$ from -3 to -7.5% (see Fig. 6). Differences in the isotopic composition of open lacustrine carbonates are consistent with sedimentary evidence of variation in organic productivity in the lake. Analyses from the entire sample suite are colinear; isotopic compositions become lighter with increasing evidence of pedogenic modification. This suggests progressive vadose zone diagenesis and influence of meteoric waters rich in soil-derived carbon dioxide (Platt 1989a; 1989d).

Environmental synthesis

An environmental model for the Rupelo Formation is provided by Fig. 7. The formation was deposited in a basin draining carbonate terrain dominated by Middle and Lower Jurassic shelf limestones (Platt, 1986; Platt & Wright, 1991). Evidence for frequent shoreline fluctuations, the presence of evaporites in the Rio

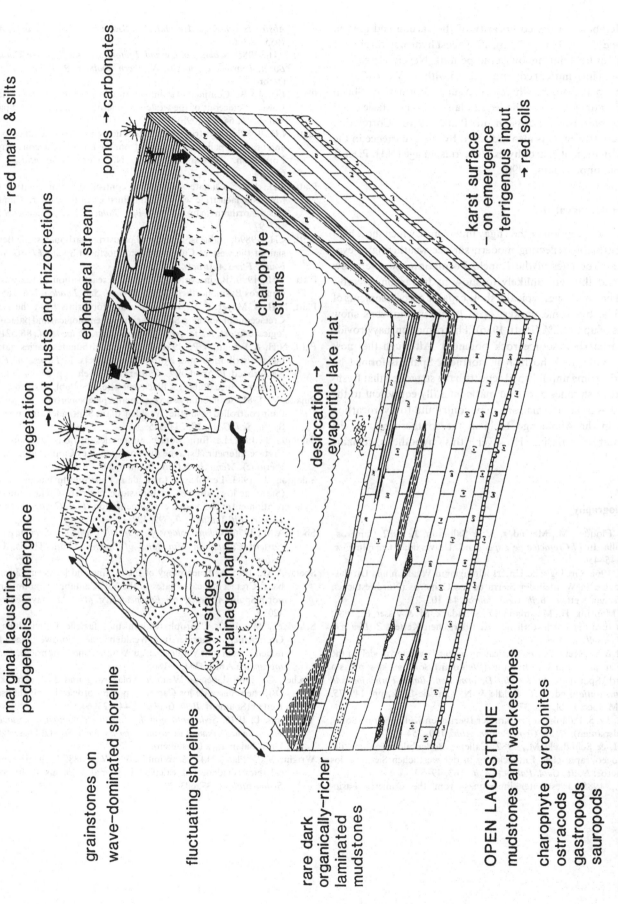

"PALUSTRINE"
marginal lacustrine
pedogenesis on emergence

periodically inundated alluvial plain
– red marls & silts

vegetation
→ root crusts and rhizocretions

ephemeral stream

ponds → carbonates

grainstones on
wave–dominated shoreline

fluctuating shorelines

low-stage
drainage channels

charophyte
stems

desiccation →
evaporitic lake flat

karst surface
– on emergence
terrigenous input
→ red soils

rare dark
organically–richer
laminated
mudstones

OPEN LACUSTRINE
mudstones and wackestones

charophyte gyrogonites
ostracods
gastropods
sauropods

Fig. 7. Sedimentological model for the Rupelo Formation (after Platt, 1989b).

Cabrera Member and the co-linearity of the carbon and oxygen stable isotope values (Platt, 1989b), all suggest hydrological closure of the basin, at least during low-stand periods. Nevertheless, there may have been intermittent communication with the Vasco–Cantabrian Basin to the north, where the Aguilar Formation (Sbeta, 1976, 1985; Pujalte, 1981) shows similar carbonate facies and lithostratigraphy (see discussion in Platt, 1986). Correlation between these two units is also indicated by the occurrence in the Aguilar Formation of charophytes of Berriasian age (N.H. Platt & P.-O. Mojon, unpub. data, 1990).

Economic potential

With the exception of the vuggy limestones, which show high porosities probably reflecting modern surface weathering of evaporites, the carbonates of the Rupelo Formation show very low porosities and thus are unlikely to show hydrocarbon reservoir potential. The limestones typically show organic carbon contents of around 0.5%, but some darker, open lacustrine horizons show TOC values of up to 1.2% (Platt, 1986). The formation may provide limited hydrocarbon source rock potential. Although the most important source rock horizon in northern Spain is formed by Liassic marls, some unpublished reports have suggested that freshwater to brackish limestones and marls laterally equivalent to the Rupelo and Aguilar Formations may contribute to hydrocarbon generation in the Ayoluengo Field in the southern part of the Vasco–Cantabrian Basin. However, this hypothesis remains controversial.

Bibliography

Alonso, R., Floquet, M., Meléndez, A. & Salomon, J., 1982. Cameros–Castilla. In *El Cretácico de España*, ed. Universidad Complutense, pp. 345–456.

Beuther, A., 1966. Geologische Untersuchungen in Wealden und Utrillas-Schichten in Westteil der Sierra de los Cameros (Nordwestlichen Iberischen Ketten). *Beih. Geol. Jahrb.*, **44**, 103–21.

Errenst, C., Mensink, H., Mertmann, D., Schudack, M. & Visser, H., 1984. Zum Jura der nordwestlichen Keltiberischen Ketten. *Z. Deutsch. geol. Gesellsch.*, **135**, 23–5.

Guiraud, M. & Seguret, M., 1986. Releasing solitary overstep model for the Late Jurassic–Early Cretaceous (Wealdian) Soria strike-slip basin (North Spain). In *Strike-Slip Deformation, Basin Formation and Sedimentation*, ed. K.T. Biddle & N. Christie-Blick, pp. 149–75. SEPM Spec. Publ. No. 37.

Mauthe, R., 1975. Paläokarst im Jura der Iberischen Ketten (Prov. Soria, Nordspanien). *N. Jb. Geol. Paläon. Abh.*, **150**, 354–72.

Mensink, H. & Schudack, M., 1982. Caliche, Bodenbildungen und die Paläogeographischen Entwicklung in der westlichen Sierra de los Cameros. *N. Jb. Geol. Paläont. Abh.*, **163**, 49–80.

Platt, N.H., 1985. Freshwater limestones from the Cameros Basin. *Abstracts, 6th Reg. Mtg., Int. Ass. Sediment., Lérida (Spain), April 1985*, 363–6.

Platt, N.H., 1986. *Sedimentology and Tectonics of the Western Cameros Basin, Province of Burgos, Northern Spain*. D.Phil. Thesis, Univ. Oxford.

Platt, N.H., 1989a. Continental sedimentation in an evolving rift basin: the Lower Cretaceous of the western Cameros Basin (northern Spain). *Sed. Geol.*, **64**, 91–109.

Platt, N.H., 1989b. Lacustrine carbonates and pedogenesis: sedimentology and origin of palustrine deposits from the Early Cretaceous Rupelo Formation, W. Cameros Basin, N. Spain. *Sedimentology*, **36**, 665–84.

Platt, N.H., 1989c. Climate and tectonic controls on sedimentation of a Mesozoic lacustrine sequence: the Purbeck of the western Cameros Basin, northern Spain. *Palaeogeogr., Palaeoclimat., Palaeoecol.*, **70**, 187–97.

Platt, N.H., 1989d. Stable isotopes from palustrine carbonates: the Berriasian of the western Cameros Basin, northern Spain. *EUBG Strasbourg, Terra Abstracts*, **1**, 220.

Platt, N.H., 1990. Basin evolution and fault reactivation in the western Cameros Basin, northern Spain. *J. Geol. Soc. London*, **146**, 165–75.

Platt, N.H. & Meyer, C.A., 1991. Dinosaur footprints from the Lower Cretaceous of northern Spain: their sedimentological and palaeoecological context. *Palaeogeogr., Palaeoclimat., Palaeoecol.*, **86**, 321–33.

Platt, N.H. & Wright, V.P., 1991. Lacustrine carbonates: facies models, facies distributions and hydrocarbon aspects. In *Lacustrine Facies Analysis*, ed. P. Anadón, L. Cabrera & K. Kelts, pp. 57–74. Int. Ass. Sediment. Spec. Publ. 13. Blackwell Scientific Publ., Oxford.

Pujalte, V., 1981. Sedimentary succession and paleoenvironments within a fault-controlled basin: the 'Wealden' of the Santander area, northern Spain. *Sed. Geol.*, **28**, 293–325.

Salomon, J., 1982. Les formations continentales du Jurassique supérieur-Crétacé inférieur (Espagne du Nord, Chaines Cantabrique de NW Ibérique). *Mém. Géol. Univ. Dijon*, **6**, 228.

Salomon, J., 1983. Les phases 'fossé' dans l'histoire du bassin de Soria (Sierra de los Cameros) au Jurassique supérieur-Crétacé inférieur: relation entre téctonique et sédimentation. *Bull. Centre. Recherche Explor., Elf Aquitaine*, **7**, 399–407.

Sbeta, A.M., 1976. *Sedimentology of a Jurassic Carbonate Sequence in the Eastern Cantabrian Mountains of Spain*. D.Phil. Thesis, Univ. Oxford.

Sbeta, A.M., 1985. Sedimentology of the marine and lacustrine Jurassic-basal Cretaceous carbonates in the SW Cantabrian mountains of north Spain. *Abstracts, 6th European Reg. Mtg., Int. Ass. Sediment.*, 420–3.

Schudack, M., 1987. Charophytenflora und fazielle Entwicklung der Grenzschichten mariner Jura-Wealden in den Nordwestlichen Iberischen Ketten (mit Vergleichen zu Asturien und Katabrien). *Paläontographica* Abt., **B204**, 1–180.

Tischer, G., 1966. Über de Wealden-Ablagerung und die Tektonik der Östlichen Sierra de los Cameros in den nordwestlichen Iberischen Ketten (Spanien). *Beih. Geol. Jb.*, **44**, 123–64.

Valladares, I., 1976. *Sedimentologia del Jurásico y Cretácico al sur de la Sierra de la Demanda (Provincias de Burgos y Soria)*. Dissertation, Univ. Salamanca (Resumen).

Wright, V.P., Platt, N.H. & Wimbledon, W., 1988. Biogenic laminar calcretes: evidence of calcified root mat horizons in paleosols. *Sedimentology*, **35**, 603–20.

The western Cameros Basin, northern Spain: Hortigüela Formation (?Valanginian-Barremian)

NIGEL H. PLATT

GECO-PRAKLA, Ltd. Schlumberger House, Buckingham Gate, Gatwick Airport West Sussex RH6 0NZ, UK

Introduction, stratigraphy, and geological setting

The Cameros Basin lies in northern Spain (Fig. 1), approximately 200 km to the north of Madrid. The Cameros was a major Mesozoic rift basin (Salomon, 1983; Platt, 1989), subsequently inverted during Pyrenean compression (Platt, 1990). A thick sequence of Late Jurassic–Early Cretaceous continental sediments was described by Salomon (1982). The stratigraphy of this succession in the western half of the basin is outlined in Fig. 2 (Platt, 1986, 1989). This contribution provides a brief sedimentological summary of the lacustrine Hortigüela Formation, which forms the lower part of the Pedroso Group.

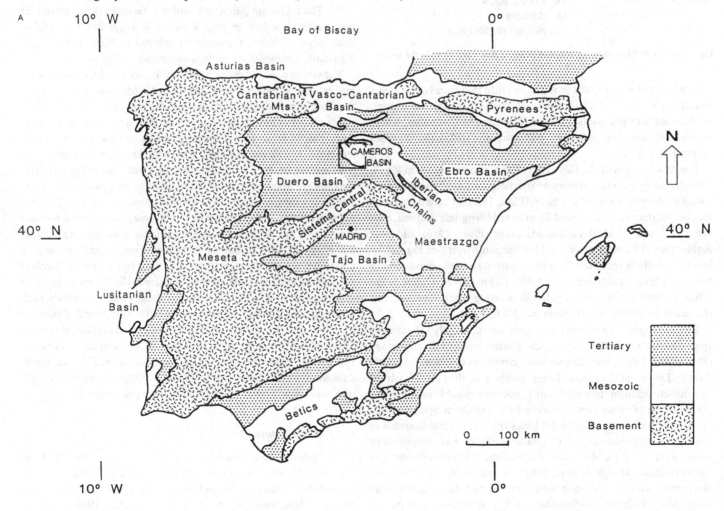

Fig. 1. Location maps. A: Simplified geological map of Iberia showing location of the western Cameros Basin (after Platt, 1990).

203

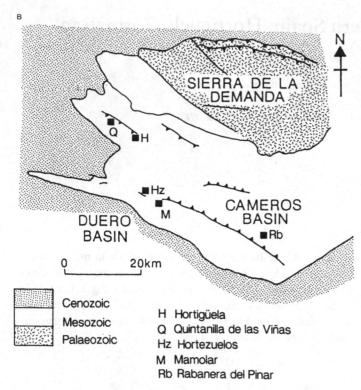

Fig. 1 (*cont.*) **B: Map showing location of localities referred to in the text.**

The Hortigüela Formation is divided into two members, the San Martín Member, which consists of distal alluvial sandstones, mudstones and paleosols, and the Zurramujeres Member, which consists of oncoidal limestones, paleosols and distal alluvial mudstones (see Fig. 2).

The distribution of the Hortigüela Formation is shown in Fig. 3. Most outcrops occur between Hortigüela and Quintanilla de las Viñas in the northwest of the basin (Platt, 1986), where it reaches a maximum thickness of around 75 m near Hortigüela, but outcrops are also found at Hortezuelos, Mamolar (Platt, 1986), and near Rabanera del Pinar (Forrer, 1991) in the southwest (see Fig. 1B for location of these sections). This distribution suggests deposition in two main lake areas, each at least 30 × 10 km in size. Outcrops are tilted, faulted, folded or thrusted depending upon location, this reflecting Pyrenean deformation (see Platt, 1990).

The Hortigüela Formation is poorly dated, largely reflecting its sparse biota. The underlying Rupelo Formation is of Berriasian age (Platt 1986, 1989), but the overlying continental strata are poorly dated. The most common dating method is that of lithostratigraphic correlation; however, this is not very reliable owing to the poor dating of other continental units in northern Spain and the reported occurrence of oncoidal limestones at several horizons in the Upper Jurassic and Lower Cretaceous. Similar lithologies occur in the Barremian La Huérguina Formation of Cuenca Province to the southeast (Monty & Mas, 1981) as well as in the Tithonian–Barremian Madero Group in the eastern part of the Cameros Basin (Salomon, 1982; see also Schudack, 1987). In the southern part of the Vasco–Cantabrian Basin, the upper part of the Aguilar Formation (Pujalte, 1981) also contains oncoidal limestones. The lower part of the Aguilar Formation exhibits charophytes of Berriasian age (N.H. Platt & P.-O. Mojon, unpub. data, 1990) and the oncoidal horizons are thus somewhat younger in age.

The Hortigüela Formation has been variably assigned to the Upper Berriasian (Salomon, 1982), the Valanginian (Platt, 1989), and the Lower Barremian (Schudack, 1987). The latter date is based on charophytes, and should thus be more reliable, but may in fact refer to reworked oncoids present in limestones of the overlying Piedrahita de Muñó Formation (see discussion in Platt, 1989). In any case, a Valanginian–Barremian age is likely for the Hortigüela Formation.

Fauna/flora

The Hortigüela Formation shows a biota consisting of cyanobacteria, rare charophytes, ostracods and gastropods.

Sedimentology

The following paragraph outlines the main facies present at outcrop (no borehole materal is currently available). For a fuller description of facies, the reader is referred to Platt (1986, 1989). Schematic sedimentological logs are shown in Fig. 4.

The terrigenous red mudstones, locally cross-bedded sandstones, and calcareous paleosols of the San Martín Member record deposition in a distal alluvial setting. The Zurramujeres Member comprises oncoidal limestones, marls, rare fine sandstones and calcareous paleosols showing circumgranular cracks and prismatic structure. The limestones are 1–6 m in thickness, and are sheet-like with erosive bases. Charophytes are rare but calcified cyanophyte oncoids are abundant. Oncoid morphology and paleoecology are similar to examples documented from freshwater settings. Oncoids are generally from 1–3 cm in size, but inverse grading of the oncoids is common, and those at the tops of some oncoid limestone beds 2.5 km north of Mambrillas de Lara (see Figs 1B and 4) reach a maximum size of 20 cm. Lateral variations in facies mostly reflect the varying thickness and number of oncoidal limestone beds (up to six beds, up to 6 m thick in the northwest; only one or two beds, each reaching a maximum thickness of 1 m in the southwest). Fenestrae are common in the northwest. Black pebbles and blackened oncoids are also locally present in this area (e.g., at Mambrillas de Lara, see Fig. 1B). In the southwest, the oncoidal limestones include abundant sand-sized detrital quartz grains, indicating increased terrigenous influence, and cross-bedding is more pronounced.

Stable isotopes

Oncoidal limestones show ^{13}C from -7 to -9%, ^{18}O from -4 to -7%; there is no discernible difference in isotopic composition between the oncoids and the limestone matrix. These values support a freshwater origin for the oncoids (Platt, 1989).

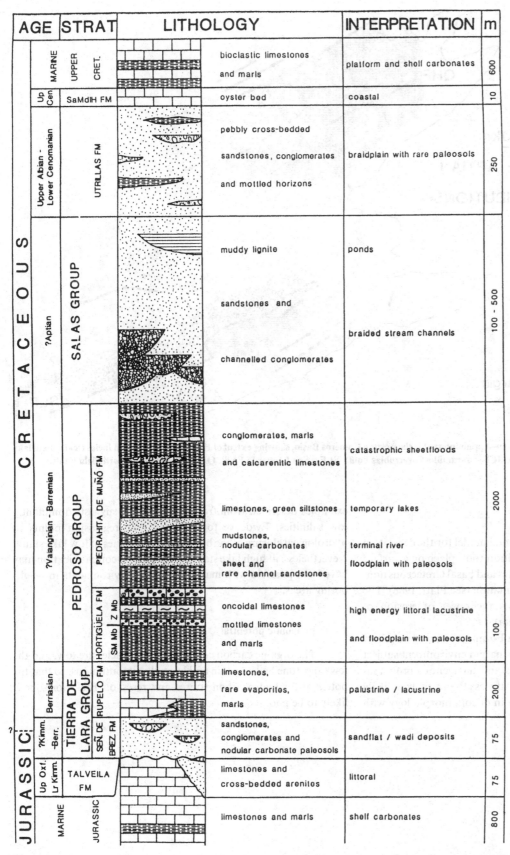

AGE	STRAT	LITHOLOGY	INTERPRETATION	m
MARINE UPPER CRET.		bioclastic limestones and marls	platform and shelf carbonates	600
Up Cen	SaMdlH FM	oyster bed	coastal	10
Upper Albian - Lower Cenomanian	UTRILLAS FM	pebbly cross-bedded sandstones, conglomerates and mottled horizons	braidplain with rare paleosols	250
?Aptian	SALAS GROUP	muddy lignite	ponds	100 - 500
		sandstones and	braided stream channels	
		channelled conglomerates		
?Valanginian - Barremian	PEDROSO GROUP / PEDRAHITA DE MUÑÓ FM	conglomerates, marls and calcarenitic limestones	catastrophic sheetfloods	2000
		limestones, green siltstones	temporary lakes	
		mudstones, nodular carbonates	terminal river	
		sheet and rare channel sandstones	floodplain with paleosols	
	HORTIGÜELA FM Z Mb / SM Mb	oncoidal limestones mottled limestones and marls	high energy littoral lacustrine and floodplain with paleosols	100
Berriasian	TIERRA DE LARA GROUP / RUPELO FM	limestones, rare evaporites, marls	palustrine / lacustrine	200
?Kimm. -Berr.	SEÑ DE BREZ FM	sandstones, conglomerates and nodular carbonate paleosols	sandflat / wadi deposits	75
Up Oxf. Lr Kimm.	TALVEILA FM	limestones and cross-bedded arenites	littoral	75
MARINE JURASSIC		limestones and marls	shelf carbonates	800

Fig. 2. General stratigraphic column for the Upper Jurassic–Lower Cretaceous of the western Cameros Basin (modified after Platt, 1989).

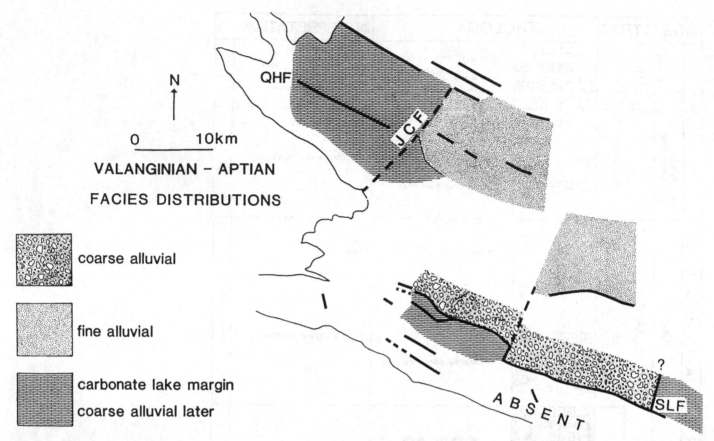

Fig. 3. Facies distributions for the Valanginian–Aptian interval in the western Cameros Basin, showing extent of Hortigüela Formation facies ('carbonate lake margin'). Modified after Platt (1989, 1990). JCF = Jaramillo-Covarrubias Fault, SLF = San Leonardo Fault, QHF = Quintanilla-Hortigüela Fault.

Environmental synthesis

Figure 5 presents an environmental model for the deposition of the Hortigüela Formation. Deposition took place on a surface consisting of Jurassic marine carbonates and basal Cretaceous non-marine carbonates of the Rupelo Formation (see Platt, 1986). The presence of clastic horizons suggests the onset of erosion of crystalline massifs (e.g., Iberian Meseta).

The facies present in the Hortigüela Formation record a shallow water, low-gradient, high-energy lake margin environment subject to occasional regression, pedogenesis and terrigenous progradation. Water depths were probably mostly less than 5 m, with likely maxima of 10 m or so. The similarity in oncoid morphology with examples described from freshwater settings suggests dominantly low salinities. Evidence for repeated lake regression points to hydrological closure of the basin, at least at times of low lake stand. Nevertheless, at high lake stands, intermittent communication may have been possible with marginal marine environments in neighboring areas.

Economic potential

The organic carbon contents of these facies are low, and the few sandstone beds are thin and laterally discontinuous, so that the potential for hydrocarbon source and reservoir rock horizons is likely to be poor.

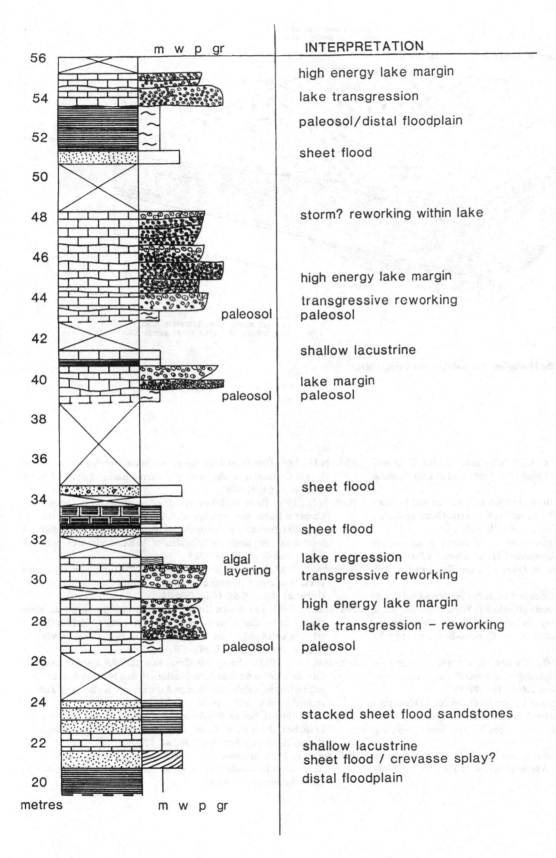

INTERPRETATION

high energy lake margin

lake transgression

paleosol/distal floodplain

sheet flood

storm? reworking within lake

high energy lake margin

transgressive reworking
paleosol

shallow lacustrine

lake margin
paleosol

sheet flood

sheet flood

lake regression
transgressive reworking

high energy lake margin

lake transgression – reworking

paleosol

stacked sheet flood sandstones

shallow lacustrine
sheet flood / crevasse splay?

distal floodplain

paleosol

algal
layering

paleosol

Fig. 4. Detailed sedimentological log through the upper part of the Hortigüela Formation, Zurramujeres Member (after Platt, 1989). Locality: 100 m to the west of sharp left-hand bend on road from Mambrillas de Lara to Campolara, 2.5 km north of Mambrillas de Lara, Province of Burgos.

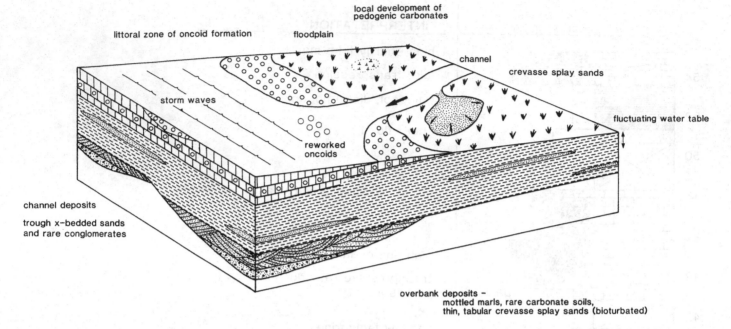

local development of
pedogenic carbonates

littoral zone of oncoid formation   floodplain

channel

crevasse splay sands

storm waves

fluctuating water table

reworked
oncoids

channel deposits

trough x-bedded sands
and rare conglomerates

overbank deposits –
mottled marls, rare carbonate soils,
thin, tabular crevasse splay sands (bioturbated)

Fig. 5. Sedimentological model for the Hortigüela Formation (after Platt, 1989).

Bibliography

Alonso, A., Floquet, M., Meléndez, A., & Salomon, J., 1982. Cameros–Castilla. In *El Cretácico de España*, ed. Universidad Complutense, 345–456.

Beuther, A., 1966. Geologische Untersuchungen in Wealden und Utrillas-Schichten in Westteil der Sierra de los Cameros (Nordwestlichen Iberischen Ketten). *Beih. Geol. Jb.*, **44**, 103–21.

Forrer, A., 1991. *Geologische Betrachtungen insbesondere der unteren Oberkreide entlang der San Leonardo-Überschiebung bei Hontoria del Pinar (NW Keltiberische Ketten, Provinz Burgos/Spanien).* Liz. Arb. Thesis, Univ. Bern.

Guiraud, M. & Seguret, M., 1986. Releasing solitary overstep model for the Late Jurassic–Early Cretaceous (Wealdian) Soria strike-slip basin (North Spain). In *Strike-Slip Deformation, Basin Formation and Sedimentation*, ed. K.T. Biddle & N. Christie-Blick, pp. 149–75. SEPM Spec. Publ. No. 37.

Mensink, H. & Schudack, M., 1982. Caliche, Bodenbildungen und die Paläogeographischen Entwicklung in der westlichen Sierra de los Cameros. *N. Jb. Geol. Paläont., Abh.*, **163**, 49–80.

Monty, C.L. & Mas, J.R., 1981. Lower Cretaceous (Wealdian) blue-green algal deposits of the Province of Valencia, Eastern Spain. In *Phanerozoic Stromatolites*. ed. C.L. Monty, pp. 85–120. Springer-Verlag, Berlin.

Platt, N.H., 1986. *Sedimentology and Tectonics of the Western Cameros Basin, Province of Burgos, Northern Spain.* D.Phil. Thesis, Univ. Oxford.

Platt, N.H., 1989. Continental sedimentation in an evolving rift basin: the Lower Cretaceous of the western Cameros Basin (northern Spain). *Sed. Geol.*, **64**, 91–109.

Platt, N.H., 1990. Basin evolution and fault reactivation in the western Cameros Basin, northern Spain. *J. Geol. Soc. London*, **146**, 165–75.

Pujalte, V., 1981. Sedimentary succession and paleoenvironments within a fault-controlled basin: the 'Wealden' of the Santander area, northern Spain. *Sed. Geol.*, **28**, 293–325.

Salomon, J., 1982. Les formations continentales du Jurassique supérieur-Crétacé inférieur (Espagne du Nord, Chaines Cantabrique de NW Ibérique). *Mém. Géol. Univ. Dijon*, **6**.

Salomon, J., 1983. Les phases 'fossé' dans l'histoire du bassin de Soria (Sierra de los Cameros) au Jurassique supérieur-Crétacé inférieur: relation entre téctonique et sédimentation. *Bull. Centre Recherche Explor., Elf Aquitaine*, **7**, 399–407.

Schudack, M., 1987. Charophytenflora und fazielle Entwicklung der Grenzschichten mariner Jura-Wealden in den Nordwestlichen Iberischen Ketten (mit Vergleichen zu Asturien und Katabrien). *Paläontographica Abt.*, **B204**, 1–180.

Tischer, G., 1966. Über de Wealden-Ablagerung und die Tektonik der Östlichen Sierra de los Cameros in den nordwestlichen Iberischen Ketten (Spanien). *Beih. Geol. Jb.*, **44**, 123–64.

Valladares, I., 1976. *Sedimentología del Jurásico y Cretácico al sur de la Sierra de la Demanda (Provincias de Burgos y Soria).* Dissertation, Univ. Salamanca (Resumen).

The eastern Cameros Basin (Kimmeridgian to Valanginian) – northern Iberian Ranges (Spain)

MARINA NORMATI

Centre des Sciences de la Terre et U.R.A., C.N.R.S. no 157, Université de Bourgogne, 6 Bd. Gabriel, 21000 Dijon, France

Introduction

The northwestern part of the Iberian Ranges (North Spain) consists of two parts: the Sierra de los Cameros and the Demanda Paleozoic Massif. During the Upper Jurassic and the Lower Cretaceous, the Sierra de los Cameros contained a continental sedimentary basin: the Cameros Basin (Fig. 1).

From Kimmeridgian to Valanginian, a graben-type structure formed in the northern and northeastern part of the basin with a thick sequence of varied continental sediments. These continental rocks, particularly those in the Berriasian to Valanginian lacustrine basin, have been studied by Normati (1991) and dated biostratigraphically by ostracods and charophytes. From Barremian to Albian times, a younger graben-type structure formed in the southwestern part of the basin. It was narrower and showed less subsidence, and the sediment infill comprises exclusively fluviatile sediments (Salomon, 1982a).

General and detailed sections

The complex lacustrine system (Fig. 2), which developed during the Berriasian to Valanginian, comprises the Sierra Matute Formation, Aguilar del Rio Alhama Formation, Inestrillas Formation, Cervera Formation and the lower-most part of the Las Casas Formation (Salomon, 1982b).

During the Lower Berriasian, a lacustrine complex developed: represented by the Sierra Matute Formation in the western part of the area and by the Aguilar del Rio Alhama Formation in the southern part of the area. Detailed study of these deposits traces the limits of four typical zones – Hinojosa, Renieblas, Sierra Matute and Sierra de Pegado – corresponding to four characteristic environments (Fig. 3) based on lithofacies, relative development, and sequential arrangements (Normati & Salomon, 1989).

Hinojosa (200 m in thickness): Vertical sequences from 0.1 to 1 m in thickness, are organized in a regular succession of facies types (Fig. 3). Each sequence includes, from the bottom to the top: (a) homogeneous mudstones with charophytes and limnic fauna, (b) clotted or nodular micrite with rootlet traces, (c) brecciated or fractured limestones, with deep cavities, and (d) laminar horizon. This sequence records a transgressive or flood stage followed by a drying out of the lake and subsequent pedogenesis.

Renieblas (70 m in thickness): These deposits are characterized by a succession of several cycles from 2 to 10 m in thickness (Fig. 3). Each cycle comprises (a) cross-bedded sandstones with oncoids, (b) sandy mudstones and siltstones with burrows and root traces, and (c) dark micritic limestones with charophytes and ostracods. Each sequence records deposition in an alluvial setting with limestones which developed in small lakes developed on a floodplain.

Sierra Matute (500 m in thickness): This facies consists mainly of dark micritic limestones with gastropods, ostracods and charophytes. The limestones are massive, commonly show clotted fabric and cracks are rare (Fig. 3). These deposits are interpreted as permanent lake sediments. (Sierra Matute Formation.)

Sierra de Pegado (50 m in thickness): Vertical sequences are organized in a regular succession of facies types from 5 to 20 cm in thickness (Fig. 3). Each sequence comprises: (a) ostracod packstones, (b) homogeneous mudstones with ostracods and charophytes, and (c) brecciated limestones or mudcracks. These sequences record a transgressive or flood stage followed by desiccation of the lake. (Aguilar del Rio Alhama Formation.)

During the Upper Berriasian, the western part of the study area (Fig. 2) was filled by fluvial clastic sediments (El Royo Formation). But, in the eastern part of the area (Sierra de Pegado), the sediment infill exclusively comprised of evaporitic carbonates (Inestrillas and Cervera Formations), interpreted as sebkha deposits. Cumulative thicknesses reach as much as 1300 m.

During the Valanginian, the northeastern part of the study area was filled by fluvial clastic sediments, but the base of the Las Casas Formation is characterized by a succession of several cycles (Fig. 3). Each cycle comprises three units: sandstones, siltstones and limestones. These limestones do not exceed 200 m in thickness and consist mainly of dark micritic limestones with ostracods, charophytes, gastropods and oncoids. The lower unit (sandstones) shows an erosional lower contact. Each cycle records deposition on a floodplain with fluvial channels and small lakes.

The boundaries between the depositional centers are prominent and well-defined. Transitional facies are rare. This is consistent with an active extensional tectonic regime, possibly related to basin-forming strike–slip movements associated with the opening of the Bay of Biscay.

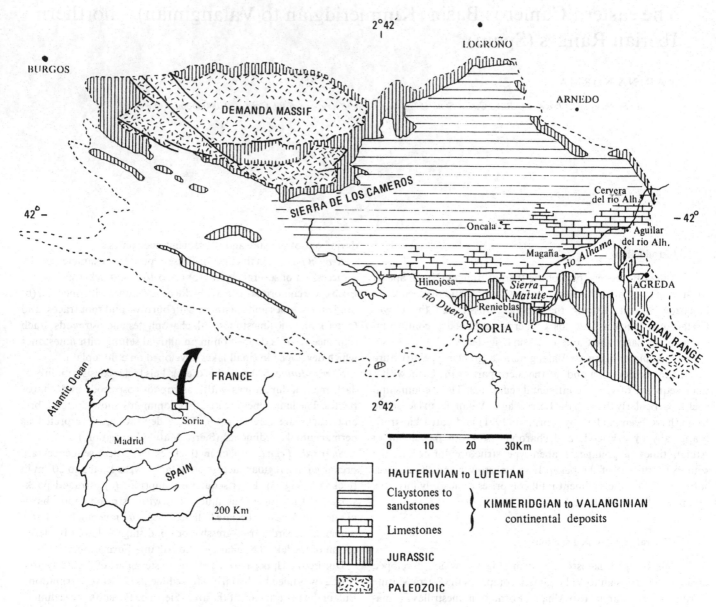

Fig. 1. Locality map. Location of the fluvio-lacustrine deposits and their extent in the Cameros Basin (Sierra de los Cameros).

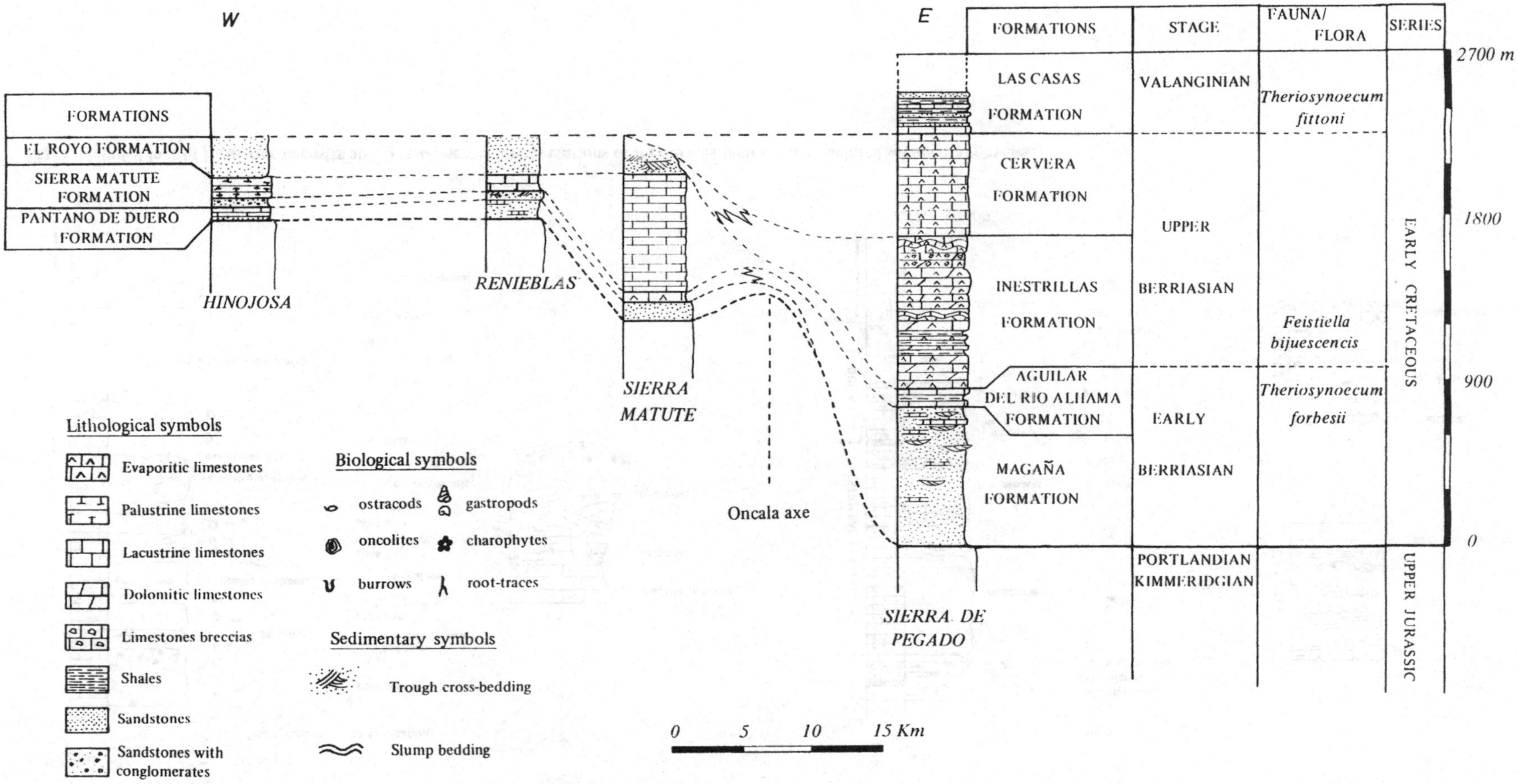

Fig. 2. Schematic view of the eastern part of the Cameros Basin, showing major formations deposited during the Upper Jurassic–Lower Cretaceous.

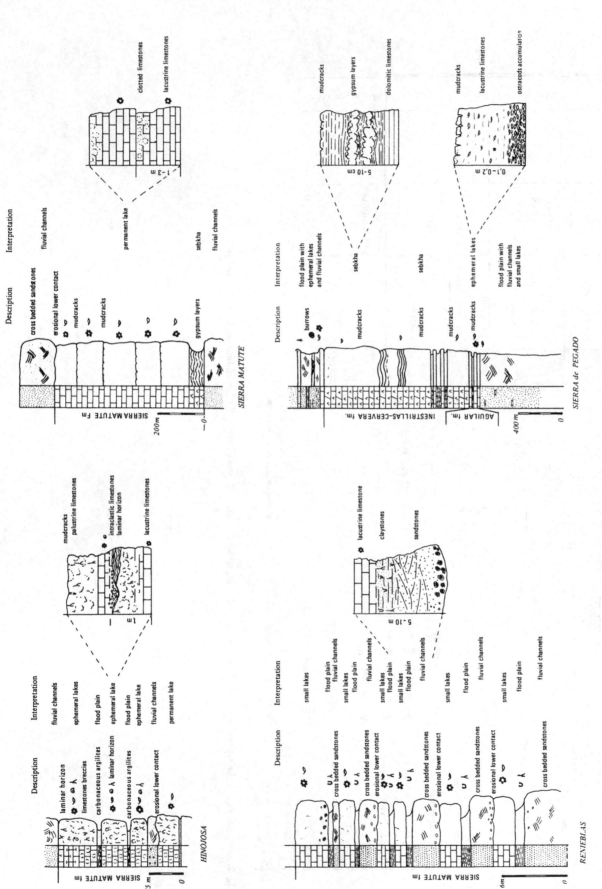

Fig. 3. Detailed logs of lacustrine deposits and environmental interpretations of the various facies and associated sedimentary cycles.

Bibliography

Arnold, M., Guillou, J.J., Michel, B. & Servajean, G., 1976. *La pyrite du wealdien de la Sierra de Los Cameros (Logrono, Espagne), sa liaison avec un environnement volcano-sédimentaire métamorphisé de type rift.* 4eme. R.A.S.T., Paris.

Beuther, M., 1966. Geologische Untersuchungen in Wealden und Utrillas-Schichten im Westteil der Sierra de Los Cameros (Nordwestliche Iberische Ketten). *Beih. Geol. Jb.*, **44**, 103–21.

Brenner, P., 1976. Ostracoden und charophyten des spanischen Wealden (systematikes, ökologie, stratigraphie, paläogeographie). *Palaeontographica*, **152** (4–6), 113–201.

Bulard, P.F., 1972. *Le Jurassique moyen et superieur de la Chaîne Iberique sur la bordure du bassin de l'Ebre (Espagne).* Thèse Fac. Sciences, Nice.

Guiraud, M., 1983. *Evolution tectono-sédimentaire du bassin wealdien (Crétacé inférieur) en relais de décrochements de Logrono-Soria (N-W Espagne).* Thèse Doct. 3éme cycle, Univ. Sciences et Techniques, Languedoc, Montpellier.

Guiraud, M. & Seguret, M., 1986. A releasing solitary overstep-model for the Late Jurassic-Early Cretaceous (Wealdian) Soria strike-slip basin (Northern Spain). In *Strike-slip Deformation, Basin Formation and Sedimentation.* ed. N. Christie-Blick & K.T. Biddle, pp. 159–75. Soc. Econ. Paleont. Min., Spec. Publ. **37**.

Kneuper-Haack, F., 1966. Ostracoden aus dem Wealden der Sierra de los Cameros (Nord-westliche Iberische Ketten). *Beih. Geol. Jb.*, **44**, 165–209.

Normati, M., 1991. *Les systèmes carbonatés du Bassin de Soria (Espagne). Expression du rifting atlantique en milieu continental.* Thèse Nouveau Doctorat., Univ. Bourgogne.

Normati, M. & Salomon, J., 1989. Reconstruction of a Berriasian lacustrine paleoenvironment in the Cameros Basin (Spain). *Palaeogeogr., Palaeoclimat., Palaeoecol.*, **70**, 215–223.

Palacios, C. & Sanchez Lozano, R., 1885. La formacion wealdense en los provincias de Soria y Logrono. *Bol. Com. Mapa. Geol. Espana*, **12**, 109–140.

Salomon, J., 1982a. *Les formations continentales du Jurassique supérieur–Crétacé inférieur en Espagne du Nord.* Thèse d'Etat. Mem. Geol., Univ. Dijon, **6**.

Salomon, J. 1982b. El Cretacico inferior, Cameros-Castilla. In *El Cretacico de Espana*, pp. 345–90. *Univ. Complutense Madrid.*

Sanchez Lozano, R., 1894. Descripcion fisica, geologica y minera de la provincia de Logrono. *Mem. Com. Mapa Geol. Espana*, **18**.

Seanz Garcia, C., 1935. Notas para el estudio de la facies wealdica espanola. *Asociation Espagnola para el Progreso de las Ciencias*, 59–76.

Tischer, G., 1966. Über die wealden-Ablagerung und die Tektonik der ostlichen Sierra de Los Cameros in den nordwestlichen Iberischen Ketten (Spanien) *Beih. Geol. Jb.*, **44**, 123–64.

Wright, V.P., Platt, N.H. & Wimbledon, W.A. 1988. Biogenic laminar calcretes: evidence of calcified root-mat horizons in paleosols. *Sedimentology*, **35**, 603–20.

La Serranía de Cuenca Basin (Lower Cretaceous), Iberian Ranges, central Spain

N. MELÉNDEZ,[1] A. MELÉNDEZ[2] AND J.C. GÓMEZ FERNÁNDEZ[1]

[1]Dpto. Estratigrafía, Universidad Complutense, E-28040, Madrid, Spain
[2]Dpto. Geología, Universidad de Zaragoza, E-50009, Zaragoza, Spain

The Serranía de Cuenca mountains are located in the western part of the Iberian Ranges in central Spain (Fig. 1). The transition period from Jurassic to Cretaceous in central Spain was dominated by distensive tectonics, coinciding with the rifting phases which took place during the Lower Cretaceous (Vilas et al., 1983). The creation of horsts and grabens allowed the development of many subbasins in which continental sedimentary sequences were deposited after a long period of erosion and karstification of the Jurassic basement.

This paper deals with the Upper Hauterivian?–Barremian depositional sequence in one of the sub-basins, the Serranía de Cuenca Basin (part of the Iberian Basin), delimited by two unconformities corresponding to the Neokimmerian unconformity at the base and the Intra-Barremian unconformity at the top (Fig. 2). This depositional sequence comprises two formations: El Collado Formation (siliciclastic and alluvial in origin) and the La Huérguina Formation (carbonate and lacustrine-palustrine in origin). The vertical juxtaposition of these two formations (Fig. 2) shows that the carbonate lacustrine system expanded over the siliciclastic alluvial system with time, probably a result of changes in basin subsidence rates.

The El Collado Formation contains coarse siliciclastic material showing cross-bedding and ripple cross-lamination within channel bodies, as well as claystone with pedogenetic features such as root traces. This formation is interpreted as floodplain deposits with a source area toward the northeast and east.

The La Huérguina Formation contains a wide variety of carbonate facies reflecting the many different subenvironments (Fig. 3):

(1) Lacustrine deposits: shallow lakes and ponds are represented by marls with ostracodes and charophytes; marly limestones with organic debris, pyrite nodules and lignites; and biomicrites (wackestones-packstones) with charophytes, ostracodes, gastropods, rare oncolites, black pebbles and intraformational intraclasts. In this same order, from base to top, these facies comprise a shallowing-upwards sequence that represents the infilling of shallow carbonate lakes. Brecciation horizons and rhizoliths can be found at the top of these sequences (Gómez Fernández, 1988; Gierlowski-Kordesch et al., 1991). Deep lakes are represented by rhythmically-laminated limestones (lacustrine carbonate varves) and turbiditic layers containing a high content of organic matter (Gómez Fernández & Meléndez, 1991). The rhythmically-laminated limestones are extremely fossiliferous with an exceptionally good quality of preservation: microphytes, crustacea, insects, fish, amphibia, reptiles (including Crocodylia) and a bird (Sanz et al., 1988a, 1990).

(2) Lacustrine deltaic deposits: deltaic lobes are represented by a coal-limestone facies association. Micrites (wackestones) contain varying amounts of coalified plant material, ostracodes, charophytes, gastropods, rare oncolites and bluegreen algae. The coalified macrophytic debris defines a relict cross-stratification in the micrites, suggesting accumulation by tractive currents. The coal, identified as a lignite, is mainly composed of allochthonous plant material. The thickest coal bed is 50 cm thick (Gierlowski-Kordesch et al., 1991). Pollen and spores from these coal beds have been dated as Early Barremian in age (Mohr, 1987). Other fossils include mammals (Crusafontia cuencana (Dryolestidae)) (Henkel & Krebs, 1969), frogs (Wealdenbatrachus jucarensis (Discoglossidae)) (Fey, 1988), crocodiles, lizards and other amphibians and reptiles (Kühne & Crusafont, 1968; Henkel, 1970).

(3) Palustrine deposits: alluvial/palustrine marginal areas are represented by carbonate conglomerates, grainstones, oncolitic sandstones, wackestones/packstones and clayey marls. Early subaerial diagenesis of these rocks is evidenced by rhizoliths (vertical prismatic structures), nodularization with clotted textures and nodular structures, brecciation, calichification and marmorization. All these diagenetic features are related to pedogenesis (Arribas et al., 1989).

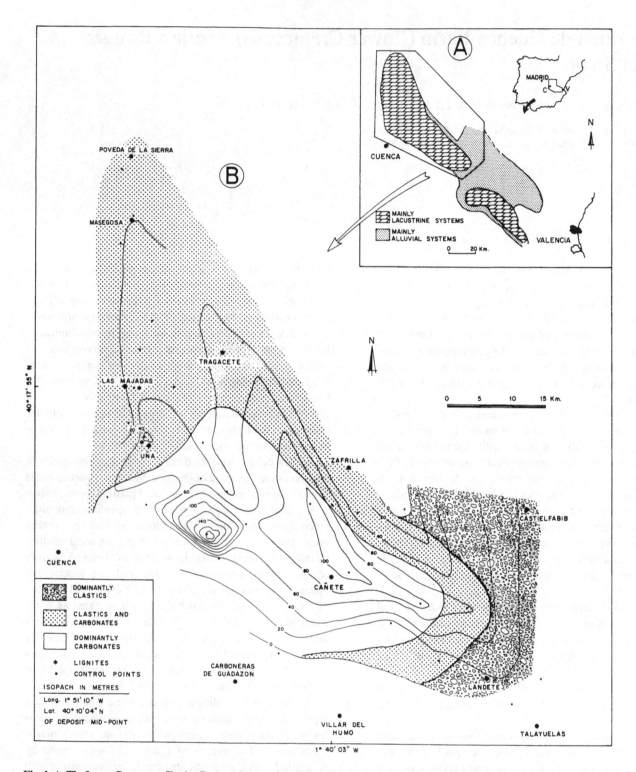

Fig. 1. A: The Lower Cretaceous Iberian Basin with the enclosed lacustrine sub-basin of La Serranía de Cuenca. B: Isopach and lithofacies assemblage map of the La Huérguina and the El Collado Formation.

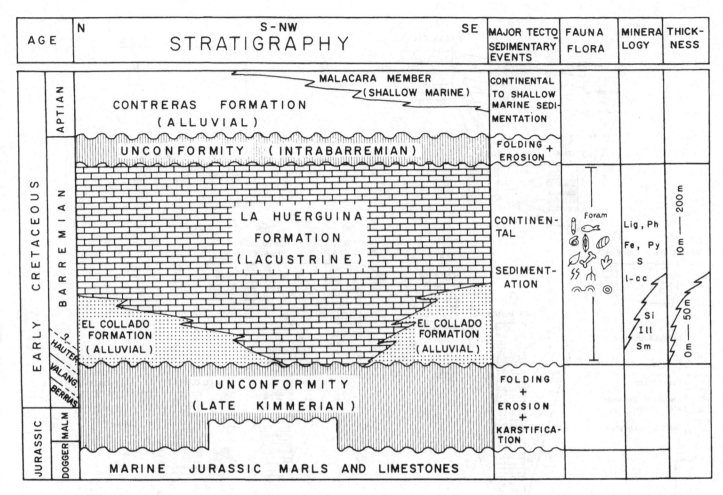

Fig. 2. Stratigraphy of the Serranía de Cuenca Basin.

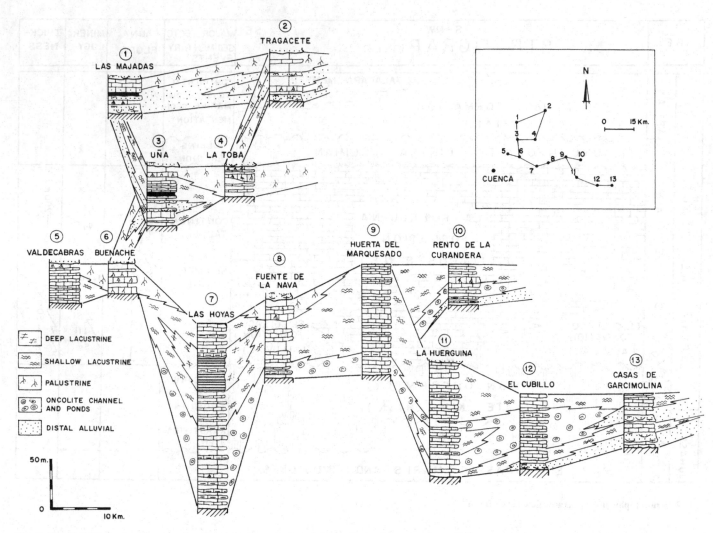

Fig. 3. Detailed lithologic columns and environmental correlation of the La Huérguina and El Collado Formations.

Bibliography

Arribas, M.E., Gómez Fernández, J.C. & Meléndez, N., 1989. Early diagenetic processes in lacustrine and in related floodplain sediments from the Lower Cretaceous (central Spain). *Abstracts, 10th Int. Ass. Sediment. Regional Mtg., Budapest, Hungary,* pp. 3–4.

Fey, B., 1988. *Die Anurenfauna aus der Unterkreide von Uña (Ostspanien).* Dissertation, Inst. Paläontologie, Freie Univ. Berlin.

Gierlowski-Kordesch, E. & Janofske, D., 1989. Paleoenvironmental reconstruction of the Weald around Uña (Serranía de Cuenca, Cuenca Province, Spain). In *Cretaceous of the Western Tethys,* ed. J. Weidmann, pp. 239–64. Proc. 3rd Int. Cretaceous Symp., Tübingen, 1987. E. Schwiezerbart'sche Verlagsbuchhandlung. Stuttgart.

Gierlowski-Kordesch, E., Gómez Fernández, J.C. & N. Meléndez, 1991. Carbonate and coal deposition in an alluvial-lacustrine setting: Lower Cretaceous Weald in the Iberian Ranges (east-central Spain). In *Lacustrine Facies Analysis,* ed. P. Anadón, L. Cabrera & K. Kelts, pp. 109–25. Int. Ass. Sediment. Special Pub. No. 13, Blackwell Sci. Pub., Oxford.

Gómez Fernández, J.C., 1988. *Estratigrafía y sedimentología del Cretácico Inferior en 'Facies Weald' de la Región Meridional de la Serranía de Cuenca* Masters Thesis, Univ. Complutense, Madrid.

Gómez Fernández, J.C. & Meléndez, N., 1991. Rhythmically laminated lacustrine carbonates of La Serranía de Cuenca Basin (Iberian Ranges, Spain). In *Lacustrine Facies Analysis,* ed. P. Anadón, L. Cabrera & K. Kelts, pp. 245–56. Int. Ass. Sediment. Special Pub. No. 13, Blackwell Sci. Pub., Oxford.

Henkel, S., 1970. Eine neue Fossillagerstätte in Ostspanien und ihre Bedeutung für die Stammgeschichte der Wirbeltiere. *Umsch. wiss. Techn.,* **8,** 247–8.

Henkel, S. & Krebs, B., 1969. Zwei Säugetier-Unterkiefer aus der unteren Kreide von Uña (Prov. Cuenca, Spanien). *N. Jb. Geol. Paläont. Mh.,* 449–63.

Kühne, W.G. & Crusafont, M. 1968. Mamíferos del Wealdense de Uña cerca de Cuenca. *Acta. Geol. Hispanica,* **3,** 133–4.

Mas F.R., Alonso, A. & Meléndez, N., 1982. El Cretácico basal 'Weald' de la Cordillera Ibérica Suroccidental (NW de la provincia de Valencia, E de la de Cuenca). *Cuad. Geol. Ibérica,* **8,** 309–35.

Meléndez Hevia, F., Villena Morales, J., Ramirez del Pozo, J., Portero García, J.M., Olivé Davó, A., Assens Caparros, J. & Sanchez Soria, P., 1975. Síntesis del Cretácico de la Zona Sur de la 'Rama Castellana' de la Cordillera Ibérica. *Actas Ier Symp. Cret. Cord. Ibérica, Cuenca,* 241–52.

Meléndez Hevia, F., Meléndez Hevia, A., Ramirez del Pozo, J., Portero García, J.M. & Gutiérrez, G., 1975. Guía de las excursiones geológicas a Tragacete-Las Majadas y Cañete-Landete. *Actas Ier Symp. Cret. Cord. Ibérica, Cuenca,* 253–73.

Meléndez, N., 1982. Presencia de una discordancia cartográfica intrabarremiense de la Cordillera Ibérica Occidental (provincia de Cuenca). *Estud. Geol.,* **38,** 51–4.

Meléndez, N., 1983. *El Cretácico de la Región de Cañete-Rincón de Ademuz (provincias de Cuenca y Valencia).* Semin. Estrat. Ser. Monograf. No. 9. Universidad Complutense, Madrid.

Meléndez, N., Mas, J.R., & Alonso, A., 1987. Lacustrine occurrences in the Lower Cretaceous of the Iberian Ranges, Spain. *Terra Cognita,* **7,** 223.

Meléndez, N., Meléndez, A. & Gómez Fernández, J.C., 1989. *Los Sistemas lacustres del Cretácico inferior de la Serranía de Cuenca (Cordillera Ibérica).* Guía de Campo, IV Reunión de Campo del Grupo Español de Trabajo, IGCP 219, Universidad Complutense, Madrid.

Mohr, B.A.R., 1987. Mikrofloren aus Vertebraten-Führenden Unterkreide-Schichten bei Galve und Uña (Ostspanien). *Berliner geowiss. Abh.,* **86A,** 69–85.

Ramírez del Pozo, J. & Meléndez Hevía, F., 1972. Nuevos datos sobre del Cretácico inferior en facies 'Weald' de la Serranía de Cuenca. *Bol. Geol. Min. (IGME),* **83**(6), 569–81.

Ramírez del Pozo, J., Portero García, J.M., Olivé Davó, A. & Meléndez Hevia, F., 1975. El Cretácico de la Serranía de Cuenca y de la región Fuentes-Vilar del Humo: correlación y cambios de facies. *Actas Ier Symp. Cret. Cord. Ibérica, Cuenca,* pp. 189–205.

Sanz, J.L., Bonaparte, J.F. & Lacasa, D., 1988a. Unusual Early Cretaceous Birds from Spain. *Nature,* **331,** 433–5.

Sanz, J.L., Diéguez, C., Fregenal-Martínez, M.A., Martínez-Delclos, X., Meléndez, N. & Poyato-Ariza, F.J., 1990. El yacimiento de Fósiles de Las Hoyas, Provincia de Cuenca (España). *Com. Runión de Tafonomía y Fosilización, Madrid,* 337–55.

Sanz, J.L., Wenz, S., Yébenes, A., Estes, R., Martínez-Delclos, X., Jiménez-Fuentes, E., Diéguez, C., Buscalioni, A.D., Barbadillo, L.J. & Vía, L., 1988b. An Early Cretaceous faunal and floral continental assemblage: Las Hoyas fossil site (Cuenca, Spain). *Geobios,* **21**(5), 611–35.

Vilas, L., Alonso, A., Arias, C., García, A., Mas, J.R., Rincon, R. & Meléndez, N., 1983. The Cretaceous of the southwestern Iberian Ranges (Spain). *Zitteliana,* **10,** 245–54.

Vilas, L., Mas, J.R., García, A., Arias, A., Alonso, A., Meléndez, N. & Rincon, R., 1982. Ibérica Suroccidental. *El Cretácico de España.* Univ. Complutense, Madrid, pp. 457–509.

Late Cretaceous

Uppermost Cretaceous and Paleogene fluvio-lacustrine basins in the northern Iberian Ranges (Spain)

MARC FLOQUET, JEAN SALOMON AND JEAN-PAUL VADOT

Centre des Sciences de la Terre et URA CNRS no 157, Université de Bourgogne, 6, Bd Gabriel 211000 Dijon, France

Geographic and geologic setting

Continental deposits of Late Cretaceous–Paleogene age occur in the Santo Domingo de Silos and Arganza-Talveila Basins (Fig. 1), at the southern border of the Northwestern Iberian Ranges of Spain. These deposits comprise fluvial claystones, sandstones, and conglomerates, as well as carbonates attributed to lacustrine–palustrine and sabkha paleoenvironments. Biostratigraphic data from dinosaur remains (Lapparent *et al.*, 1957), from gastropods, ostracods and charophytes (Floquet, 1990; Vadot, 1992) indicate a Maastrichtian to Ypresian age. These deposits form the Santibañez del Val Formation *sensu* Floquet *et al.*, (1982) and were previously studied by Valladares (1976, Unit IV), Pol Mendez (1985), Floquet *et al.* (1985), Floquet (1990) and Vadot (1992).

In the Santo Domingo de Silos Basin, the Santibañez del Val Formation generally rests on marginal marine carbonates of the Santo Domingo de Silos Formation (Fig. 2) of uppermost Campanian age according to the rudistid fauna (Floquet, 1990). In the Arganza-Talveila Basin, the Santibañez del Val Formation rests on marginal marine and palustrine carbonates of the Sierra de la Pica Formation (Fig. 2), attributed to the uppermost Campanian by reason of its lateral equivalence with the top of the Silos Formation (Floquet, 1990).

General and detailed sections

Santo Domingo de Silos Basin

The Santibañez del Val Formation is approximately 500 m thick in this area (Fig. 2) and is divided into seven Members which correspond to the successive terrigenous and carbonate facies.

The carbonate facies consist mainly of micritic limestones with various features arranged in a depositional and diagenetic sequence. An example of such a sequence (Fig. 3) is taken from the Santa Cecilia Member (Fig. 2). Each sequence shows an evolution of environments from very shallow lacustrine (deposition of homogeneous micrites) to marginal lacustrine, characterized by alternating subaerial exposure and submergence (formation of nodular micrites with microkarst cavities), and finally to settings with

prolonged subaerial exposure (formation of laminar calcrete and rhizoliths). There is also a large scale upward trend in environments evident through the whole formation: from shallow lakes, subject to only temporary emergence (La Yecla Member of Maastrichtian age (*Lychnus* fauna)), to palustrine settings with prolonged subaerial exposure of marshy marginal areas (Santa Cecilia Member of Ypresian age as indicated by the *Planorbis, Lymnaea* fauna) (Vadot, 1992).

The dolomitic limestones of the Retuerta Member record marine-influenced sabkha environments. Deposition of these rocks probably correlates with the transgression of Dano-Montian age documented from the Northern Castilian and Cantabrian marine platform (Floquet, 1990).

The terrigenous facies consist predominantly of red silty claystones and scattered lenticular sandstone or conglomerate-sandstone bodies. This suggests a distal fluvial environment. First, in the lower terrigenous Members (Castroceniza and Las Porqueras), much of the run-off was in the form of unchannelized surface flow: the few channels were unstable and switched freely over the laterally extensive floodplain. Then, in the upper terrigenous members (Quintanilla del Coco and La Cañada), facies become progressively coarser with larger sandstone and conglomerate-sandstone bodies and with a higher degree of organization up sequence.

The general coarsening upward of the terrigenous deposits, the increase in organization towards the top of the Santibañez del Val Formation and the progressive upward evolution from lacustrine to palustrine carbonates probably all reflect tectonic uplift (Floquet *et al.*, 1985, Alonso *et al.*, 1987; Floquet, 1990), which resulted in increasing continental influence in the Santo Domingo de Silos area between the Maastrichtian and the Ypresian.

Arganza–Talveila Basin

The Santibañez del Val Formation is less than 250 m thick in this area and is divided into two Members (Fig. 2). Carbonates dominate the Valdelacasa Member of Maastrichtian age (*Lychnus* fauna). These deposits indicate lacustrine–palustrine environments with frequent subaerial exposure. Spectacular oncoids, which

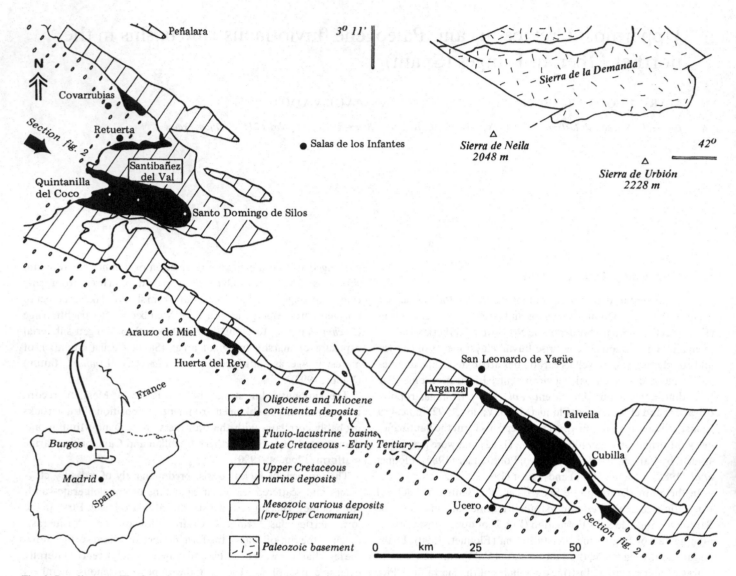

Fig. 1. Locality map. The fluvio-lacustrine basins of Santo Domingo de Silos and Arganza–Talveila on the southwestern border of the Northwestern Iberian Ranges.

rarely encrust gastropods and reptile bone splinters, fill the small fluvial channels.

The El Colmenar Member, attributed to the lowermost Tertiary, is entirely terrigenous, comprising muddy sandstones and sandstones with intercalated coarse conglomerates. These deposits record a fluvial environment, proximal in comparison with that of the Santo Domingo de Silos Basin. Conglomerates at the top of the formation may have been deposited in an alluvial fan.

Evolution of the fluvio-lacustrine basins related to the northern Castilian and Cantabrian marine platform

Comparison of the Late Cretaceous–Paleogene sedimentary series exposed in the study area with that present in areas of the Northern Castilian and Cantabrian Platform 50–100 km to the north (Floquet, 1990) suggests that deposition of the lacustrine–palustrine or sabkha carbonates of the Santibañez del Val Forma-

tion may have correlated with periods of marine transgression on the Northern Castilian Platform.

During the Upper Maastrichtian, lacustrine–palustrine limestones were deposited in both the Santo Domingo de Silos and Arganza–Talveila Basins. Subtidal limestones with *Orbitoides* and supratidal dolomicritic limestones with gypsum were laid down on the marine platform to the north at this time (Figs 2 and 4). During the Dano-Montian, coastal sabkha and palustrine facies were laid down in the Santo Domingo de Silos Basin, whereas reefal and bioclastic dolomitic limestones were deposited on the marine platform. During the Ypresian, lacustrine–palustrine limestones were again deposited in the Santo Domingo de Silos Basin, while subtidal limestones with Alveolinids were deposited on the marine platform. The formation of lakes and marshes may have been related to a rise in base level and/or to damming of fluvial outlets at the time of absolute or relative sea level rises.

Conversely, the deposition of fluvial deposits in the study area

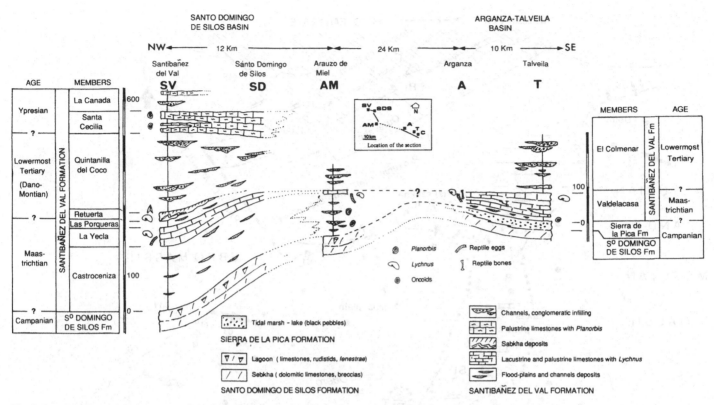

Fig. 2. Stratigraphy and environments of the fluvio-lacustrine series in the Santo Domingo de Silos and Arganza–Talveila basins.

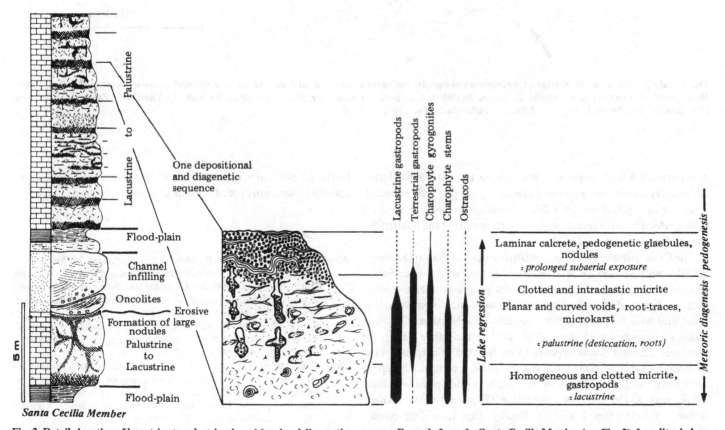

Fig. 3. Detailed section of lacustrine to palustrine depositional and diagenetic sequences. Example from the Santa Cecilia Member (see Fig. 2). Locality: below the Ermita Santa Cecilia, south of the Santibañez del Val village. From Vadot (1992).

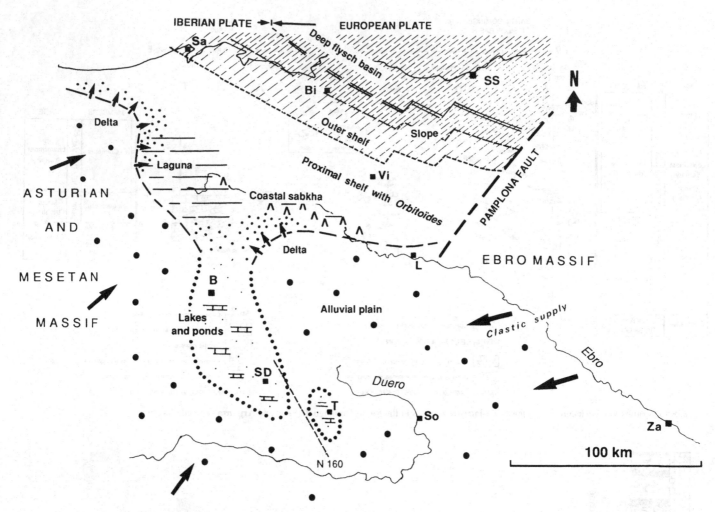

Fig. 4. Paleogeography of the Northern Castilian coastal complex during the Upper Maastrichtian and location within of the fluvio-lacustrine basins of Santo Domingo de Silos and Arganza–Talveila. *B.* Burgos, *Bi.* Bilbao, *L.* Logroño, *Sa.* Santander, *SS.* San Sebastian, *So.* Soria, *Vi.* Vitoria, *Za.* Zaragoza, *SD.* Santo Domingo de Silos Basin, *T.* Arganza–Talveila–Cubilla Basin. From Floquet (1990).

was correlated with times of regression on the marine platform. Marine regression and increased clastic supply probably reflected absolute or relative sea-level falls as well as periods of tectonic uplift, which became increasingly important during the Maastrichtian to Ypresian time-interval.

The Covarrubias–Retuerta, Santibañez del Val–Santo Domingo de Silos and Arauzo de Miel outcrops belong to the same Santo Domingo de Silos 'Basin' according to the progressive sedimentological evolution of all facies. The deposits were laid down over a low relief and wide homoclinal ramp linked to the Northern Castilian coastal complex as a whole (Fig. 4). On the other hand, the Arganza–Talveila Basin appears to be isolated and formed in a developing syncline. This Basin may have been tectonically separated from the Santo Domingo de Silos by an incipient anticlinal ridge, with a 150°–160° N orientation (Eastern Iberian Ranges tectonic trend) (Figs 2 and 4). Finally, this Late Cretaceous–Paleogene fluvio-lacustrine sedimentation occurred in 'proto-

foreland basins' in which development was related to compression associated with northward movement of Iberia towards France.

Bibliography

Alonso, A., Floquet, M., Mas, R., Meléndez, A., Meléndez, N., Salomon, J. & Vadot, J.P., 1987. Modalités de la régression marine sur le détroit ibérique (Espagne) à la fin du Crétacé. In *Transgressions et régressions au Crétacé en France et dans les pays limitrophes. Mém. Géol. Univ. Dijon* **11**, 91–102.

Floquet, M., Alonso, A. & Meléndez, A., 1982. El Cretácico superior–Cameros–Castilla. In *El Cretácico de España*, pp. 387–456. Universidad Complutense, Madrid, 387–456.

Floquet, M., Salomon, J., & Vadot, J.P., 1985. Evolution of a continental basin according to carbonate facies, sequences and diagenesis: example in the Late Cretaceous–Early Tertiary of Santo Domingo de Silos (Province of Burgos, Spain). *Abstracts 6th European Regional Meeting Sedimentology I.A.S.*, Lleida, pp. 566–9.

Floquet, M., 1990. La plate-forme nord-castillane au Crétacé supérieur

(Espagne). Arrière-pays ibérique de la marge passive basco-cantabrique. Sédimentation et vie. Thèse d'Etat, Université de Bourgogne, Dijon. (*Mém. géol. Univ. Dijon*, **14**, 1–925, 1991.)

Lapparent, A.F. de, Quintero, I., & Trigueros, E., 1957. Descubrimiento de huesos de dinosaurios en el Cretácico terminal de Cubilla (Soria). *Not. Comunic. Inst. Geol. Min. España*, **45**, 61–3.

Pol Mendez, C., 1985. *Estratigrafía y paleogeografía de los sedimentos Cretácicos–Paleogenos y Miocenos del Este de la Cuenca del Duero.* Tesis Doctoral, Univ. Oviedo.

Vadot, J.P., 1992. *Bassins fluvio-lacustres au Crétacé final et au Paléogène dans les chaînes ibériques septentrionales (Espagne).* Thèse Doctorat, Univ. Bourgogne.

Valladares, I., 1976. Estratigrafía del Cretácico superior calcáreo en el borde occidental de la Cuenca de Cameros (Burgos-Soria). *Stud. Geol.*, **XI**, 93–108.

The Upper Cretaceous lacustrine record in the central part of the Iberian Ranges, Spain

A. MELÉNDEZ[1] AND N. MELÉNDEZ[2]

[1] Dpto. Geología, Universidad de Zaragoza, E-50009, Zaragoza, Spain
[2] Dpto. Estratigrafía, Universidad Complutense, E-28040, Madrid, Spain

During the uppermost Cretaceous, a marine regression occurred in the Iberian trough (Fig. 1) allowing the spread of continental conditions throughout most of the Iberian Ranges (Alonso et al., 1987). This regression was controlled by uplift during the Upper Santonian–Campanian and the first compressive movements during the Maastrichtian (Floquet & Meléndez, 1982), which are related to the movement of the Iberian Plate towards the northeast (Boilot et al., 1984; Alonso et al., 1987, 1989). From a paleogeographic point of view, basement lineaments (direction: N 60° E and N 140° E) controlled the location and development of sub-basins in the area, including the Soria lineation, Ateca High, Alto Tajo Fault, Priego-Cuenca lineation, the Gallocanta High, Ateca–Castellón Fault, and Montalban–Oropesa Fault (Fig. 1). The relative movements of the different blocks created three sub-basins: the Ibérica Central, the Serranía de Cuenca and the El Maestrazgo. Climatic conditions in these isolated continental sub-basins were probably arid and warm with strong, sporadic rainfall (Alonso et al., 1987).

The Ibérica Central area contains the Sierra de La Pica Formation (Figs 2 and 3). This sequence includes black pebble limestones, which have channelized geometries and mud mounds with blue-green algae. Lychnus (gastropod), charophytes, ostracods and rhizoliths are associated with limestones containing features related to the cyclic saturation and desiccation of a lake margin. This formation is interpreted as an extensive lacustrine–palustrine environment with marshes, coastal ponds and alluvial channels.

In the Serranía de Cuenca area, the uppermost Cretaceous lacustrine sedimentation is represented by the Villalba de la Sierra Formation. This formation is composed of sandstones, claystones, charophyte limestones and gypsum. The rock sequence is interpreted as a transition from a littoral sabkha environment (Campanian) to an alluvial continental environment (Maastrichtian) (Figs. 2 and 3). The continental floodplains, though siliciclastic in nature, also contained large carbonate ponds where plant colonization occurred.

The Fortanete Formation represents the Maastrichtian sequence in the El Maestrazgo area. The main portion of this formation is comprised of micritic limestones with Lychnus and black pebbles and is interpreted as sediments of a shallow carbonate lacustrine environment surrounded by wide palustrine margins with plant colonization. The upper part of the Fortanete Formation contains abundant nodular gypsum structures, pointing toward evaporitic episodes associated with a littoral sabkha environment.

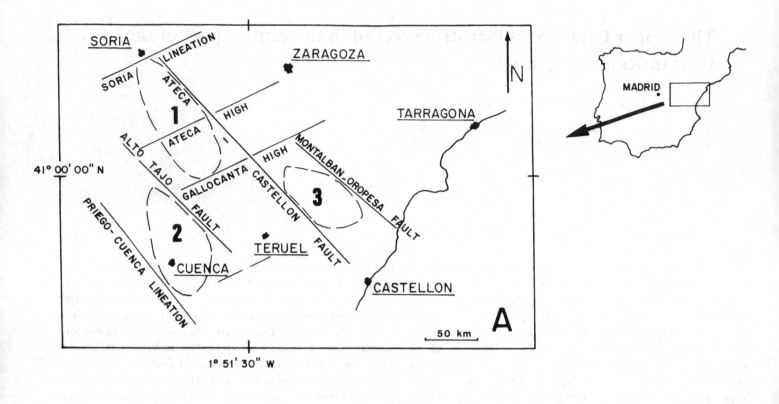

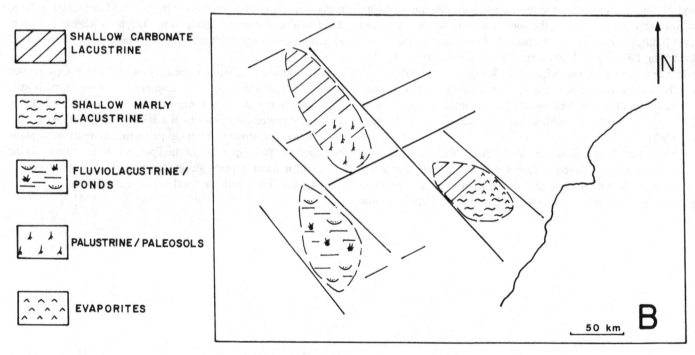

Fig. 1. A: Location map showing the tentative subdivision into three main paleogeographic domains for the Upper Cretaceous (late Senonian) in the Iberian Ranges. (1) Ibérica Central, (2) Serranía de Cuenca, and (3) Maestrazgo. B: Paleogeographic distribution of lacustrine deposits.

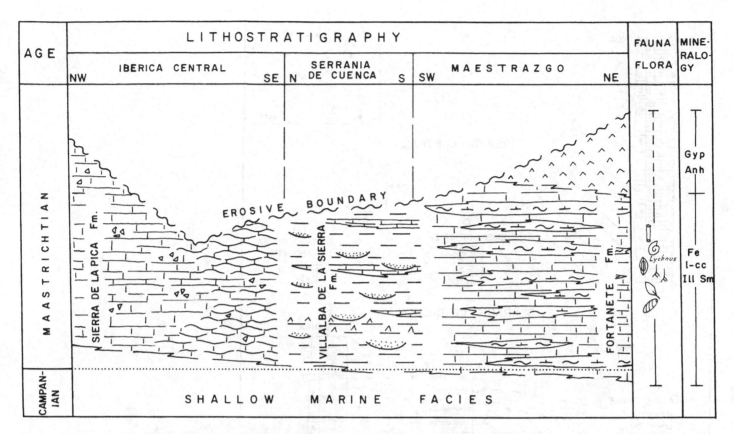

Fig. 2. Schematic stratigraphic diagram of the Upper Cretaceous lacustrine facies of the central part of the Iberian Ranges. See Fig. 3 for thickness variations in the three paleogeographic domains and Fig. 1 for lithologic symbols.

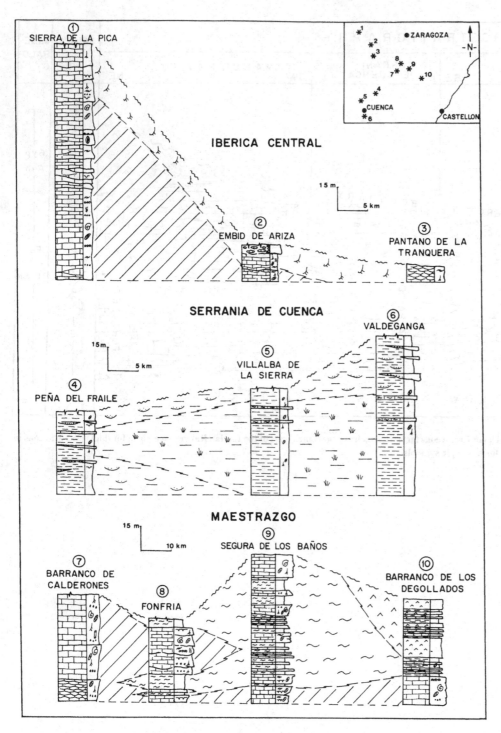

Fig. 3. Lithologic columns and environmental correlation in the three paleogeographic domains of the Upper Cretaceous. See Fig. 1 for lithologic symbols.

Bibliography

Almunia, A., 1984. *Estratigrafía y Sedimentología del Cretácico Superior Carbonatado al Oeste del Maestrazgo.* Tesis de Licenciatura, Univ. Zaragoza.

Almunia, A., Arqued, V. & Meléndez, A., 1985. Caracerísticas sedimentológicas durante el ciclo Senoniense en el Maestrazgo. *Trabajos Geol.,* **15,** 159–67.

Alonso, A., Floquet, M., Mas, J.R. & Meléndez, A., 1983. Evolution paléogeographique des plates-formes de la Meseta Nord-Castillane et de la Cordillère Ibérique (Espagne) au Senonien. *Géol. Méditerranéenne* **3–4,** 361–7.

Alonso, A., Floquet, M., Mas, J.R., Meléndez, A., Meléndez, N., Salomon, J. & Vadot, J.P., 1987. Modalités de la regression marine sur de détroit Ibérique (Espagne) à la fin du Crétacé. *Mem. Géol. Univ. Dijon,* **11,** 91–102.

Alonso, A., Floquet, M., Mas, J.R. & Meléndez, A., 1989. Origin and evolution of an epeiric carbonate platform, Upper Cretaceous, Spain. *XII Congr. Español Sediment., Simposios y Conferencias,* 21–31.

Arqued, V., 1984. *La serie carbonatada del Cretácico superior en el Maestrazgo spetentrional: estratigrafía, sedimentología y paleogeografía.* Tesis de Licenciatura, Univ. Zaragoza.

Boillot, G., Montadert, L., Lemoine, M. & Biju-Duval, B., 1984. *Les Marges Continentales Actuelles et Fossiles Autour de La France.* Paris: Masson.

Canerot, J., Cugny, P., Pardo, G., Salas, R. & Villena, J., 1982. Ibérica Central-Maestrazgo. In *El Cretácico de España,* pp. 273–344. Universidad Complutense, Madrid.

Floquet, M., Alonso, A. & Meléndez, A., 1982. Cameros-Castilla. In *El Cretácico de España,* pp. 387–456. Universidad Complutense, Madrid.

Floquet, M. & Meléndez, A., 1982. Características sedimentarias y paleogeográficas de la regresión finincretácica en el sector central de la Cordillera Ibérica. *Cuad. Geol. Ibérica,* **8,** 237–57.

Floquet, M., Salomon, J. & Vadot, J.P., 1985. Evolution of a continental basin according to carbonate facies, sequences, diagenesis: example in the Late Cretaceous-Early Tertiary of Santo Domingo de Silos (Province of Burgos, Spain). *Abstracts, 6th Int. Ass. Sediment. Eur. Reg. Mtg., Lleida, Spain,* pp. 566–9.

Meléndez, A. & Meléndez, N., 1989. Depósitos lacustres del Cretácico terminal en la Cordillera Ibérica Central y Maestrazgo. *Geogaceta,* **6,** 99.

Meléndez, A., Meléndez, F., Portero, J. & Ramirez, J., 1985. Stratigraphy, sedimentology, and paleogeography of Upper Cretaceous evaporitic-carbonate platform in the central part of the Sierra Ibérica. *Excursion Guidebook, 6th Int. Ass. Sediment. Eur. Reg. Mtg.,* pp. 188–213.

Vilas, L., Alonso, A., Arias, C., García, A., Mas, J.R., Rincón, R. & Meléndez, N., 1983. The Cretaceous of the southwestern Iberian Ranges (Spain). *Zitteliana,* **10,** 245–54.

Vilas, L., Mas, J.R., Garcá, A., Arias, A., Alonso, A., Meléndez, N. & Rincón, R., 1982. Ibérica Suroccidental. In *El Cretácico de España,* pp. 457–509. Universidad Complutense, Madrid.

The 'Lignites du Sarladais' Basin, Dordogne, southwest France

J.-P. COLIN

Esso Rep, 213 Cours Victor Hugo, 33323 Bègles, France

Age

Upper Cretaceous, Cenomanian. essentially based on palynology (Colin & Médus, 1972) and Charophyta (Colin, 1973a) with *Atopochara multivolvis*.

Underlying sediments are of Late Jurassic (Portlandian) age, dated by foramimifera (Arnaud, 1865; Séronie-Vivien, 1959).

Overlying sediments in the east are marine Early Turonian micritic limestones, with marine late Cenomanian clays in the West (Colin, 1973a).

Lithology

In the Sarladais (Dordogne, southwest France), the Cenomanian is often represented by a non-marine facies called 'Lignite du Sarladais'. It is known in outcrops in two distinct areas separated by about 30 km (Fig. 1):

(1) The so-called zone of La Chapelle Péchaud, near the village of Saint-Cyprien. The following succession of facies can be observed (Fig. 2):
   (i) lignite, overlying karstified Late Jurassic limestones;
   (ii) lignitic clays often gypsiferous toward the top;
   (iii) marine oyster marls with foraminifera and ostracodes.
(2) The Sarladais *sensu stricto* Zone, northeast of the town of Sarlat. In this area, the Cenomanian is typically lacustrine, represented by a succession of lignites, lacustrine limestones very rich in gastropods, ostracodes and charophytes, and horizons of bituminous shales and boghead (Fig. 3).

Paleontology

Gastropods (Répelin, 1902)

Helix cenomanensis, H. (Xerophila) petrocoriensis, Bulimus sp., *Cylindrogyra varians, Nisopsis fluviatilis, Auricula* sp., *Limnaea conica, L. munieri, L. acuta, Planorbis cretaceus, Physa simeyrolensis, P. minima, P. granum, P. subcylindrica, Chilina olivula, Melania sulcorugata, M. nitida, M. quadricostata, M. tricostata, M. costulata, Hamtkenia subovoidea, H. munieri, Hydrobia moureti, Bithinia primigenia, B. pisum, Paludina dordoniensis, Valvata arnaudi, Ampulllopsis faujasi, Aperostoma primigenia, Ampullina cureti, Neritina primordialis, Cyclotus primigenius, Nisopsis* sp., *Potamides (Bittium) malviensis.*

Pelecypods (Répelin, 1902)

Unio sp., *Corbula zurcheri.*

Ostracodes (Colin, 1973a, 1974a)

Metacypris xestoleberiformis, Rosacythere grekoffi.

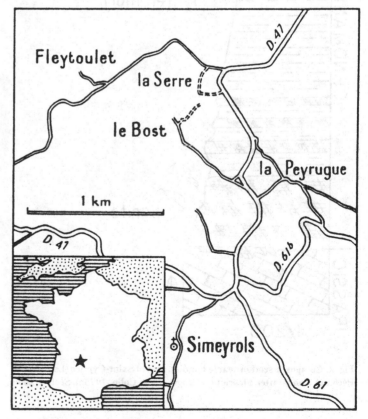

Fig. 1. Situation map of the type area of the 'Lignites du Sarladais' (Sarladais s.s. Zone).

235

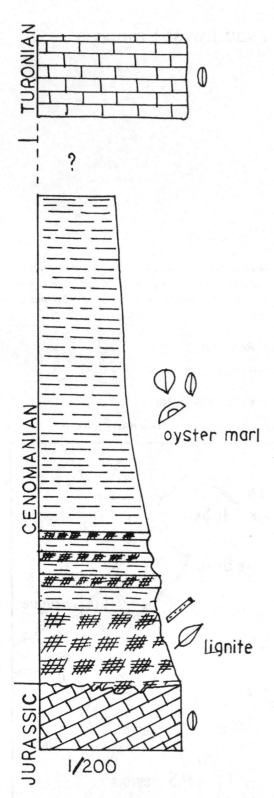

Fig. 2. **Composite section near le Fournet, area of Saint-Cyprien (La Chapelle Péchaud Zone) (after Fleuriot de Langle, 1964; Colin, 1973a). Scale 1/200.**

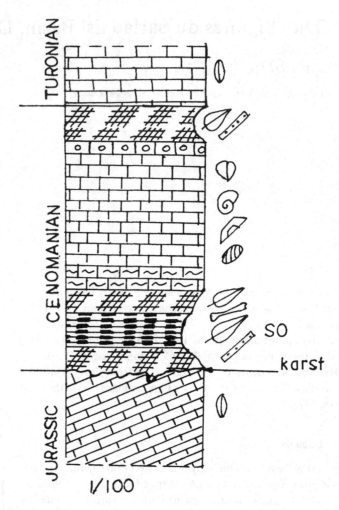

Fig. 3. **Composite section near La Serre, area of Simeyrols–Sarlat (Sarladais *s.s.* Zone) (after Arnaud, 1877; Colin, 1973 b). Scale 1/100.**

Algae (Colin, 1973a; Colin & Vachard, 1977)

Atopochara multivolvis (Charophyta), *Girvanella? palustris*, *Botryococcus* sp., *Munieria* cf. *baconica*.

Spores and pollen (Colin & Médus, 1972)

Retricolpites vulgaris, Tricolpites cf. *reticulatus, T.* cf. *erugatus, Tricolpopollenites micromunus, Taxodiaceapollenites hiatus, Liliacidites* cf. *variegatus, Cyathydites* sp., *Camarazonosporites* sp., *Costatoperforosporites* sp., *Gleicheniidites* sp., *Appendicisporites tricuspidatus, A. stylosus, Cicatricosisporites dorogensis, Vadazisporites* sp.

Megaspores and other botanical microfossils (Colin, 1973a, b; 1974b)

Microcarpolithes hexagonalis, Spermatites ellipticus, Costatheca dakotaensis, C. halli, Verrutriletes aff. *dubius, V.* aff. *compositipunctatus, Striatriletes?* sp.

Plants (Zeiller, 1887)

Sequoia aliena, S. reichenbachi, aff. *Myrica* sp.

Vertebrates

Remains of Saurians and Chelonians.

Bibliography

Arnaud, H., 1865. Des argiles lignitifères du Sarladais. *Bull. Soc. Géol. France*, Ser. 2, **23**, 59–63.

Arnaud, H., 1877. Mémoire sur le terrain crétacé du Sud-Ouest de la France. *Mém. Soc. Géol. France* Ser. 2, **10** (4), 1–110.

Arnaud, H., 1879. Les lignites de Saint-Cyprien. *Bull. Soc. Géol. France,* Ser. 3, **8**, 32–3.

Arnaud, H., 1887. Réunion de la Société Géologique de France dans la Charente-Maritime et la Dordogne. *Bull. Soc. Géol. France* Ser. 3, **15**, 809.

Colin, J.-P., 1973a. *Etude stratigraphique et micropaléontologique du Crétacé supérieur de la région de Saint-Cyprien (Dordogne).* PhD Thesis, Univ. Paris.

Colin, J.-P., 1973b. Microfossiles végétaux dans le Cénomanien et le Turonien de Dordogne (S.O. France). *Palaeontographica*, **143B** (5–6), 106–19.

Colin, J.-P., 1974a. Nouvelles espèces des genres *Metacypris* et *Theriosynoe-cum* (ostracodes lacustres) dans le Cénomanien de Dordogne (S.O. France). *Rev. Espań. Micropaleontol.*, **5** (2), 183–9.

Colin, J.-P., 1974b. Quelques mégaspores du Cénomanien et du Turonien supérieur du Sarladais (Dordogne, S.O. France) *Rev. espań. Micropaleontol.*, **7** (1), 15–23.

Colin, J.-P., & Médus, J., 1972. Un gisement de lignite du Sarladais d'âge Cénomanien: données palynostratigraphiques. *C. R. Soc. Géol. France, Paris*, **1**, 1–22.

Colin, J.-P. & Vachard, D., 1977. Une 'Girvanelle' dulçacquicole du Cénomanien du Sud-Ouest de la France: *Girvanella(?) palustris* Colin et Vachard n.sp. *Rev. Palaeobot. Palynol.*, **23**, 293–302.

Fleuriot de Langle, P. 1964. *Analyse stratigraphique du Cénomanien et évolution en bordure Nord Aquitaine.* Doctoral Dissertation, Univ. Bordeaux.

Harlé, 1863. Note sur le niveau géologique des calcaires crétacés de Sarlat (Dordogne). *Bull. Soc. Géol. France*, Ser. 2, **20**, 120–2.

Meugy, P., 1865. Observations sur l'âge des lignites de Cimeyrols et des environs de Sarlat (Dordogne). *Bull. Soc. Géol France*, Ser. 2, **23**, 89–93.

Mouret, 1867. Compte rendu de l'excursion du 16 septembre aux mines de Simeyrols. *Bull. Soc. Géol. France*, Ser. 2, **15**, 875–81.

Répelin, J., 1902. Description des faunes et des gisements du Cénomanien saumâtre ou d'eau douce du Midi de la France. *Ann. Mus. Hist. Nat. Marseille*, **7**.

Séronie-Vivien, R.M., 1959. Etude geologique de l'antidinal de Saint-Cyprien (Dordogne). *8h Congrès Soc. Sav. Dison.* pp. 571–578.

Zeiller, R., 1887. Note sur la flore des lignites de Simeyrols. *Bull. Soc. Géol. France*, Ser. 3, **15**, 882–3.

Paleocene to Eocene

The Eocene lacustrine–palustrine system of Peguera (Mallorca Island, western Mediterranean)

E. RAMOS-GUERRERO, L. CABRERA AND M. MARZO

Dpto. Geologia Dinàmica, Geofisica i Paleontologia, Facultat de Geologia, Zona Universitaria de Pedralbes, 08028 – Barcelona, Spain

Geological setting

Mallorca is the largest of the Balearic Islands in the western Mediterranean Sea (Fig. 1.A). Its rocks belong to the Alpine–Betic thrust-fold belt. From Paleogene until Middle Miocene, this area underwent a dominant compressive tectonism caused by the convergence between the European and African plates (Decourt *et al.*, 1986).

Mallorca Island consists of a Neogene horst and graben system developed after Langhian (Middle Miocene) times. The grabens are infilled by Middle Miocene to Quaternary post-compressive rocks, which are more than 1000 m thick. The Neogene horsts make up the major reliefs on the island and include Paleozoic to Middle Miocene rocks. These rocks are structured (Fig. 1.B) in a set of northwest-directed pre-Langhian thrust sheets (Fallot, 1922; Darder, 1925; Alvaro, 1987; Sàbat *et al.*, 1988).

Paleogene rocks on Mallorca unconformably overlie the Mesozoic substratum, usually Lower Cretaceous in age. There is a Paleocene and Lower Eocene sedimentation gap, and the older Paleogene rocks are Lutetian in age (Middle Eocene). The Paleogene sedimentary record on the island has been split into several lithostratigrafic units (Fig. 2) and assembled in two depositional sequences bounded by erosive surfaces (Ramos-Guerrero *et al.*, 1989b). Both depositional sequences include platform and near-shore marine deposits in the southern depositional area. Alluvial and lacustrine deposits occur in the northern study area. Lacustrine sedimentary rocks include the Peguera Limestone Formation (Bartonian–Priabonian in age, Depositional Sequence I), and the mainly siliciclastic Cala Blanca Formation (Oligocene in age, Depositional Sequence II) also described in this volume.

The Eocene lacustrine system (Sequence I)

The Peguera Limestone Formation

This formation crops out along the southern edge of the Serra de Tramuntana (Fig. 1.B). Good exposures are available at the Peguera, Alaró and Biniamar localities (PE, AL and BI in Fig. 1.B). This unit is up to 140 m thick and bounded by erosive surfaces (Fig. 3). The Bartonian (Middle Eocene) age of this unit has been established on the basis of charophytes (Ramos-Guerrero, 1988) and rodent fossil assemblages (De Bruijn *et al.*, 1978; Hugueney & Adrover, 1982). Work currently in progress suggests that the uppermost levels of the unit may be Priabonian.

The Peguera Formation is made up by two kinds of facies assemblages: alluvial and lacustrine. The alluvial red-bed facies assemblage (mudstones, sandstones and minor conglomerates) crops out only locally (Platja d'es Morts Member at Peguera Section, PE in Fig. 3) and wedges out towards the northeast. The lacustrine facies assemblage consists nearly exclusively of carbonates.

Lacustrine facies in the Peguera Formation

The carbonate assemblages of the Peguera Formation record the development of a lacustrine basin filling in a fault-related basin located in a foreland platform (Ramos-Guerrero *et al.*, 1989a, b). The minimal areal extension of this lacustrine basin was larger than 115 km^2.

Three major lacustrine facies assemblages are recognized in the Peguera Formation:

(1) *Lower shallow lacustrine assemblage*. The lowermost lacustrine deposits in the Peguera Formation consist of shallow lacustrine facies which include organic-rich marls and oncolitic beds. This facies assemblage has been recorded in the Alaró and Biniamar sections (AL and BI in Fig. 3).

(2) *Inner lacustrine facies assemblage*. This assemblage consists of alternating coal and carbonate beds, a few dm thick. The carbonates are darkly-laminated, bioclastic wackestones and packstones with a high organic matter content. The major components are reworked cyanobacteria filaments. The organic matter contents in the carbonates range between 2 and 12%, with mean values about 8%. The geochemical characteristics allow us to attribute the recorded kerogen to the II and III types, pointing to an important macrophytic contribution. Nevertheless minor algal contributions can also be envisaged (Ramos-Guerrero *et al.*, 1989a).

The coal is a poorly matured ($R_0 = 0.37$), humic sub-bituminous coal. Cutinite, resinite and alginite macerals are included in a

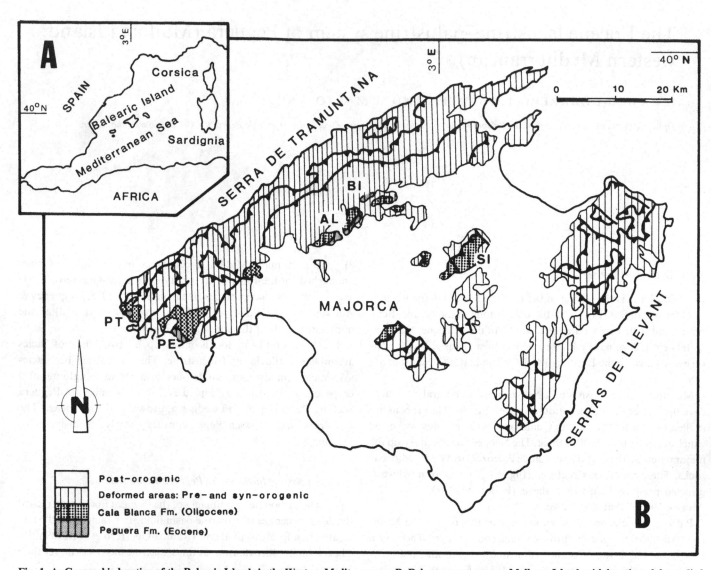

Fig. 1. A: Geographic location of the Balearic Islands in the Western Mediterranean. B: Paleogene outcrops on Mallorca Island, with location of the studied sections. PT = Puig d'en Tió, PE = Peguera, BI = Biniamar and SI = Sineu.

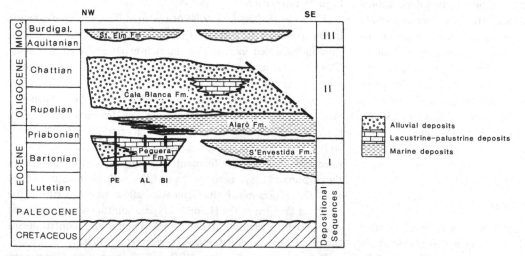

Fig. 2. Stratigraphic framework for the Paleogene on Mallorca. The location of the studied sections is shown by the same captions used in Fig. 1. After Ramos-Guerrero *et al.*, 1989 a.

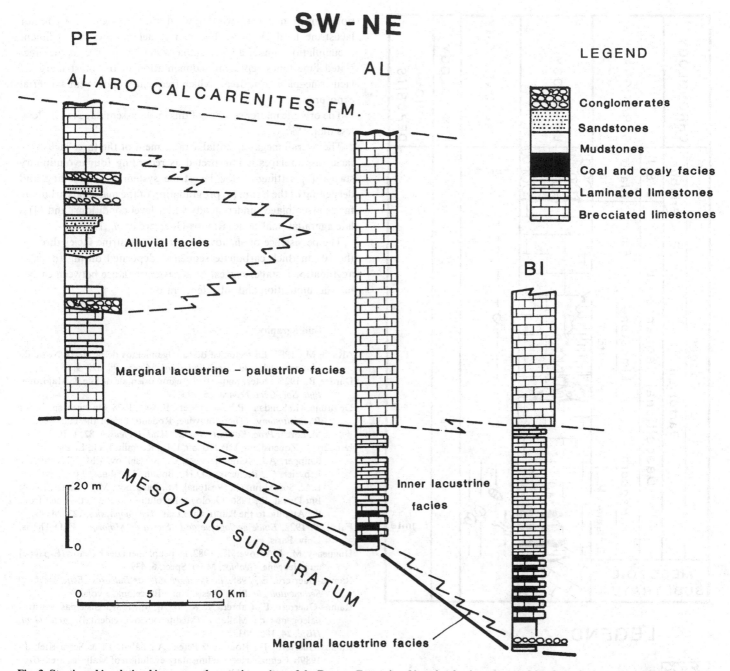

Fig. 3. Stratigraphic relationships among three main sections of the Peguera Formation. Note the clastic wedge resulting from the interrelationships between alluvial and lacustrine facies as well as the pinching out against the substratum of the lowermost shallow lacustrine facies and the inner lacustrine facies. See location of the sections in Fig. 1. (After Ramos-Guerrero *et al.*, 1989 a).

huminitic matrix. The main inorganic components are carbonates, but the sulfur content is rather high (13–15% as calculated from pure coal samples).

This facies assemblage only occurs in the northeastern localities (AL and BI in Fig. 3).

(3) *Upper marginal lacustrine facies assemblage.* This facies assemblage is recorded in all the studied sections (PE, AL and BI, in Fig. 3). It consists of thick successions of lacustrine–palustrine limestones which overlie the inner lacustrine assemblage.

These limestones are pale-colored and show massive lenticular

beds, up to 1 m thick, with minor marly interbeds. The limestones are peloidal and algal mudstones and wackestones, with minor and small-size stromatolitic bioconstructions, and bioclastic accumulations of ostracodes, gastropods and charophytes. They are commonly brecciated and often display an intensive root bioturbation. *Microcodium* also occurs rarely.

The thin marl interbeds contain many fossils consisting of charophyte and vertebrate remains, which include cocodrilidae and freshwater fish teeth.

Limestones and marls in the upper levels of the Peguera Forma-

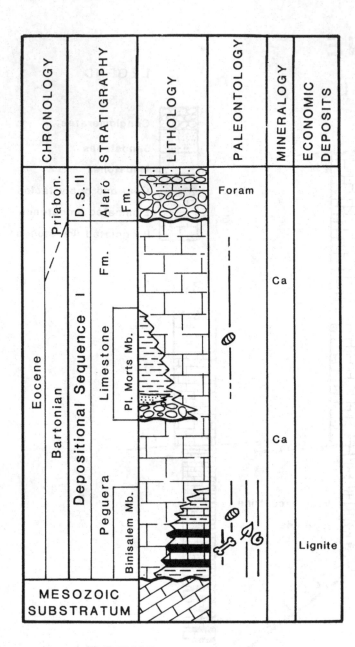

LEGEND

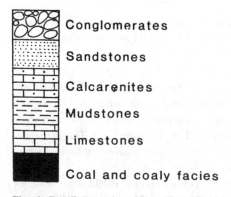

Conglomerates

Sandstones

Calcarenites

Mudstones

Limestones

Coal and coaly facies

Fig. 4. Detailed stratigraphic section and characteristics of the Peguera Limestone Formation. See thickness variations in Fig. 3.

tion form minor shallowing-upward sequences caused by minor lacustrine level changes. The marly facies represents sediment accumulations under a more stable water level, whereas the brecciated limestones represent sedimentation in the most marginal lacustrine–palustrine areas, which frequently underwent subaerial exposure.

The organic matter content of this facies assemblage is very low, less than 1%.

The overall megasequential arrangement of the three lacustrine facies assemblages is interpreted as recording four evolutionary stages: (1) settling of the lacustrine system, (2) spreading and deepening of the basin, (3) progradation of the marginal carbonate facies assemblages under steady water-level conditions, and (4) a late aggradational stage (Ramos-Guerrero *et al.*, 1989a).

The persistence of shallow lacustrine–palustrine facies through the 100 m thick carbonate sequence, deposited during the final aggradational stage, suggests a persistent balance between carbonate accumulation and subsidence rates.

Bibliography

Alvaro, M., 1987. La tectonica de cabalgamientos de la Sierra Norte de Mallorca (Islas Baleares). *Bol. Geol. Minero*, **98**, 622–9.

Darder, B., 1925. La tectonique de la region orientale de l'ile de Majorque. *Bull. Soc. Géol. France*, **25**, 245–78.

De Bruijn, H., Sondaar, P.Y. & Sanders, E.A.C., 1978. On a new species of *Pseudoltinomys* (Theridomydae, Rodentia) from the Paleogene of Mallorca. *Proc. Koninkl. Nederl. Akad. Wetensch*, **82**, 1–10.

Decourt, J., Zonenshain, L.P., Ricou, L.E., Kazmin, V.G., Le Pichon, X., Knipper, A.L., Grandjacquet, C., Sortshikov, I.M., Geysant, J., Lepvrier, C., Pechersky, D.H., Boulin, J., Sibuet, J.C., Savostin, L.A., Sorokhtin, O., Westphal, M., Bazhenov, M.L., Lauer, J.P. & Biu-Duval, B., 1986. Geological evolution of the Tethys belt from the Atlantic to the Pamir since Lias. *Tectonophysics*, **123**, 241–315.

Fallot, P. 1922. *Etude geologique de la Sierra de Majorque*. PhD Thesis, Univ. Paris.

Hugueney, M., & Adrover, R., 1982. Le peuplement des baleares (Espagne) au paléogène. *Geobios*, Mem. Spec., **6**, 439–49.

Ramos-Guerrero, E., 1988. *El Paleógeno de las Baleares: Estratigrafia y Sedimentologia*. PhD Thesis, Univ. Barcelona, 3 vols.

Ramos-Guerrero, E., Cabrera, L. & Marzo, M., 1989a. Sistemas lacustres paleógenos de Mallorca (Mediterráneo occidental). *Acta Geol. Hisp.*, **24**, 185–203.

Ramos-Guerrero, E., Rodriguez-Perea, A., Sábat, F. & Serra-Kiel, J., 1989b. Cenozoic tectosedimentary evolution of Mallorca area. *Geodinamica Acta*, **3**, 53–72.

Sàbat, F., Muñoz, J.A. & Santanach, P., 1988. Transversal and oblique structures at the Serres de Llevant thrust belt (Mallorca Island). *Geol. Rundschau*, **77**, 529–38.

The lacustrine deposits of the Eocene Pontils Group (eastern Ebro Basin, northeast Spain)

P. ANADÓN

Institut de Ciències de la Terra (J. Almera), CSIC, c. Martí i Franqués s.n., E-08028 Barcelona, Spain

Geological setting

The Tertiary Ebro Basin (northeast Spain) is the intact autochthonous part of the southern foreland basin of the Pyrenees (Fig. 1). The Eocene sequence along the central SE margin of the Ebro Basin (Igualada area) consists of alternating marine and non-marine deposits up to 2000 m thick. Non-marine conditions persisted during the Cuisian (late Ypresian) to early Bartonian and resulted in the deposition of the Pontils Group.

General stratigraphy of the Pontils Group

The Pontils Group (Fig. 2) comprises the following units:

Santa Càndia Formation: Palustrine and lagoonal carbonates, sandstones and mudstones.
Carme Formation: Mudflat deposits composed of red mudstones with interstratified sandstone and gypsum beds.
Fontanelles Formation: Palustrine and floodplain deposits consisting of limestones, mudstones and marls.
Valldeperes Formation: Calcareous mudstones, limestones, dolostones and gypsum interpreted as mudflat to saline lake (playa or sabkha) deposits.
Bosc d'En Borràs Formation: Shallow lake deposits composed of nodular limestones, calcareous mudstones and lignites.
La Portella Formation: Sandstones and mudstones deposited in a lagoonal environment related to the Bartonian transgression.

The Santa Càndia Formation records the palustrine and lagoonal environments which were established in the Igualada area after the Cuisian regression. The overlying Carme Formation reflects the widespread development of dry mudflats during the late Cuisian–early Lutetian. During the late Lutetian and early Bartonian, a complex of closely related environments developed, forming the upper Pontils Group. This includes deposits of distal alluvial fans (Pobla de Claramunt Formation), floodplain to palustrine environments (Fontanelles Formation), saline lakes and sabkhas (playa) (Valldeperes Formation), and shallow lakes (Bosc d'En Borràs Formation). Later, the Biarritzian transgression (early Bartonian) led to the formation of lagoonal environments (La Portella Formation) which were precursors of the widespread marine conditions which led to the deposition of the Santa Maria Group.

Representative detailed sections

Lacustrine deposits are the main constituents of some units of the Pontils Group. Figure 3 shows detailed sections of some of the most representative lacustrine deposits. Column 1 corresponds to limestone-dominated shallow lacustrine deposits (lower part of the Bosc d'En Borràs Formation). These deposits are mainly formed by nodular to brecciated limestones which display abundant soil features (pedogenic imprints): nodulisation, fissuration, mottling and pseudo-microkarst related to root traces. These features were formed when the original carbonate muds in littoral zones of a lake underwent repeated desiccation and flooding.

Column 2 corresponds to a typical sequence of a shallow saline lake (playa or sabkha) and surrounding saline mud flats. The deposits comprise dolostone, mudstone and nodular (chicken-wire) gypsum beds. The gypsum nodules (ancient anhydrite) range from less than 1 mm to 1 m in diameter. Some macronodules of gypsum and massive gypsum beds were formed by the amalgamation of small-sized gypsum nodules.

Column 3 shows a typical succession of the palustrine deposits of the Fontanelles Formation. The red and variegated mudstones display abundant soil features formed in a ponded mudflat or floodplain which underwent frequent water table fluctuations. The limestones, which originated from shallow lakes, show abundant soil features that evidence lake-level oscillations.

Column 4 also shows shallow lacustrine deposits related to mudflats of the Santa Càndia Formation. The limestones display similar soil features to those observed in columns 1 and 3. In addition, dolomite and evaporite (mainly gypsum) casts are present.

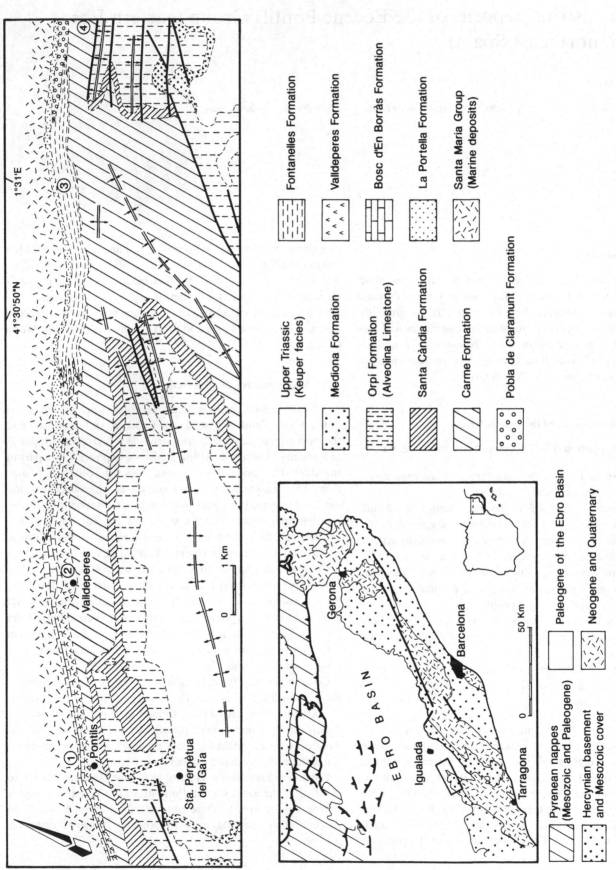

Fontanelles Formation

Valldeperes Formation

Bosc d'En Borràs Formation

La Portella Formation

Santa Maria Group
(Marine deposits)

Upper Triassic
(Keuper facies)

Mediona Formation

Orpí Formation
(Alveolina Limestone)

Santa Càndia Formation

Carme Formation

Pobla de Claramunt Formation

Pyrenean nappes
(Mesozoic and Paleogene)

Hercynian basement
and Mesozoic cover

Paleogene of the Ebro Basin

Neogene and Quaternary

Gerona

Barcelona

EBRO BASIN

Igualada

Tarragona

Pontils

Valldeperes

Sta. Perpètua del Gaià

1°31'E

41°30'50"N

Fig. 1. Geological map of the eastern Ebro Basin, in the Igualada Area. Circled numbers refer to the location of the detailed lithologic logs (Fig. 3).

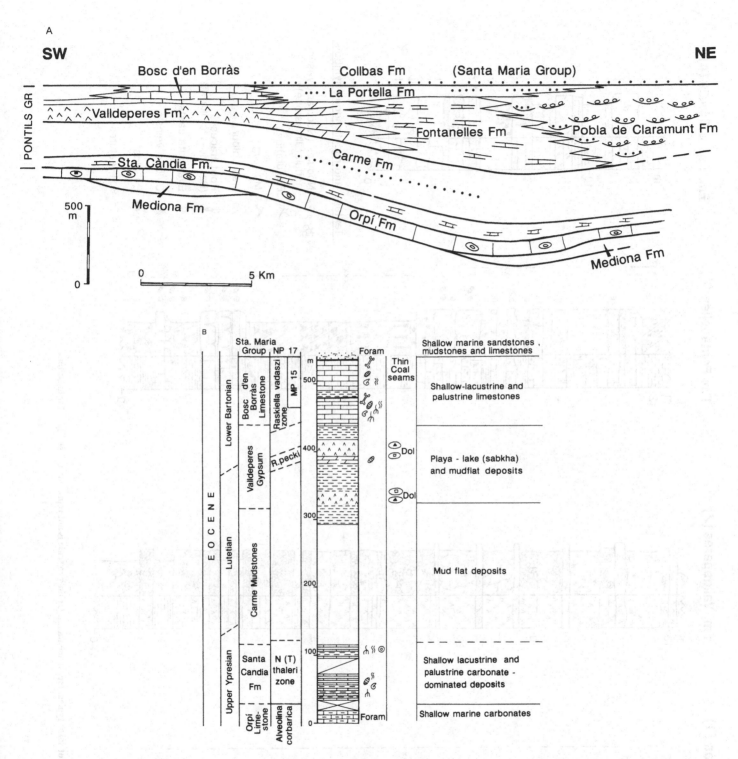

Fig. 2. A: Stratigraphic framework of the Pontils Group (Igualada Area) and B: general stratigraphic column for the Pontils type section. Reliability index A–B.

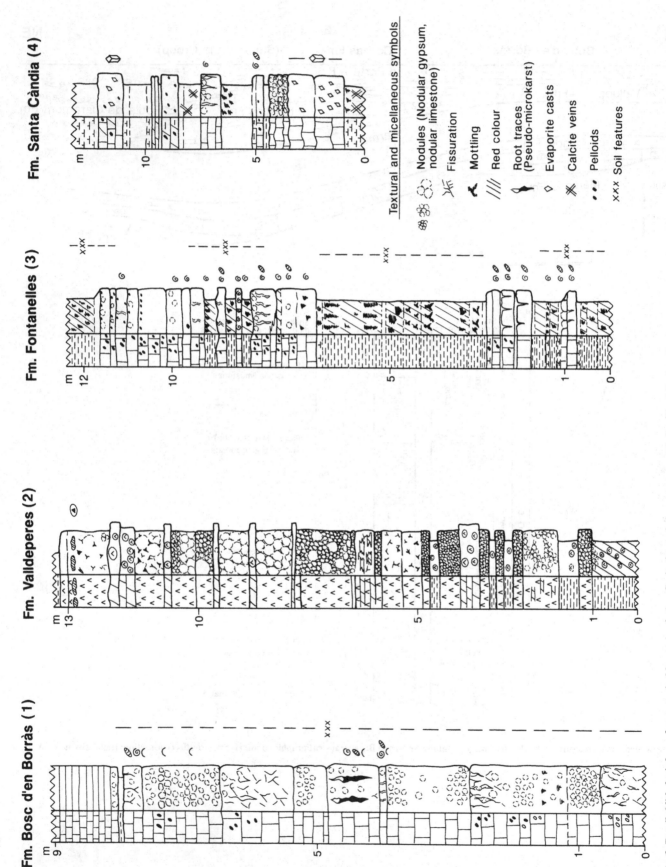

Fm. Bosc d'en Borràs (1)

Fm. Valldeperes (2)

Fm. Fontanelles (3)

Fm. Santa Càndia (4)

Textural and micellaneous symbols

Nodules (Nodular gypsum, nodular limestone)

Fissuration

Mottling

Red colour

Root traces (Pseudo-microkarst)

Evaporite casts

Calcite veins

Pelloids

Soil features

Fig. 3. Detailed lithologic logs of some typical lacustrine facies. Grain size scale is not represented. See Fig. 1 for location.

Bibliography

Anadón, P., 1978. El Paleógeno continental anterior a la transgresión biarritziense (Eoceno medio) entre los ríos Gaià y Ripoll (provincias de Tarragona y Barcelona). *Estudios Geol.*, **34**, 431–40.

Anadón, P., Cabrera, L., Guimerà, J. & Santanach, P., 1985. Paleogene strike slip deformations and sedimentation along the southeastern margin of the Ebro Basin. In *Strike–Slip Deformation, Basin Formation and Sedimentation.* ed. K.T. Biddle & N. Christie-Blick, pp. 303–18. SEPM Spec. Publ. No. 37.

Anadón, P., Colombo, F., Esteban, M., Marzo, M., Robles, S., Santanach, P. & Solé-Sugrañes, L., 1979. Evolución tectonoestratigráfica de los Catalánides. *Acta Geol. Hispánica*, **14**, 242–70.

Anadón, P. & Feist, M., 1981. Charophytes et biostratigraphie du paléogène inférieur du Bassin de l'Ebre oriental. *Palaeontographica Abt.*, **B178**, 143–68.

Anadón, P., Feist, M., Hartenberger, J-L., Müller, C. & Villalta, J., 1983. Un example de corrélation biostratigraphique entre échelles marines et continentales dans l'Eocène: la coupe de Pontils (bassin de l'Ebre, Espagne). *Bull. Soc. Géol. France*, Ser 7, **25**, 747–55.

Anadón, P. & Marzo, M., 1986. Sistemas deposicionales eocenos del margen oriental de la Cuenca del Ebro: Sector Igualada-Montscrrat. *Guía de las Excursiones, XI Congr. Español Sedimentología Barcelona* pp. 4.1–4.59.

Anadón, P. & Zamarreño, I., 1981. Paleogene nonmarine algal deposits of the Ebro Basin, Northeastern Spain. In *Phanerozoic Stromatolites.* ed. C. Monty, pp. 140–54. Springer-Verlag, Berlin.

Ferrer, J., 1971. El Paleoceno y Eoceno del borde suroriental de la depresión del Ebro (Cataluña). *Mém. Suisses Paléontologie*, **90**, 1–70.

Hottinger, L., 1962. Recherches sur les Alvéolines du Paléogène et de l'Eocène. *Mém. Suisses Paleontologie*, **75**, 1–76.

Inglés, M. & Anadón, P., 1987. Clay-mineral distribution in marine and non-marine eocene deposits of the SE Ebro Basin (Spain). *Sixth Meeting, European Clay Group, Summaries – Proceedings*, pp. 304–6.

Inglés, M. & Anadón, P., 1991. Relationship of clay minerals to depositional environment in the non-marine Eocene Pontils Group. SE Ebro Basin (Spain). *J. Sed. Petrology*, **61**, 926–39.

Pi, M.D. & Buurman, P., 1987. Authigenic palygorskite and smectite in early Paleogene paleosols of the SE Ebro Basin (Catalonia, NE Spain). *Geol. en Mijnbow*, **65**, 287–96.

Rosell, J., Julià, R. & Ferrer, J., 1966. Nota sobre la estratigrafía de unos niveles con carofitas existentes en el tramo rojo de la base del Eoceno al S de los Catalánides (Provincia de Barcelona). *Acts Geol. Hispánica*, **1** (5), 17–20.

Lacustrine deposits of the kaolinitic Argiles Plastiques Formation, southern Paris Basin, France

MÉDARD THIRY

Ecole des Mines de Paris, 35 rue St. Honoré, F-77305 Fontainebleau, France

General section

On the southern margin of the Paris Basin, the Argiles Plastiques Formation overlies the Cretaceous chalk and comprises the first deposits after the withdrawal of the Cretaceous sea. Two distinct deposits can be recognized: (1) a relatively early unit in the west consisting of mottled clays and nodular carbonates, interpreted as floodplain deposits and (2) a later unit in the east consisting of kaolinitic clays and sands, interpreted as lacustrine deposits, which are described herein. The kaolinitic Argiles Plasti-

ques Formation crops out along the edge of the Tertiary Paris Basin, and on an anticlinal high (Breuillet area) (Fig. 1).

This formation is Early Eocene in age: it is overlain by Middle Eocene marine deposits and correlated lithologically with other formations containing Lower Eocene fossils. Pollen related to the Sparnacian associations of Ypresian age are present.

The average thickness of this formation is 10 m, but it can be as thick as 25 m. Claystones, siltstones and sandstones are the major lithologies. The claystones are mined for refractory and ceramic uses: the annual production being around 300,000 tonnes.

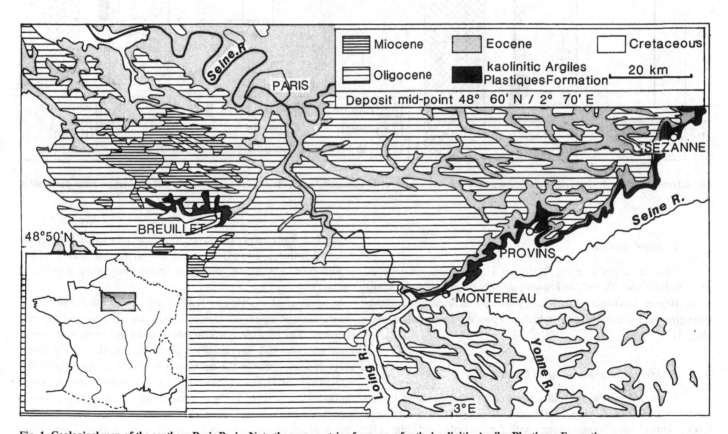

Fig. 1. Geological map of the southern Paris Basin. Note the narrow strip of exposure for the kaolinitic Argiles Plastiques Formation.

251

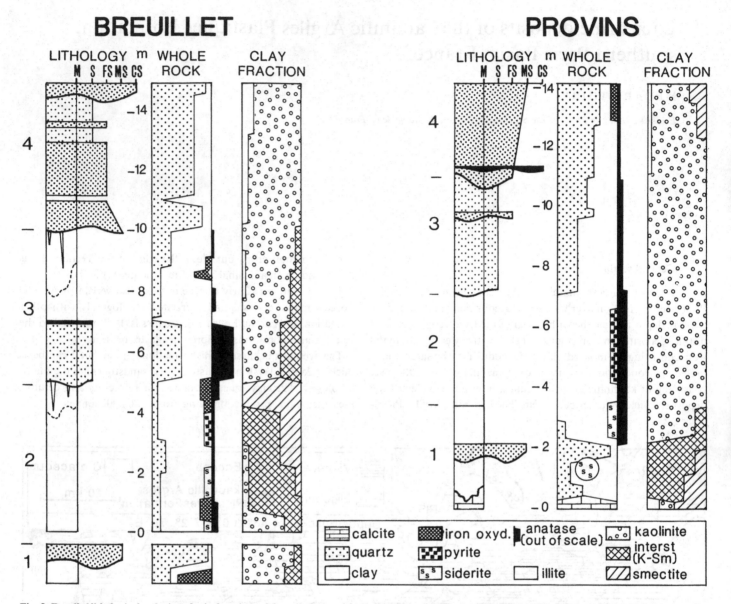

Fig. 2. Detailed lithological and mineralogical sections of the main facies of the kaolinitic Argiles Plastiques Formation. The unit numbers are those described in the text.

Detailed sections

The kaolinitic Argiles Plastiques Formation displays two main sedimentary facies: the brown and grey facies which crop out in the Brie region (near Montereau, Provins and Sézanne), and the ferruginous and mottled facies which is exposed in the Breuillet area (Fig. 1).

The ferruginous and mottled facies

Four different sedimentary units can be distinguished in the Breuillet area (Fig. 2): (1) a basal iron crust capping the underlying Cretaceous chalk and sandstones with flint pebbles and chips, (2) grey to brown claystones, truncated at the top by an oxidation horizon containing mudcracks, (3) argillaceous siltstone and clay-

stone containing an organic layer, topped by an oxidized zone with paleosol features, and (4) an upper unit of fine to coarse, white, argillaceous sandstone overlying claystone units (erosional contact).

Mineralogical analyses by XRD indicate that clays and quartz are the main components, iron compounds are minor, and feldspar is absent. The clay fraction consists of three major phases: aluminum smectite and interstratified kaolinite/smectite in the lower units and kaolinite in the upper units. Micas are present in the uppermost sandstone. Anatase (TiO_2) is present in the claystone units and can be up to 3–6% in some samples. Regionally, there is a succession of smectite-rich, anatase-rich and carbon-rich (lignite) layers which can be used as a mineralogical marker. Correlation among sections shows that the claystone/sandstone boundary (between units 3 and 4) is diachronous (Fig. 3).

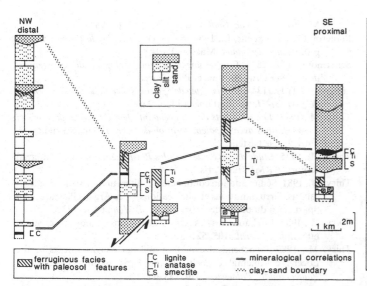

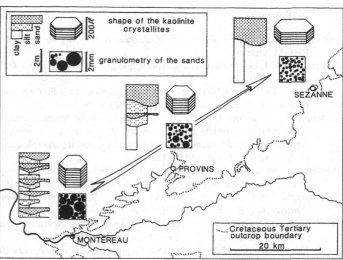

Fig. 3. Correlation among sections of the ferruginous and mottled facies on the edge of the lacustrine basin (Breuillet area). Note the lateral gradation from sand to clay.

Fig. 4. Progressive sedimentary sorting of the sandstones, claystones and kaolinite in the lacustrine basin.

The brown and grey facies

The claystones of the central part of the Argiles Plastiques Formation are brown and grey at the base, light-colored toward the top, and generally lacking in sedimentary structures. A typical section near Provins contains four units (Fig. 2): (1) a basal colluvium of reworked chalk and flint pebbles with coarse sandstone and claystone lenses which infill irregularities at the top of the Cretaceous chalk, (2) a unit of pure claystone which ranges from one to five meters in thickness, (3) lenticular beds of silty kaolinitic claystone, and (4) sandstone with variable clay content interbedded with lignite lenses.

Numerous exploratory drill holes provide precise data on the stratigraphy. The pure claystone unit forms a continuous layer covered by lenses of silty claystone, between which a transitional sandstone layer occurs. These units are truncated by channel sandstones. There is a progressive sorting of deposits from southwest to northeast (Fig. 4). In the southwestern part of the study area, there are several thin claystone layers, and the associated sandstones are coarse-grained. In the northeastern study area, only one or two thick claystone layers exist and the sandstones are fine-grained and well-sorted.

The clay mineralogy of the deposits includes smectite and interstratified kaolinite restricted to the basal unit and only kaolinite in the overlying claystone and sandstone units. Detailed XRD analysis shows variations in the crystallinity and size of the kaolinite (Fig. 4): they decrease from southwest to northeast in the basin.

Interpretation

The lacustrine basin of the kaolinitic Argiles Plastique Formation is surrounded by several anticlines of Variscian and Hercynian age (Fig. 5). The kaolinitic clays and sands were deposited in deltas that prograded into a lake with euxinic conditions. Sands

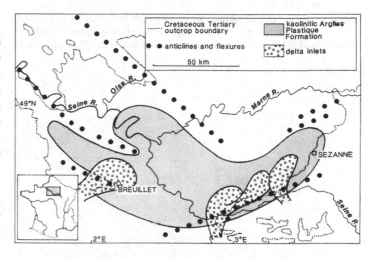

Fig. 5. Paleogeography of the kaolinitic Argiles Plastiques Formation. Note the importance of the structural framework for the partition of the clay deposits in the basin.

brought in by streams were deposited in distributary channels, whereas the kaolinitic clays collected in the lake. The silty clays were deposited by mud flows on the delta front, caused by the breaching of natural levees along the channel banks. On the edge of the basin, tectonic pulses led to successive exposure events resulting in the development of mottled horizons and paleosols.

Smectite and interstratified kaolinite/smectite are derived from the erosion of highly weathered soils established on the flint-bearing Cretaceous chalk. Kaolinite is derived from the erosion of soils established on more acidic parent rocks, such as the argillaceous and sandy rocks of the Triassic and Lower Cretaceous (well-exposed in the eastern portion of the Paris Basin).

Bibliography

Bricon, C., 1969. *Notice de la carte géologique à 1/50,000 de Dourdan.* BRGM, Orléans.

Colson, C., 1979. Les carrières d'argile plastique du bassin de Provins. *Ind. Minérale*, 207–29.

Demarcq, G., 1955. Le problème du 'Sparnacien' dans le Sud-Est du bassin parisien. *Bull. Soc. Géol. France*, sér. 6, **5**, 155–67.

Ducreux, J.L., Michoux, D. & Wyns, R., 1984. Contrôle climatique de la sédimentation yprésienne (Eocène inférieur) en Brie et en champagne (Est du Bassin de Paris, France): conséquences stratigraphiques. *C.R. Acad. Sci. Paris*, **299**(II), 1283–6.

Duplaix, S., 1948. Contribution à l'étude pétrographique des sables sparnaciens du Bassin de Paris. *Bull. Soc. Géol. France*, sér. 5, **18**, 493–510.

Feugueur, L., 1963. *L'Yprésien du Bassin de Paris – Essais de monographie stratigraphique.* Mém. Carte Géol. France.

Fritel, P.H., 1910. Etude sur les végétaux fossiles de l'étage Sparnacien du Bassin de Paris. *Mém. Soc. Géol. France*, sér. 4, **40**, 1–37.

Gruas-Cavagnetto, C., 1968. Etude palynologique des divers gisements du Sparnacien du Bassin de Paris. *Mém. Soc. Géol. France*, **110**, 1–144.

Guettard, J., 1756. Description minéralogique des environs de Paris. *Hist. Acad. Roy. Sci*, 217–58.

Hebert, E., 1854. Observations dur l'argile plastique et les assises qui l'accompagnent dans la partie méridonale du Bassin de Paris et leurs relations avec les couches tertiaires du Nord. *Bull. Soc. Géol. France*, sér. 2, **11**, 418–42.

Janet, L., 1903. Stratigraphique, pétrographique et tectonique de la roche du Breuillet (Seine-et-Oise). *Bull. Soc. Géol. France*, sér. 4, **3**, 622–9.

Poirier, G., 1884. Sur l'allure de la composition de l'Argile Plastique dans le Montois, *Bull. Soc. Géol. France*, sér. 3, **13**, 68–76.

Pomerol, C. & Feugueur, L., 1968. *Bassin de Paris, Ile-de-France. Guides géologiques régionaux.* Masson, Paris.

Sénarmont, H.H., 1844. *Essais d'une description géologique du département de Seine-et-Oise.* Béthune et Plon, Paris.

Sénarmont, H.H., 1844. *Essais d'une description géologique du département de Seine-et-Marne.* Béthune et Plon, Paris.

Thiry, M., 1973. *Les sédiments de L'Eocène inférieur du Bassin de Paris et leurs relations avec la paléoaltération de la craie.* Thèse 3ème cycle, Univ. Louis Pasteur.

Thiry, M., 1975. *Etude du bassin des Argiles Plastiques de Provins.* Internal Report, ARMINES, Paris.

Thiry, M., 1981. Sédimentation continentale et altérations associées: calcitisations, ferruginisations et silicifications – Les Argiles Plastiques du Sparnacien du bassin de Paris. *Sci. Géol. Mém.* **64**, 1–173.

Thiry, M., 1982. Les kaolinites des Argiles de Provins: géologie et cristallinité. *Bull. Minéral.*, **105**, 521–6.

Thiry, M., 1989. Geochemical evolution and paleoenvironments of the Eocene continental deposits in the Paris Basin. *Palaeogeogr., Palaeoclimat., Palaeoecol.*, **70**, 153–63.

Thiry, M., Cavelier, C. & Trauth, N., 1977. Les sédiments de L'Eocène inférieur de Bassin de Paris et leurs relations avec la paléoaltération de la craie. *Sci. Géol. Bull.*, **30**, 113–28.

Thomas, H., 1900. Contribution à la géologie des environs de Provins. *Bull. Soc. Géol. France*, sér. 3, **28**, 72–85.

Tourenq, J., 1968. Etude sédimentologique comparée des formations yprésiennes de Fosses (Val d'Oise) et de Breuillet (Essonne). Implications paléogéographiques. *C.R. Somm. Soc. Géol. France*, sér. 7, **10**, 177–8.

Paleogene of the Madrid Basin (northeast sector), Spain

M.E. ARRÍBAS

Dpto. Petrología y Geoquímica, Universidad Complutense, E-28040, Madrid, Spain

The Madrid Basin is located in central Spain (Fig. 1). This basin is limited at its northwest edge by the Central System and at the northeast edge by the Iberian Range. The Central System is a large exposure of Hercynian granites surrounded by low to high-rank metamorphic rocks. The Iberian Range is a mountain belt developed from a depositional trough of the aulacogen-type which was filled with Mesozoic sediments (Alvaro *et al.*, 1979). The Madrid Basin is a foreland basin formed during the compressive phase of the Alpine build-up, which also caused the formation of the Iberian Range and Central System. The Paleogene succession within the Madrid Basin is synorogenic with this orogeny.

Paleogene deposits are scattered along the northeast border of the Madrid Basin. The northern Paleogene outcrops are nearest to the area of convergence between the Iberian Range and the Central System (Fig. 1). The best exposure for the Paleogene succession occurs in the province of Guadalajara in these localities: Beleña de Sorbe, Membrillera, Torremocha de Jadraque, Negredo and Baides (Fig. 1).

The Paleogene succession (Fig. 2) contains a faunal association of macro- and micro-mammals which corresponds, in its lower part, to the MP-17 and MP-20 zones (Upper Eocene) and, in its upper part (Arríbas *et al.*, 1983; Lopez-Martinez, 1983), to the MP-21 zone (Lower Oligocene) (Schmidt-Kittler, 1987). Other fossil faunal and floral components include ostracodes, charophytes, gastropods, macrophytic debris and blue-green algae.

Within the Paleogene succession (Fig. 2), two lithological units are differentiated (Arríbas, 1986a): a carbonate unit and a siliciclastic unit. The carbonate unit, with a thickness variation of 200–500 m, contains a variety of facies interpreted as deposits of a lacustrine–paludal paleoenvironment (Arríbas, 1986a, b). The carbonate facies are divided into two main groups: hard carbonates and powdery carbonates. Within the hard carbonate group, two groups of limestones are distinguished: (1) lacustrine limestones, which contain bioclastic limestones, intraclastic limestones, oncolitic limestones, sandy limestones, laminated limestones and dolomitic limestones/dolostones, and (2) limestones with pedogenetic features (nodular limestones, nodular sandy limestones, bioturbated limestones, limestones with vertical prismatic structures, vuggy limestones and limestones with deformed lamination and fenestral porosity). Within the powdery carbonate group, four facies are distinguished: marls, dolomitic marls, chalks and dolomitic chalks. The siliciclastic unit, which grades into the carbonate unit, contains several facies ascribed to prograding alluvial fans (such as lobes, channels, sheets and massive lutites) (Arríbas *et al.*, 1983).

The lithological facies associations and their sequence patterns exhibit the sedimentary evolution of the lacustrine basin deposits as the lacustrine and alluvial-fan environments expanded and contracted through time (see Fig. 3). The Paleogene succession reflects an evolution from a lacustrine carbonate environment (carbonate unit) to a prograding alluvial-fan environment (siliciclastic unit).

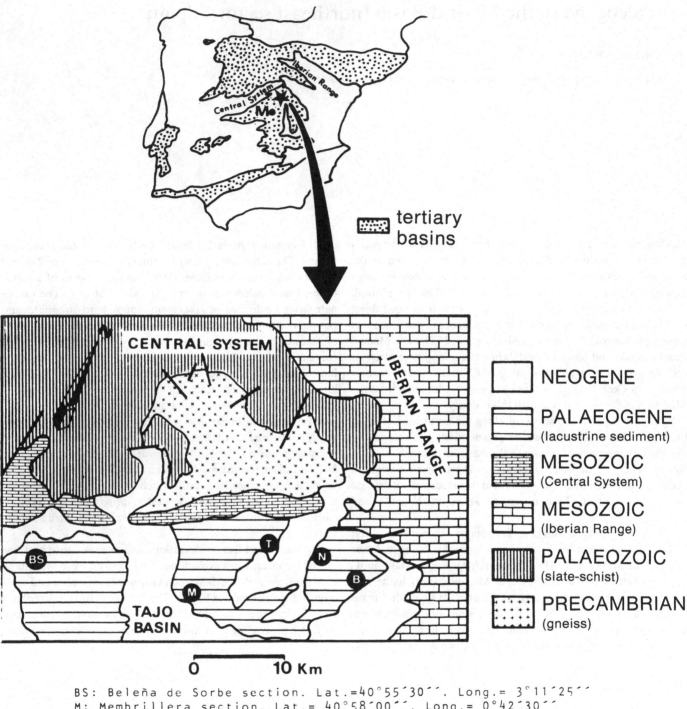

BS: Beleña de Sorbe section. Lat.=40°55´30´´. Long.= 3°11´25´´
M: Membrillera section. Lat.= 40°58´00´´. Long.= 0°42´30´´
T: Torremocha de Jadraque.Lat.= 41°01´20´´. Long.= 2°53´30´´
N: Negredo section. Lat.= 41°01´25´´. Long.= 0°51´40´´
B: Baides section. Lat.= 41°01´10´´. Long.= 0°55´06´´

Fig. 1. Locality map of Paleogene of Madrid Basin (NE sector), Spain.

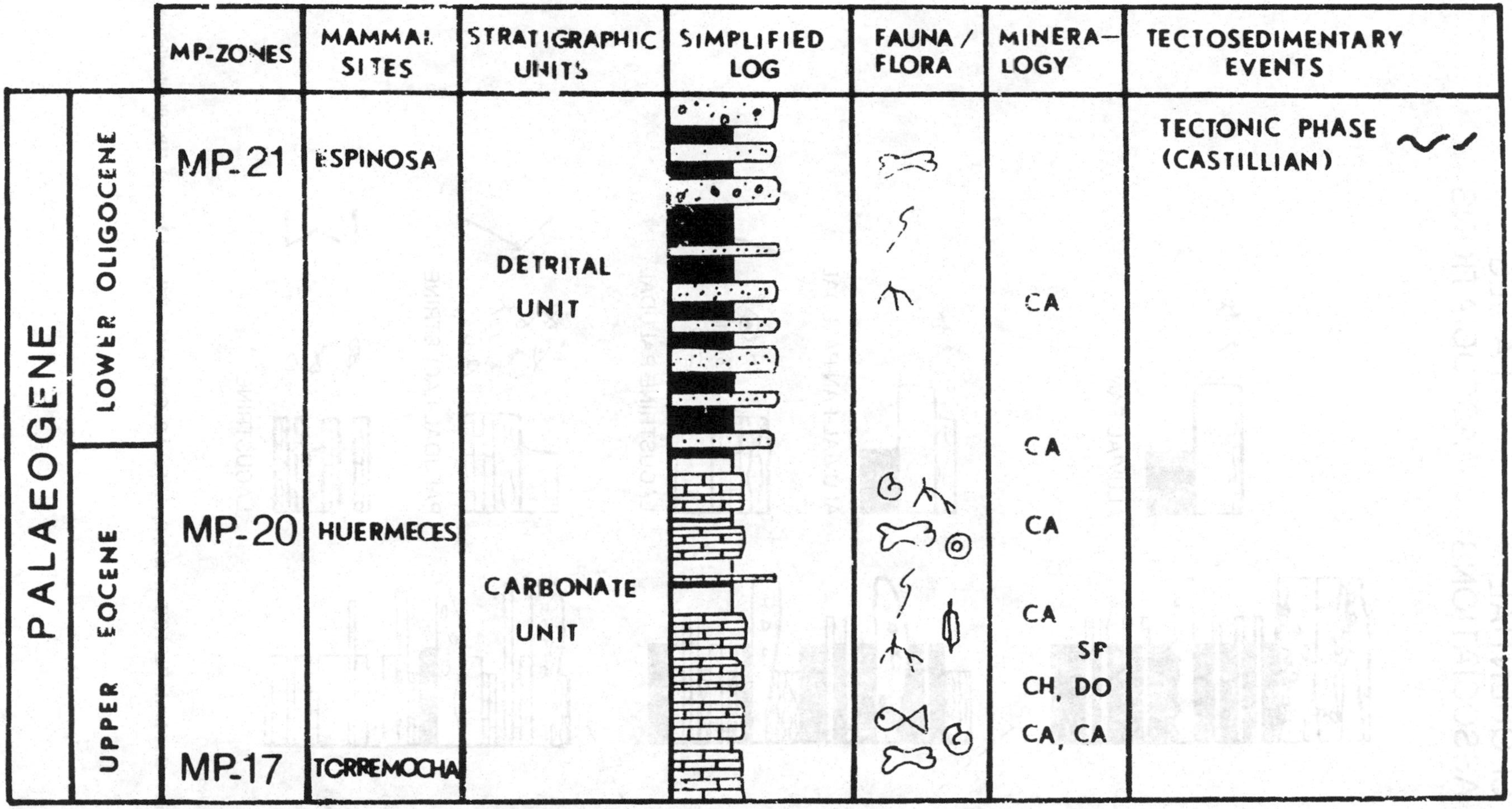

Fig. 2. Paleogene of Madrid Basin (NE sector), Spain. Dating of macromammal faunas (MP-zones) from Schmidt-Kittler (1987), López-Martínez (1983), and Arribas *et al.* (1983). Stratigraphic units from Arribas (1986 a, b). Sedimentology and diagenesis from Arribas & Bustillo (1985) and Arribas (1986b). Fauna and flora include vertebrates (mammals and reptiles), ostracodes, charophytes, gastropods, plant debris and blue-green algae. Reliability index: some biostratigraphic information with some zonal fossils.

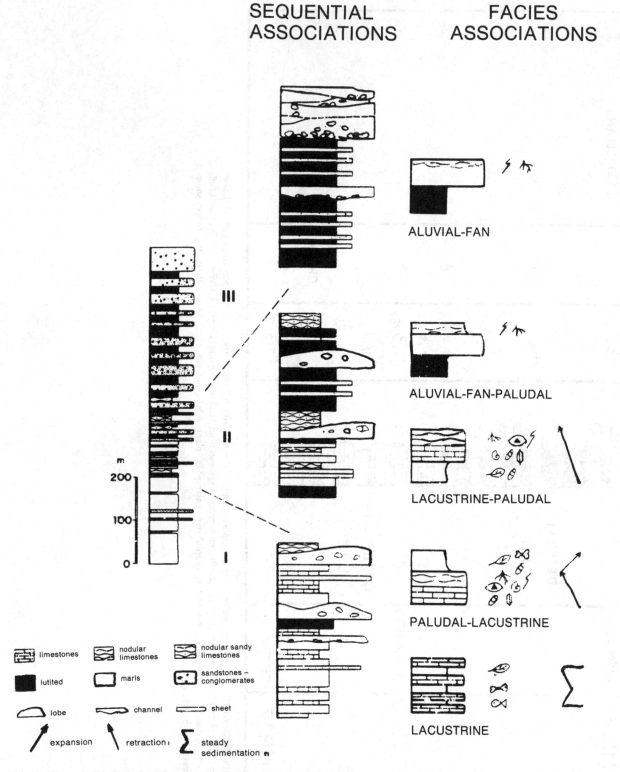

SEQUENTIAL ASSOCIATIONS

FACIES ASSOCIATIONS

ALUVIAL-FAN

ALUVIAL-FAN-PALUDAL

LACUSTRINE-PALUDAL

PALUDAL-LACUSTRINE

LACUSTRINE

limestones

nodular limestones

nodular sandy limestones

lutited

marls

sandstones – conglomerates

lobe

channel

sheet

expansion

retraction

steady sedimentation

Fig. 3. Schematic lithological section of the Paleogene in the Madrid Basin (NE sector), Spain. I and II: Carbonate unit (Upper Eocene). III: Siliciclastic unit (Lower Oligocene).

Bibliography

Alvaro, M., Capote, R. & Vegas, R., 1979. Un modelo de evolución geotectónica para la Cadena Celtibérica. *Acta Geol. Hisp.*, **14**, 172–7.

Arríbas, M.E., 1982. Petrología y sedimentología de las facies carbonáticas del Paleógeno de la Alcarria (Sector NW). *Estudios Geol.*, **38**, 27–41.

Arríbas, M.E., 1984. Facies and sequences in lacustrine carbonates: Paleogene of NE sector of the Tertiary Tajo Basin (Spain). *5th European Congress on Sedimentology, Marseilles, Abstracts*, pp. 489–90.

Arríbas, M.E., 1985. *Sedimentología y diaqénesis de las facies carbonáticas del Paleógeno del sector NW de la Cuenca del Tajo.* PhD Dissertation, Univ. Complutense.

Arríbas, M.E., 1986a. Estudio litoestratigráfico de una unidad de edad paleógena. Sector N de la Cuenca Terciaria del Tajo (Prov. Guadalajara). *Estudios Geol.*, **42**, 103–16.

Arríbas, M.E., 1986b. Petrología y análisis secuencial de los carbonatos lacustres del Paleógeno del sector N de la Cuenca Terciaria del Tajo (provincia de Guadalajara). *Cuad. Geol. Ibérica*, **10**, 295–334.

Arríbas, M.E. & Arríbas, J., 1989. Petrographic evidence of different provenance in two alluvial fan sequences (Palaeogene of the N. Tajo Basin, Spain). *Geological Society of London*, Spec. Pub. No. 57, 263–79. Blackwell Scientific Publishers, Oxford.

Arríbas, M.E. & Bustillo, M.A., 1985. Modelos de silicificación en los carbonatos lacustres-palustres del Paleógeno del borde NE de la Cuenca del Tajo. *Bol. Geol. Min.*, **96** (3), 325–43.

Arríbas, M.E., Diaz, M., Lopez, N. & Portero, J., 1983. El abanico aluvial Paleógeno de Beleña de Sorbe (Cuenca del Tajo): facies, relaciones espaciales y evolución. *X Congreso Nacional de Sedimentología, Menorca, Abstracts*, pp. 134–9.

Crusafont, M., Meléndez, B. & Truyols, J., 1960. El yacimiento de Vertebrados de Huérmeces del Cerro (Guadalajara) y su significado cronoestratigráfico. *Estudios Geol.*, **16**, 243–54.

Lopez-Martinez, N., 1983. *Los Micoromamíferos fósiles de la Cuenca del Tajo.* Unpublished report, Instituto Geológico y Minero de España.

Portero, J.M. & Aznar, J.M., 1984. Evolución morfotectónica y sedimentación terciarias en el sistema central y cuencas limítrofes (Duero y Tajo). *I Congreso Español de Geología*, **3**, 253–63.

Schmidt-Kittler, N. (Ed.), 1987. *10th International Symposium on Mammalian Biostratigraphy and Palaeoecology of European Palaeogene.* Münchner Geowissenschaftliche Abhandlung, Mainz.

The Bighorn Basin, Montana–Wyoming, USA

R. YURETICH

Department of Geology and Geography, University of Massachusetts, Amherst, Massachusetts, USA

The Bighorn Basin of northwestern Wyoming and adjacent Montana (Fig. 1) contains a thick sequence of non-marine deposits ranging in age from Late Cretaceous through Eocene. Much of the Paleocene strata are part of the Fort Union Formation, which contains the sedimentary record of the principal uplift of the mountain ranges around the Bighorn Basin. This formation reaches thickness of approximately 3000 m with much of it in the subsurface. Rea & Barlow (1975), on the basis of well-log data, postulated that, in the Paleocene, the Bighorn Basin was divided into two major depocenters of lacustrine and palustrine sedimentation. The center of this structure called the 'Shoshone River Arch' (Gingerich, 1983), separates the northern or Clark's Fork Basin from the larger southern basin (Fig. 1). It is in the northern part of the basin where the Fort Union Formation is exceptionally well exposed, facilitating detailed facies analysis.

An initial study of the Fort Union Formation by Hickey (1980) identified fluvial, lacustrine, swamp and alluvial-fan facies. Subsequent investigation of the lacustrine lithologies led to a broad definition of the physical and chemical characteristics of the lake and a definition of the Belfry Member, which encompasses largely lacustrine lithofacies (Yuretich *et al.*, 1984). Field efforts have measured 24 major sections and correlated these by tracing individual sandstone or limestone beds and measuring shorter sections as necessary. This has permitted a detailed sedimentological interpretation of the Belfry Member (Yuretich & Hicks, 1986).

Lithologies

The lacustrine deposits of the northern Bighorn Basin are part of a complex of non-marine sedimentary rocks of alluvial, deltaic, lacustrine and palustrine origin (Fig. 2). The lacustrine interval is recognized by the lateral continuity of the beds, by the calcareous nature of the deposits and by the floral, faunal and sedimentological aspects indicative of extensive ponded water.

The characteristics of the Belfry Member indicate that the sediments were deposited in shallow water (< 15 m?), presumably near shore. Sandstone and lignite are often of greater volumetric importance than mudstone or limestone. This reflects the significance of deltaic and palustrine environments in shaping the 'Lake Belfry' basin.

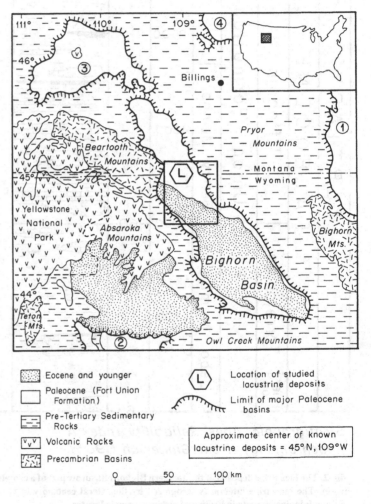

Eocene and younger

Paleocene (Fort Union Formation)

Pre-Tertiary Sedimentary Rocks

Volcanic Rocks

Precambrian Basins

Location of studied lacustrine deposits

Limit of major Paleocene basins

Approximate center of known lacustrine deposits = 45° N, 109° W

0   50   100 km

Fig. 1. The Bighorn Basin straddles the state line between northwestern Wyoming and southwestern Montana. Lacustrine deposits of Paleocene age are developed primarily in this area and Eocene lake deposits of the Tatman Formation also crop out in the west-central part of the Bighorn Basin (Van Houten, 1944). Other basins illustrated are: (1) Powder River Basin; (2) Wind River Basin; (3) Crazy Mountain Basin; (4) Musselshell Basin. Paleocene lake deposits have been reported from the Powder River Basin (Ayers, 1986) and the Wind River Basin (Keefer, 1961), although these are somewhat older and with different sedimentological characteristics. For summaries of Paleocene lacustrine rocks in the Rocky Mountain basins, see Yuretich (1989).

261

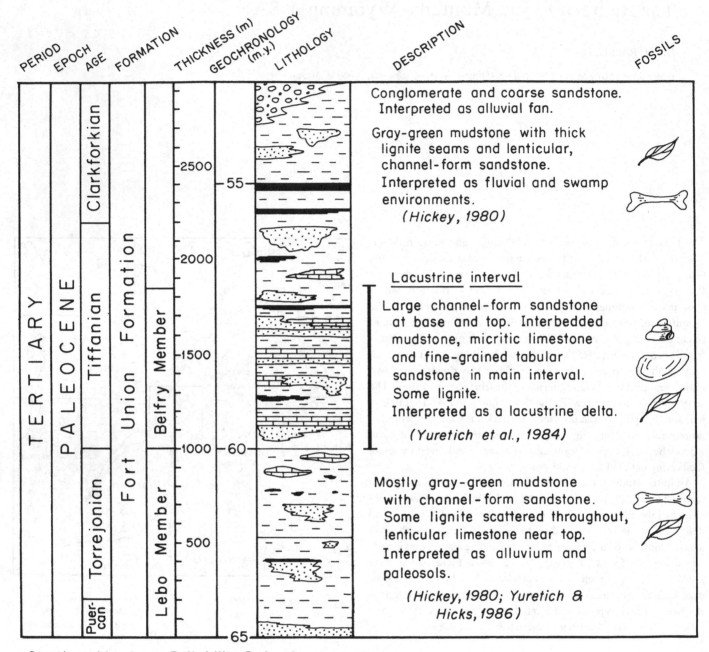

Stratigraphic Ages Reliability Index A
Geochronology from Gingerich (1983)

Fig. 2. The lacustrine deposits of the northern Bighorn Basin are part of a complex of non-marine sedimentary rocks of alluvial, deltaic, lacustrine and palustrine origin. The lacustrine interval is recognized by the lateral continuity of the beds, by the calcareous nature of the deposits and by the floral, faunal and sedimentological aspects indicative of extensive ponded water.

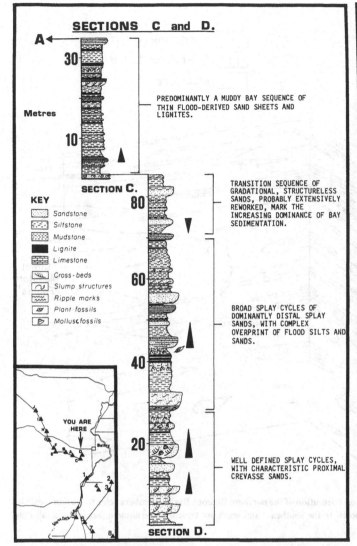

Fig. 3. Stratigraphic section through the lacustrine sequence typical of the 'bay and splay' facies with fine-grained and calcareous intervals punctuated with crevasse-splay sandstones from the nearby distributary channels. Inset shows relative location of the section.

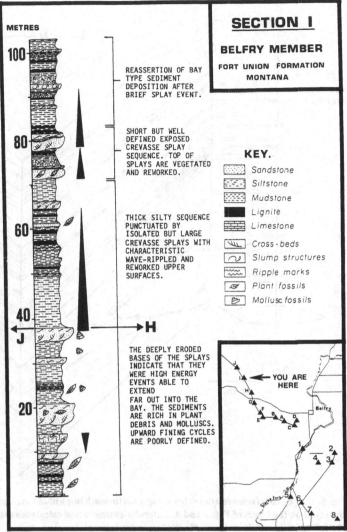

Fig. 4. Stratigraphic section of the lacustrine sequence farther away from lake basin and containing more channel-form sandstone, coals and rooted mudstones. Inset shows relative location of the section.

The various lithologies often occur in discrete packages which aid in their environmental interpretation (Figs 3 and 4). The most easily identifiable of these comprise cross-bedded lenticular sandstone which grades upward into siltstone, mudstone and lignite. These prominent, fining-upward cycles resemble typical channel-to-floodplain transitions or waning flood phases in a fluvial setting (Yuretich et al., 1984).

Other recurring sequences can also be recognized in the rocks. Interbedded siltstone, sandstone and mudstone, each bed about 30 cm thick, occur frequently in many outcrops. The sandstone is usually poorly indurated, and ripple-bedding is common throughout. Fossils of any kind are rare. These deposits probably correspond to a proximal overbank environment, such as natural levees. The repetitive nature and relatively thick (5 m) occurrences of these sequences suggest deposition within an aggradational system.

A third major sequence is dominated by fine-grained lithologies. Mudstone with thin lignites, plant fossils and scattered lenticular lime mudstone form relatively thick sequences (several meters). These are thought to be marsh and pond deposits, and commonly overlie the levee and channel lithologies previously described, thus completing a fining-upward sequence.

Finally, we noted a calcareous sequence, which included both coarse and fine-grained lithologies. The fine-grained portions are mudstone, often containing mollusc fossils, and some tabular micritic limestone. The coarse-grained portion consists of thin (<30 cm) ripple-bedded tabular sandstone. These are more representative of semipermanent bodies of ponded water, with the sandstones being formed from crevasse splays spilling over from nearby channels. It is the prevalence of this sequence which gives the Belfry Member its distinctive characteristics (Fig. 3).

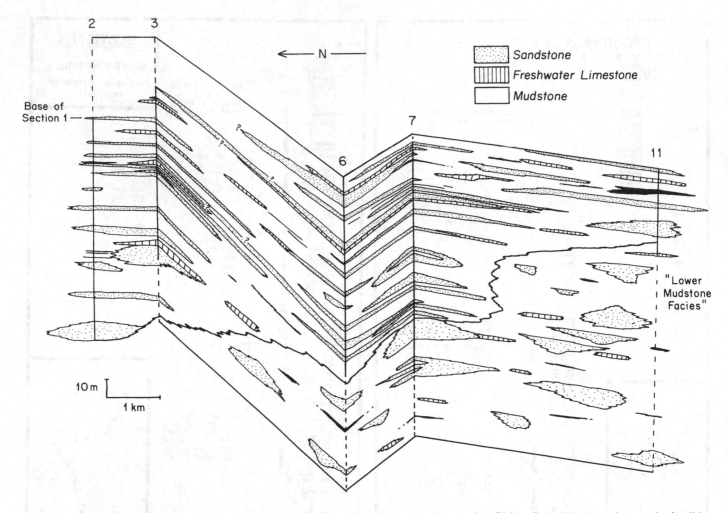

Fig. 5. Large-scale facies relationships along a north-south line within Fort Union Formation of the northern Bighorn Basin. Numbers refer to section localities displayed on the insets of Figs 3 and 4. Laterally extensive and calcareous deposits to the southeast and south are typical of permanent lacustrine conditions. Fluvial deposits overlie and underlie the lacustrine interval.

Regional facies

The lithologic relationships in the Belfry Member support a lacustrine delta-distributary lobe origin for this unit. In addition, facies relationships based on the measured sections reveal clearly that the Belfry Member undergoes pronounced thinning to the west of the town of Belfry and again south near the Wyoming border. In both directions it grades into mudstone-dominated lithofacies (Figs 5 and 6) which have been called the 'lower' and 'upper mudstone facies'.

There are, however, some noticable differences between the lower and upper mudstones. Although both are characterized by somber, dark grey-green and carbonaceous mudstone, the lower facies has a greater abundance of isolated large channel sandstones 5 to 10 m in thickness. Plant fossils are common, but there are few molluscan accumulations, with most of these in channel lags. In contrast, the upper mudstone facies has much less sandstone, with Unionid bivalve shells often abundant. These sandstones are usually thin

(less than 1 m), and are not very extensive. Freshwater carbonate occurs regularly within the mudstone sequence.

It seems clear that the lower mudstone is an alluvial deposit, based upon the preponderance of rooted, soil-like horizons similar to those described from the Eocene (Neasham & Vondra, 1972; Bown & Kraus, 1981). The origin of the upper mudstone is much less certain. Part of the sequence may also be fluvial, but tabular carbonate beds become more prominent towards the south of the area studied, suggesting that 'deep-water' (i.e., non-deltaic) lake beds lie in this direction.

Facies relationships to the east of the study area are still obscure. The Fort Union Formation has been deformed into a series of gentle anticlines and synclines related to the uplift of the surrounding mountain blocks. East of the study area lies an anticlinal fold axis, and most of the facies correlative with the Belfry Member have been removed by erosion. The general paleocurrent data show clearly that the source areas lay to the west, so one might also expect more 'deep-water' sediments to be in the opposite direction.

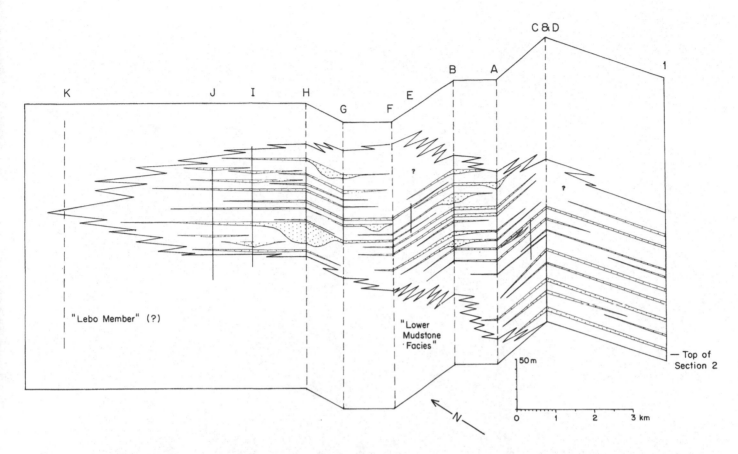

Fig. 6. Large-scale facies relationships along an east-west line within the Fort Union Formation in the northern part of the Bighorn Basin. Letters refer to section localities displayed on insets of Figs 3 and 4. Deltaic deposits to the west grade laterally into permanent lakes toward the east.

Acknowledgments

The research on the Fort Union Formation was supported by the National Science Foundation (EAR83-06153).

Bibliography

Ayers, W.B. Jr., 1986. Lacustrine and fluvial–deltaic depositional systems, Fort Union Formation (Paleocene), Powder River Basin, Wyoming and Montana. *Am. Assoc. Petrol. Geol. Bull.*, **70**, 1651–73.

Bown, T.M. & Kraus, M.J., 1981. Lower Eocene alluvial paleosols, (Willwood Formation, northwest Wyoming, USA) and their significance for paleoecology, paleoclimatology and basin analysis; *Palaeogeog., Palaeoclimat., Palaeoecol.*, **34**, 1–30.

Gingerich, P.D., 1983. Paleocene-Eocene faunal zones and a preliminary analysis of Laramide structural deformation in the Clark's Fork Basin, Wyoming. In *Geology of the Bighorn Basin*, ed. W.W. Boberg, pp. 185–96. Wyoming Geol. Assoc. Guidebook, 34th Field Conference, Billings, Montana.

Hickey, L.J., 1980. Paleocene stratigraphy and flora of the Clark's Fork Basin. In *Early Cenozoic Paleontology and Stratigraphy of the Bighorn Basin, Wyoming*, ed. P.D. Gingerich, pp. 33–49. Univ. Michigan Papers on Paleontology, 24.

Hickey, L.J., Johnson, K.R. & Yuretich, R.F., 1986. Field trip through the facies of the Fort Union. In *Geology of the Beartooth Uplift and Adjacent Basins*, ed. P. Garrison, pp. 279–90. Montana Geological Soc.-Yellowstone-Bighorn Research Assoc. Joint Field Conference and Symposium, Red Lodge, Montana.

Keefer, W.H., 1961. Waltman Shale and Shotgun Members of the Fort Union Formation (Paleocene) in the Wind River Basin, Wyoming. *Am. Assoc. Petrol. Geol. Bull.*, **45**, 1310–23.

Neasham, J.W. & Vondra, C.F., 1972. Stratigraphy and petrology of the Lower Eocene Willwood Formation, Bighorn Basin, Wyoming. *Geol. Soc. Am. Bull.*, **83**, 2167–80.

Rea, B.D. & Barlow, J.A., 1975. Upper Cretaceous and Tertiary rocks, northern part of the Bighorn Basin, Wyoming and Montana. In *Geology and Mineral Resources of the Bighorn Basin*. ed. F.A. Exum & G.R. George, pp. 63–71. Wyoming Geol. Assoc. Guidebook, 27th Field Conference, Cody, Wyoming.

Van Houten, F.B., 1944. Stratigraphy of the Willwood and Tatman Formations in northwestern Wyoming. *Geol. Soc. Am. Bull.*, **55**, 165–210.

Yuretich, R.F., 1989. Paleocene lakes of the central Rocky Mountains, western USA. *Palaeogeog., Palaeoclimat., Palaeoecol.*, **70**, 53–63.

Yuretich, R.F. & Hicks, J.F., 1986. Sedimentology and facies relationships of the Belfry Member, Fort Union Formation, northern Bighorn Basin. In *Geology of the Beartooth Uplift and Adjacent Basins*, ed. P. Garrison, pp. 53–69. Montana Geological Soc.-Yellowstone-Bighorn Research Assoc. Joint Field Conference and Symposium, Red Lodge, Montana.

Yuretich, R.F., Hickey, L.J., Gregson, B.P. & Hsia, Y.L., 1984. Lacustrine deposits in the Paleocene Fort Union Formation, northern Bighorn Basin, Montana. *J. Sed. Petrol.*, **54**, 836–52.

Oligocene

Lacustrine deposits in the Tertiary Daban Basin (northern Somalia)

MARIO SAGRI, ERNESTO ABBATE AND PIERO BRUNI

Dipartimento Scienze della Terra, via La Pira 4, 50121 Firenze, Italy

The Daban Basin, located 25 km southeast of Berbera, northern Somalia (Fig. 1), is filled with Middle Eocene to Oligocene deposits. More than 2700 m of clastic sediments (of which 2000 m are lacustrine) accumulated in the Daban Basin during 20–25 Ma, giving an average sedimentation rate of 10.8 to 13.5 cm/1000 year.

A broad syncline is recognized in the Daban deposits; the northern limb has a gentle southerly dip and rests conformably on the Middle Eocene Taleh Evaporites, while the beds of the southern margin have a steeper dip and are truncated by the Dagah Shabelle fault zone against the Mesozoic and the crystalline basement, or are covered by Plio-Quaternary deposits (Fig. 1). The Daban Basin fill sediments rest conformably on the Middle Eocene Taleh Evaporites and are overlain unconformably by fanglomerates (Boulder Beds) of Pliocene age (Macfadyen, 1933).

In the sedimentary succession of the Daban Basin, the environments listed below can be distinguished (Abbate *et al.*, 1983; Bruni *et al.*, 1987) from the base of the sequence upward: restricted lagoon, delta, lagoon, alluvial plain, ephemeral lake and perennial lake (Fig. 2). The perennial lake sediments are, westward, laterally transitional to a thick fan-delta sequence (Figs 1 and 2).

Lacustrine deposits occur in the middle and upper portion of the Daban sequence (Fig. 2) and are represented by ephemeral lake sediments and perennial lake deposits (Sagri *et al.*, 1989). The lacustrine sediments contain only few probable Oligocene fossils, such as freshwater fishes (cichlids, Van Couvering, 1982), gastropods, ostracods and silicified trees (Macfadyen, 1933).

Ephemeral lake

The lower portion of the lacustrine sequence consists of playa mudflat sediments with intercalated evaporitic and clastic sequences (Fig. 3) deposited in a saline playa or inland sabkha (Sagri *et al.*, 1989). The ephemeral lacustrine series rest conformably on alluvial sediments and the inception of lacustrine deposition is marked by a 3 m thick key bed consisting of micritic chalk-like limestone containing ostracods, algal remains and ooids.

The main facies recognized in the playa mudflat are (Fig. 4, A, C):

(1) Red and green massive siltstones and mudstones, intensely bioturbated, with abundant gypsum and manganese concretions and desiccation cracks.

(2) Red and brown, horizontally or cross-stratified pebbly sandstones and imbricated, channelized, fine-grained conglomerates in beds 1–5 m thick.

(3) Green mudstones and dark lignitic shales.

The saline playa sediments consist of three facies arranged in cycles of about 5–15 m thick (Fig. 4, A, B) reflecting lacustrine expansion and contraction. From bottom to top:

(1) Red massive siltstones intensely bioturbated, containing gypsum veins and calcrete.

(2) Green massive, mottled and bioturbated siltstones.

(3) Gypsum-green shale layers, up to 2 m thick, that consist of alternating gypsum laminae (0.5–2 cm thick) and green shales.

Perennial lake

Perennial lake deposits occur in the upper portion of the Daban sequence (Figs 2 and 3) and consist of terrigenous sediments with subordinate amounts of limestones and marls.

Central and marginal lacustrine facies can be distinguished (Sagri *et al.*, 1989). Marginal facies outcrop only in the western side of the basin and are represented by fan-delta deposits consisting of thick beds of sandstone and conglomerates, green siltstones and nodular chalky limestones. Central lake facies consist of (Fig. 5):

(1) Green massive siltstones and mudstones with rare dark lignitic shales. This facies is volumetrically the most significant.

(2) Paper-thin rhythmic laminations of green marls and siltstones, faintly burrowed.

(3) Laminated to poorly-laminated chalky limestones in units up to 10 m thick, containing well-preserved, freshwater fishes, ostracods and chert nodules. These layers are traceable across the entire basin and are good stratigraphic markers (Figs 1 and 2).

(4) Horizontal or cross-laminated fine to coarse greenish-grey sandstones in beds up to 4 m thick with current-rippled and wave-rippled fine-grained sandstones generally arranged in thickening and coarsening upward sequences.

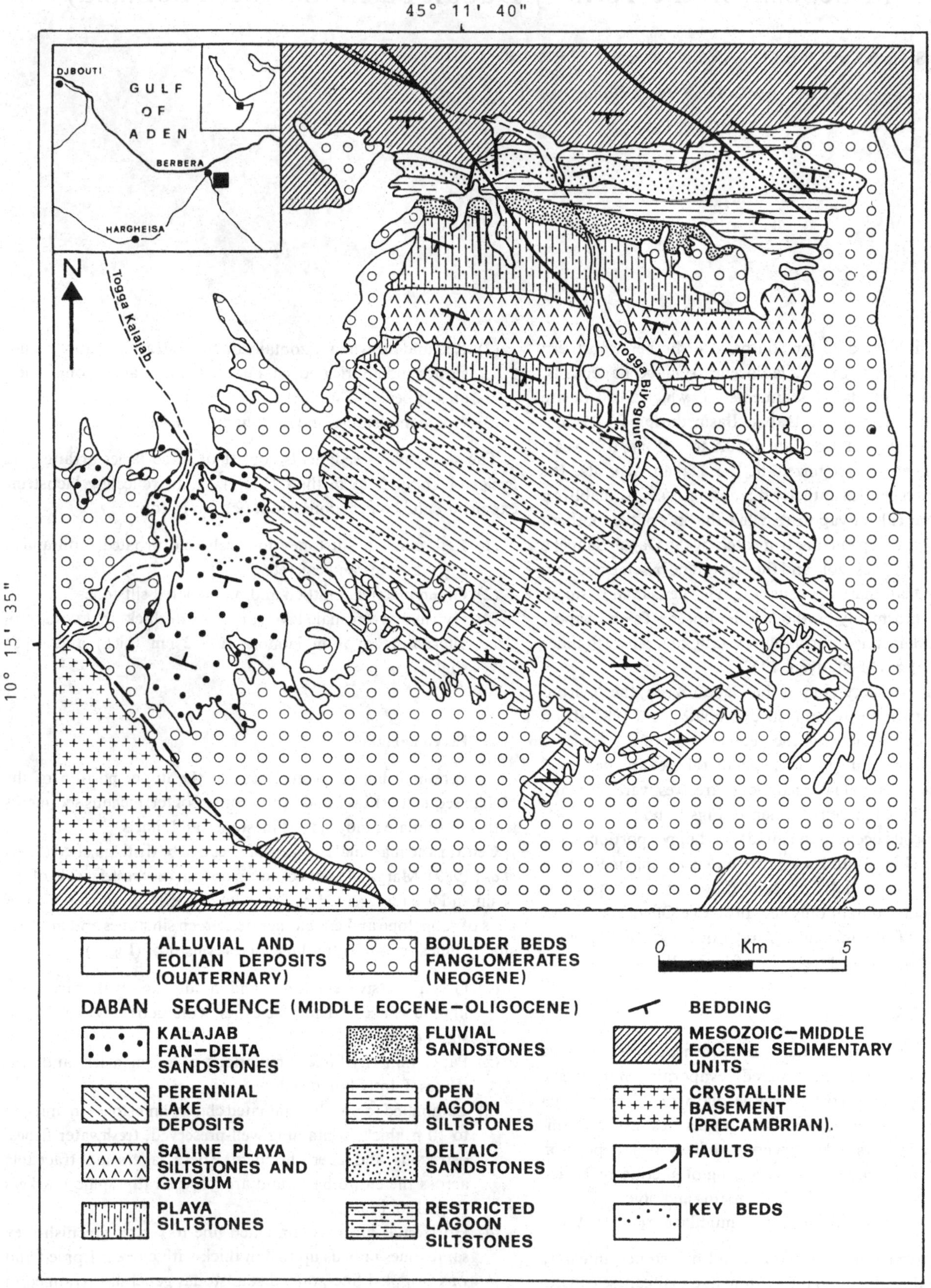

Fig. 1. Simplified geology of the Daban Basin (after a 1:100,000 geological map of Bruni *et al.*, 1987).

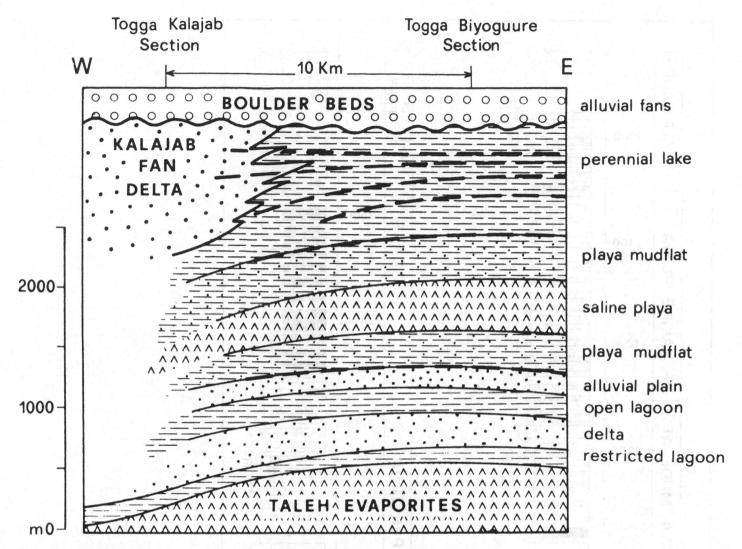

Fig. 2. Schematic stratigraphy and paleoenvironments of the Daban Basin fill. Dashed lines indicate key beds.

(5) Graded sandstones, in beds up to 1 m thick, with well developed Bouma subdivisions.

(6) Basin-wide, whitish tuffs up to 7 m thick are locally interbedded in the lacustrine deposits. They permit good correlation between the lacustrine and fan-delta sequences (Figs 1 and 2).

(7) Massive red siltstones and mudstones with silicified wood and gastropod remains occur in the upper portion of the perennial lacustrine deposits.

The lacustrine sequence of the Daban Basin was deposited in a rapidly subsiding half-graben, parallel to the Gulf of Aden (Abbate *et al.*, 1988). The tectonic and sedimentary evolution of this basin is therefore influenced by the breakup of the Somali-Arabian continental block and the rifting of the Gulf of Aden (Sagri *et al.*, 1989).

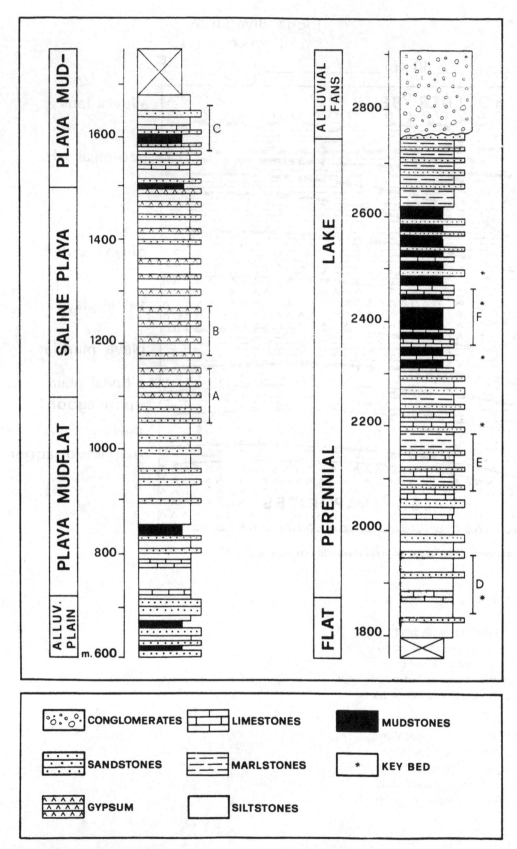

CONGLOMERATES   **LIMESTONES**   **MUDSTONES**

SANDSTONES   **MARLSTONES**   **KEY BED**

GYPSUM   **SILTSTONES**

Fig. 3. Lithology of the lacustrine sequence in the Biyoguure section. Position of the sedimentological logs of Figs 4 and 5 are indicated with capital letters on the right side of the columns.

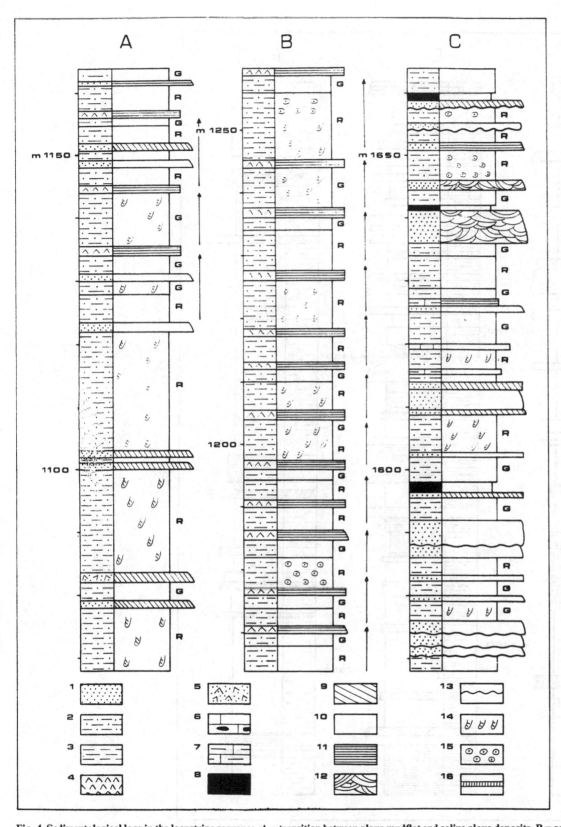

Fig. 4. Sedimentological logs in the lacustrine sequence. A = transition between playa mudflat and saline playa deposits. B = saline playa deposits characterized by cyclic sedimentation. Cycles are marked by arrows. C = playa mudflat deposits. 1 = sandstones. 2 = siltstones. 3 = mudstones. 4 = gypsum. 5 = gypsum arenites. 6 = limestones and cherty limestones. 7 = marlstones. 8 = lignitic mudstones. 9 = cross-lamination. 10 = massive sediments. 11 = horizontal lamination. 12 = trough cross-lamination. 13 = erosional surfaces. 14 = bioturbation. 15 = caliches. 16 = tuffs. R = red. G = green. Logs position in the lacustrine sequence see Fig. 3 (from Sagri et al., 1989).

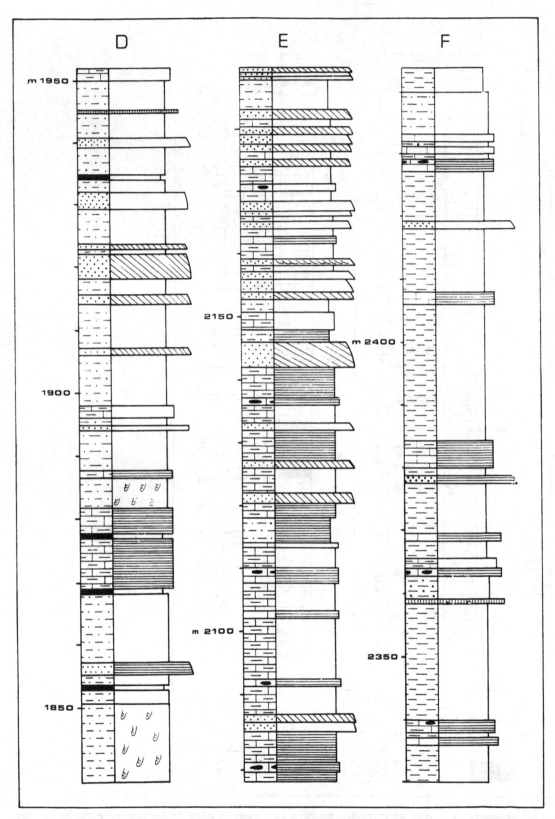

Fig. 5. Sedimentological logs in the lacustrine deposits. D = playa mudflat-perennial lake transition marked by a thick laminated marlstone key bed. E = perennial lake deposits composed of marlstones, limestones, sandstones and siltstones. F = perennial lake deposits composed of mudstones, marlstones and limestones.

The legend is the same as in Fig. 4. For log positions in the lacustrine sequence, see Fig. 3 (from Sagri *et al.*, 1989).

Bibliography

Abbate, E., Bruni, P., Fazzuoli, M. & Sagri, M., 1983. Le facies di transizione e continentali nel bacino terziario del Daban, Somalia settentrionale. Dati preliminari. *Quad. Geol. Somalia*, **7**, 7–32.

Abbate, E., Bruni, P., Fazzuoli, M. & Sagri, M., 1988. The Gulf of Aden continental margin of Northern Somalia: Tertiary sedimentation, rifting and drifting. *Mem. Soc. Geol. Ital.*, **31**, 427–45.

Bruni, P., Abbate, E., Abdi Salah Hussein, Fazzuoli, M. & Sagri, M., 1987. *Geological Map of the Daban Basin, Northern Somalia*. S.E.L.C.A., Firenze.

Macfadyen, W.A., 1933. *Geology of British Somaliland*. Crown Agents, London.

Sagri, M., Abbate, E. & Bruni, P., 1989. Deposits of ephemeral and perennial lakes in the Tertiary Daban Basin (Northern Somalia). *Palaeogeogr., Palaeoclimatol., Palaeoecol.*, **70**, 225–33.

Van Couvering, J.A., 1982. *Fossil Cichlid Fish of Africa*. Spec. Paper Paleontol., Paleontol. Assoc. London, 29.

The Lough Neagh Basin, Northern Ireland, UK

JOHN PARNELL AND BALVINDER SHUKLA

Department of Geology, Queens' University, Belfast BT7 1NN, UK

The Tertiary Lough Neagh Basin probably consisted of a number of fault-controlled sub-basins (Fig. 1), delimited by northeast–southwest and north-northwest–south-southeast fractures in an extensional regime. The deposits of the basin represent a range of sedimentary environments associated with a lacustrine system (Figs 2 and 3). The lake waters appear to have been fresh with a low sulfur content, and the iron mineralogy is therefore dominated by siderite

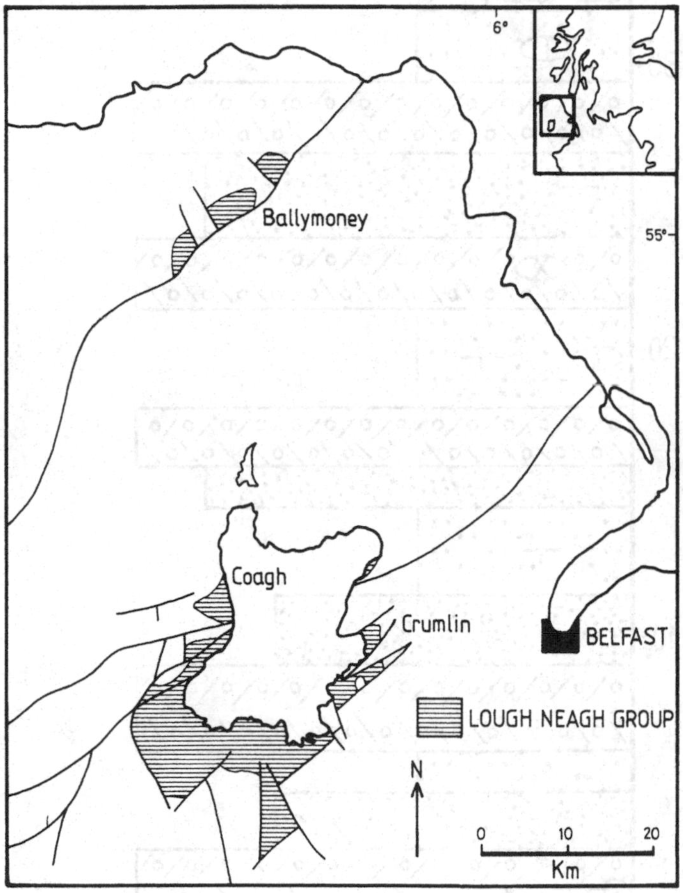

Fig. 1. Location of Oligocene Lough Neagh Group rocks, Northern Ireland, UK.

rather than pyrite. The sediments have been palynologically dated as Upper Oligocene (Chattian) age (Wilkinson *et al.*, 1980; Wilkinson & Boulter, 1980).

The rock types that constitute the Lough Neagh Group can be subdivided into four main lithologies according to grain size and fabric:

(1) Conglomerates (generally matrix-supported): these occur particularly near the base of the sequence, where they consist largely of basalt and chalk fragments, and pass downward into brecciated and *in situ* regolithic basalt. Thin conglomerate layers higher in the sequence additionally contain pebbles of dolerite, quartzite, Carboniferous limestone and other basement rocks.

(2) Sandstones: these occur at several levels in the sequence and can be divided into (i) poorly sorted and (ii) well sorted types. The poorly sorted sandstones consist of fine to coarse angular sand grains in an abundant clay matrix. The well sorted sandstones consist of grains which are more rounded and more quartzose, that is, lower contents of feldspar, mica and other unstable grains. Both types of sandstones exhibit low-angle cross-lamination.

(3) Mudrocks: these rock types constitute a significant proportion of the sequence. They are variably colored green, brown, grey and black. The green/black mudstones are organic-rich and all shades include finely disseminated plant matter. Some mudrocks also include plant roots, and more rarely stems in growth position, and a limited fauna (see below). The mudrocks are mostly massive; lamination, bedding and grading are exceptional.

In addition to plant debris, the biota in the mudstones includes gastropods, bivalves, ostracods and diatoms. The gastropods *Vivaparus lentus* and *Bernicia sp.* are common throughout the sequence, and three specimens of *Unio sp.* have been collected: all are freshwater genera. Extraction of ostracods was attempted using two samples, and both yielded ostracod carapaces, but the genus could not be determined. Diatom fragments have been recorded in several samples, and in a borehole from the Coagh District, a bed

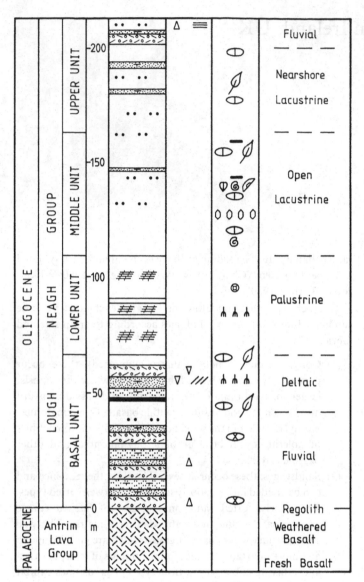

Fig. 2. Summary stratigraphic column for the Lough Neagh Group, Coagh District.

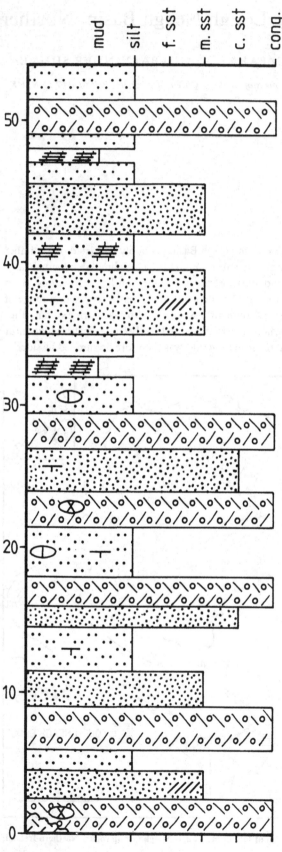

Fig. 3. Detailed lithological succession for lower part of sequence in Coagh District, determined from borehole records.

several centimeters thick was found to consist largely of diatom frustules. Fragments of salamander teeth have also been obtained from mudstone in the Coagh District. Possible bioturbation has been noted in several siltstones and sandstones.

(4) Lignite: this occurs interbedded with all of the lithologies above, but particularly grades into dark mudrocks. The lignite is divisible into (i) woody and (ii) non-woody forms. The woody form consists of clearly identifiable large plant fragments, in which the cell structure may be particularly well preserved by silicification. Non-woody lignite is massive and blocky, and does not exhibit recognizable botanical components. The lignites contain up to 75% organic carbon; the woody lignites are the most pure. Organic-rich mudstones typically contain 10–20% organic carbon.

The palynology of the Lough Neagh Group has been summarized by Wilkinson *et al.*, (1980), who concluded from the presence of ferns, conifers, palms and swamp cypress that the vegetation represents a low-lying, warm, frost-free environment. More recent work by one of us (B.S.) has shown that vertical changes in spore assemblages can be recognized. The topmost assemblage in borehole cores from the Coagh region contains *Sequoia*, which suggests a change to a drier climate.

Lignites and organic-rich mudrocks represent a swampy environment at the margin of a lake. Much of the plant matter is more or less *in situ*, although it may have suffered bacterial breakdown to a gyttja in which the plant structure is no longer recognizable. Thick lignites with low ash contents (less than 10%) imply isolation from clastic input. This is most likely to be engendered by peat deposition in floating or raised swamps which occur above flood level and never receive water-transported mineral matter. These swamps were readily susceptible to degradation unless they become drowned or rapidly buried. The thick sequences of mudrock found above some lignites suggest drowning of the peats by lake waters. Towards the deeper parts of the basins, the lignite seams split up into thinner seams as they become interbedded with mudrocks.

Some seams are silty at the base and become more purely lignitic towards the top. Once a swamp had been established, vegetation at the margin of the swamp would filter out any silt and prevent it from flowing into the center.

The lignite seam are in places interbedded with poorly sorted sandstones and conglomerates, which represent rivers draining into the lakes as small deltas. The sandstones which are well sorted were probably continually reworked sediment through wave action around the edge of the lake. Progressively further out into the lake, finer-grained sediment was deposited along with some thin lignite seams produced by erosion and limited transport of material from lake margin swamps (parautochthonous lignite). The mudstones contain many small fragments of eroded lignite. The lack of bedding and grading in the mudstones, and the occurrence of a mud matrix in some sandstones suggest rapid deposition of clays by flocculation.

Bibliography

Fowler, A., & Robbie, J.A., 1961. *Geology of the Country around Dungannon*. Memoir Geol. Surv., Northern Ireland.

Griffith, A.E., Legg, I.C. & Mitchell, W.I., 1987. Mineral Resources. In *Province, City and People: Belfast and its Region*, eds. R.H. Buchanan & B.M. Walker, pp. 43–58. Greystone Books, Belfast.

Manning, P.I., Robbie, J.A. & Wilson, H.E., 1970. *Geology of Belfast and the Lagan Valley*. Memoir Geol. Surv., Northern Ireland.

Parnell, J. & Meighan, I.G., 1989. Lignite and associated deposits of the Tertiary Lough Neagh Basin, Northern Ireland (conference report). *J. Geol. Soc. Lond.*, **146**, 351–2.

Parnell, J., Shukla, B. & Meighan, I.G., 1989. The lignite and associated sediments of the Tertiary Lough Neagh Basin. *Irish J. Earth Sci.*, **10**, 67–88.

Wilkinson, G.C., Bazley, R.A.B. & Boulter, M.C., 1980. The geology and palynology of the Oligocene Lough Neagh Clays, Northern Ireland. *J. Geol. Soc. Lond.*, **137**, 65–75.

Wilkinson, G.C. & Boulter, M.C., 1980. Oligocene pollen and spores from the western part of the British Isles. *Palaeontographica* B, **175**, 27–83.

Wright, W.B., 1924. Age and origin of the Lough Neagh Clays. *Quart. J. Geol. Soc. Lond.*, **80**, 468–88.

The Oligocene Campins Basin (northeast Spain)

P. ANADÓN

Institut de Ciències de la Terra (J. Almera), CSIC, c. Martí i Franqués s.n., E-08028 Barcelona, Spain

Geological setting – facies distribution

The Oligocene Campins Basin is located in the Paleogene strike–slip fault system of the Catalan Coastal Ranges, northeast Spain (Fig. 1). The present outcrops (a few km²) correspond to the remnants of an Oligocene basin of unknown extent which were overlain by Miocene and Quaternary alluvial deposits.

General stratigraphy

The Oligocene sequence is formed by three units (Fig. 2). The lower and upper units are formed by alluvial deposits (arkosic sandstones and conglomerates). The intermediate, or lacustrine unit, consists of shallow lacustrine deposits (mudstones, limestones and minor coals, and travertines) which are overlain by deep lacustrine deposits. The deep lacustrine deposits comprise terrigenous facies and carbonate-rich facies. The terrigenous facies crop out in the northeast part of the basin and are composed of sandstones, pebbly sandstones and siltstones. These facies grade laterally to the southwest into the carbonate-rich facies which are formed by a complex arrangement of thin-bedded limestones and

dolostones, massive and laminated mudstones and marls (carbonate mudstones). In some places, the laminated mudstones have a high organic matter content (oil shales). The upper part of the lacustrine sequence is composed of carbonate mudstones which were formed in a shallow lacustrine environment.

The overall basin-fill sequences record: (1) a basin-formation phase with predominant alluvial sedimentation, (2) an early shallow-phase which evolved into a deep meromictic lake-episode and finished with a new shallow phase and (3) a late alluvial sedimentation phase.

The Campins basin-fill sequence is Upper Oligocene (Lower

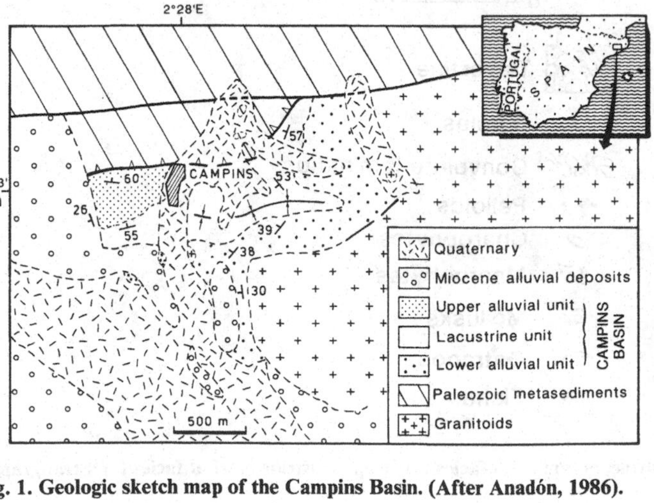

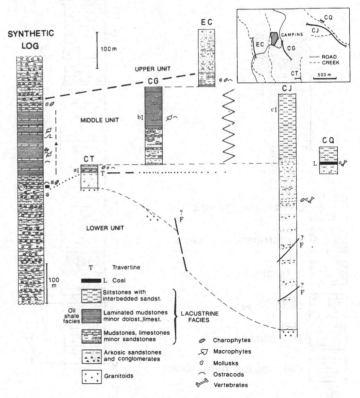

Fig. 2. Stratigraphic pattern of the Campins Basin. Modified after Anadón (1986).

Fig. 1. Geologic sketch map of the Campins Basin. (After Anadón, 1986).

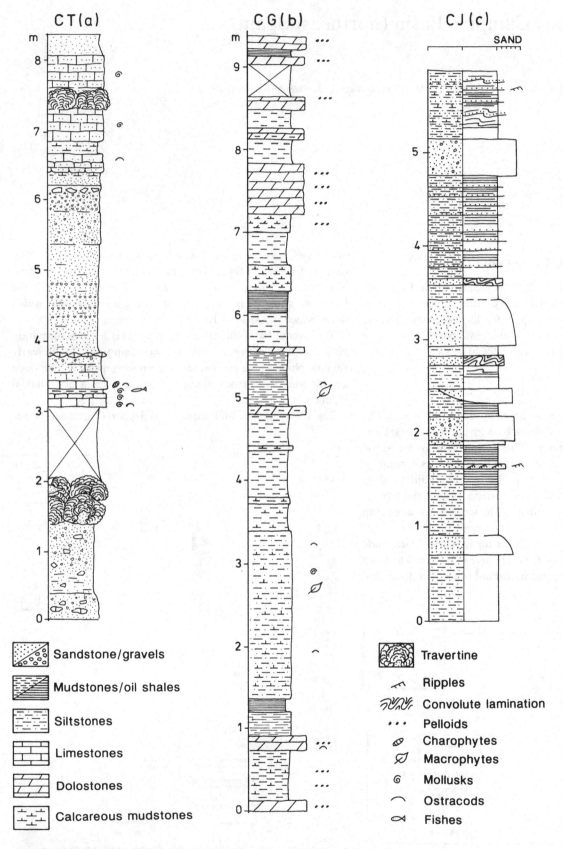

Fig. 3. Detailed lithologic columns. Shallow lacustrine deposits (a), deep lacustrine, organic-rich facies (b), deep lacustrine detrital facies (c). Stratigraphic position in Fig. 2.

Chattian). The age determination is based on two fossil mammal sites (*Theridomys* aff. *major* zone of the biozonation proposed by Agustí *et al.*, 1987; reliability index B). The mammal sites are located at the base of the lacustrine unit. Some Upper Oligocene charophytes have been recorded from the top of the lacustrine deposits.

Detailed representative sections

A typical column of the shallow lacustrine deposits is shown in Fig. 3a. These facies include variegated arkosic sandstones with interbedded limestones and marls which contain fossils of freshwater organisms. Travertines are composed of cauliflower-like bodies up to 1 m in diameter; they probably formed in spring-related settings of the lake shore due to bacterial-algal action.

The deep carbonate-rich facies (Fig. 3b) consists of an alternation of laminated, organic-rich mudstones, siltstones and thin dolostones. Varves (calcite or aragonite and organic-rich clay couplets) are locally present. In some horizons the laminated mudstones contain fossils of abundant plant leaves and, less commonly, insects and fishes. The laminated mudstones are organic-rich (up to 11.5% TOC), with HI index up to 500 mg HC/g TOC. The organic matter was derived from terrestrial plants with subordinate algal/bacterial input.

The deep, terrigenous facies (Fig. 3c) are formed by an alternation of siltstones, arkosic sandstones and small-pebble conglomerates. The siltstone packets, up to 5 m thick, frequently display lamination and contain minor, interbedded thin sandstones that in some places exhibit ripple and convolute lamination. The coarse sandstones and small pebble conglomerates display massive or graded bedding. The deep, terrigenous facies show the effects of several sedimentary processes: (1) mass flows and high-density turbidity currents from which the coarser sandstones and small-pebble conglomerates originated, (2) settling from suspension, leading to siltstones and (3) underflows and low-density turbidity

currents from which formed the thin, rippled sandstones. The sedimentary features, high organic matter content, and lack of bioturbation of the deep lacustrine facies may be interpreted as characteristic of deposits formed under the anoxic bottom of a permanent, water-stratified lake.

Bibliography

Agustí, J., Anadón, P., Arbiol, G., Cabrera, L., Colombo, F. & Sáez, A., 1987. Biostratigraphical characteristics of the Oligocene sequences of North-Eastern Spain (Ebro and Campins Basins). *München Geowiss. Abh* (A)**10**, 35–42.

Almera, J., 1883. Excursión al Montseny. *Mem. R. Acad. Ciencias Artes Barcelona*, **1**, 435–60.

Almera, J., 1907. Estudio de un lago oligocénico en Campins. *Mem. R. Acad. Ciencias Artes Barcelona*, **6**, 11–20.

Anadón, P., 1973. *Estudio estratigráfico y sedimentológico de los afloramientos terciarios de Campins (Barcelona)*. Thesis, Univ. Barcelona.

Anadón, P., 1986. Las facies lacustres del Oligoceno de Campins (Vallés oriental, Provincia de Barcelona). *Cuadernos Geol. Ibérica*, **10**, 271–94.

Anadón, P., Cabrera, L., Juliá, R., Roca, E. & Rosell, L., 1989. Lacustrine oil-shale basins in Tertiary grabens from NE Spain (Western European Rift System). *Palaeogeogr., Palaeoclimatol., Palaeoecol.*, **70**, 7–28.

Anadón, P., Cawley, J.J. & Juliá, R., 1988. Oil source rocks in lacustrine sequences from Tertiary grabens, western Mediterranean Rift System, Northeast Spain. *A.A.P.G. Bull.*, **72**, 983.

Anadón, P. & Utrilla, R., 1993. Sedimentology and isotope geochemistry of lacustrine carbonates of the Oligocene Campins Basin, north-east Spain. *Sedimentology*, **40**, 699–720.

Anadón, P. & Villalta, J.F., 1975. Caracterización de terrenos de edad estampiense en Campins (Vallés Oriental). *Acta Geol. Hisp.*, **10**, 6–9.

De las Heras, X., 1989. Geoquímica orgánica de conques lacustres fòssils. PhD Thesis, Univ. Barcelona.

Permanyer, A. & García-Vallès, M., 1986/1987. Generación de hidrocarburos en sedimentos lacustres terciarios: las cuencas de Ribesalbes y Campins (NE de España). *Rev. Inv. Geol.*, **42/43**, 23–44.

Oligocene lacustrine deposits in the Cala Blanca Formation (Mallorca Island, western Mediterranean)

E. RAMOS-GUERRERO, L. CABRERA AND M. MARZO

Dpto. Geología Dinàmica, Geofìsica i Paleontologia, Facultad de Geología. Universidad de Barcelona, E-08028 Barcelona, Spain

Geological setting

Mallorca is the largest of the Balearic Islands in the western Mediterranean Sea (Fig. 1). These islands are the emerged part of a major morphostructural feature (Balearic Promontory) located between the Valencia Trough and the Algerian Basin. The Balearic Promontory makes up the eastern part of the Betic thrust-fold belt and displays a Middle Miocene to Quaternary horst and graben structure superimposed on a previously developed Lower Miocene (Langhian) thrust-and fold-dominated structure.

The horsts and grabens of Mallorca developed after the Langhian. Paleozoic to Middle Miocene rocks crop out in the horsts, where a set of northwest-oriented thrust sheets (Burdigalian–Langhian in age) are recognized (Fallot, 1922; Darder, 1925; Alvaro, 1987; Sábat *et al.*, 1988). Younger Serravalian to Quaternary deposits, up to 1000 m thick, are recognized in the grabens.

The Paleogene sedimentary record in Mallorca has been divided into two major depositional sequences (Fig. 2) bounded by erosion surfaces (Ramos-Guerrero *et al.*, 1989b). These sequences include marine shelf and nearshore deposits in the southern depositional area, while alluvial and lacustrine deposits dominate in the northern area. Lacustrine sedimentary rocks include the Peguera Limestone Formation (Bartonian–Priabonian in age, Depositional Sequence I) described in this volume, and the mainly siliciclastic Cala Blanca Formation (Oligocene in age, Depositional Sequence II).

The Oligocene lacustrine (Sequence II)

The Cala Blanca Formation:

This formation crops out extensively in the Serra de Tramuntana as well as in the central part of the island (Fig. 1,B). The Puig d'en Tió, Peguera, Alaró and Sineu Sections (PT, PE, AL and SI, in Figs 1–3) have been studied in detail. This unit is up to 200 m thick and it is bounded by erosive surfaces (Fig. 3). The Oligocene age of this formation has been established on the basis of fossil charophytes (Ramos-Guerrero, 1988) and mammal assemblages (Forsyt-Major, 1904; Vidal, 1905; Hugueney & Adrover, 1982) Palynological assemblages, which indicate a warm, tropical humid climate, have also been documented (Alvarez-Ramis *et al.*, 1987).

The Cala Blanca Formation records the activity of an alluvial complex originating from the erosion of emerged lands located to the north-northwest of Mallorca (Ramos-Guerrero *et al.*, 1989b). The generation of the paleoreliefs could be related to the activity of a northwest–southwest strike–slip fault system.

The paleocurrent trends show dominant southward and southeastward transport directions. Major lateral changes from proximal to distal facies are also observed in the same direction (Ramos-Guerrero & Marzo, 1989).

Lacustrine facies in the Cala Blanca Formation

Minor lacustrine sequences occur in the Peguera and Alaró sections (PE and AL in Figs 1–3) related to the fluvial facies, whereas a major clastic-carbonate lacustrine sequence is developed in Sineu (SI in Fig. 3), in relation to the distal alluvial facies. The Sineu lacustrine succession is composed of two facies assemblages (see Fig. 4).

Internal lacustrine facies (Deflá limestone Member)

The internal lacustrine facies assemblage is mainly composed of a lower rhythmic limestone unit which grades upwards to marls. The rhythmic limestones are composed of alternating carbonate and clay laminae and thin beds. The carbonate laminae are 2 to 6 mm thick and consist of micrites with algal filaments, diatoms, charophytes and ostracodes; scarce chert nodules occur at the lower part of the section. The clay laminae range from 2 to 4 mm in thickness, and show higher organic matter contents.

Marginal lacustrine facies (Sineu marls Member)

The marginal lacustrine facies assemblage is more terrigenous-influenced and consists of marls and finely laminated rhythmites made up of black clays and coal.

The marls, which are the major lithology, are grey, massive and display changing organic matter contents (0.5 to 5% TOC). The black clay-coal couplets form beds up to 10 m thick with common turtle, crocodile, mammal and gastropod fossil remains. These laminated beds alternate with massive, brown-yellowish marls which show paleosol features.

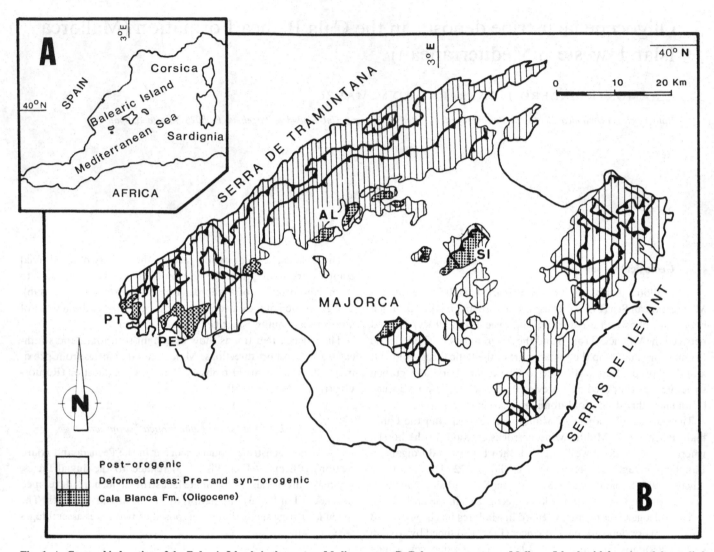

Fig. 1. A: Geographic location of the Balearic Islands in the western Mediterranean. B: Paleogene outcrops on Mallorca Island, with location of the studied sections. PT = Puig d'en Tió, PE = Peguera and SI = Sineu.

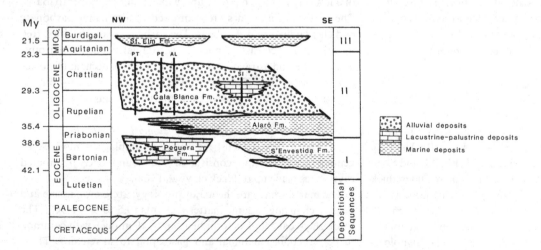

Fig. 2. Stratigraphic framework for the Paleogene on Mallorca. The location of the studied sections is represented by the same letters in Fig. 1. (After Ramos-Guerrero *et al.*, 1989a).

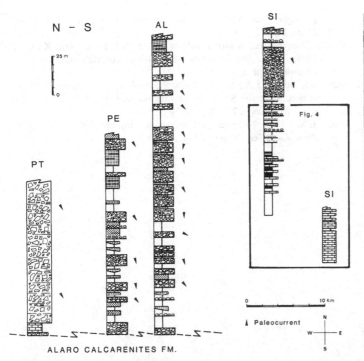

Fig. 3. Schematic stratigraphic logs in the Cala Blanca Formation. Same legend as in Fig. 4. See Fig. 1 for location of the sections. (After Ramos-Guerrero et al., 1989a.)

The coal facies occur in seams up to a few cm in thickness. These coals are moderate rank ($R_0 = 0.28$) humic lignites, which are made up of a huminite matrix including semifusinite and liptinite macerals. Clay and carbonate contents are very high. The sulfur content is also high (11–12%).

The relationship between internal and marginal facies assemblages is not directly observable in the field, but the marginal lacustrine facies probably prograded over the internal facies (Ramos-Guerrero et al., 1989a). The sequential arrangement of the facies assemblages shows a final progradation of the fluviatile environments over the lacustrine system with thin fine conglomerate and sandstone beds occurring at the top of the lacustrine sequence (Figs 3 and 4).

Bibliography

Alvarez-Ramis, C., Ramos-Guerrero, E. & Fernandez-Marron, T., 1987. Estudio paleobotánico del Cenozoico de la zona central de Mallorca: Yacimiento de Son Ferragut. *Bol. Geol. Minero*, **98**, 349–56.

Alvaro, M., 1987. La tectónica de cabalgamientos de la Sierra Norte de Mallorca (Islas Baleares). *Bol. Geol. Minero*, **98**, 622–9.

Darder, B., 1925, La tectonique de la region orientale de l'île de Majorque. *Bull. Soc. Géol. France*, **25**, 245–78.

Decourt, J., Zonenshain, L.P., Ricou, L.E., Kazmin, V.G., Le Pichon, X., Knipper, A.L., Grandjacquet, C., Sortshikov, I.M., Geysant, J., Lepvrier, C., Pechersky, D.H., Boulin, J., Sibuet, J.C., Savostin, L.A., Sorokhtin, O., Westphal, M., Bazhenov, M.L., Lauer, J.P. & Biu-Duval, B., 1986. Geological evolution of the Tethys belt from the Atlantic to the Pamir since Lias. *Tectonophysics*, **123**, 241–315.

Fallot, P. 1922. *Etude geologique de la Sierra de Majorque*. PhD Dissertation, Univ. Paris.

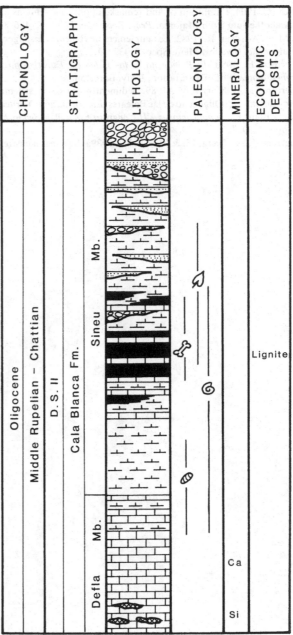

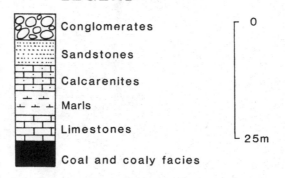

LEGEND

- Conglomerates
- Sandstones
- Calcarenites
- Marls
- Limestones
- Coal and coaly facies

0

25m

Fig. 4. Detailed stratigraphic log and main characteristics of the lacustrine deposits in the Cala Blanca Formation (see location in Fig. 3 inset). D.S. = Depositional Sequence.

Forsyt-Major, C. 1904. Exhibition of and remarks upon some remains of Anthracotherium from Majorca. *Proc. Zool. Soc. London* I, 456–8.

Hugueney, M., & Adrover, R., 1982. Le peuplement des baleares (Espagne) au Paléogène. *Geobios*, Mem. Spec. 6, 439–49.

Ramos-Guerrero, E., 1988. *El Paleógeno de las Baleares: Estratigrafia y Sedimentologia.* PhD Dissertation, Univ. Barcelona.

Ramos-Guerrero, E. & Marzo, M., 1989. Sedimentologia de un sistema fluvio-aluvial en el Oligoceno de las Baleares: La Formación Detrítca de Cala Blanca. *Comunicaciones XII Congreso Español de Sedimentol.,* pp. 47–50.

Ramos-Guerrero, E., Cabrera, Ll. & Marzo, M., 1989a. Sistemas lacustres paleógenos de Mallorca (Mediterráneo occidental). *Acta Geol. Hisp.*, **24**, 185–203.

Ramos-Guerrero, E., Rodriguez-Perea, A., Sábat, F. & Serra-Kiel, J., 1989b. Cenozoic tectosedimentary evolution of Mallorca area. *Geodinamica Acta*, **3**, 53–72.

Sábat, F., Muñoz, J.A., & Santanach, P., 1988. Transversal and oblique structures at the Serres de Llevant thrust belt (Mallorca Island). *Geol. Rundschau*, **77**, 529–38.

Vidal, L.M., 1905. Note sur l'Oligocene de Majorque. *Bull. Soc. Geol. France*, 4th Ser., V, pp. 651–4.

Swiss Molasse Basin (Chattian)

NIGEL H. PLATT

GECO-PRAKLA, Ltd. Schlumberger House, Buckingham Gate, Gatwick Airport West Sussex RH6 0NZ, UK

Introduction

This short contribution describes lacustrine limestones and dolomites present within the lower part of the Upper Oligocene–Lower Miocene Lower Freshwater Molasse of western Switzerland (Fig. 1). This area forms part of the Molasse Basin, the northern foreland basin of the Alpine Chain. The Molasse Basin contains a thick sedimentary succession of Tertiary age (Fig. 2). This sequence, termed the Molasse, is dominated by continental and marine clastic deposits, but carbonates also occur at several horizons. Two distinct units are discussed here: the Lower Chattian 'Calcaires Inférieurs' and the Upper Chattian 'Calcaires d'eau douce et dolomie'.

The Lower Chattian Calcaires Inférieurs comprise palustrine carbonates resting on karstified Mesozoic limestones at the southern margin of the Jura Mountains. The Calcaires Inférieurs contain a shallow freshwater fauna/flora of ostracodes and charophytes, but also display clear evidence of subaerial exposure (glaebule/pisoid formation, circumgranular cracking, rhizoliths and microkarstic cavities). These rocks record deposition in shallow, ephemeral lakes at the northern margin of the Molasse Basin. Repeated desiccation led to locally intense modification of the lacustrine carbonates by pedogenesis. Stable isotope analyses reveal typical freshwater compositions. Weak covariance of $\delta^{13}C$ and $\delta^{18}O$ is consistent with deposition in hydrologically closed lakes, although the range of $\delta^{13}C$ values may alternatively record pedogenetic overprinting.

The Upper Chattian Calcaires d'eau douce et dolomie include laminated basinal facies and bioturbated shallow lake charophyte-ostracod mudstones and wackestones. Detrital clastic contents are higher towards the Alps in the south. A white dolomite marker bed caps the sequence in the north. These rocks record deposition in an extensive, possibly deeper lake complex; later, increased aridity led to the onset of evaporitic conditions. Variations in carbon isotopic composition may reflect changes in organic productivity or areal differences in the proportions of biogenic, authigenic and detrital carbonate. The higher $\delta^{18}O$ values shown by the dolomites are consistent with formation from evolved, evaporitic lake waters.

Stratigraphy and geological setting

The Molasse Basin is a major foreland basin of Tertiary age lying to the north of the Alps (Fig. 1). The basin contains a thick sequence of continental and shallow marine deposits (Fig. 2), mostly clastics derived from erosion of the uplifting Alpine Chain, but also including some carbonate deposits which have previously received relatively little attention. This paper describes lacustrine carbonates of Late Oligocene age which crop out in western Switzerland between Lake Geneva and Lake Neuchâtel, to the north of Lausanne (Fig. 1). During Late Oligocene times, this area was fed by streams draining a source area to the north dominated by Jurassic marine carbonates of the Jura Mountains, as well as by streams draining the more distant and varied terrains of the Alps to the south, where exposed lithologies at this time included Mesozoic carbonates and clastics as well as basement rocks.

The Molasse sequence is divided into four main lithostratigraphic units, two of them laid down under dominantly marine and two under mainly continental conditions (Fig. 2). The deposits described here lie within the lower part of the Upper Oligocene–Lower Miocene Lower Freshwater Molasse. This thick succession is dominated by alluvial clastics (see e.g., Platt & Keller, 1992), and shows a pronounced proximal-distal facies change from alluvial-fan conglomerates along the Alpine front in the south to fluvial sandstones and mudstones with interbedded lacustrine carbonates and evaporites further to the north. Recent biostratigraphic work on charophytes and micromammals (Berger, 1986; Engesser & Mayo, 1987) has distinguished two separate horizons, the Calcaires Inférieurs in the Lower Chattian and the Calcaires d'eau douce et dolomie in the Upper Chattian (Fig. 3).

Calcaires Inférieurs

The Calcaires Inférieurs (Fig. 3) crop out at the northern edge of the Molasse Basin along the southern foot of the Jura Mountains, where they form a laterally persistent horizon approximately 5 m in thickness resting on the Mesozoic substrate and the thin cover of Eocene karstic deposits (the Siderolithikum or Sidérolithique).

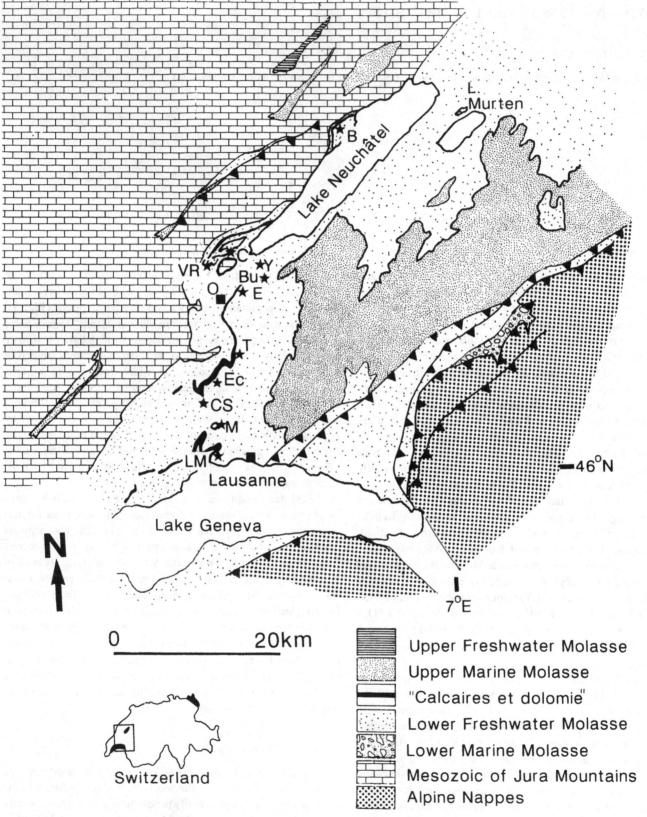

Fig. 1. Simplified geological map of western Switzerland, showing location of study area and of important localities mentioned in the text.

Upper Freshwater Molasse
Upper Marine Molasse
"Calcaires et dolomie"
Lower Freshwater Molasse
Lower Marine Molasse
Mesozoic of Jura Mountains
Alpine Nappes

B = Boudry,   Y = Yverdon,   VR = Valeyres-sous-Rances,   O = Orbe,
C = Champvent,   CS = Cossonay,   Bu = R.   Buron,   E = Essertines,
T = R. Talent, Ec = Eclépens, M = Mex, LM = La Morges.

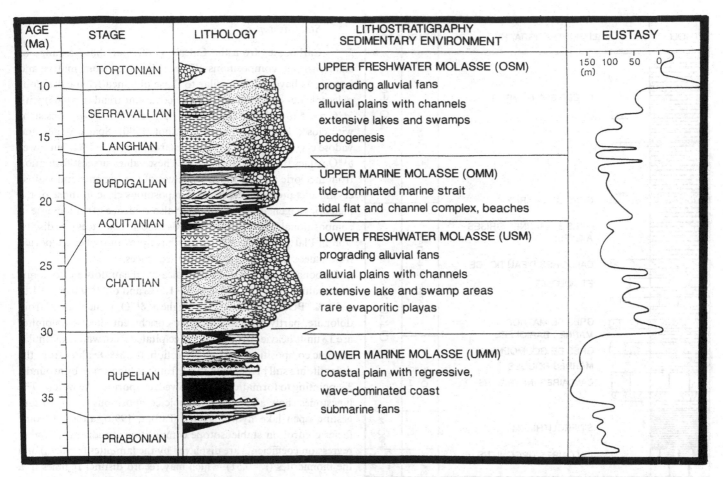

Fig. 2. Generalized stratigraphy of the Molasse Basin (after Keller, 1989, 1990). Thicknesses highly variable; in the study area the Lower Freshwater Molasse is probably from 500–1000 m thick. Figure courtesy of Beat Keller.

The Calcaires Inférieurs are generally unlaminated and contain a freshwater fauna/flora of charophytes, ostracods, gastropods and bivalves. Some horizons show a wide variety of features pointing to subaerial exposure, including rhizoliths, circumgranular cracks, glaebules, pedogenetic brecciation and microkarstic cavities. This association of facies indicates deposition in a shallow, oxygenated lake subject to periodic desiccation and pedogenetic modification of the carbonate muds during emergence.

Vadose crystal silt and layered internal sediment occur in micro-karst cavities. Coarser, blocky calcite is interpreted as fresh-water phreatic cement; rare ferroan calcite is a late void-filling phase present in fissures and bioclasts. Vadose cements are non-luminescent; blocky calcite cements include both early non-luminescent and later complexly zoned, brightly luminescent generations.

The carbonates of the Calcaires Inférieurs are arranged in 1–2 m thick regressive cycles and appear similar to 'palustrine' deposits previously described from the Mesozoic and Tertiary of southern Europe and the western USA. These deposits are thought to have been laid down in low-gradient, low-energy settings at the margins of shallow lakes (Platt & Wright, 1991) or in extensive freshwater carbonate marsh complexes similar to the modern environments of the Florida Everglades (see Platt & Wright 1992).

Calcaires d'eau douce et dolomie

The Calcaires d'eau douce et dolomie (Fig. 3) form part of a carbonate-bearing interval up to 120 m in thickness, comprising a series of thin carbonate beds alternating with alluvial sandstones, overbank mudstones, grey lacustrine marls and paleosols. The carbonates occur in a variety of facies, including bioturbated charophyte-ostracod mudstones and wackestones, as well as lami-nated ostracod-algal limestones. Some of the laminae are graded and show sharp erosional bases. Detrital contents are higher in the south, where the carbonates contain abundant 0.5 to 1 mm quartz grains. One or two thin beds of dolomite are present at several localities, where they form a valuable marker at the top of the unit (Fig. 3). The dolomites are locally cut by vertical cracks and are overlain by gypsiferous marls ('Marnes grises à gypse').

Although the charophyte-ostracod mudstones are likely to have

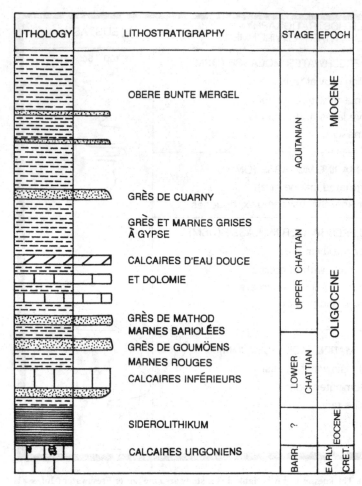

LITHOLOGY	LITHOSTRATIGRAPHY	STAGE	EPOCH
	OBERE BUNTE MERGEL	AQUITANIAN	MIOCENE
	GRÈS DE CUARNY		
	GRÈS ET MARNES GRISES À GYPSE	UPPER CHATTIAN	OLIGOCENE
	CALCAIRES D'EAU DOUCE ET DOLOMIE		
	GRÈS DE MATHOD MARNES BARIOLÉES		
	GRÈS DE GOUMÖENS MARNES ROUGES	LOWER CHATTIAN	
	CALCAIRES INFÉRIEURS		
	SIDEROLITHIKUM	?	EOCENE
	CALCAIRES URGONIENS	BARR.	EARLY CRET.

Fig. 3. Detailed stratigraphy of the study area (after Kissling, 1974). Not to scale; Calcaires Inférieurs are less than 10 m in thickness, Calcaires d'eau douce et dolomie make up to 50% of a 120 m thick interval.

been deposited in a shallow lake, with oxygenated bottom waters, the presence of laminated facies points to the periodic development of stratification. The presence of grading and erosional bases suggests an origin through resedimentation, which would be consistent with the presence of a significant slope and with deposition in relatively deep water. Evidence for the development of stratification is likewise consistent with deposition in deeper water or during times of higher lake productivity (see Kelts & Hsü, 1978). The higher detrital quartz contents in the south record clastic input from the Alpine Chains. The presence of the cracks within the dolomite points to root action or desiccation; a shift to a more arid climate is supported by the presence of evaporites in the Marnes grises à gypse above.

Figure 4 presents a schematic reconstruction of western Switzerland during Late Oligocene times, showing the sites of carbonate deposition in relation to the paleogeography.

Stable isotopes

Carbonates from the Calcaires Inférieurs have a range of stable isotope compositions (Fig. 5). Samples from matrix and intraclasts have values similar to those obtained from calcareous paleosols but also exhibit a weakly covariant trend (r = 0.65) with $\delta^{13}C = -5$ to -7.5 ‰ and $\delta^{18}C = -5.5$ to -9 ‰ PDB, consistent with closed lake hydrology (cf. Talbot, 1990). Spar cements and vadose crystal silts have $\delta^{13}C$ values from -6 to -7 ‰, but lower $\delta^{18}O$ values from -10 to -11 ‰. These values suggest formation from meteoric waters; the $\delta^{18}O$ values of the cements and vadose crystal silts probably represent compositions close to those of the meteoric diagenetic fluids. There is also evidence of an increase in compositional heterogeneity towards exposure surfaces (see discussion in Platt, 1992); similar trends have previously been reported from paleosols in marginal marine sequences.

Carbonate samples from the Calcaires d'eau douce et dolomie show values of $\delta^{13}C$ from -5 to -1.5 ‰ and of $\delta^{18}O$ from -1 to -10 ‰ PDB (Fig. 6A). The highest $\delta^{18}O$ values come from dolomite, partly reflecting isotopic enrichment effects (dolomites are 3 δ units heavier than calcites precipitated from waters of similar isotopic compositions). After correction, the $\delta^{18}O$ values from the dolomite are still higher than those from the limestones by around 2 ‰, pointing to formation from evolved, evaporitic lake waters. The limestones show a spread of isotopic compositions (r = 0.35) suggesting open lake hydrology (cf. Talbot, 1990), although some facies control on stable isotope composition is indicated. Higher regression coefficients are displayed by the laminites (r = 0.73) and the biomicrites (r = 0.51), which may record distinct regimes in a lake showing intermittent hydrological closure; the higher $\delta^{13}C$ values shown by the laminites may reflect higher organic productivity during times of high lake stand.

Evidence of areal variations in carbon isotope composition (see Fig. 6B) may be the result of differences in the proportions of authigenic and detrital carbonate. Marine limestones typically show heavier carbon isotope compositions than lacustrine carbonates, with $\delta^{13}C$ values near to 0 ‰ PDB, so that input of detrital carbonate derived from the erosion of older marine formations favors heavier carbon isotope compositions. In this case, samples from the central part of the basin, further from sources of detrital carbonate, show lower $\delta^{13}C$ values than those from the southern area, fed by streams draining the mixed lithologies of the Alps, and those from the northern and northwestern areas, where the drainage area was comprised of the Mesozoic carbonates of the Jura Mountains.

Economic potential

The Chattian carbonates show total organic carbon values of up to 0.8%, indicating moderate to poor economic potential as hydrocarbon source rocks; nevertheless, laterally equivalent strata of the Molasse à Charbon (Fasel, 1986; Homewood et al., 1986) include sub-economic coal horizons which may form potential gas-prone hydrocarbon source rock targets in proximal basin areas.

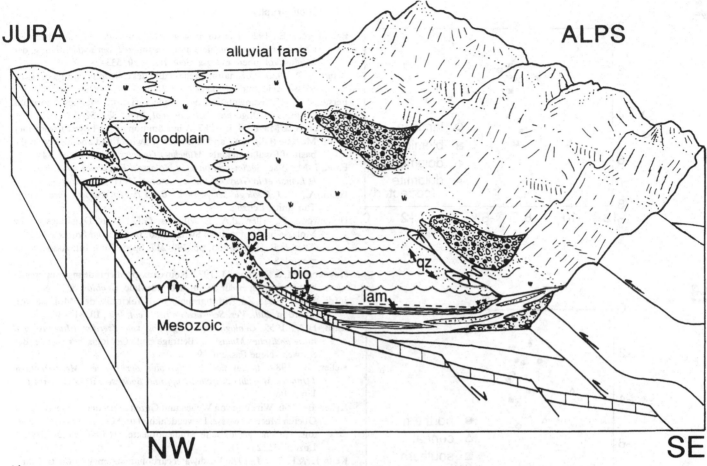

JURA

ALPS

alluvial fans

floodplain

pal

bio

qz

lam

Mesozoic

NW

SE

1) Lower Chattian -
   palustrine / lacustrine
   carbonates at
   Jura margin (pal)

2) Upper Chattian - extensive lake complex
   N lake margin: biomicrites (bio)
   basin center: laminites (lam)
   S clastic fringe: sandy limestones (qz)

Fig. 4. Summary diagram for carbonate sedimentation in the Swiss Molasse Basin during the Chattian. Modified after a figure by D. Rigassi (in Weidmann 1987). Area of the Molasse Basin shown approximately 40 km from NW–SE and 60 km from SW–NE.

1) Calcaires Inférieurs (Lower Chattian) shallow lacustrine/palustrine limestones (*pal*) were deposited on karstified Mesozoic substratum at the northern basin margin. Repeated periods of emergence are recorded by the development of pedogenetic fabrics and by the spectacular meteoric cementation history.

2) Calcaires d'eau douce et dolomie (Upper Chattian) were deposited in a more extensive lake complex. In the northwest, only shallow water lacustrine biomicrites (*bio*) were laid down; central and northern basin areas also saw the development of laminated carbonates (*lam*), recording deposition beneath a stratified water column. Local evidence for re-sedimentation may point to the presence of a high-gradient, bench-type lake margin in the south. Sandy limestones (*qz*) were deposited on the south of the basin. Higher clastic supply in this area probably reflected distal alluvial input at the front of alluvial fans draining the Alps.

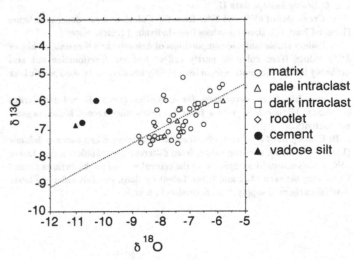

○ matrix
△ pale intraclast
□ dark intraclast
◇ rootlet
● cement
▲ vadose silt

Fig. 5. Stable isotope data: I.
Cross plot of $\delta^{13}C$ and $\delta^{18}O$ for the Calcaires Inférieurs (Lower Chattian). Note that data from matrix and intraclast samples may show weak covariance (r = 0.65).

A

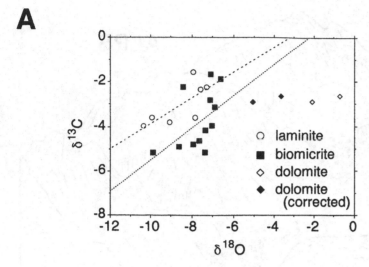

B

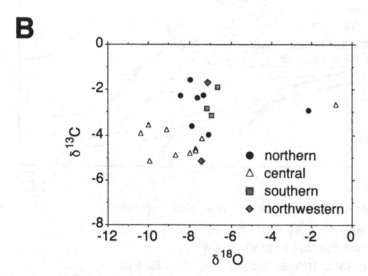

Fig. 6. Stable Isotope data II.

A: Cross plot of δ¹³C and δ¹⁸O from the Calcaires d'eau douce et dolomie (Upper Chattian), showing values from individual facies. Note:

1) distinct stable isotopic compositions of dolomite and limestones – higher δ¹⁸O values from dolomite partly reflect isotopic fractionation but also probably indicate formation from isotopically heavier, evolved evaporitic lake waters.

2) possible weak covariance for laminites (r = 0.73) and biomicrites (r = 0.51). Higher δ¹³C values from laminites may record higher organic productivity.

B: Cross plot of δ¹³C and δ¹⁸O from the Calcaires d'eau douce et dolomie (Upper Chattian), showing values from different geographical areas. Lower δ¹³C values shown by samples from the central part of the basin (area around Cossonay, between Mex and River Talent section), possibly reflecting lower detrital carbonate supply. See discussion in text.

Bibliography

Baumberger, E., 1927. Die stampischen Bildungen der Nordwestschweiz und ihrer Nachbargebiete mit besonderer Berücksichtigung der Molluskenfaunen. *Eclogae Geol. Helv.*, **20**, 533–78.

Berger, J.-P., 1986. Biozonation préliminaire des charophytes oligocènes de Suisse occidentale. *Eclogae Geol. Helv.*, **79**, 897–912.

Bersier, A., 1945. Sédimentation molassique: variations latérales et horizons continus à l'Oligocène. *Eclogae Geol. Helv.*, **38**, 452–8.

Engesser, B. & Mayo, N.A., 1987. A biozonation of the Lower Freshwater Molasse (Oligocene and Agenian) of Switzerland and Savoy on the basis of fossil mammals. *Münchner Geowiss Abh.* A10, 67–84.

Fasel, J.-M., 1986. *Sédimentation de la Molasse d'Eau douce subalpine entre le Leman et la Gruyère.* Thèse doctorat, Univ. Fribourg.

Heim, A., 1919. *Geologie der Schweiz. Band I: Molasse und Juragebirge.* Tauchnitz, Leipzig.

Homewood, P., Allen, P.A. & Williams, G.D., 1986. Dynamics of the Molasse basin of western Switzerland. In *Foreland Basins*, eds. P.A. Allen & P. Homewood, pp. 199–217. Spec. Publ. Int. Assoc. Sedimentol. 8.

Hugueney, M. & Kissling, D., 1972. Nouveaux gisements de mammifères de l'Oligocène supérieur de Suisse occidentale. *Géobios*, **5**, 55–66.

Jordi, H.A., 1951. Zur Stratigraphie und Tektonik der Molasse von Yverdon. *Bull. Ver. Schweiz. Petrol.-Geol. Ing.*, **18**, 817–36.

Jordi, H.A., 1955. *Geologie der Umgebung von Yverdon (Jurafuss und mitteländische Molasse).* Beitrage zur Geologischer Karte der Schweiz, Neue Fassung 99.

Keller, B., 1989. *Fazies und Stratigraphie der Oberen Meeresmolasse (Unteres Miozän) zwischen Napf und Bodensee.* PhD Dissertation, Univ. Bern.

Keller, B., 1990. Wirkung von Wellen und Gezeiten bei der Ablagerung der Oberen Meeresmolasse. Löwendenkmal und Gletschergarten – zwei anschauliche geologische Studienobjekte. *Mitteil. natur. Gessell.* Luzern, **31**, 245–71.

Kelts, K. & Hsü, K.J., 1978. Freshwater carbonate sedimentation. In *Lakes, Chemistry, Geology, Physics*, ed. A. Lerman, pp. 295–323. Springer Verlag, Berlin.

Kissling, D., 1974. *L'oligocène de l'extrémité occidentale du bassin molassique suisse. Stratigraphie et aperçu sédimentologique.* Thèse doctorat, Univ. Genéve.

Matter, A., Homewood, P., Caron, C., van Stuyvenburg, J., Weidmann, M. & Winkler, W., 1980. Flysch and molasse of western and central Switzerland. In *Geology of Switzerland, a Guide Book*, ed. R. Trümpy, pp. 261–93. Schweizerische Geologische Kommission.

Mojon, P.-O., Engesser, B., Berger, J.-P., Bucher, H. & Weidmann, M., 1985. Sur l'age de la Molasse d'Eau douce inférieure de Boudry NE. *Eclogae Geol. Helv.*, **78**, 631–67.

Platt, N.H., 1989. Lacustrine carbonates and pedogenesis: sedimentology and origin of palustrine deposits from the Early Cretaceous Rupelo Formation, W. Cameros Basin, N. Spain. *Sedimentology*, **36**, 665–84.

Platt, N.H., 1992. Freshwater limestones from the Lower Freshwater Molasse (Oligocene), western Switzerland: sedimentology and stable isotopes. *Sedim. Geol*, **78**, 81–99.

Platt, N.H. & Keller, B., 1992. Distal alluvial deposits in a foreland basin setting: the Lower Freshwater Molasse (Lower Miocene), Switzerland: sedimentology, architecture and paleosols. *Sedimentology*, **39**, 545–65

Platt, N.H. & Wright, V.P., 1991. Lacustrine carbonates: facies models, facies distributions and hydrocarbon aspects. In *Lacustrine Facies Analysis*, ed. P. Anadón, Ll. Cabrera & K. Kelts, pp. 55–73. Spec. Publ. Int. Assoc. Sedimentol. 13.

Platt, N.H. & Wright, V.P., 1992. Palustrine carbonates and the Florida Everglades: Towards an exposure index for the fresh-water environment? *J. Sedim. Petrol.*, **62**, 1058–71.

Reggiani, L., 1989. Faciés lacustres et dynamique sédimentaire dans la Molasse d'eau douce inférieure Oligocène (USM) de la Savoie. *Eclogae geol. Helv.*, **82**, 325–50.

Rigassi, D.A., 1977. Subdivision et datation de la molasse 'd'eau douce inférieure' du plateau suisse. *Paleolab. News*, 1, Nyon.

Talbot, M.R., 1990. A review of the paleohydrological interpretation of carbon and oxygen isotopic ratios in primary lacustrine carbonates. *Chem. Geol. (Isotope Geosci. Sectn.)*, **80**, 261–79.

Trümpy, R., 1980. *Geology of Switzerland: a Guide Book. Part A: an Outline of the Geology of Switzerland.* Schweizerische Geologische Kommission, Wepf & Co., Basel.

Weidmann, M., 1984. Le sidérolithique et la molasse basale d'Orbe (VD). *Bull. Soc. vaud. Sci. nat.* (Lausanne), **77**(366), 135–41.

Weidmann, M., 1987. *Les dessous d'une ville: Petite géologie Lausannoise.* Les Cahiers de la Forêt Lausannoise, 2, Direction des finances de la ville de Lausanne, service des forêts, domaines et vignobles.

Miocene to Pliocene

Neogene lacustrine deposits of karstic origin (Ardenne Massif, Belgium)

CHRISTIAN DUPUIS[1] AND RONAN ERTUS[2]

[1]*Geologie fondamentale et appliquée, Faculté Polytechnique, rue de Houdain, 9, B-7000 Mons, Belgium*
[2]*Laboratoire de géochimie des roches sédimentaires, UA. 723, Université de PARIS-sud, F-91405 Orsay Cedex, France*

Neogene lacustrine deposits occupy karstic holes (cryptolapiaz) scattered on the carbonate rocks of the folded and peneplained Variscan Belt exposed in the Ardenne Massif (Fig. 1). The carbonate rocks are restricted to two main large tectonic units: the Synclinorium of Namur (S.N.) and of Dinant (S.D.). The Syncli-

norium of Namur contains Frasnian and Dinantian limestones interbedded with Famennian siliciclastics. The Dinantian limestones (Tournaisian and Visean age) are covered by Namurian and Westphalian coal measures. To the south, the Synclinorium of Dinant exposes Givetian, Frasnian and Dinantian limestones. They

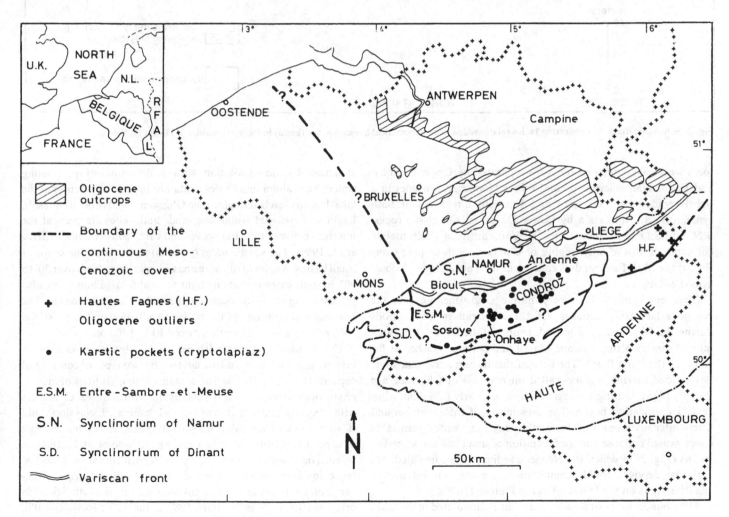

Fig. 1. Location map of the Neogene karstic lakes of the Ardenne Massif.

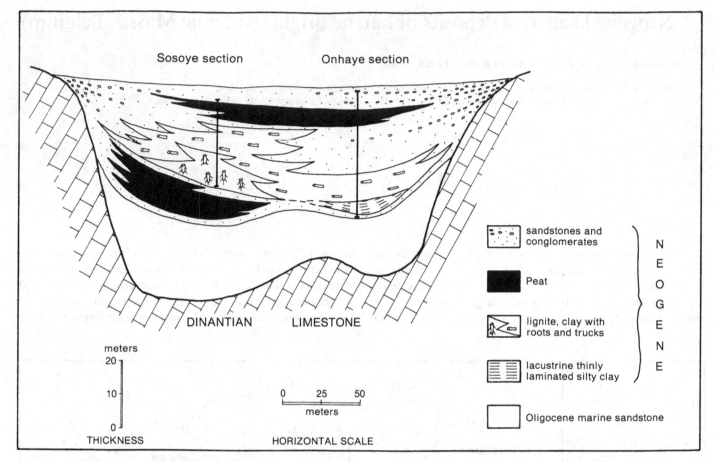

Fig. 2. Schematic diagram illustrating facies relationships in an hypothetical karstic pocket (basin) in the synclinorium of Dinant (E.S.M.).

form west–east elongated outcrops separated by Upper Frasnian and Famennian siliciclastics and minor Namurian synclines in a tightly folded belt. Givetian and Frasnian limestones of both synclinoria are underlain by Lower Devonian siliciclastic rocks. S.N. and S.D. are separated by the Variscan front which mainly corresponds to the outcropping of a major south-dipping over-thrust: the Midi Fault and its equivalent tectonic features (Robas-zynski & Dupuis, 1983).

Oligocene (–Early Miocene?) marine sandstones underlie the Neogene lacustrine deposits. Of the two transgressive periods during the Oligocene, the second, probably the largest, deposited marine sands on the Ardenne Massif and perhaps reached as far south as the Paris Basin. The base of this transgressive sandstone has yielded marine organic-walled microfossils of Oligocene age (Soyer, 1978). During the regression of the Early (–Middle) Mio-cene, karstification began. The dissolution of the Devonian and Carboniferous limestones below the Oligocene sandy permeable cover caused collapse and the formation of small holes (pockets) or basins (Fig. 2) in which the Neogene sediments accumulated. No Neogene deposits could accumulate where the substratum was siliciclastic, as on the Hautes–Fagnes Plateau (H.F.).

The abundance of organic matter that accumulated in the lakes within the karstic pockets and the weathering of pyrite probably enhanced karstic dissolution with acidification of percolating water. Also, aluminum oxides contained in the clay fraction of the lake deposits leached out into the Oligocene sands and, as a result, kaolinite developed within the sands and halloysite grew at the interface between the sandy cover and calcareous basement (Ertus *et al.*, 1989). The karstic holes (pockets) slowly took the shape of small basins. Angles of dip of the basin walls range from over 30° to 90°, in some cases resembling tight vertical folds. Faults were also present along and in the vicinity of the wall of the karstic basins. The horizontal dimensions of the basins vary between about 100 and 1000 m with a vertical depth between 10 and 100 m.

The marine Oligocene sandstones contain *Ophiomorpha* and very rare glauconitic layers, and are covered by two types of continental deposits (Fig. 2, 3, 4). The first contains high quantities of macro-phytic organic matter: sandy peat, lignite with Sequoia, clay layers with Sequoia roots and trunks, and laminated claystone and siltstone (pseudo-varves) with fossil leaves and seeds. This first type of deposit is interpreted to be from a palustrine or/and lacustrine paleoenvironment. The second type of deposit comprises siliciclas-tic sedimentary rocks interpreted to be of alluvial origin. Pale-oenvironmental, sedimentological and mineralogical studies are in progress (Boxus, 1989; Mottart, 1989; Ertus, 1990; Russo Ermolli, 1990).

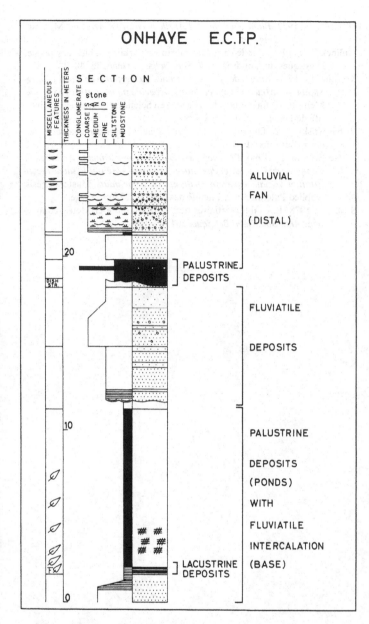

Fig. 3. Detailed lithologic column of the Onhaye E.C.T.P. section in a karst basin in the Synclinorium of Dinant (E.S.M.). Location of outcrop indicated in Figs 1 and 2.

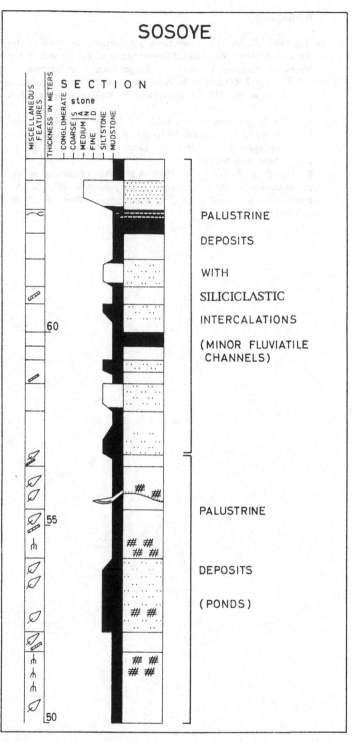

Fig. 4. Detailed lithologic column of the Sosye section in a karst basin in the Synclinorium of Dinant (E.S.M.). Location of outcrop indicated in Figs 1 and 2.

Bibliography

Boxus, P., 1989. *Palynologie quantitative du Tertiaire de Bioul (Entre-Sambre-et-Meuse, Belgique)*. Memoire de fin d'étude de la licence en Sciences botaniques. Université de Liège, Faculté des Sciences.

Calembert, L., 1954. L'Oligocène. In *Prodrome d'une description géologique de la Belgique*, pp. 510–32. Imprimerie H. Vaillant-Carmanne S.A. Liège.

Ertus, R., 1990. *Les néoformations d'halloysite dans les cryptokarsts oligo-miocènes de l'Entre-Sambre-et-Meuse. Approche sédimentologique, pétrographique et minéralogique*. Doctoral dissertation, Faculté Polytechnique de Mons.

Ertus, R., Dupuis, C. & Trauth, N. 1989. A new type of surficial mineral accumulation of meteoric origin: halloysitisation of a silicified limestone in a covered karst. *C.R. Acad. Sciences Paris*, sér. II, **309**, 595–601.

Gilkinet, A., 1922. *Plantes fossiles de l'Argile plastique d'Ardenne*. Mémoire Soc. géol. Belgique, Liège.

Gulinck, M., 1966. Sur le caractere marin de certains sables des poches karstiques du Condroz. *Bull. Soc. Belge de Géol.*, **75**, 48–9.

Mottart, F., 1989. *Etudes des fruits et graines du Tertiaire de Bioul (Entre-Sambre-et-Meuse): implications paléoécologique et stratigraphique*. Mémoire de fin d'étude de la licence en Sciences botaniques, Université de Liège, Faculté des Sciences.

Robaszynski, F. & Dupuis, C., 1983. *Belgique. Guides géologiques régionaux*. Editions Masson, Paris.

Russo Ermolli, E., 1990. *Datation palynologique de gisements tertiaires de Campine, du Condroz et de l'Entre-Sambre-et-Meuse. Essai de reconstitution des paléoenvironnements et des paléoclimats*. Masters Thesis, Applied Paleontology, Faculté des Sciences, Univ. Liège.

Soyer, J., 1978. Les sables tertiaires de l'Entre-Sambre-et-Meuse condrusien. *Ann. Soc. Géol. Belgique*, **101**, 93–100.

Madrid Basin (Neogene), Spain

JOSE P. CALVO,[1] SALVADOR ORDOÑEZ,[1] M. ANGELES GARCIA DEL CURA,[2] MANUEL HOYOS[3] AND ANA M. ALONSO ZARZA[1]

[1] Dpto. Petrología y Geoquímica, Facultad C.C. Geológicas, Universidad Complutense E-28040 Madrid, Spain
[2] Institute of Economic Geology, Consejo Superior de Investigaciones Científicas, E-28040 Madrid, Spain
[3] National Museum of Natural Sciences, Consejo Superior de Investigaciones Científicas, E-28006 Madrid, Spain

The Madrid Basin of central Spain is a fault-bounded intracratonic basin with a complex deformational history. The basin is bordered by Paleozoic granitoids and metamorphic rocks (northern and southern margins), as well as Mesozoic sedimentary formations (eastern margin) (Fig. 1). All basin margins are limited by reverse faults with variable dip angles. Alpine tectonic evolution of the Madrid Basin was characterized by E-W to N 115° E shortening during Early and Middle Miocene times. By contrast, a period of uniaxial (N-S) extension resulting in strike–slip faults is recorded during the Late Miocene and Pliocene. The thickness of Tertiary continental rocks in the basin ranges from 2000 to 3500 m. The Neogene sedimentary record has been divided into three Miocene stratigraphic units (Alberdi et al., 1984; Junco & Calvo, 1983) and two Upper Miocene–Pliocene sedimentary cycles (Calvo et al., 1990), as shown in Fig. 2.

This contribution deals with the three Miocene units. Each Miocene unit is composed of deposits interpreted as alluvial and lacustrine sediments. The lake facies are widespread in central parts of the basin. During the Early and Middle Miocene, the lakes occupied up to 7000 km² in area. The lacustrine facies show remarkable lithological variations from unit to unit, indicating an evolution of lake complexes through time (Calvo et al., 1989b).

Figure 3 summarizes the common sequences and facies associations of the paleolakes recognized in the Miocene units. The Miocene Lower Unit is characterized by an evaporite facies that consists of a variety of saline minerals (Ordóñez et al., 1991). Glauberite and thenardite deposits are exploited from this unit.

The Miocene Intermediate Unit shows remarkable lateral facies transitions from peripheral alluvial systems to marginal lacustrine (paludal) and basinal facies (Calvo et al., 1989b). Towards the top of the unit, an episode of sedimentation in a freshwater lake is recorded (Alonso Zarza et al., 1992; Bellanca et al., 1992). Sepiolite and bentonite deposits are extensively exploited from this unit (Calvo et al., 1986; Ordóñez et al., 1991).

The lacustrine facies of the Miocene Upper Unit consists mainly of freshwater carbonates (biomicrites, oncoidal limestone, tufa) and marls, which were deposited in shallow lakes and fluvio-lacustrine paleoenvironments. These Upper Miocene lakes represent a fresher evolutionary stage in the development of lake complexes through the remainder of the Neogene in the Madrid Basin. The Miocene–Pliocene sedimentary cycles represent short-lived carbonate lakes limited in areal extent.

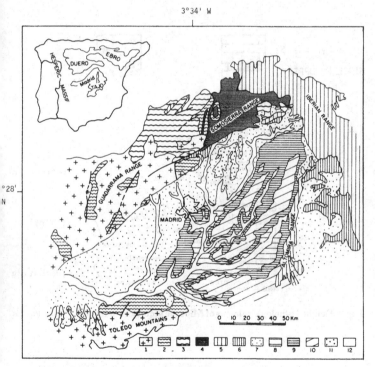

Fig. 1. Locality Map and Facies Distribution
1. plutonic rocks
2. shales, marbles, quartzites and gneisses
3. shales and metagreywackes
4. shales, quartzites and metavulcanites
5. Mesozoic terranes (mainly carbonates)
6. Paleogene (terrigenous and carbonates)
7. Lower to Upper Miocene terrigenous sediments
8. Lower Unit, Miocene
9. Intermediate Unit, Miocene
10. Upper Unit, Miocene
11. Pliocene
12. Quaternary

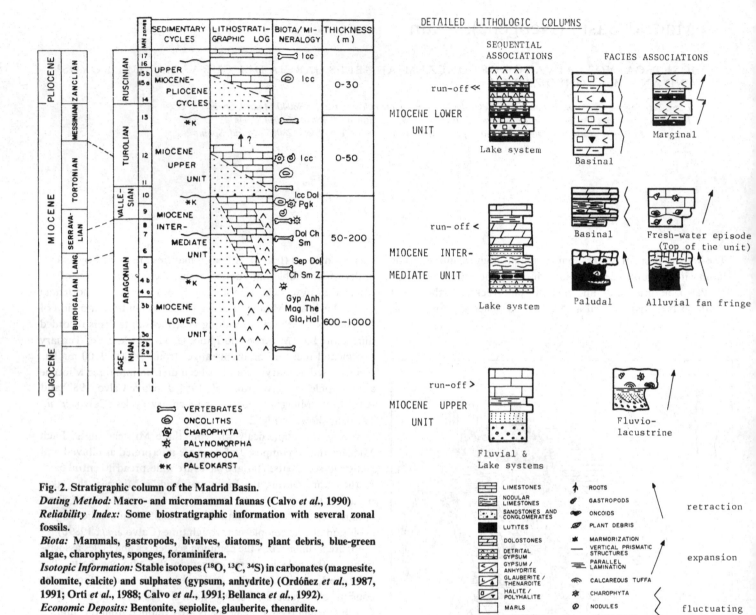

Fig. 2. Stratigraphic column of the Madrid Basin.

Dating Method: Macro- and micromammal faunas (Calvo *et al.*, 1990)

Reliability Index: Some biostratigraphic information with several zonal fossils.

Biota: Mammals, gastropods, bivalves, diatoms, plant debris, blue-green algae, charophytes, sponges, foraminifera.

Isotopic Information: Stable isotopes (^{18}O, ^{13}C, ^{34}S) in carbonates (magnesite, dolomite, calcite) and sulphates (gypsum, anhydrite) (Ordóñez *et al.*, 1987, 1991; Orti *et al.*, 1988; Calvo *et al.*, 1991; Bellanca *et al.*, 1992).

Economic Deposits: Bentonite, sepiolite, glauberite, thenardite.

Environmental Interpretation:

Miocene Lower Unit: Closed perennial saline lake (Ordóñez *et al.*, 1991)

Miocene Intermediate Unit: Closed perennial, slightly saline to freshwater lake complexes (Calvo *et al.*, 1989a, b).

Miocene Upper Unit: Freshwater shallow lake (Calvo *et al.*, 1989b).

Fig. 3. Detailed lithologic columns of Miocene sequences in the Madrid Basin.

Bibliography

Alberdi, M.T., Hoyos, M., Junco, F., López-Martínez, N., Morales, J., Sesé, C. & Soria, D., 1984. Biostratigraphy and sedimentary evolution of continental Neogene in the Madrid area. *Paléobio. Continent.*, **14**, 47–68.

Alonso Zarza, A.M., Calvo, J.P. & García del Cura, M.A., 1986. Sedimentología y petrología de los abanicos aluviales y facies adyacentes en el Neógeno de Paracuellos del Jarama (Madrid). *Estudios geol.*, **42**, 79–101.

Alonso Zarza, A.M., Calvo, J.P. & García del Cura, M.A., 1992. Palustrine sedimentation and associated features – grainification and pseudomicrokarst in the Middle Miocene (Intermediate Unit) of the Madrid Basin, Spain. *Sed. Geol.*, **76**, 43–61.

Bellanca, A., Calvo, J.P., Censi, P., Neri, R., & Pozo, M., 1992. Recognition of lake-level changes in Miocene lacustrine units of the Madrid Basin, Spain. Evidence from facies analysis, isotope geochemistry, and clay mineralogy. *Sed. Geol.*, **76**, 135–53.

Calvo, J.P., Alonso Zarza, A.M. & García del Cura, M.A., 1986. Depositional sedimentary controls on sepiolite occurrence in Paracuellos de Jarama, Madrid Basin. *Geogaceta*, **1**, 25–8.

Calvo, J.P., Alonso Zarza, A.M. & García del Cura, M.A., 1989a. Models of Miocene marginal lacustrine sedimentation in response to varied depositional regimes and source areas in the Madrid Basin (central Spain). *Palaeogeogr., Palaeoclimatol., Palaeoecol.*, **70**, 199–214.

Calvo, J.P., Hoyos, M., Morales, J. & Ordóñez, S., 1990. Neogene stratigraphy, sedimentology and raw materials of the Madrid Basin. *Paleont. Evol., Spec. Mem.*, **2**, 61–95.

Calvo, J.P., Ordóñez, S., García del Cura, M.A., Hoyos, M., Alonso Zarza, A.M., Rodríguez Aranda, J.P., & Sanz, E., 1989b. Sedimentología de los complejos lacustres miocenos de la Cuenca de Madrid. *Acta Geol. Hisp.*, **24**, 281–98.

Calvo, J.P., Ordóñez, S., García del Cura, M.A., Alonso Zarza, A.M., Bustillo, M., & Fort, R., 1991. Composición isotópica de carbonatos en las sucesiones neógenas de la Cuenca de Madrid: valoración de los datos existentes. *I Congr. Grupo Español Terciario, Comunicaciones*, pp. 59–62.

Calvo, J.P., Pozo, M. & Servant-Vildary, S., 1988. Lacustrine diatomite deposits in the Madrid Basin (central Spain). *Geogaceta*, **4**, 14–17.

García del Cura, M.A., Ordóñez, S. & López Aguayo, F., 1979. Estudio petrológico de la 'Unidad Salina' de la Cuenca del Tajo. *Estudios Geol.*, **35**, 325–39.

Junco, F. & Calvo, J.P., 1983. Cuenca de Madrid. In *Geología de España*, vol. 2, pp. 534–43. Instituto Geológico y Minero de España, Madrid.

Megías, A.G., Ordóñez, S., Calvo, J.P. & García del Cura, M.A., 1982. Sedimentos de flujo gravitacional yesíferos y facies asociadas en la cuenca neógena de Madrid, España. *V Congr. Latinoam. Geología, Buenos Aires*, **2**, 311–28.

Ordóñez, S., Calvo, J.P., García del Cura, M.A., Alonso Zarza, A.M. & Hoyos, M., 1991. Sedimentology of sodium sulphate and special clays from the Tertiary Madrid Basin (Spain). In *Lacustrine Facies Analysis*, ed. P. Anadón, L. Cabrera & K. Kelts, pp. 39–55. IAS Special Publication No. 13. Blackwell Scientific Publ., Oxford.

Ordóñez, S., Fontes, J.Ch. & García del Cura, M.A., 1987. Estudio isotópico de la paragénesis sulfatada sódica, calcosódica y cálcica de la Unidad Salina de la Cuenca de Madrid. *II Cong. Geoquímica España, Soria, Abstracts*, pp. 90–8.

Orti, F., Rosell, L., Utrilla, R., Inglés, M., Pueyo, J.J., & Pierre, C., 1988. Reciclaje de evaporítas en la peninsula Iberíca durante el ciclo alpino. *II Congreso Geol. España, Granada, Comunicaciones*, **1**, 421–4.

Pliocene lacustrine basin of Villarroya (Iberian Ranges, northern Spain)

ARSENIO MUÑOZ, ANTONIO PEREZ AND JOAQUIN VILLENA

Facultad de Ciencias, Departamento de Geologia, Universidad de Zaragoza, E-50009 Zaragoza, Spain

The Villarroya Basin is located in the southeastern part of La Rioja Province in northern Spain (Fig. 1) in a zone called the 'Tectonized Fringe' (Durantez *et al.*, 1982). This zone separates the Cameros Mountains (northwestern sector of the Iberian Ranges) from the Tertiary Ebro Depression. The basin is a half-graben (Fig. 2) which was generated by distensive movement.

The Villarroya Basin is quite small (6 km long by 2 km wide) and filled with continental sedimentary rocks (Brinkmann, 1957), which are interpreted as alluvial to lacustrine–palustrine deposits (Figs 3, A, B and 4). These deposits have been dated as Upper Pliocene

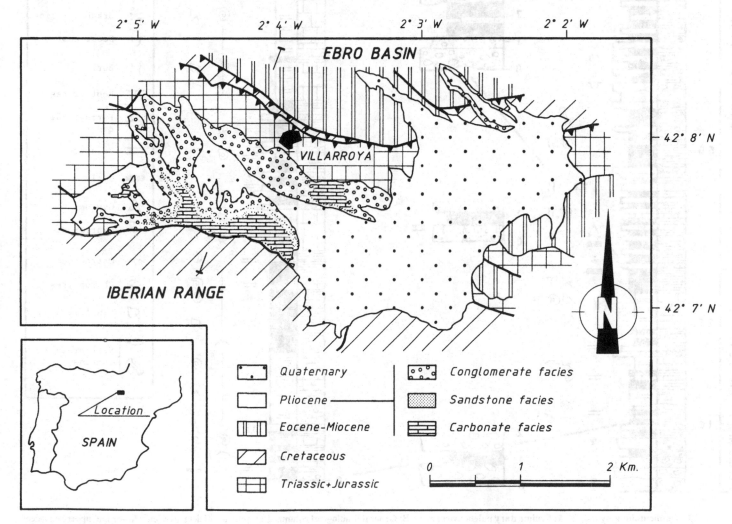

Fig. 1. Locality map and facies distribution of the Pliocene Villarroya Basin.

307

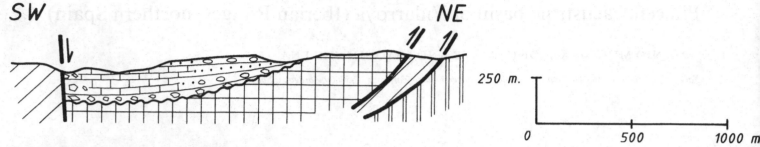

Fig. 2. Schematic cross-section of the Villarroya Basin showing geometry and facies distribution (see Fig. 1 for location of cross-section and legend).

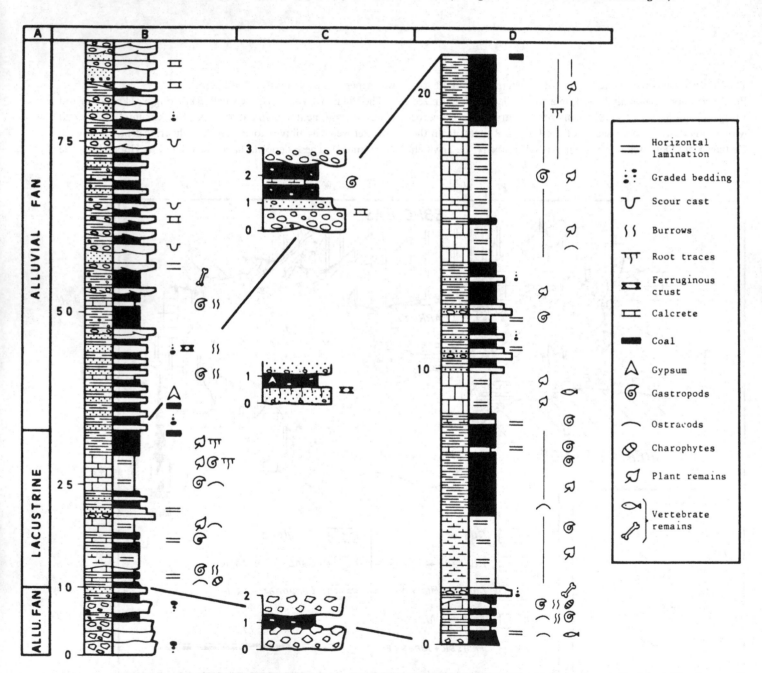

Fig. 3. Sedimentologic synthesis. **A:** Sedimentary paleoenvironment. **B:** General lithological column. **C:** Cycles in siliciclastic facies: lower and upper sequences interpreted as alluvial fan deposits; middle sequence interpreted as interbedded high energy (turbidites) and low energy deposits in the lacustrine environment. **D:** Detailed section of lacustrine facies. Scales in meters.

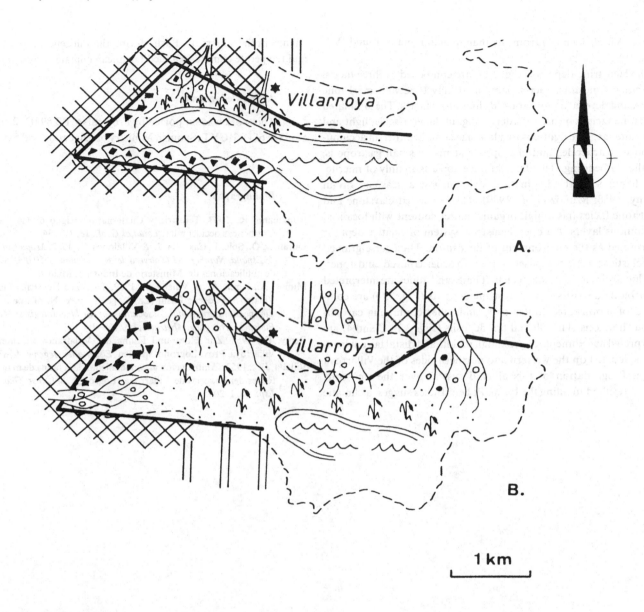

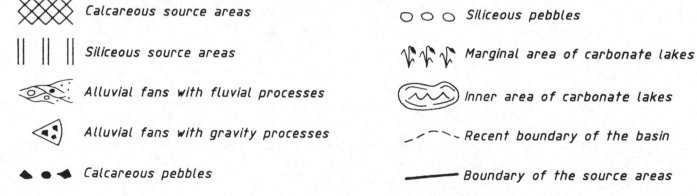

Fig. 4. Schematic paleogeographic representations of the Pliocene Villarroya Basin. A: Lower part of basin sequence. B: Upper part of basin sequence.

(Lower Villafranchian) from the mammalian remains found (Villalta, 1952).

The lacustrine deposits (Fig. 3,D) are composed of three facies: laminated limestone facies, laminated claystone and marlstone facies, and detrital coarse-grained siliciclastic facies. The laminated limestone facies consists of alternating, millimeter-scale, light and dark laminae. The light-colored laminae contain either an accumulation of ostracodes and charophyte stems or small gastropods, parallel to bedding. The dark laminae consists mainly of micrite. The features of the lamination style suggest a seasonal origin (Remy, 1958; Muñoz *et al.*, 1989). The laminated claystone and marlstone facies has a high organic matter content with localized bituminous layers. An open lacustrine system of relative depth is interpreted as the environment of deposition. The coarse-grained siliciclastic facies is composed of grey- to ocher-colored sandstones, that locally include angular clasts. These sandstones are interpreted as turbidite deposits. The palustrine deposits (Fig. 3,D) are composed of bioturbated lutites and sandstones containing carbonaceous horizons. The alluvial fan deposits (Figs 3,B, C and 4) are interpreted as sediments of proximal, middle and distal fan environments, located on the western and northern sides of the Villarroya Basin. Progradation of this alluvial system over the lacustrine system resulted in filling the basin. Pollen studies suggest a climatic change in the Villarroya Basin during the Pliocene from a humid and temperate climate to a mediterranean climate (Remy, 1958).

Acknowledgements

This work was supported by Project No. PB 89-0342 and PB 89-0344 of DGICYT (Spanish Ministry of Education and Science).

Bibliography

Brinkmann, R., 1957. Terciario y Cuaternario antiguo de las Cadenas Celtibéricas occidentales. *Estudios Geol.*, **13**, 123–34.

Durantez, O., Solé, J., Castiella, J., & Villalobos, L., 1982. *Mapa Geológico de España. Number 281 (Cervera del Rio Alhama) 1:50,000.* Servicio de publicaciones del Ministerio de Industria, Madrid.

Muñoz, A., Pérez, A., & Villena, J., 1989. The open lacustrine facies of Villarroya Pliocene Basin (La Rioja Province, N. of Spain). *10th European International Association of Sedimentologists Meeting, Budapest, Hungary, Abstracts*, pp. 168–9.

Remy, H., 1958. Zur Flora und Fauna der Villafranca-schichten von Villarroya, Prov. Logroño/Spanien. *Eiszeitalter Gegenw.*, **9**, 83–103.

Villalta, J.F., 1952. Contribución al conocimiento de la fauna de mamíferos fósiles del Plioceno de Villarroya (Logroño). *Bol. Inst. Geol. Min. Esp.*, **64**, 1–204.

The Miocene lacustrine evaporite system of La Bureba (western Ebro Basin, Spain)

P. ANADÓN

Institut de Ciències de la Terra 'J. Almera', (CSIC), C. Martí y Franquès s. n., E-08028 Barcelona, Spain

Geological setting – facies distribution

The Miocene lacustrine evaporites of La Bureba are located in the western Ebro Basin, northeastern Spain. The Tertiary Ebro Basin is the southern foreland basin of the Pyrenees. The western sector of the Ebro Basin is bounded by the southern Pyrenean nappes to the north (Sierra de Cantabria) and the Sierra de la Demanda to the south. This range belongs to the Iberian Chain, an intraplate Tertiary fold-thrust system. The non-marine depositional systems in the western Ebro Basin were related to the basin margin history of both the Pyrenees and the Iberian Chain. The depositional systems evolved simultaneously with the overthrust and nappe emplacement of the western Pyrenees and the Sierra de la Demanda (Fig. 1). The depositional framework during the Early to Middle Miocene was formed by alluvial systems related to the active basin margins (La Demanda and La Bureba alluvial systems) and a lacustrine system which was located in the center of the latest depositional trough (i.e., La Bureba lacustrine system, Figs 1 and 2). The dating of the La Bureba lacustrine deposits (Middle to Upper Miocene) has been obtained by stratigraphic interpolation and lithological correlation with some mammal sites (Crusafont *et al.*, 1966; Santafé *et al.*, 1982).

General stratigraphy of the La Bureba Neogene lacustrine system

The lacustrine evaporite system of La Bureba consists of several facies associations. The central lacustrine facies are mainly constituted by sulfate-dominated evaporites (gypsum, anhydrite, glauberite). These facies are bounded to the north and to the south by mud- and sand-flat facies. To the west, the evaporite lacustrine facies change laterally into carbonate-dominated shallow lacustrine facies (Riba, 1955a; Portero *et al.*, 1979).

The evaporite facies, the Cerezo de Rio Tirón Gypsum, is in the center of the basin and is about 250 m thick (Fig. 3). The lower part of this unit, over 70 m thick, is formed by an alternation of evaporite packets, each up to 8 m thick, and siltstone-dominated intervals (Ríos, 1963). These evaporite packets are composed of massive and banded glauberite and nodular anhydrite. The mudstone-dominated intervals contain grey mudstones and nodular anhydrite (Ordóñez *et al.*, 1982; Menduiña *et al.*, 1984). Chloride evaporites have not been reported from the La Bureba lacustrine evaporite system (Ortí, 1982).

The upper part of the evaporite succession, which crops out extensively, is formed by diverse gypsum lithofacies and interbedded mudstone beds (Anadón, 1990). In the upper parts of the evaporite succession, the gypsum is mainly primary (i.e., does not replace pre-existing evaporites), whereas in the lower and intermediate parts, the gypsum that occurs in outcrop is secondary (i.e., after hydration of anhydrite or replacement of glauberite). In some outcrops of the intermediate parts of the evaporite unit, both primary and secondary gypsum coexist.

Detailed representative sections

Gypsum deposits are the main constituents in outcrop of the evaporite lacustrine lithofacies of La Bureba. Glauberite deposits are typically found in open cast mines and, in general, are not found in outcrop due to their transformation by weathering into gypsum. Figure 4 shows detailed sections of some of the most representative gypsum-dominated lacustrine sequences. Column 1 corresponds to the mudstone- and sandstone-dominated successions of the transition between distal facies of the alluvial system of La Bureba and the evaporite lacustrine facies. In this section, siliciclastic, alluvial deposits predominate and are composed of an alternation of red, green and grey mudstones, and sandstones. The sandstone beds show sedimentary structures indicative of traction and loading. The gypsum is secondary, after anhydrite, and the nodules were formed in a sabkha-like environment which surrounded the central areas of the evaporite lacustrine system.

Column 2 is located in a more central position in the basin framework than column 1. Column 2 corresponds to a typical sabkha-saline lake sequence formed by a cyclic succession of: (1) massive, or nodular alabastrine gypsum, (2) laminated (alabastrine) gypsum and (3) grey to brown thin-bedded mudstones, which in some places include enterolithic gypsum horizons. Alabastrine gypsum (secondary after anhydrite) has a microcrystalline texture, although some beds have crystalline fabric (crystal size 1–2 mm in

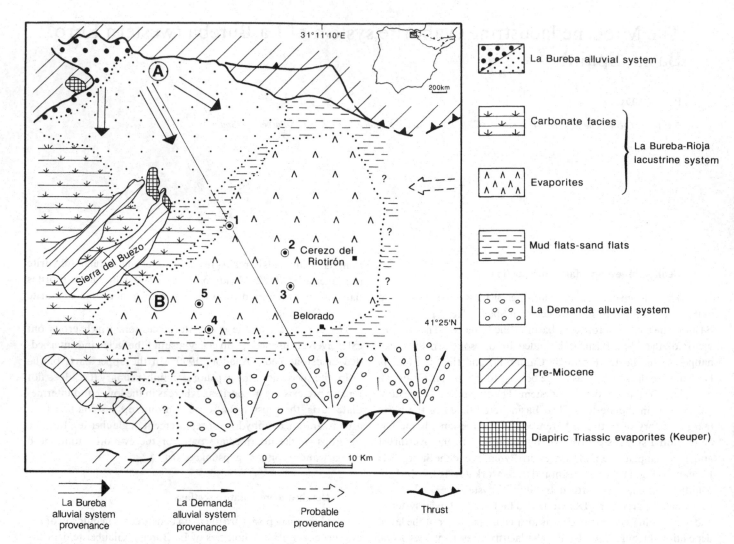

Fig. 1. Schematic facies distribution and paleogeography of the Neogene non-marine deposits of the western Ebro Basin. Modified after Anadón (1990). Location of the detailed lithologic columns of Fig. 4, and schematic cross-sections of Fig. 2, are indicated.

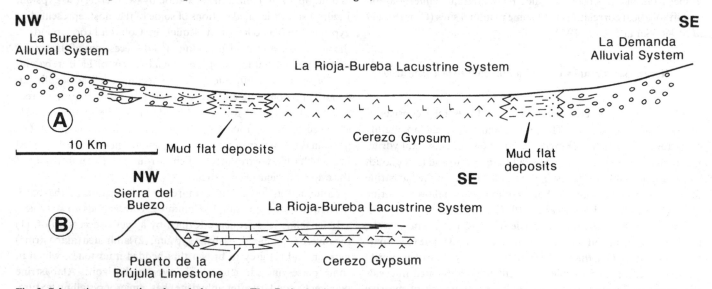

Fig. 2. Schematic cross-sections through the western Ebro Basin showing the relationship among the diverse depositional systems. After Anadón (1990). See location in Fig. 1. Vertical exaggeration not to scale.

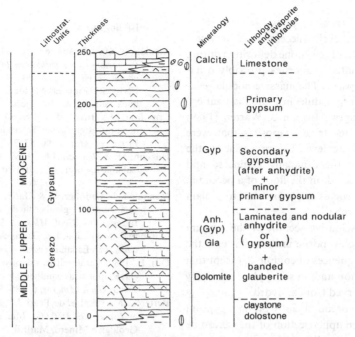

Fig. 3. General stratigraphic log of La Bureba lacustrine evaporite deposits. The age assignment is based on several mammal sites located in the alluvial facies near the basin margin. The mammal sites are stratigraphically above and below the lacustrine facies of the La Bureba lacustrine system (Crusafont *et al.*, 1966; Santafé *et al.*, 1982).

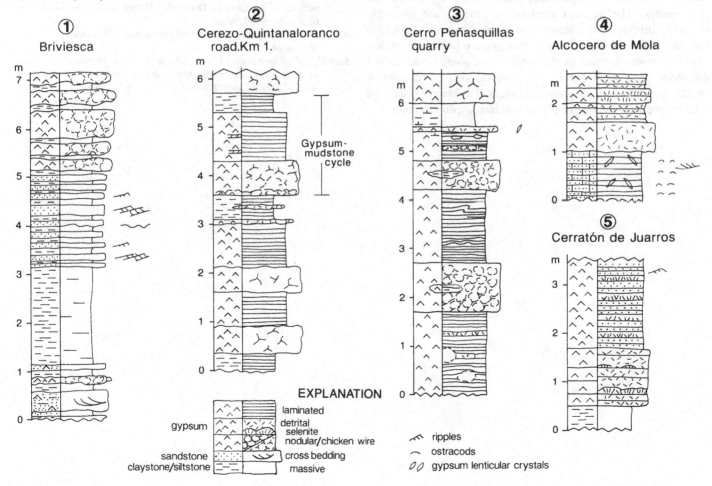

Fig. 4. Detailed lithologic columns of the La Bureba lacustrine evaporite deposits at five different sites. (See Fig. 1.)

diameter). In thin sections, tiny relics of anhydrite have been commonly observed. The described cyclic succession can be interpreted as resulting from evaporite and mudstone deposition during lake-level oscillations. The laminated evaporites probably originated as subaqueous, primary gypsum. The massive-nodular evaporites probably formed as anhydrite nodules in a 'sabkha-tization' process during low lake-level stages. Hussain & Warren (1989) describe Pleistocene laminated lacustrine evaporites that were 'sabkhatized' by a fall in the lake water level. This process led to the growth of nodular evaporites in the originally laminated and layered sediments that were deposited on the floor of a perennial Pleistocene brine lake. A similar process can be invoked to explain some gypsum sequences of La Bureba.

Column 3 is a detailed representative section of the gypsum deposits of the middle part of the evaporite unit. In this case the succession resembles the cyclic sequences of column 2, except that the mudstone beds are less common and, in some places, primary gypsum horizons have been preserved from anhydritization.

Columns 4 and 5 correspond to predominantly primary gypsum lithofacies which are typical of the upper section of the Cerezo de Rio Tirón Gypsum unit. These gypsum lithofacies crop out predominantly in the western part of the evaporite lacustrine basin. Column 4 represents gypsrudites (clastic selenite gypsum) and gypsarenites. The gypsum crystals, in general, are randomly oriented, although some horizons of vertically-growing selenites also have been observed. Some detrital gypsum beds show ripples.

In the uppermost part of the gypsum unit, some limestone and dolostone beds occur (column 5). The carbonate beds contain bioclastic accumulations of ostracod shells, that in some places, are dolomitized and cemented by large poikilitic gypsum crystals.

Bibliography

Anadón, P., 1990. Yesos de Cerezo. In *Guía práctica para el estudio de las formaciones evaporíticas en la Cuenca del Ebro y sistemas circundantes, y en Levante*, ed. F. Ortí & J.M. Salvany, pp. 127–31. ENRESA (Empresa Nacional de Residuos S.A.), Depto. de Geoquímica, Petrología y Prospección Geológica. Universitat de Barcelona.

Crusafont, M., Truyols, J. & Riba, O., 1966. Contribución al conocimiento de la estratigrafía del Terciario de Navarra y Rioja. *Notas y Com. Inst. Geol. Minero*, **90**, 53–76.

Hussain, M. & Warren, J.K., 1989. Nodular and enterolitic gypsum: the 'sabkha-tization' of Salt Flat playa, west Texas. *Sed. Geol.*, **64**, 13–24.

Menduiña, J., Ordóñez, S. & García del Cura, M.A., 1984. Geología del yacimiento de glauberita de Cerezo del Rio Tirón (provincia de Burgos). *Bol. Geol. Minero*, **95**, 33–51.

Ordóñez, S., Menduiña, J. & García del Cura, M.A., 1982. El sulfato sódico natural en España. *Tecniterrae*, **46**, 16–32.

Ortí, F., 1982. Características deposicionales y petrológicas de las secuencias evaporíticas continentales en las cuencas terciarias peninsulares. *Temas Geol. Mineros*, **6**, 485–508.

Portero, J.M., Ramírez del Pozo, J. & Hernández, A., 1979. *Mapa Geológico de España, E. 1:50,000*. Memoria y Hoja no. 168. Briviesca. Inst. Geológico Minero, Madrid.

Riba, O., 1955a. Sur le type de sedimentation du tertiaire continental de la partie ouest du bassin de L'Ebre. *Geol. Rundschau*, **43**, 363–70.

Riba, O., 1955b. Sobre la edad de los conglomerados terciarios del borde norte de las Sierras de La Demanda y Cameros. *Notas y Comun. Inst. Geol. Minero*, **39**, 39–50.

Ríos, J.M., 1963. *Materiales salinos del suelo español*. Memorias Inst. Geol. Minero 64.

Santafé, J.V., Casonovas, M.L. & Alférez, F., 1982. Presencia del Vallesiense en el Mioceno continental de la Depresión del Ebro. *Rev. Real Acad. Ciencias Exact. Fís. Nat. Madrid*, **76**, 277–84.

The Miocene Ribesalbes Basin (eastern Spain)

P. ANADÓN

Institut de Ciències de la Terra (J. Almera), CSIC, c. Martí i Franqués s.n., E-08028 Barcelona, Spain

Geological setting – facies distribution

The Ribesalbes Basin (Fig. 1) is a fault-bounded Neogene basin located in the southern Maestrat (eastern Iberian Chain, Spain). The southern Maestrat comprises a carbonate-dominated Mesozoic cover up to 1000 m thick, and an Hercynian (pre-Upper Carboniferous) basement. This area was affected during the Neogene by the widespread rifting event which formed the Western European Rift System. The present structure of the Eastern Iberian Chain corresponds to a Neogene system of horsts and grabens bounded by northeast–southwest extensional faults. These extensional structures are clearly superimposed on the Paleogene compressive structures of the eastern Iberian Chain.

The Miocene Ribesalbes–Alcora Basin, 150 km² in area, is bounded by east-northeast–west-southwest to north-northeast–south-southwest normal faults. Several Mesozoic fault-bounded blocks crop out within the graben. The Neogene deposits in the Ribesalbes-Alcora area comprise two main sequences: the Lower to Middle Miocene Ribesalbes sequence (up to 600 m thick) and the Upper Miocene Alcora sequence (up to 200 m thick). The Alcora sequence is interpreted as alluvial deposits, whereas the Ribesalbes sequence contains lacustrine and alluvial deposits. The Ribesalbes sequence unconformably overlies the Mesozoic succession, and itself is conformably overlain by the Alcora sequence which occasionally onlaps the Mesozoic rocks of surrounding horsts.

General stratigraphy of the Ribesalbes sequence

The most complete succession of the Ribesalbes sequence is best observed close to the village of Ribesalbes. In this zone, the Ribesalbes Neogene sequence is composed of five units (Fig. 2).

Unit A. Maximum thickness 300 m, consists of pebble to boulder breccias of Mesozoic limestone clasts with minor interbedded red mudstones and sandstones. The breccias are poorly sorted and display disorganized fabric, and lack internal stratification. The unit has been interpreted as mass flow dominated alluvial fan deposits.

Unit B (Ribesalbes Dolostones). Maximum thickness 100 m comprises laminated dolostones with interbedded massive or laminated brown mudstones and sandstones. Organic-rich beds (oilshales) are abundant. The existence of slump horizons and olistholists of Cretaceous rocks are probably related to synsedimentary tectonism. This unit is interpreted as resulting from deposition in an anoxic bottom of a meromictic lake (see next section).

Unit C. Maximum thickness 90 m, formed by massive and laminated grey mudstones with interbedded dolostone and sandstone packets. Unit C records an alternation of shallow-water sedimentation phases (sandstone-rich intervals) and deeper sedimentation phases (laminated mudstones and dolostones) in a lake basin (see next section).

Unit D. Maximum thickness 70 m comprises an olisthostrome of disorganized, heterometric large blocks of Cretaceous limestones.

Unit E. Maximum thickness 25 m, comprised of thin-bedded limestones with thin interbedded mudstones. The limestones contain ostracodes and charophytes. This unit has been interpreted as having formed in a shallow offshore lacustrine environment.

The Ribesalbes megasequence has been interpreted as recording a graben formation phase, with related coarse alluvial fan deposits (unit A) overlain by a carbonate-dominated lacustrine sequence originating from a meromictic lake (unit B). The lacustrine system experienced significant water-level fluctuations recorded in unit C. The olisthostrome unit (D) evidences a renewed tectonic activity in the area previous to a later shallow water sedimentation phase (unit E).

The age determinations are based on several mammal sites located in C unit mudstones near Araya, and fossil plant leaves in the Ribesalbes unit near the village of Ribesalbes (reliability index B). The fossil mammals indicate a MN 4 biozone (Lower Aragonian = Lower to Middle Miocene; Agustí *et al.*, 1988).

Representative detailed sections

The dolostones in Ribesalbes unit B are mainly formed by Mg-poor dolomite. Some dolostone beds contain minor, variable amounts of calcite and aragonite. Early diagenetic CT-opal has also been found. Figure 3 shows a representative section of the Ribesalbes Dolostone (unit B), formed by an alternation of dolostone, dolomitic marls and sandstones. Organic-rich horizons are present

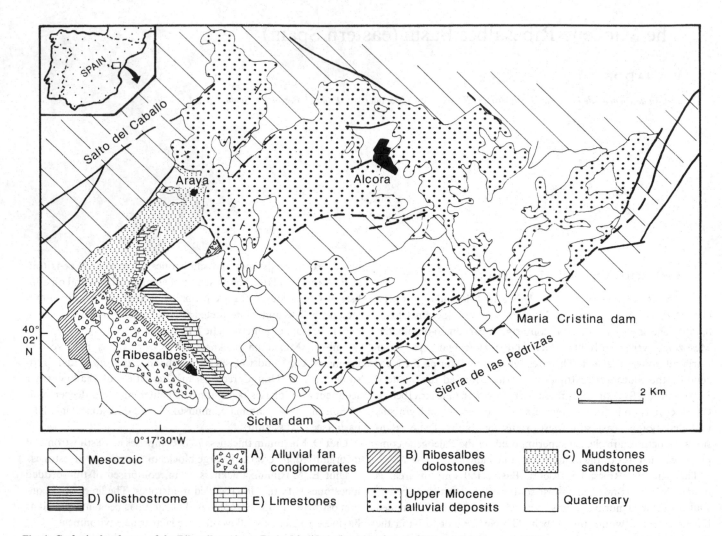

Fig. 1. Geologic sketch map of the Ribesalbes-Alcora Basin. Modified after Agustí *et al.*, 1988 and Anadón *et al.*, 1989.

in the dolostone and marl packets. The dolostones display fine laminations, well-preserved plant leaves, insects, and amphibian skeletons, and lack benthonic fauna and bioturbation. These features suggest that these deposits originated under the anoxic bottom of a meromictic lake. The organic-rich facies consist of two rock-types: laminated mudstones and dolostones. These facies have an oil yield up to 250 1/MT. The dolostones display total organic carbon (TOC) contents ranging from 1% to 15% and HI > 700 mgHC/gTOC. The organic matter has a mixed algal/plant origin.

The representative section for Unit C (Fig. 3) shows grey mudstones with interbedded dolostones and sandstones. The dolo-

stones show similar features to those observed in the Ribesalbes Dolostone (Unit B). Sandstone-dominated intervals are comprised of an alternation of sandstone beds up to 25 cm with thin mudstones. The sandstone beds show horizontal- and cross-lamination and have wave-rippled tops. Some sandstone beds are amalgamated and flaser structures are common. Horizontal burrows are also present. This section records an alternation of development under: (1) relatively deep, anoxic bottom conditions (laminated mudstones and dolostones) and (2) shallower, oxic bottoms under wave and current actions (sandstone beds).

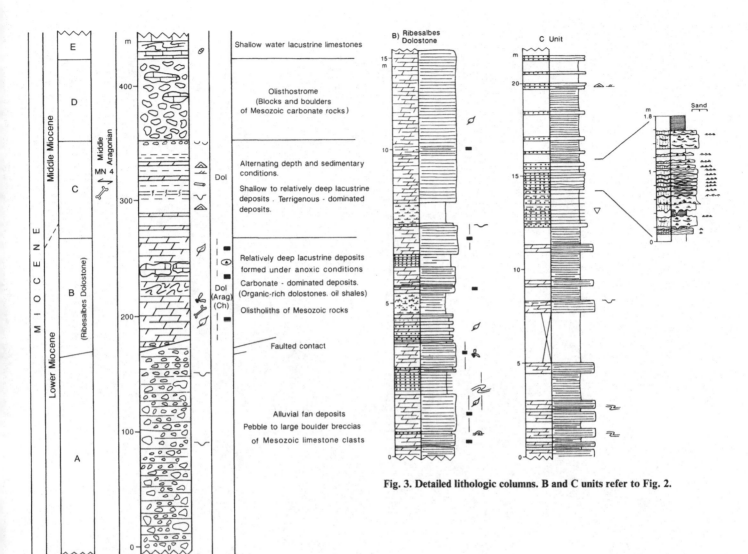

Fig. 3. Detailed lithologic columns. B and C units refer to Fig. 2.

Fig. 2. General stratigraphic log of the Ribesalbes sequence.

Bibliography

Agustí, J., Anadón, P., Ginsburg, L., Mein, P. & Moissenet, E., 1988. Araya et Mira: Nouveaux gisements de Mammifères dans le Miocène infèrieur-moyen des chaines iberiques orientales et mediterra-neennes. Consequences stratigraphiques et tectoniques. *Paleontol. Evol.*, **22**, 83–101.

Anadón, P., Cabrera, L., Julià, R., Roca, E. & Rosell, L., 1989. Lacustrine oil-shale basins in Tertiary grabens from NE Spain (Western European rift system). *Palaeogeogr., Palaeoclimatol., Palaeoecol.*, **70**, 7–28.

Anadón, P., Cawley, S.J. & Julià, R., 1988. Oil source rocks in lacustrine sequences from Tertiary grabens, Western Mediterranean Rift System, Northeast Spain. *Am. Asoc. Petrol. Geol. Bull.*, **72**, 983.

Bastida, J., 1979. *Arcillas cerámicas de Araya y Villar. Geología, Mineralogía y Físico-Química*. Cerámica Azuvi-Dept. Cristalografía, Univ. Auton. Barcelona.

De las Heras, X., 1989. *Geoquímica orgánica de conques lacustres fòssils*. PhD Thesis, University of Barcelona.

Faura y Sans, M., 1914. *Informe sobre la cuenca petrolífera de Ribesalbes (Provincia de Castellón), y en particular de las minas de Disodila que en San Chils explota la 'Compañía de Aceites de Esquisto, S.A.'* Unpublished report.

Fernández Marrón, M.T. & Alvarez-Ramis, C., 1967. Contribución al estudio de las gimnospermas fósiles del Oligoceno de Ribesalbes (Castellón). *Estudios Geol.*, **23**, 155–61.

Hernández-Sampelayo, P. and Cincunegui, M., 1926. Esquistos bitumino-sos de Ribesalbes. *Bol. Inst. Geol. Minero España*, **46** (6), 1–86.

I.G.M.E., 1981. *Investigación de pizarras bituminosas en la cuenca terciaria de Ribesalbes (Castellón)*. Unpublished Report, Inst. Geol. Minero España.

Permanyer, A. & García-Vallés, M., 1986–87. Generación de hidrocarburos en sedimentos lacustres terciarios: las cuencas de Ribesalbes y Campins (NE de España). *Rev. Inv. Geol.*, **42/43**, 23–44.

The Cenajo and Las Minas-Camarillas basins (Miocene), southeastern Spain

JOSÉ P. CALVO[1] AND EMILIO ELIZAGA[2]

[1] Dpto. Petrología y Geoquímica, Facultad C. C. Geológicas, Universidad Complutense E-28040, Madrid, Spain
[2] Dpto. Geológicas, ITGE, Ríos Rosas 23, E-28003 Madrid, Spain

The Cenajo and Las Minas–Camarillas basins are small, fault-bounded basins which developed in southeastern Spain during the Late Miocene (Fig. 1). The sedimentary fill is similar in the two basins, being composed of two tectono-sedimentary units (Fig. 2). The episode of breakdown of the lacustrine platforms (base of the upper unit) was closely related to the extrusion of ultrabasic volcanic rocks (lamproites) (Elizaga & Calvo, 1989).

Figure 3A contains the generalized cross-section of the Las Minas–Camarillas basins showing the main sedimentary units, faults and associated volcanism. The lower unit is interpreted as basal fan-delta and turbidite clastic deposits below an alternation of finer-grained channel fills, marls and dolostones (Figs 2, 3B). This sequence grades upwards into gypsum deposits that contain diagenetic sulfur. Figure 3C shows a section of the gypsum facies showing the various types of gypsum. The top of the lower unit consists of a monotonous succession of marl-limestone sequences which intercalate with thin lenses of sandstone. Diatomaceous marls become more abundant towards the top of this lower unit.

The lower part of the upper unit is interpreted as turbidites and megaslumps (up to 50 m in thickness). Re-sedimented diatomite slices form most of the megaslumps, that in turn is overlain by a monotonous succession of laminated diatomites (currently economically exploited) and limestones (Fig. 3D). Porcelanites occur as recrystallized diatomites. The numbers in Fig. 3D correspond to samples analyzed in Bellanca et al. (1989). These samples evidence a change from heavier δ18 values from aragonite-rich mixtures in the diatomites to higher values in calcite-rich mixtures of limestones toward the top of the sequence. In some areas, the uppermost part of the upper unit consists of deltaic deposits (Fig. 3E) that in places are composed of coarsening-upwards sandstone bodies, mudstones and paludal carbonates (Fig. 3F).

Dating method

(1) Isotopic determination from volcanic rocks interbedded with lacustrine sequences (Bellon et al., 1981).
(2) Mammal faunas (Calvo et al., 1978).

Isotopic information

Stable isotopes (^{18}O, ^{13}C) in carbonate-diatomite cycles and evaporite facies (Bellanca et al., 1989; Pierre et al., 1989).

Economic deposits

Diatomite (currently exploited), sulfur (abandoned exploitation), and oil shales (under investigation).

Environmental interpretation

Shallow (evaporite facies) to moderately deep, meromictic lake (carbonate-diatomite cycles).

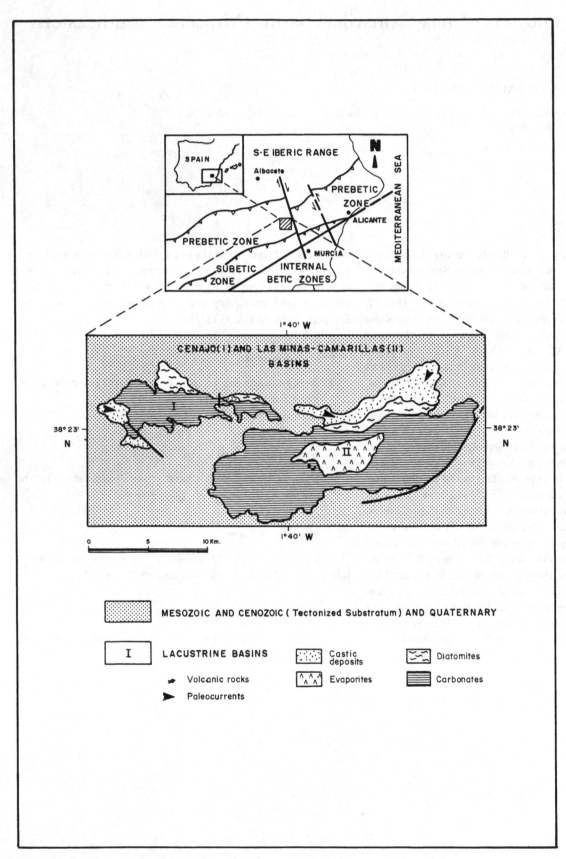

Fig. 1. Locality map and facies distribution.

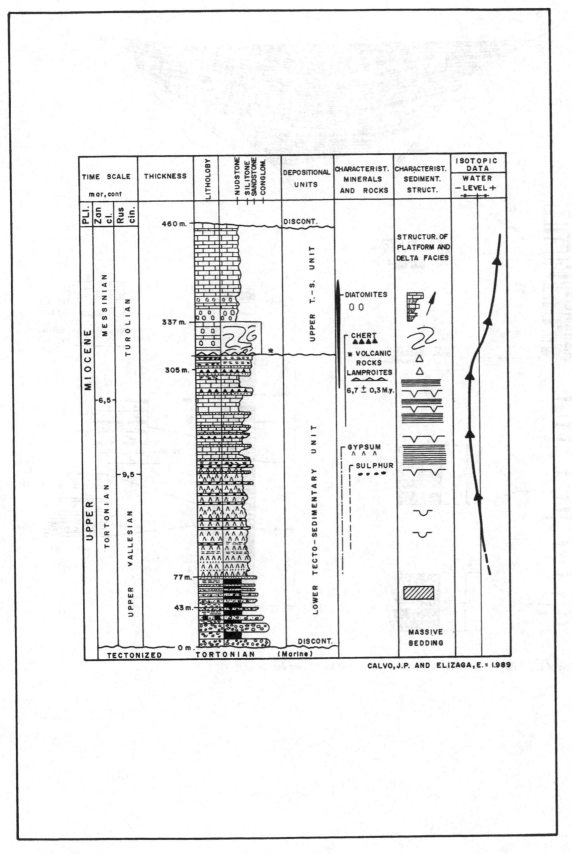

Fig. 2. General stratigraphic column of Cenajo and Las Minas Camarilla Basins.

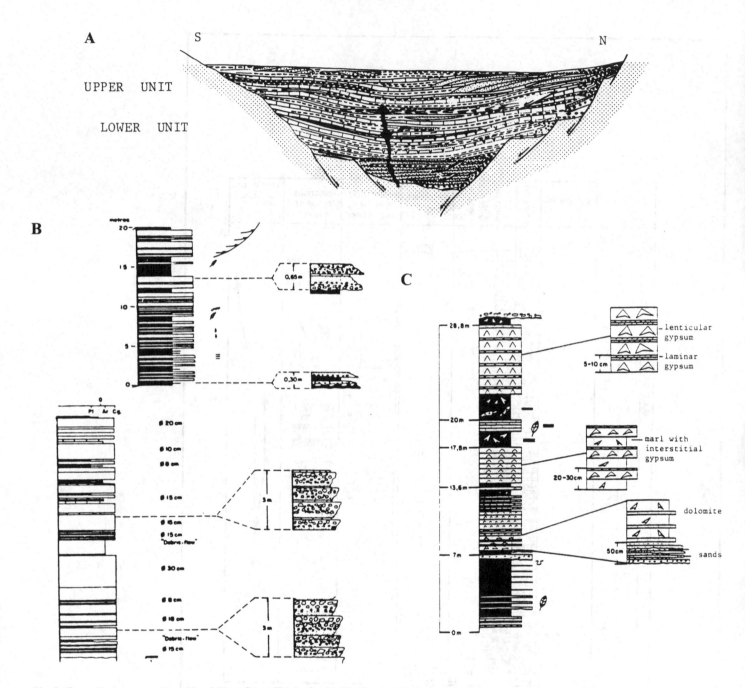

Fig. 3. Generalized cross-section of Las Minas–Camarillas basins (A–F). (See text for further details.)

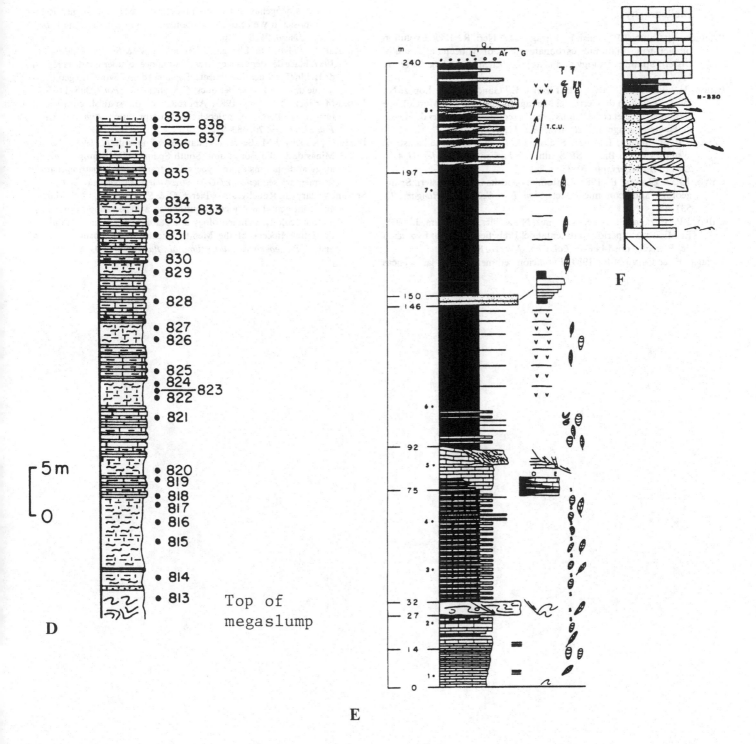

Top of
megaslump

Fig. 3 (*cont.*)

Bibliography

Bellanca, A., Calvo, J.P., Censi, P., Elizaga, E. & Neri, R., 1989. Evolution of lacustrine diatomite carbonate cycles of Miocene age, southeastern Spain: petrology and isotope geochemistry. *J. Sed. Petrol.*, **59**, 45–52.

Bellon, H., Bizon, G., Calvo, J.P. Elizaga, E., Gaudant, J. & Lopez, N., 1981. Le volcan du Cerro del Monagrillo (province de Murcie): age radiométrique et corrélations avec les sediments néogénes du Bassin de Hellin (Espagne). *C.R. Acad. Sci. Paris*, **292**, 1035–8.

Calvo, J.P. & Elizaga, E., 1985. Sedimentology of a Neogene lacustrine system, Cenajo Basin, SE Spain. *6th European Reg. Meet. I.A.S., Lleida, Abstracts*, pp. 70–3.

Calvo, J.P. & Elizaga, E., 1987. Diatomite deposits in southeastern Spain: geologic and economic aspects. *Ann. Inst. Geol. Publ. Hungary*, **70**, 537–43.

Calvo, J.P., Elizaga, E., Lopez-Martínez, N., Robles, F. & Usera, J., 1978. El Mioceno superior continental del Prebético Externo: Evolución del Estrecho nord-bético. *Bol. Geol. Minero*, **89**, 9–28.

Elizaga, E. & Calvo, J.P., 1989. Evolución sedimentaria de las cuencas lacustres neógenas de la zona prebética (Albacete, Espana). Relación, posición y efectos del vulcanismo durante la evolución. *Bol. Geol. Minero*, **96**, 837–46.

Foucault, A., Calvo, J.P., Elizaga, E., Rouchy, J.M. & Servant-Vildary, S., 1987. Place des dépots lacustres d'age miocéne supérieur de la région de Hellin (Province de Albacete, Espagne) dans l'évolution géodynamique des Cordilléres bétiques. *C.R. Acad. Sci. Paris*, **305**, 1163–6.

López-Martínez, N. *et al.*, 1987. Approach to the spanish continental neogene synthesis and paleoclimatic interpretation. *Ann. Inst. Geol. Publ. Hungary*, **70**, 383–91.

Pierre, C., Rouchy, J.M., Servant-Vildary, S. & Foucault, A., 1989. The Las Minas de Hellín Formation (South Spain): stable isotopes, mineralogy and diatoms of an hypersaline transitonal lacustrine-marine sedimentary sequence. *EUG V, Strasbourg, Abstracts*, p. 219.

Servant Vildary, S., Rouchy, J.M., Pierre, C. & Foucault, A., 1990. Marine and continental water contributions to a hypersaline basin using diatom ecology, sedimentology and stable isotopes: an example in the Late Miocene of the Mediterranean (Hellín Basin, southern Spain). *Palaeogeogr., Palaeoclimatol., Palaeoecol.*, **79**, 189–204.

Lacustrine Neogene deposits of the Ebro Basin (southern margin), northeastern Spain

ANTONIO PÉREZ, ARSENIO MUÑOZ, GONZALO PARDO AND JOAQUIN VILLENA

Facultad de Ciencias, Departamento de Geologia, Universidad de Zaragoza, E-50009 Zaragoza, Spain

Four Neogene tectosedimentary units consisting of the deposits of alluvial fans and lacustrine systems have been established along the southern margin of the Ebro Basin (Fig. 1) (Pérez *et al.*, 1988b). From the various lacustrine facies assemblages, four types of lacustrine systems were defined:

(A) *Marginal freshwater lacustrine system* containing (1) marginal palustrine environments (limestone facies with bioturbation and nodules with evidence of nearly constant vegetative cover) and (2) inner zones characterized by massive bioclastic limestone facies with charophytes, gastropods and ostracodes, and limestones facies with rhizoliths. This system is interpreted as shallow freshwater lake deposits of inner to marginal conditions depending on water table fluctuations.

(B) *Shallow saline lake system* characterized by a diverse facies assemblage. This assemblage consists of lutites and brecciated limestones containing gypsum pseudomorphs, as well as high Mg-calcite-bearing laminated limestones. This system is interpreted as shallow lake deposits associated with high salinity waters and periodic desiccation leading to evaporitic pumping and deposition of salts.

(C) *Open freshwater lake system* with high biological activity and strong surges. Sedimentary structures in sandstones include wave ripples and hummocky cross-stratification. Large bioclastic accumulations in limestones contain gastropods, ostracodes, pelecypods, and foraminifera (*Ammonia beccarii tepida*).

(D) *High salinity playa-lake system* containing saline lutitic facies with halite and nodular, selenitic to microcrystalline gypsum.

All four of these lake systems show lateral relationships with adjacent alluvial fan deposits. Based on these relationships, an evolution of lake environments based on tectosedimentary units is surmised (Figs 2, 3, 4); lacustrine systems evolve from playa lake complexes to saline shallow lakes to freshwater lakes.

N1 (tectosedimentary unit 1) – Extensive development of a playa-lake complex (type D) containing a large central lake (100 km in length) surrounded by some smaller shallow marginal lakes (type A). Alluvial systems cover areas about 25 km in length.

N2 (tectosedimentary unit 2) – Existence of shallow saline carbonate lakes (type B) and small freshwater lakes (type A). The areal extent of the saline lakes was around 25 × 75 km, while the smaller freshwater lakes contained diverse macrophytes in their marginal zones. The associated alluvial system ranged from 15 to 30 km in length.

N3 (tectosedimentary unit 3) – Development of open freshwater lakes (type C) and shallow marginal lakes (type A). The areal extent of this lacustrine system is approximately 75 × 140 km. Alluvial systems are restricted to the western sector.

N4 (tectosedimentary unit 4) – Existence of dominantly alluvial systems containing conglomerates and oncolitic limestones.

Acknowledgements

This work has been supported by PB89–0344 project of DGICYT.

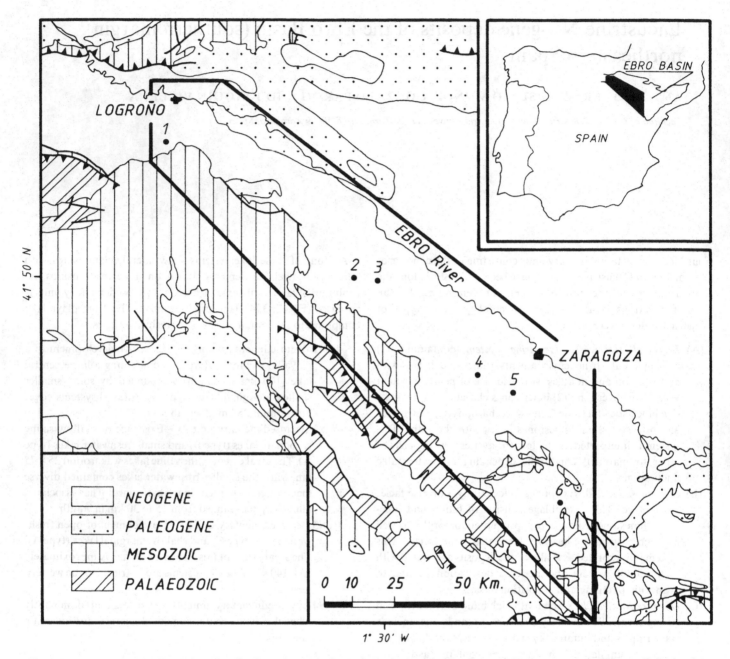

Fig. 1. Location map and stratigraphic sections for the study area in the southern margin of the Ebro Basin. (1) Western sector, (2) El Buste, (3) Borja, (4) La Muela, (5) La Plana, (6) Moyuela.

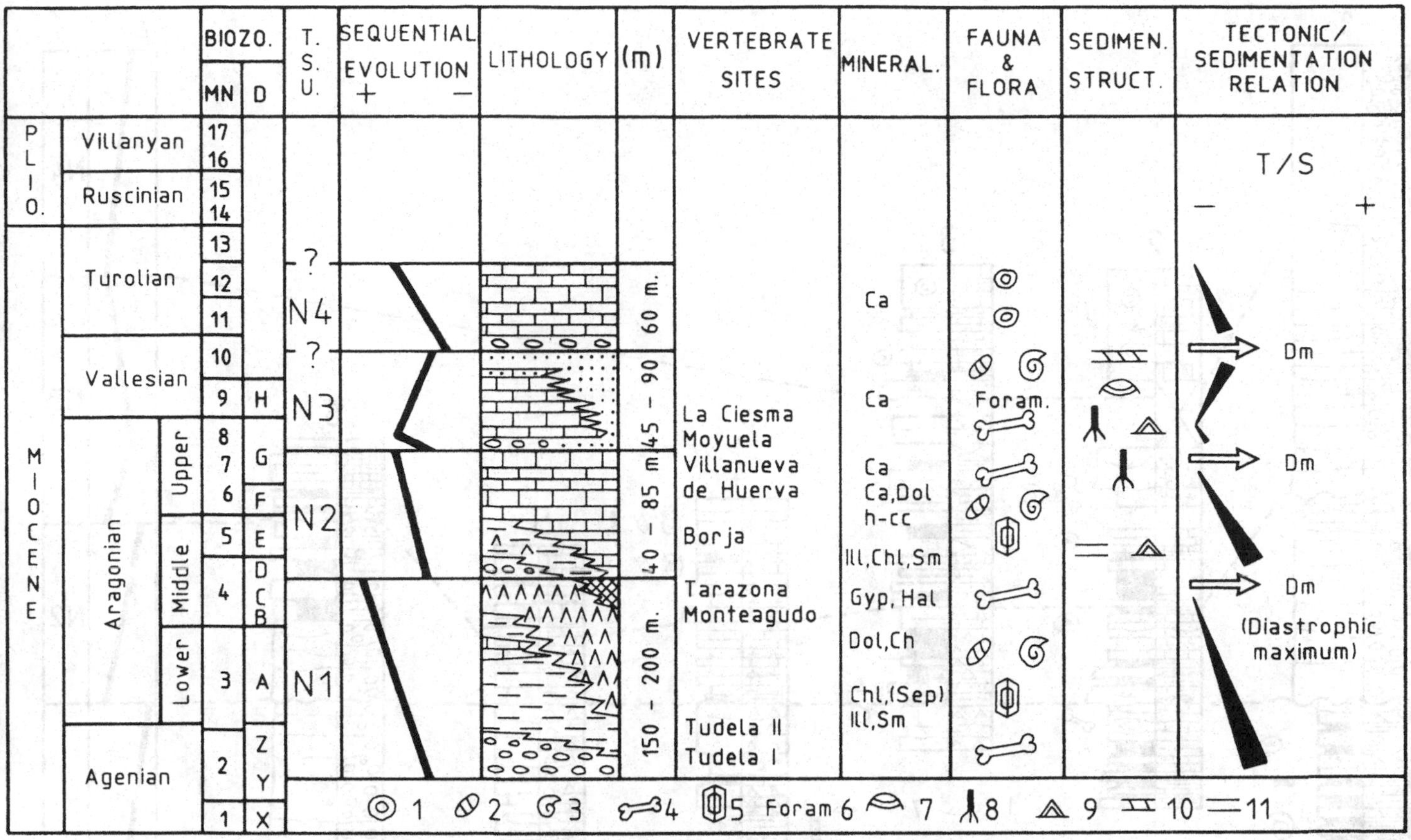

Fig. 2. Stratigraphic framework. (1) oncolites, (2) charophytes, (3) gastropods, (4) vertebrate remains, (5) ostracodes, (6) foraminifera, (7) hummocky cross-stratification, (8) root traces, (9) symmetrical ripple cross-lamination, (10) megaripples, (11) horizontal lamination. T.S.U. = tectosedimentary unit. See Fig. 4 for lithofacies legend.

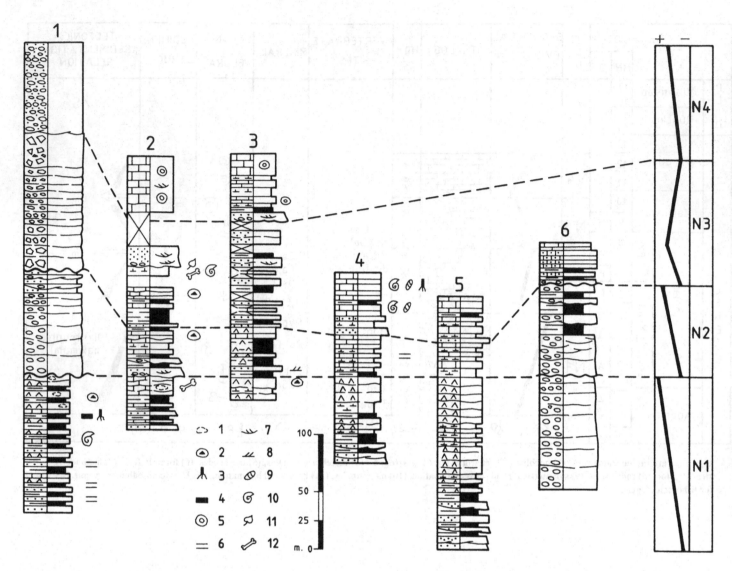

Fig. 3. Representative stratigraphic sections of the tectosedimentary units (see Fig. 1 for location of sections in study area). (1) nodularization, (2) chert nodules, (3) root traces, (4) coal, (5) oncolites, (6) horizontal lamination, (7) trough cross-stratification, (8) ripple cross-lamination, (9) charophytes, (10) gastropods, (11) macrophytic remains, (12) vertebrate remains. See Fig. 4 for lithofacies legend.

TECTOSEDIMENTARY UNIT (T.S.U.) **N1** *(Early to middle Miocene)*

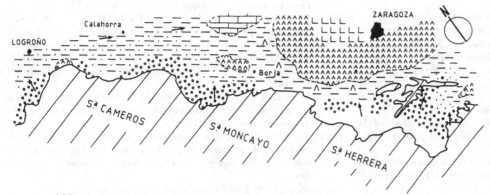

T.S.U. **N2** *(Middle Miocene)*

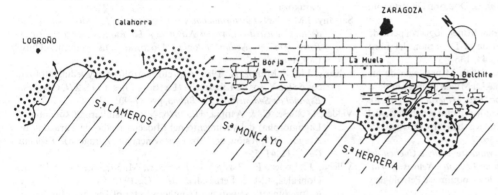

T.S.U. **N3** *(Middle to late Miocene)*

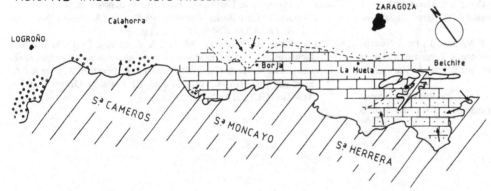

T.S.U. **N4** *(Late Miocene)*

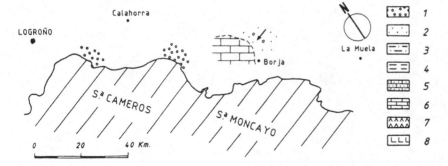

Fig. 4. Facies distribution and depositional systems of the four tectosedimentary units through time. (1) conglomeratic lithofacies, (2) sandy lithofacies, (3) sandy-lutitic lithofacies, (4) lutitic lithofacies, (5) calcarenitic lithofacies, (6) micritic limestone lithofacies, (7) gypsum lithofacies, (8) halite.

Bibliography

Astibia, H., Mazo, A., Morales, J., Sese, C., Soria, D. & Valdes, G., 1984. Mamíferos del Mioceno medio de Tarazona de Aragón (Zaragoza). *I. Congreso Español de Geología, Simposios*, pp. 383–90.

Azanza, B., Canudo, I., & Cuenca, G., 1988. Nuevos datos bioestratigráficos del Terciario continental de la Cuenca del Ebro (sector centro-occidental). *II Congreso Geológico de España, Simposios*, pp. 261–4.

Birnbaum, S.J., 1976. *Non-marine Evaporite and Carbonate Deposition.* Doctoral Dissertation, Univ. Cambridge.

Crusafont, M., Truyols, J. & Riba, O., 1966. Contribución al estudio de la estratigrafía del Terciario continental de Navarra y Rioja. *Notas Commun. I.G.M.E.*, **90**, 53–76.

González, I. & Galan, E., 1984. Minerlogía de los materiales terciarios del área de Tarazona-Borja-Ablitas (Depresión del Ebro). *Estudios Geol.*, **40**, 115–28.

Mandado, J., 1987. *Litofacies yesíferas del sector aragonés de la Cuenca terciaria del Ebro. Petrogénesis y Geoquímica.* Doctoral Dissertation, Univ. Zaragoza.

Mata, M.P., Pérez, A., & Lopez Aguayo, F., 1988. Mineralogía del perfil de La Muela; Terciario del sector central de la Depresión del Ebro (provincia de Zaragoza). *Estudios Geol.*, **44**, 135–43.

Mata, M.P., Pérez, A., & Lopez Aguayo, F., 1989. Mineralogía de los depósitos lacustres del Terciario de Borja-La Muela (Borde Sur de la Depresión del Ebro, Zaragoza). *Bol. Soc. Esp. Mineral.*, **12**, 213–20.

Muñoz, A., Pardo, G. & Villena, J., 1986–87. Análisis tectosedimentario del Terciario de la Depresión de Arnedo (Cuenca del Ebro, provincia de La Rioja). *Acta Geol. Hisp.*, **21–2**, 427–36.

Orti, F. & Salvany, J.M., 1990. *Formaciones evaporiticas de la Cuenca del Ebro y cadenas periféricas, y de la zona de Levante. Neuvas aportaciones y guia de superficie.* Departament de Geoquímica, Petrología y Prospecció Geológica, Barcelona.

Orti, F., Salvany, J.M., Rossell, L., Pueyo, J.J. & Ingles, M., 1986. Evaporitas antiguas (Navarra) y actuales (Los Monegros) de la Cuenca del Ebro. *Guia de las Excursiones del XI Congreso Español de Sedimentología*, pp. 2.1–2.36.

Pérez, A., Azanza, B., Cuenca, G., Pardo, G. & Villena, J., 1985. Nuevos datos estratigráficos y paleontológicos sobre el Terciario del borde meridional de la Depresión del Ebro (provincia de Zaragoza). *Estudios Geol.*, **41**, 405–11.

Pérez, A., Muñoz, A., Pardo, G., Arenas, C. & Villena, J., 1988a. Características de los sistemas lacustres en la transversal Tarazone–Tudela (sector Navarro-Aragonés de la Cuenca terciaria del Ebro). *II Congreso Geológico de España, Simposios*, pp. 519–27.

Pérez, A., Muñoz, A., Pardo, G., Arenas, C., & Villena, J., 1988b. Las unidades tectosedimentarias del Neógeno del borde ibérico de la Depresión del Ebro (sector central). In *Sistemas lacustres neógenos del margen ibérico de la Cuenca del Ebro*, ed. A. Pérez, A. Muñoz, & J.A. Sánchez, pp. 7–20. Guía Campo III Reunión Grupo Español Trabajo IGCP 219.

Pérez, A., Villena, J. & Pardo, G., 1986–87. Presencia de estratificación cruzada hummocky en depósitos lacustres de la Depresión del Ebro. *Acta Geol. Hisp.*, **21–2**, 27–34.

Pérez, A., 1989. *Estratigrafía y sedimentología del Terciario del borde meridional de la Depresión del Ebro (sector riojano-aragonés) y cubetas de Muniesa y Montalbán.* Doctoral Dissertation, Univ. Zaragoza.

Quirantes, J., 1978. *Estudio sedimentológico y estratigráfico del Terciario continental de Los Monegros.* Doctoral Dissertation, Inst. Católico, Zaragoza.

Salvany, J.M., 1989. *Las formaciones evaporíticas del Terciario continental de la Cuenca del Ebro en Navarra y La Rioja. Litoestratigrafía, Petrología y Sedimentología.* Doctoral Dissertation, Univ. Barcelona.

Salvany, J.M. & Muñoz, A., 1989. Aspectos petrológicos y sedimentológicos de los Yesos de Ribafrecha (La Rioja). *Res. Com. XII. Congreso Español de Sedimentología, Bilbao*, pp. 87–90.

Valdes, G., Sese, C. & Astibia, H., 1986. Micromamíferos (Rodentia Y Lagomorpha) del yacimiento del Mioceno medio de Tarazona de Aragón (Depresión del Ebro, provincia de Zaragoza). *Estudios Geol.*, **42**, 41–55.

Villena, J., Lopez, F., Pardo, G., Pérez, A., Muñoz, A., Gonzalez, J., Gonzalez, J.M. & Fernández Nieto, C., 1987. Clay mineralogy in tectosedimentary analysis of southern sector of Tertiary Ebro Basin. *6th Meeting of European Clay Group, Sevilla*, pp. 572–4.

Villena, J., Pérez, A. & Pardo, G., 1987. Storm deposits in the lacustrine Miocene of Ebro Basin. *Abstracts, 8th Regional Meeting of Sedimentology, Int. Assoc. Sediment. Tunis*, pp. 494–5.

Villena, J., Pérez, A., Pardo, G., Muñoz, A. & Arenas, C., 1988. Sistemas lacustres de la región de Moyuela. In *Sistemas lacustres neógenos del margen ibérico de la Cuenca del Ebro*. ed. A. Pérez, A. Muñoz & J.A. Sánchez, pp. 81–104. Guía Campo III Reunión Grupo Español Trabajo IGCP, 219.

The Neogene Cabriel Basin (eastern Spain)

P. ANADÓN

Institut de Ciències de la Terra (J. Almera), CSIC, c. Martí i Franqués s.n., E-08028 Barcelona, Spain

Geological setting – facies distribution

The Neogene Cabriel Basin (Upper Miocene–Lower Pliocene?) is located in the southeastern part of the Iberian Chain (eastern Spain) (Fig. 1). During the Neogene, widespread rifting affected the northeastern Iberian Peninsula resulting in extensional structures (horsts, half-grabens and tilted fault blocks). They superimposed pre-existing Paleogene compressive features. In the southern Iberian Chain, the Neogene extensional phase also led to the opening of Triassic-floored depressions which are the result of extension combined with diapirism and dissolution of Triassic evaporites. The Cabriel Basin may be regarded as one of these depressions filled with alluvial and lacustrine deposits. The alluvial-dominated succession in the basin margins, in particular the Requena–Utiel area, pass laterally to lacustrine carbonates and evaporites towards inner basin areas.

General stratigraphy

The basin fill in the Cabriel River area (Fig. 2) is mainly formed by 400 m of terrigenous fluvial deposits (red to yellow mudstones and interbedded sandstones and conglomerates) with minor shallow lacustrine carbonates and gypsum. The gypsum-dominated lacustrine deposits are found at the base (La Rambla Gypsum) and towards the middle part of the basin-fill sequence (Los Ruices Gypsum). These evaporites were formed in shallow saline lakes fed by waters from Triassic evaporite source areas.

Limestone-dominated lacustrine deposits constitute two main units: The Fuente Podrida Limestone, in the lower basin-fill

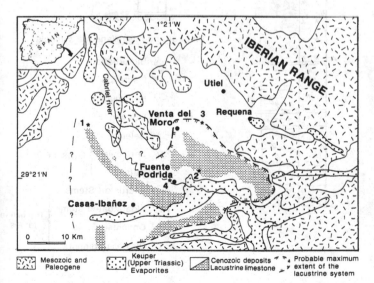

Fig. 1. Geologic sketch map of the Cabriel Basin. Numbers 1 to 4 indicate the location of the detailed lithologic logs (Fig. 3).

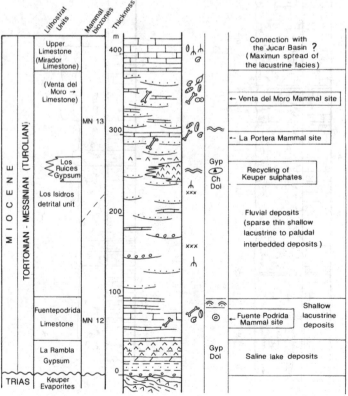

Fig. 2. General stratigraphic log.

331

sequence, and the Upper Limestone (or Mirador Limestone) at the top. Other minor lacustrine deposits formed by limestones and lignite-bearing marls are also present (i.e., Venta del Moro Limestone).

The age determinations are based on several mammal sites within the basin-fill (reliability index A–B). The fossil mammals indicate an age of Upper Miocene (Turolian mammal stage). A lowermost Pliocene age cannot, however, be excluded for the uppermost beds of the Upper Limestone unit.

Detailed representative sections

A detailed column of a limestone-dominated shallow lacustrine sequence is shown in Fig. 3 (Rambla de San Pedro). The lower

part of the section consists of bioturbated limestones (mudstones and pelloid packstones) with abundant oncoid fragments. The upper part of the section, in addition to these features, shows pedogenic structures (e.g., nodulisation, mottling). These are related to water-level fluctuations in littoral zones.

The Sardinero section shows a lower part of the section to be formed by white and grey mudstones. Mottling and nodulisation are present in some carbonate-rich horizons, indicating soil overprints. The upper part of the section comprises bioturbated limestones with abundant neomorphic features. This sequence records the vertical transition from floodplain deposits to shallow, littoral lacustrine deposits.

Lacustrine evaporite deposits are shown in the Los Ruices section. In the lower part of the section lenticular gypsum beds

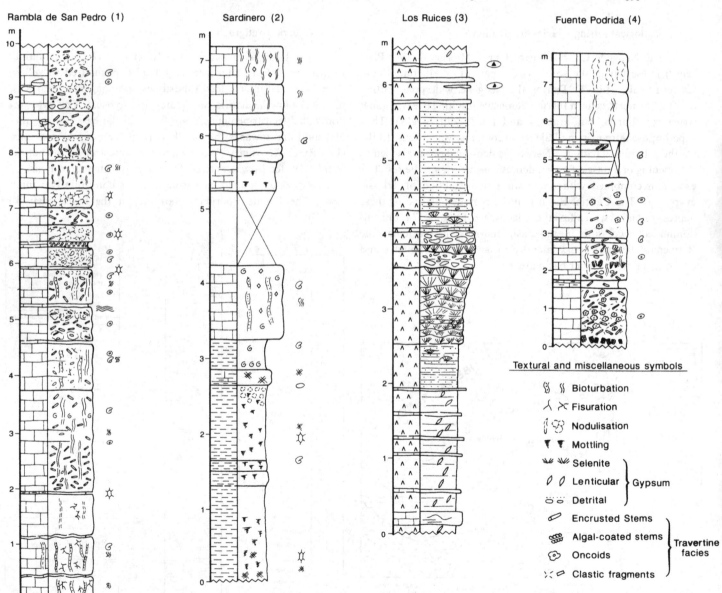

Fig. 3. Detailed lithologic logs of some typical lacustrine facies. See Fig. 1 for location (Anadón, unpublished data.)

predominate, with thin interbedded carbonates. They are overlain by gypsarenites and a selenitic gypsum unit. The selenite beds alternate with gypsarenite and gypsrudite beds. The upper part of the sequence comprises gypsarenite and lenticular gypsum beds. Carbonate replacement of gypsum deposits (individual crystals and thin beds) is evidenced. The described successions were formed in a shallow, saline lake and surrounding saline mudflats.

The Fuente Podrida section is representative of the oncolite-travertine facies. The limestones (rudstones) are formed by oncoids and fragments of algal-travertine encrusted stems. In some places, encrusted stems are preserved in their vertical position and constitute discrete horizons. These deposits were formed in shallow, high energy, littoral lacustrine areas with very active carbonate deposition.

Bibliography

Anadón, P., 1981. *Trabajo estratigráfico, sedimentológico y paleogeográfico del Terciario del Sector Suroriental de la Cordillera Ibérica.* Inst. Geol. Min. España. (Unpublished report.)

Anadón, P., 1985. *Terciario. Memoria del Mapa Geol. España 1:200,000, 2ª serie, Hoja 55, Llíria.* Inst. Geol. Minero España, pp. 80–97.

Anadón, P., Rosell, L. & Talbot, M.R., 1989. Carbonate replacement of gypsum in Upper Miocene lacustrine deposits, Eastern Spain. *Terra Abstracts*, **1**, 219–20.

Anadón, P., Rosell, L. & Talbot, M.R., 1992. Carbonate replacement of lacustrine gypsum in two Neogene continental basins, eastern Spain. *Sedim. Geol.*, **78**, 201–16.

Assens, J., Ramírez del Pozo, J., Giannini, G., Riba, O., Villena, J. & Reguant, S., 1973. *Mapa Geológico de España 1:50,000, 2ª serie, MAGNA, Memoria y Hoja no 719.* Venta del Moro. Inst. Geol. Minero España.

Lacomba, J.I., Morales, J., Robles, F., Santisteban, C. & Alberdi, M.T., 1986. Sedimentología y paleontología del yacimiento finimiocene de La Portera (Valencia). *Estudios Geol.*, **42**, 167–80.

Mathissen, M. y Morales, J., 1981. Stratigraphy, facies and depositional environments of the Venta del Moro vertebrate locality, Valencia, Spain. *Estudios Geol.*, **37**, 199–207.

Mein, P., Moissenet, E. & Truc, G., 1978. Les formations continentales du Neogène superieur des vallées du Jucar et du Cabriel au NE d'Albacete (Espagne). Biostratigraphie et environment. *Doc. Lab. Géol. Fac. Sci. Lyon*, **72**, 99–147.

Moissenet, E., 1985. Les dépressions tarditectoniques des Chaines Ibériques méridionales: distension, diapirisme et dépots néogènes associés. *C.R. Acad. Sc. Paris* **300** (II), 523–8.

Opdyke, N., Mein, P., Moissenet, E., Pérez González, A., Lindsay, E. & Petko, M., 1990. The magnetic stratigraphy of the Late Miocene sediments of the Cabriel Basin, Spain. In *European Neogene Mammal Chronology*, ed. E.H. Lindsay, pp. 507–14. Plenum Press, New York. (NATO ASI Series A, 180.)

Robles, F., 1979. *Estudio estratigráfico y paleontológico del Neógeno continental de la cuenca del Río Júcar.* PhD Thesis, Univ. Valencia.

Robles, F., Torrens, J., Aguirre, E., Ordoñez, S., Calvo, J.P., Santos, J.A., 1974. Levante. Libro Guía del Coloquio Int. sobre Bioestratigrafía continental Neógeno superior y Cuaternario inferior. *Guía*, 4.10, 87–133.

A high-altitude Late Tertiary intermontane basin: the Wakka Chu Group, Ladakh, northwestern India

MICHAEL E. BROOKFIELD

Land Resource Science, Guelph University, Guelph, Ontario N1G 2W1, Canada

The Wakka Chu Group is a Miocene–Pliocene late orogenic clastic sequence. It fills an intermontane basin developed along the Indus Suture Zone during uplift of the High Himalaya to the south (Fig. 1). The sediments consist of coarse clastic alluvial fan deposits

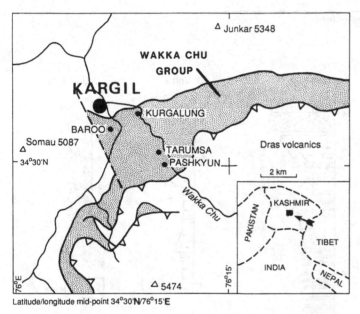

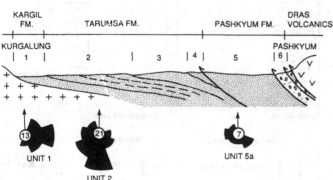

Fig. 1. Location map and cross section along Wakka Chu from Kurgalung to Pashkyum. Numbers on cross-section refer to 'units' shown in Fig. 2 and 3. Current roses are below with number of measurements in center.

passing upwards into braided and meandering stream deposits and finally into lake sediments. The basin was deformed and overthrust from the south in probably Pliocene times. This ubiquitous late Tertiary backthrusting, north of the High Himalaya, has destroyed many late Tertiary intermontane lake deposits along the Indus Suture Zone.

The Wakka Chu Group is one of the most thoroughly studied late Tertiary units along the Indus Suture Zone. The sections were described by Bhandari *et al.* (1977), Brookfield & Andrews-Speed (1984) and Sahni & Bhatnagar (1958), and noted by Frank *et al.* (1977) and Shah *et al.* (1976). Faunas and floras have been recorded, with varying degrees of accuracy, by Bhandari *et al.* (1977), De Terra (1935), Lydekker (1883), Sahni & Bhatnagar (1958), Tewari & Dixit (1972) and Tewari & Sharma (1972).

The sedimentary units distinguished can be related to the three formations proposed by Bhandari *et al.* (1977) (Fig. 2).

Of the six lithostratigraphic units, the lowermost three units form a gradational sequence unconformable on the Ladakh batholith: unit 4 is bounded above and below by thrusts: units 5 and 6 appear gradational although Frank *et al.* (1977) noted a minor thrust between them.

Units 1 to 3 represent ephemeral semi-arid stream deposits passing upwards into large braided and meandering streams and finally into lacustrine and lacustrine delta deposits.

Unit 1 unconformably overlies, and fills depressions in, the Ladakh batholith complex. It is mostly unfossiliferous and consists of tabular immature sand-bodies with low-angle cross-bedding, red siltstones, mudcracks and calcrete horizons. These features point to a semi-arid ephemeral stream system environment. Current directions show an easterly flow (Fig. 1). About 70 m above its base, a dark grey silty shale contains freshwater molluscs, arthropods, fish and plant remains of late Miocene to late Pliocene age (Mathur, 1983). Abundant microvertebrates have recently been found and indicate roughly the same age (A. Sahni, pers. comm., 1985). In contrast, an early Miocene or older age for unit 3 was based on freshwater and land plants (Tewari & Sharma, 1972). The vertebrates are probably more accurate than the plants for dating. Thus, a late Miocene beginning is likely.

Unit 1 gradually passes up into unit 2. In contrast to the random

AGE	UNITS here others[1]	FORMATION	THICKNESS (metres)	LITHOLOGY	FOSSILS	ENVIRONMENT
?	6 10	U. PASHKYUM	400	Coarse conglomerate red silty shale.		? Alluvial fan
MIOCENE	5 9 L.-M.		1100	Purple, red, green mudstone, gypsiferous and thin felspathic ssts. passing up into grey shale and greywacke ssts.	vertebrate	Ephemeral stream
_/_	Thrust _/_					
?	4 8	U.	30	Thin-bedded graded greywacke ssts. thin grey siltst. and shale.		Lacustrine turbites
_/_	Thrust _/_					
	3 7 6	M. TARUMSA	200	grey carbonaceous mdst. and olive graded siltst.	Charophytes[3] Subzebrinus	shallow water
MIOCENE	2 5 4	L.	300-400 (1500)[2]	Thick buff ssts. alternating with thin ssts. and grey-buff siltst. and shales.	Abundant plant remains	Large braided meandering streams
MIOCENE	1 3 2	KARGIL	0-200[4]	Red, grey, green shale, siltst. with thin white ssts. and conglomerate ssts. Caliche nodules: leaves, fish remains.	Unio, Viviparus Planorbis palm leaves vertebrates	Ephemeral stream
	Unconformity					
EOCENE-OLIGOCENE	LADAKH BATHOLITH COMPLEX			Granodiorite, tonalite		

[1]Sahni and Bhatnager (1958) [2]DeTerra (1935) [3]Tewari and Sharma (1972) [4]Tewari (1964) [5]Bhandari et al. (1977).

Fig. 2. Stratigraphic classification, age and environment.

succession of lithologies in unit 1, unit 2 contains regular cyclical repetitions of facies characteristic of large braided and meandering streams passing up into lacustrine river mouth bar and delta deposits (Fig. 3,A). If the thickness of epsilon cross-bedded point bar sandstones is taken to approximate bank-full depth of stream (cf. Cherven, 1978), then the streams were up to 14 m deep in places – comparable to the present Indus River. Current directions are more variable than unit 1 and show generally southerly to south-westerly flow (Fig. 1). Plant remains are abundant in the upper beds, where they occur as large fragments in channel sandstones, as finely disseminated material in crevace splay and overbank sandstones, and as well-preserved material in overbank clays and lake clays. Angiosperm pollen suggests a temperate climate (Bhandari et al., 1977) and the gastropod Subzebrinus suggests a high altitude of at least 2000 m (Tewari & Dixit, 1972).

Unit 3 consists of carbonaceous grey to black upward-fining micaceous graded siltstones and silty clays (Fig. 3,B) with rare fine-grained graded sandstones. It lies above lacustrine delta deposits suggesting a prodeltaic lacustrine environment (Cherven, 1978; Hyne et al., 1979).

Unit 4 consists of medium-bedded volcaniclastic graded sandstones (Fig. 3,C) which resemble lacustrine turbidites (Sturm & Matter, 1978). Although separated from unit 3 by a thrust, they fit the general facies passage from units 1 to 3 as a more distal lake facies.

Units 5 and 6 are tectonically emplaced over unit 4 and are the same units as 1 and 2. Both sequences have almost identical petrographies and similar faunas.

Unit 5 has a lower, red conglomerate unit which passes gradually upwards into an upper, grey unit composed of fluvial sandstone-

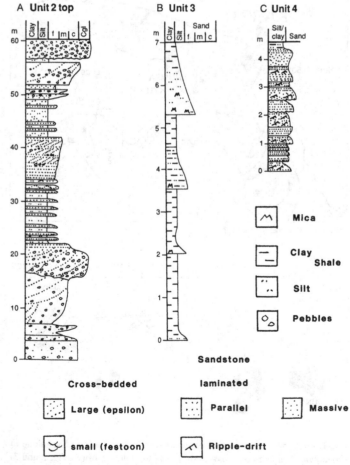

Fig. 3. Representative sedimentary sections of units 2 (top), 3 and 4.

shale cycles. They are equivalent to units 1 and 2 respectively. Current directions are variable in unit 5a (Fig. 1). Unit 6 contains massive, moderately sorted conglomerates with rounded pebbles, alternating with red silty shales and thin sandstones. The pebbles are dominantly of felsic volcanics and limestones (containing Eocene foraminifera), with less frequent cherts, siltstones, quartzites, marbles and fine-grained diorites and aplite granites. These indicate sources both to the south (Indus Suture Zone) and to the north (volcanics originally overlying the Ladakh Batholith Complex). The thickness of the silty shales and thinness of the conglomerates suggest rapid transport of coarse clastics onto a plain with little permanent stream activity. Unit 6 was probably laid down by flash floods on an arid alluvial fan system. Perhaps this arid depositional phase was the last event before late Tertiary backthrusting.

Summary

The Wakka Chu Group was deposited in an intermontane basin lying between the Zanskar Range to the south and the Ladakh Range to the north. The high elevation of the basin by unit 2 period may be related to accelerating uplift of Himalaya and Tibet in Miocene times. The development of a lake during the deposition of units 2–4 may be due to blocking of the ancestral Indus by large landslides – downcutting by the river would undoubtedly be capable of keeping pace with any gradual uplift. The change to a semi-arid climate during the deposition of unit 6 may be related to breaching of the lake dam and a consequent return to less humid, more semi-arid conditions.

Studies of Himalayan intermontane basins with lacustrine sequences like the Wakka Chu Group have great potential for relating climate, depositional environment, uplift, physiography and drainage during the development of the Himalaya.

Local oil and gas deposits may also be preserved in favorable tectonic situations: these would be extremely valuable considering the present expensive import of these materials from the south.

Bibliography

Bhandari, L.L., Venkatachalaya, B.S. & Pratap Singh, 1977. Stratigraphy, palynology and paleontology of Ladakh Molasse Group. *Proc. 4th Colloquium on Indian Micropaleontology and Stratigraphy 1974–75*, pp. 127–33. Oil and Gas Commission and Institute of Petroleum Exploration, Dehra Dun, India.

Brookfield, M.E. & Andrews-Speed, C.P. 1984. Sedimentation in a high-altitude intermontane basin – the Wakka Chu Molasse (mid-Tertiary, northwestern India). *Bull. Indian Geol. Assoc.*, **17**, 175–93.

Cherven, V.B., 1978. Fluvial and deltaic facies in the Sentinel Butte Formation, Central Williston Basin. *J. Sed. Petrol.*, **48**, 159–70.

De Terra, H., 1935. Geological studies in the northwest Himalaya between the Kashmir and Indus valleys. *Memoir Connecticut Acad. Arts & Sci.*, 7, 18–76.

Frank, W., Gansser, A. & Trommsdorf, V., 1977. Geological observations in the Ladakh area (Himalayas), a preliminary report. *Schweiz. mineral. petrogr. Mitt.*, **57**, 89–113.

Hyne, N.J., Cooper, W.A. & Dickey, P.A., 1979. Stratigraphy of an intermontane lacustrine delta, Catacumbo river, Lake Maracaibo, Venezuala. *Bull. Am. Assoc, Petrol. Geol.*, **63**, 2042–57.

Lydekker, R., 1883. *The Geology of the Kashmir and Chamba Territories and the British District of Khagan.* Geol. Surv. India Memoir 22.

Mathur, N.S., 1983. Age of the Kargil Formation, Ladakh Himalaya. In *Geology of the Indus Suture Zone of Kadakh, Wadia Inst.* ed. V.C. Thakur & K.K. Sharma, pp. 145–50. Himalayan Geol.

Sahni, M.R. & Bhatnagar, N.C., 1958. Freshwater mollusca and plant remains from the Tertiaries of Kargil, Kashmir. *Records Geol. Surv. India*, **87**, 467–76.

Shah, S.K., Sharma, M.L., Gergan, J.T. & Tara, C.S., 1976. Stratigraphy and structure of the western part of the Indus Suture belt, Ladakh, northwestern Himalaya. *Himalayan Geol*, **6**, 534–56.

Sturm, A. & Matter, A., 1978. Turbidites and varves in Lake Brienz, Switzerland: deposition of clastic detritus by density currents. *Internat. Assoc. Sediment.*, Spec. Publ. No. 2, 147–68.

Tewari, B.S. & Dixit, P.C., 1972. A new terrestrial gastropod from freshwater beds of Kargil, Ladakh, J. & K. State. *Bull. Indian Geol. Assoc.*, **4**, 61–7.

Tewari, B.S. & Sharma, S.P., 1972. Charophytes from Wakka River Formation, Kargil, Ladakh. *Bull. Indian Geol. Assoc.*, **5**, 52–62.

Quaternary

Lake Turkana and its precursors in the Turkana Basin, East Africa (Kenya and Ethiopia)

T.E. CERLING

Department of Geology and Geophysics, University of Utah, Salt Lake City, Utah 84112, USA

Tectonic and climatic setting

Lake Turkana occupies a structural depression associated with the development of the East African Rift. It is a closed basin lake, being fed principally by the Omo River, and secondarily by the combined flow of the Turkwel and Kerio Rivers (Fig. 1). The lake has a surface area of about 7560 km² with an average depth of only 31 m. The deepest part of the lake is about 114 m (1973). Because the lake is closed, the lake level fluctuates by almost a meter per year. The level of the lake has been dropping since the late-1960s when anthropological field work commenced in a large way in the northeastern part of the lake basin. The region has a mean annual temperature of about 29 °C and receives about 280 mm annual precipitation, which comes principally in April and May.

Chemistry of the modern lake

Lake Turkana is a brackish Na-HCO_3 lake with a total dissolved solids content of about 2500 mg/l. Details of the chemistry of modern Lake Turkana have been worked out by Yuretich and Cerling (1983), Cerling (1986) and Barton *et al.* (1987). They were able to do a detailed mass balance on the sediment, water and dissolved species in the basin. They showed that of the incoming ions, approximately 90% are removed annually by mineral precipitation or buried as pore water. Of particular significance is the sink identified for magnesium. Lake Turkana, like all East African lakes, is deficient in magnesium, having only about 2 ppm Mg^{+2}. Their detailed mass balance calculations show that magnesium is most likely removed as a silicate phase, probably as a tri-octahedral smectite. Barton *et al.* (1987) developed a numerical model for the evolution of conservative tracers, such as chloride, in transient closed basin lakes. The chemistry of the modern lake is compatible with basin closure taking place sometime between about 4000 and 10,000 years ago. Pore water chemistry profiles show that the lake recently was split into two basins when the lake level fell below the sill depth of about 30 m. The carbon isotopic composition of dissolved inorganic carbon shows that methanogenesis is taking place in the sediments.

Biology of the modern lake

The most comprehensive work on the modern lake is that of Hopson (1982), a compilation of biological observations made during a three-year survey to determine the suitability of developing a large-scale fishing industry at Lake Turkana. Cohen (1984) has studied the modern benthic fauna.

Geological record of sedimentation in the basin in the Plio-Pleistocene and Holocene

There is about one kilometer of sediments exposed in the Turkana Basin. Prior to 4.5 Ma ago a series of basalts, rhyolites and lahars were deposited in the Turkana Basin. Well developed fluvial and lacustrine sediments were present by 4.1 Ma ago and were deposited on a landscape with 20 + meters of topography. From 4.1 to about 2.3 Ma the Omo River fed the basin from the north and flowed out to the south and thence to the Indian Ocean. During that time, sediments were predominantly fluvial with some intervals of lacustrine sedimentation which often have abundant diatoms, ostracodes, fish fossils and invertebrate fauna. These lacustrine intervals are generally less than 10 m in thickness. The fauna, flora and mineralogy indicate that the lakes of this time interval were freshwater. At about 2.3 Ma ago, extensive volcanic activity at Mt. Kulal blocked the southward flowing Omo River, causing a large lake to form. The drainage blockage resulted in the development of a deep lake in the Koobi For and Shungura regions. This deep lake lasted for several hundred thousand years when the basin became filled with sediment. Geochemical arguments suggest that this deep lake was more saline than most earlier lakes, but still considerably fresher than the modern lake (Cerling, 1979), which is too alkaline to support a robust molluscan fauna. With the basin full of sediment, and the Omo River still carrying its large detrital load, an acute mass balance problem is apparent. It is likely that for the last 1.9 Ma or so, the Omo River has shifted between emptying into the Turkana Basin and going directly to the Nile River. Lake deposits of the last 1.9 Ma indicate more alkaline conditions than were ever present previously in the basin. Analcime, chabazite, clinoptilonite

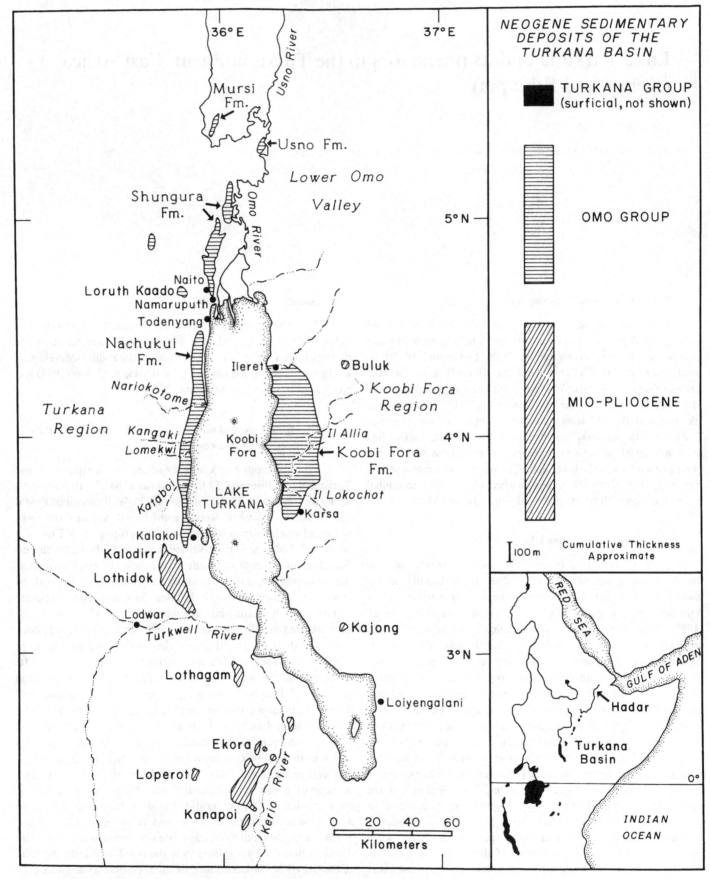

Fig. 1. Lake Turkana, showing principal inflows and sedimentary deposits in the Turkana Basin.

and dolomite are important minerals in these deposits. Based on the modeling of Cerling (1986) and Barton *et al.* (1987) the present lake probably filled between 4000 and 10,000 years ago. Since that time it has been a closed basin lake whose chemistry has been evolving to the present composition. Recent high resolution multi-channel seismic work shows that the lake has had lowstands at least 60 m below the present lake's surface, most likely within the last 10,000 years (Johnson *et al.*, 1987). Fluvial sediments throughout the entire sequence have abundant mammalian remains, including hominid fossils.

The combined thickness of the Pliocene and Pleistocene deposits in the Turkana Basin is about 600 m. Paleomagnetic studies and extensive tephrachronology studies of the Shungura, Nachukui and Koobi Fora Formations have provided an extremely well-dated sequence with many tuffs that can be correlated between different sections (McDougall, 1985; Brown *et al.*, 1985; Hillhouse *et al.*, 1986). This provides the detailed framework for the studies of mammalian fauna which has been the focus of research in the Turkana Basin.

Acknowledgements

Most of the work in the Turkana Basin was done with the cooperation of the National Museums of Kenya with the encouragement and direct help of the Director, R.E.F. Leakey. Without his encouragement, little of the work cited would have been done.

Bibliography

Lake Turkana (Lake Rudolf, Galana Boi) was virtually unknown in the scientific literature before hominid fossils were discovered in the Plio-Pleistocene sediments surrounding the lake in the late 1960s. Its importance to the evolutionary history of man has sparked a tremendous interest in the lake basin and numerous paleontologists, biologists, anthropologists, geologists, chemists and physicists have studied various aspects of the modern basin and its geologic record. The list below cites only a few of the most important papers published on geologic aspects of the basin. Many additional papers are to be found in the reference lists of the papers cited below.

Barton, C.E., Solomon, D.K., Bowman, J.R., Cerling, T.E. & Sayer, M.D., 1987. Chloride budgets in transient lakes: Lakes Baringo, Naivasha, and Turkana, Kenya. *Limnol. Oceanogr.*, **32**, 745–51.

Brown, F.H. & Cerling, T.E., 1982. Stratigraphical significance of the Tulu Bor Tuff on the Koobi Fora Formation. *Nature*, **299**, 212–15.

Brown, F.H. & Feibel, C.S., 1986. Revision of lithostratigraphic nomenclature in the Koobi Fora region, Kenya. *J. Geol. Soc. London*, **143**, 297–310.

Brown, F.H. & Feibel, C.S., 1989. 'Robust' hominids and Plio-Pleistocene paleogeography of the Turkana Basin, Kenya and Ethiopa. In *The Evolutionary History of the 'Robust' Autralopithecines*, ed. F.E. Grine, pp. 325–41. Aldine deGruyter, New York.

Brown, F.H., McDougall, I., Davies, T. & Maier, R., 1985. An integrated Plio-Pleistocene chronology for the Turkana Basin. In *Ancestors: The Hard Evidence*, ed. E. Delson, pp. 82–90. A.R. Liss, New York.

Cerling, T.E., 1979. Paleochemistry of Plio-Pleistocene Lake Turkana, Kenya. *Palaeogeogr., Palaeoclimatol., Palaeoecol.*, **27**, 247–85.

Cerling, T.E., 1986. A mass balance approach to basin sedimentation: constraints on the recent history of the Turkana Basin. *Palaeogeogr., Palaeoclimatol., Palaeoecol.*, **54**, 63–86.

Cerling, T.E., & Brown, F.H., 1982. Tuffaceous marker horizons in the Koobi Fora region and the lower Omo Valley. *Nature*, **299**, 216–21.

Cerling, T.E., Bowman, J.R. & O'Neil, J.R., 1988. An isotopic study of a fluvial-lacustrine sequence: the Plio-Pleistocene Koobi Fora Formation, East Africa. *Palaeogeogr., Palaeoclimatol., Palaeoecol.*, **63**, 335–56.

Cohen, A.S., 1984. Effect of zoobenthic standing crop on laminae preservation in tropical lake sediment, Lake Turkana, East Africa. *J. Paleontol.*, **58**, 499–510.

Feibel, C.S., 1987. Fossil fish nests from the Koobi Fora Formation (Plio-Pleistocene) of northern Kenya. *J. Paleontol.*, **61**, 130–4.

Johnson, T.C., Halfman, J.D., Rosendahl, B.R. & Lister, G.S., 1987. Climatic and tectonic effects on sedimentation in a rift valley lake: evidence from lake Turkana, Kenya. *Geol. Soc. Am. Bull.*, **98**, 439–47.

Halfman, J.D. & Johnson, T.C., 1988. High-resolution record of cyclic climatic change during the past 4 ka from Lake Turkana, Kenya. *Geology*, **16**, 496–500.

Hillhouse, J.W., Cerling, T.E. & Brown, F.H., 1986. Magnetostratigraphy of the Koobi Fora Formation, Lake Turkana, Kenya. *J. Geophys. Res.*, **91**, 11581–95.

Hopson, A.J. (Ed.), 1982. *Lake Turkana: A Report on the Findings of the Lake Turkana Project, 1972–1975*. Overseas Development Administration, London.

McDougall, I., 1985. K-Ar and ^{40}Ar/^{39}Ar dating of the hominid bearing Plio-Pleistocene sequence at Koobi Fora, Lake Turkana, northern Kenya. *Geol. Soc. Am. Bull.*, **96**, 159–75.

Owen, R.B., Barthelme, J.W., Renaut, R.W. & Vincens, A., 1982. Palaeo-limnology and archeology of Holocene deposits north-east of Lake Turkana, Kenya. *Nature*, **198**, 523–9.

Yuretich, R.F., 1979. Modern sediments and sedimentary processes in Lake Rudolf (Lake Turkana), Eastern Rift Valley, Kenya. *Sedimentology*, **26**, 313–31.

Yuretich, R.F. & Cerling, T.E., 1983. Hydrogeochemistry of Lake Turkana, Kenya: mass balance and mineral reactions in an alkaline lake. *Geochim. Cosmochim. Acta*, **47**, 1099–109.

Lake Manyara Basin, Tanzania

THOMAS SCHLÜTER

Department of Geology, University of Makerere, P.O. Box 7062, Kampala, Uganda

Lake Manyara is the southernmost lake of the eastern section of the Great Rift Valley (Gregory Rift) and presently covers almost 480 km² (Fig. 1). The west side of the lake is flanked by a deep scarp, while in the east, an undulating plain with isolated volcanic cones rises gently and grades into the undisturbed peneplain surface. Ten kilometers southeast of Lake Manyara is Lake Burungi, a much smaller and almost dry lake. The average water depth of Lake Manyara is now only 3 m. A series of terraces, however, indicates much higher lake levels in the past (Kent, 1942; Stoffers & Holdship, 1975; Vaidyanathan & Dixit, 1987). However, the maximum water extent of Lake Manyara is still debated. Kent (1942) and Stoffers & Holdship (1975) postulated a 100 m higher lake level, based on the analysis of a 56 m core sample and evidence for a corresponding strand terrace. Based on reef-building stromatolites, only a 20 m higher lake level is postulated during the Pleistocene (Dixit, 1984; Casanova, 1986). Lake Manyara waters are alkaline saline. Hydrochemistry is presented in Table 1. For comparison with other lakes in the East African Rift Valley, see Cerling (1979).

A summary stratigraphic column of what is known about the sediments of Lake Manyara is given in Fig. 2. Phosphate-bearing

Table 1. *Chemical composition of the waters of Lake Manyara*

	Na	K	Ca	Mg	Cl	SO$_4$	SiO$_2$	pH	Alk
Mg/1	2500– 21500	8–94	10	1–30	1173– 8670	230– 2280	16.3– 19	9.8	806

Source: Adapted from Stoffers & Holdship (1975) and Cerling (1979).

Pliocene or Pleistocene sediments occur at Minjingu, about 4 km east of Lake Manyara. These deposits host one of the few exploitable reserves of fertilizer in East Africa. Their position in relation to the lake beds is shown in Fig. 3. The phosphorites are mainly composed of bone fragments of birds (*Phalacrocorax kuehneanus*) and fish (*Tilapia*? sp.) (Schlüter, 1991; Schlüter *et al.*, 1992). An increased paleoalkalinity is theorized for the Plio-Pleistocene? lacustrine environment of Lake Manyara (Schlüter, 1986, 1987a, b; Singh, 1980).

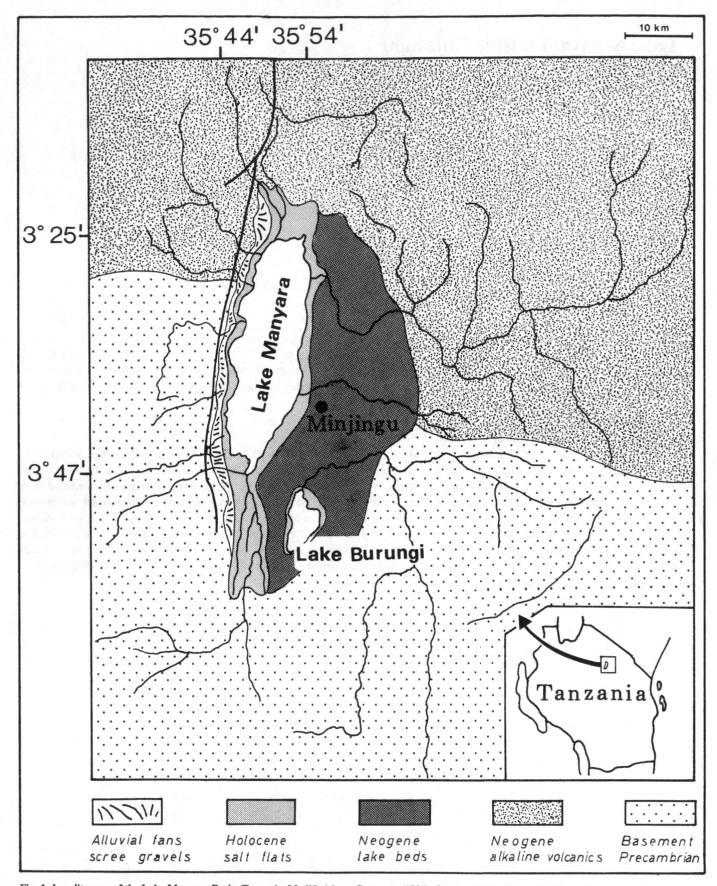

Fig. 1. Locality map of the Lake Manyara Basin, Tanzania. Modified from Casanova (1986), Orridge (1963–65), and Vaidyanathan & Dixit (1987).

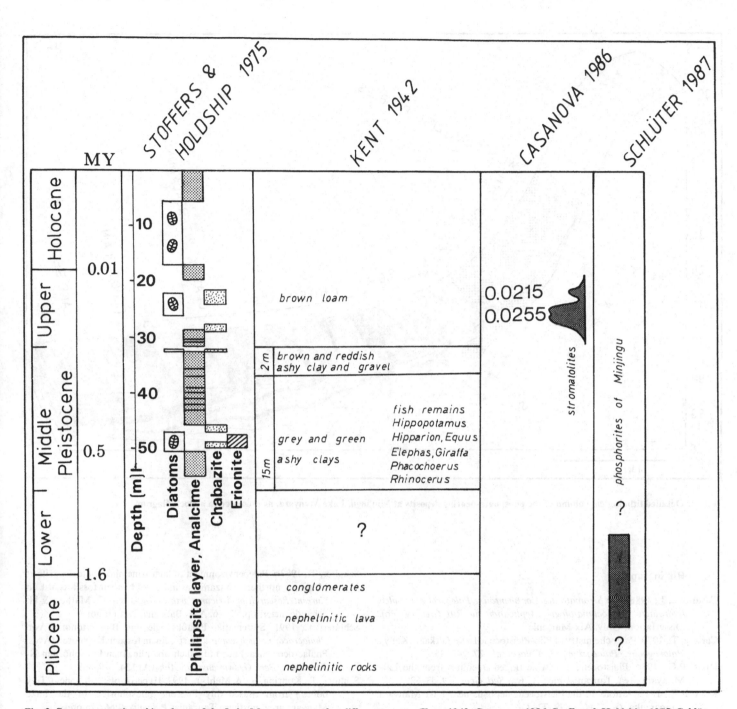

Fig. 2. Summary stratigraphic column of the Lake Manyara area, after different sources (Kent, 1942; Casanova, 1986; Stoffers & Holdship, 1975; Schlüter, 1987a, b). Stromatolite histogram refers to the frequency of C[14] ages of stromatolite samples of second and third generation, see Casanova (1986) for number of samples. 0.0215 = 21,500 BP and 0.0255 = 25,500 BP. The probable stratigraphic extent of the phosphorites is shown by the bar at right.

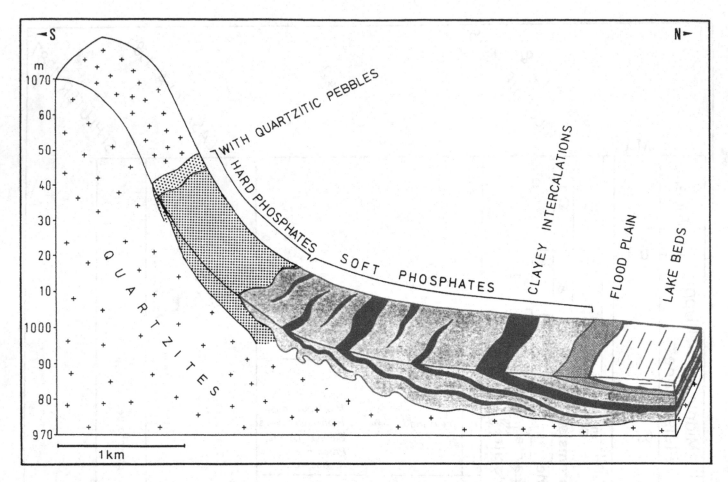

Fig. 3. Detailed lithological column of the phosphate-bearing deposits at Minjingu, Lake Manyara, as presented in a block diagram.

Bibliography

Casanova, J., 1986. *Les Stromatolites Continentaux: Paleoecologie, Paleo-hydrologie, Paleoclimatologie. Application au Rift Gregory.* PhD Dissertation, Univ. Aix-Marseille.

Cerling, T., 1979. Paleochemistry of Plio-Pleistocene Lake Turkana, Kenya. *Palaeogeogr., Palaeoclimatol., Palaeoecol.*, **27**, 247–85.

Dixit, P.C., 1984. Pleistocene lacustrine ridged oncolites from the Lake Manyara area, Tanzania, East Africa. *Sed. Geol.*, **39**, 53–62.

Kent, P.E., 1942. A note on the Pleistocene deposits near Lake Manyara, Tanganyika. *Geol. Mag.*, **79**, 72–7.

Orridge, G.R., 1963–65. *Brief Explanation of the Geology.* Quarter Degree Sheet 69, Mbulu, Mineral Resource Division, Dodoma, Tanzania.

Schlüter, T., 1986. Eine neue Fundstelle pleistozäner Kormorane (*Phalacrocorax* sp.) in Nord-Tanzania. *J. Ornithol.*, **127**, 65–91.

Schlüter, T., 1987a. A cross-section through the lacustrine environment in the Pleistocene Lake Manyara beds at Minjingu, northern Tanzania. In *Global Change in Africa during Quaterny – Past, Present, Future*, ed. H. Faure, L. Faure & E.S. Diop, pp. 419–21. Dakar.

Schlüter, T., 1987b. Paleoenvironment of lacustrine phosphate deposits at Minjingu, northern Tanzania, as indicated by their fossil record. In *Current Research in African Earth Sciences*, ed. G. Matheis & H. Schandelmeier, pp. 223–6. A.A. Balkema, Rotterdam.

Schlüter, T., 1991. Systematik, Paläökologie und Biostratonomie von *Phalacrocorax kuehneanus* nov. sp., einem fossilen Kormoran (Aves: Phalacrocoradicae) aus nutasslich oberpliozänen Phosphoriten N-Tanzanias. *Berl. Geowissenschaf. Abh.* (A) **134**, 279–309.

Schlüter, T., Kohring, R., & Mehl, J., 1992. Hyperostotic fish bones ('Tilly bones') from presumably Pliocene phosphorites of the Lake Manyara area, northern Tanzania. *Paläont. Z.* **66**, (In press.)

Singh, K., 1980. Determination of phosphate in deposits around Lake Manyara. *Univ. Sci. J. Dar Es Salaam Univ.*, **6**, 107–9.

Stoffers, P., & Holdship, S., 1975. Diagenesis of sediments in an alkaline lake: Lake Manyara, Tanzania. *Proceedings, IX Congress, Int. Sediment. Nice*, **7**, 211–18.

Vaidyanathan, R. & Dixit, P.C., 1987. Geomorphic evolution and sedimentation of Lake Manyara environs, NW Tanzania. *14th Colloquium on African Geology, Tech. Univ. Berlin, Abstracts*, pp. 129–30.

The Quaternary Katmandu Basin, Nepal

GOPAL M.S. DONGOL[1] AND MICHAEL E. BROOKFIELD[2]

[1]Geological Survey, His Majesty's Government of Nepal, Katmandu, Nepal
[2]Dept. of Land Resource Science, Guelph University, Guelph, Ontario N1G 2W1, Canada

The Katmandu basin is Nepal's largest Quaternary intermontane basin, and like the Kashmir Basin in India, evolved in response to tectonic movements associated with southward thrusting on the Main Boundary Thrust and with climatic changes due to Quaternary glaciation (see Fig. 1). The basin filled and drained several times during its development. The maximum exposed thickness is 280 m.

The basal Tarebhir Gravel (60 m) is the oldest exposed unit (see Fig. 2). It consists of stream flood and debris flow breccias of alluvial fans, derived from the south (Mahabharat granite clasts), interfingering near the top with better sorted braided stream deposits with northern source petrographies. This deposit inaugurated the Quaternary Katmandu Basin as uplift of the Mahabharat Range was occurring to the south. Dating of the lignites above indicate a Pliocene age of inauguration.

The following three formations (see Fig. 2) are found on three distinct terraces along the Bagmati River. The terraces are tilted slightly northwards indicating post-depositional uplift to the south.

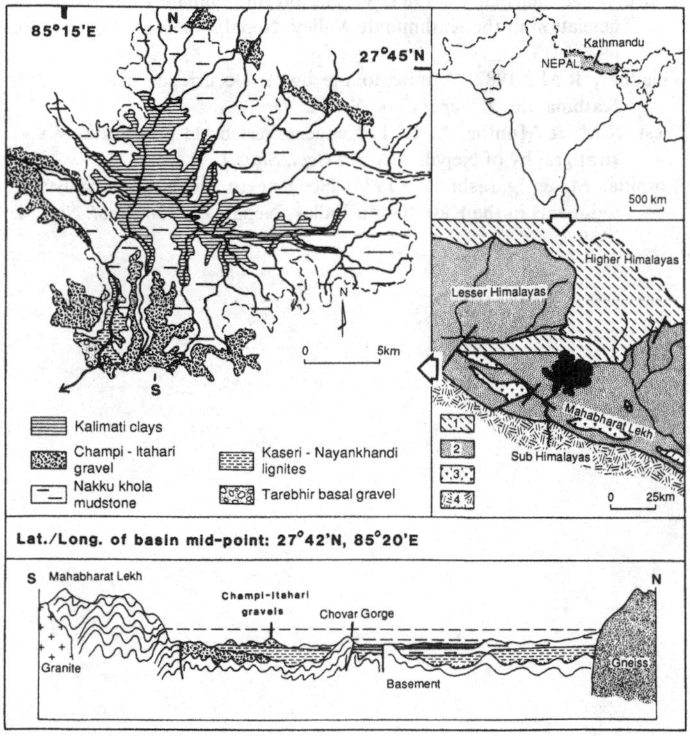

Fig. 1. Location maps and cross section of the Quaternary Katmandu basin. On inset map, 1: gneiss; 2: metasediments; 3: Mahabharat Range granite; 4: Upper Tertiary Siwalik Molasse.

On cross-section, dotted lines show successive lake levels at Lignite (upper) and Kalimati Clay (lower) times.

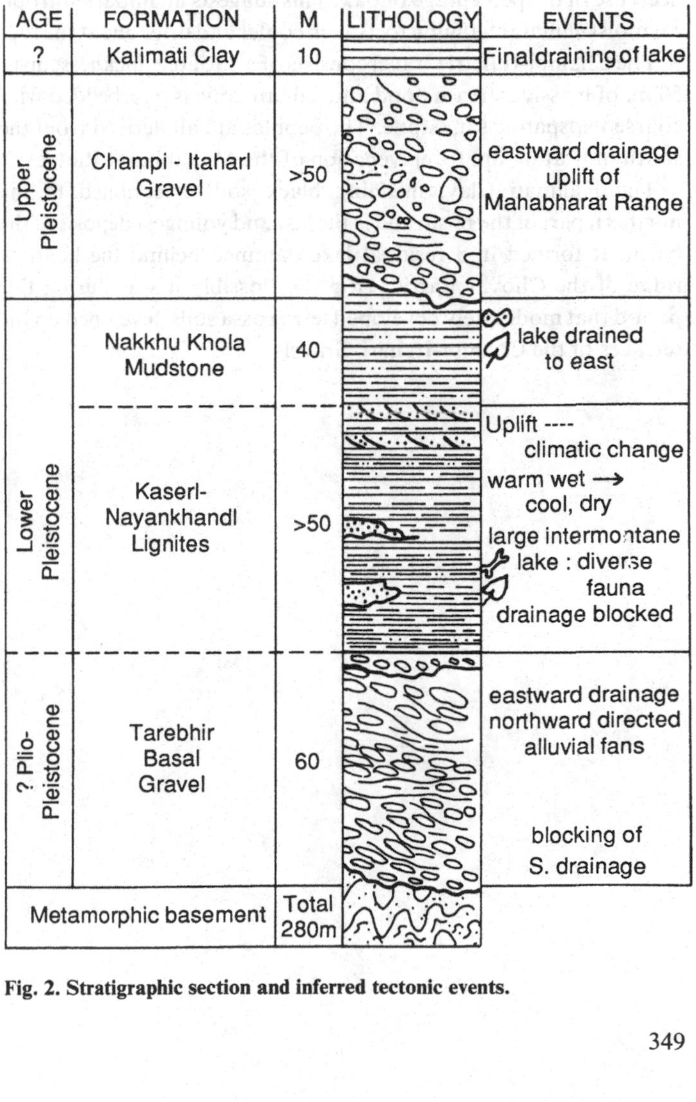

AGE	FORMATION	M	LITHOLOGY	EVENTS
?	Kalimati Clay	10		Final draining of lake
Upper Pleistocene	Champi - Itahari Gravel	>50		eastward drainage uplift of Mahabharat Range
Lower Pleistocene	Nakkhu Khola Mudstone	40		lake drained to east
Lower Pleistocene	Kaseri-Nayankhandi Lignites	>50		Uplift ---- climatic change warm wet → cool, dry large intermontane lake : diverse fauna drainage blocked
? Plio-Pleistocene	Tarebhir Basal Gravel	60		eastward drainage northward directed alluvial fans blocking of S. drainage
	Metamorphic basement	Total 280m		

Fig. 2. Stratigraphic section and inferred tectonic events.

349

The Kayseri–Nayakhandi Lignites overlie the basal gravels disconformably. They consist of a 50 m thick deltaic-lacustrine sequence of alternating sandstones, siltstones, and mudstones with interbedded thin lignites (2 cm to 1.5 m). The lignite seams form sheets and developed as *in situ* hydrophytic vegetation of swampy areas on shallow lake, river channel and crevasse splay deposits: some lignites are lensitic log-jam deposits.

The lignites contain the only Quaternary vertebrate fauna recorded from the Katmandu Basin. This Plio-Pleistocene fauna has elephant, rhinoceros, water buffalo, swamp deer, river hog and bush pig. These suggest paleoenvironments varying from reed swamp to grass jungles to humid jungle to open woodland and light forest around the lake (Dongol, 1985). The plants have not yet been studied.

The Nakkhu Khola Mudstones transitionally overlie the lignites. They consist of 40 m of massive to laminated mudstones with rare leaf impressions. They represent a more offshore, deeper euxinic lake environment. Northern exposures show more variable, shallower water conditions including rapid deposition from flooding streams. Yoshida & Igarashi (1984) identified pollen of dominantly pine, oak and alder. The upper levels of the Nakkhu Khola Mudstones show a dominance of pine, fir and spruce and a sudden decrease in the percentage of oak. This suggests an initial subtropical moist climate changing to a much cooler and drier one at the top.

The Champi-Itahari Gravel consists of a variable thickness, up to 50 m, of massive stream flood alluvial fan gravels interbedded with coarse feldspathic sandstone. The pebbles are all derived from the south indicating uplift rejuvenation of the Mahabharat Range.

The Kalimati Clay, meaning black soil, is confined to the northern part of the basin and is the last and youngest deposit of the basin. It formed in a residual lake dammed behind the bedrock ridge of the Chovar Barrier (Fig. 1). Possibly it was during this period that moderately developed terra rossa soils developed on the terraces of the Champi-Itahari Gravels.

Thus, the Katmandu Basin shows two phases of uplift of the southern Mahabharat Range, due to thrust movements on the Main Boundary Thrust. The first ponded back a large lowland Plio-Pleistocene lake in which the Kaseri–Nayakhandi Lignites and Nakkhu Khola Mudstones accumulated. This lake drained, probably to the east (Dongol, 1987), before the second uplift led to rapid deposition of the Champi–Itahari Gravels in a higher intermontane basin, followed by local tectonic ponding and deposition of the Kalimati Clay. These movements controlled the sedimentary sequences in the Lesser Himalaya Quaternary basins, and might be synchronous along the Himalaya (Burbank, 1983). With further fossil and paleomagnetic dating, and paleoenvironmental studies (some in progress) it might be possible to integrate the histories of all Himalayan Quaternary basins and relate them to tectonic phases at the mountain front.

Bibliography

Burbank, D.W., 1983. The chronology of intermontane basin development in the northwestern Himalaya and the evolution of the Northwest Syntaxis. *Earth Planet. Sci. Letters*, **64**, 77–92.

Dongol, G.M.S., 1985. Geology of the Katmandu fluviatile lacustrine sediments in the light of new vertebrate fossil occurences. *J. Nepal Geol. Soc.*, **3**, 43–57.

Dongol, G.M.S., 1987. The stratigraphic significance of vertebrate fossils from the Quaternary deposits of the Katmandu Basin, Nepal. *Newsletters in Strat.*, **18**, 21–9.

Mittre, V. & Sharma, C., 1984. Vegetation and climate during the last glaciation in the Kathmandu Valley, Nepal. *Pollen and Spores*, **26**, 69–94.

Tuladhar, R.M., 1982. A note to the lignite occurrence in Lukundol, Kathmandu. *J. Nepal Geol. Soc.*, **2**, 47–51.

West, R.M. & Munthe, J., 1981. Neogene vertebrate paleontology and stratigraphy of Nepal. *J. Nepal Geol. Soc.*, **1**, 71–14.

Yoshida, M. & Igarashi, Y., 1984. Neogene to Quaternary lacustrine sediments in the Kathmandu Valley, Nepal. *J. Nepal Geol. Soc.*, **4**, 73–97.

The proglacial lacustrine formation of the Combe d'Ain (Quaternary, Jura, France)

ROSELINE LAMY AU ROUSSEAU

Centre des Sciences de la Terre, U.R.A. C.N.R.S. no 157, 6, Bd Gabriel, F-21100 Dijon, France

Geographic and geological setting

The Combe d'Ain Basin is situated in the French Jura. The Combe d'Ain is a large basin 30 km long in the south-southwest north-northeast direction and is 2 to 6 km wide. It is bounded on the west by the Côte de l'Euthe Ridge and towards east by the Champagnole Plateau (Fig. 1). It is underlain by Jurassic rocks.

General section

During the latest Wurm (Campy, 1982), the Combe d'Ain lake was supplied by the meltwaters of the Jura glacier, which was situated on the high range to the east of the Combe d'Ain.

The basin fill, 10 to 60 m thick, is essentially composed of lacustrine laminites in the central part of the basin, and of pebbles, gravels and coarse sands at the margin. No pollen or fossils have been found, and to date, there are no radiometric dates.

Facies

The coarse facies

The coarse facies are located at the lake margin and form deltas, and can also be found farther out into the basin.

The deltas

The deltas are Gilbert-type braid deltas. Their longitudinal extension is not greater than a few hundred meters with foresets of 20 to 30 m in thickness. The topset has a 3° slope (maximum) and was covered by a braided river system. Deltaic sedimentation was dominated by gravity flow mechanisms (Fig. 2A). Matrix- to clast-supported conglomerates are the predominant facies which occur on the foreset slope. These conglomerates correspond to debris flow deposits. Openwork facies with a normal graded bedding also occur.

The coarse lenses into the basin

There are many coarse sand and gravel lenses (8 m thick by 200–300 m long) within the finer offshore deposits. The coarse lenses contain five facies which are noted (1), (2), (3), (4) and (5) on Fig. 2A. This facies association corresponds to a gravity flow deposit (Lamy Au Rousseau, 1987, 1990) and can be compared to the 'Massive Granule Bar' described in the Annot Sandstone Series (Ravenne & Beghin, 1983). These coarse lenses are interpreted as catastrophic gravity flow deposits caused, for example, by the destabilization of deltas during a fall in water level (Fig. 2B).

The fine facies

Fine material was transported by turbidity currents generated by gravity flows on the deltas. So, these rhythmic fine deposits are not 'varves'. They are turbidites dependent on the deltaic sedimentation (Lamy au Rousseau, 1990).

In the proximal area, the fine facies corresponds to an accumulation of graded beds with erosive bases (fine sand to fine silt). They are called proximal rhythmites. These proximal rhythmites gradually change to distal rhythmites in the distal area.

A distal rhythmite is characterized (in the field) by one dark and one light part (detail in Fig. 3). Each part is composed of several laminae, showing an erosive base and normal graded bedding. One lamina corresponds to the deposit of one turbidity current.

Evolution of the lake

The study of these sedimentary units and of their distribution has shown that the Combe d'Ain Formation is composed of three superimposed deltaic systems (Fig. 2A). Evolution of the lacustrine basin of Combe d'Ain depended on successive water level rises and on the topography of the basin. Lakes first occurred in low-lying areas and when the water level rose, these small lakes combined into one large lake.

During each water level rise, a transgressive system tract occurred. Deltas built up at each high water stage (Fig. 4). Falls in the water level are marked by catastrophic gravity flows.

Climate may have in part controlled lake level because a glacial tongue acted as a dam. Each delta may correspond to a stillstand during glacial advance (Fig. 5).

351

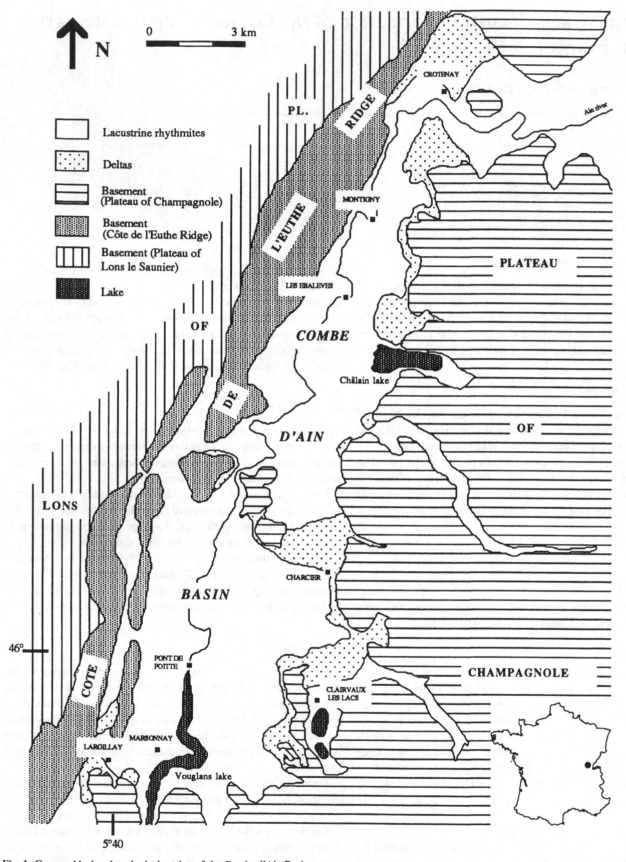

Fig. 1. Geographical and geological setting of the Combe d'Ain Basin.

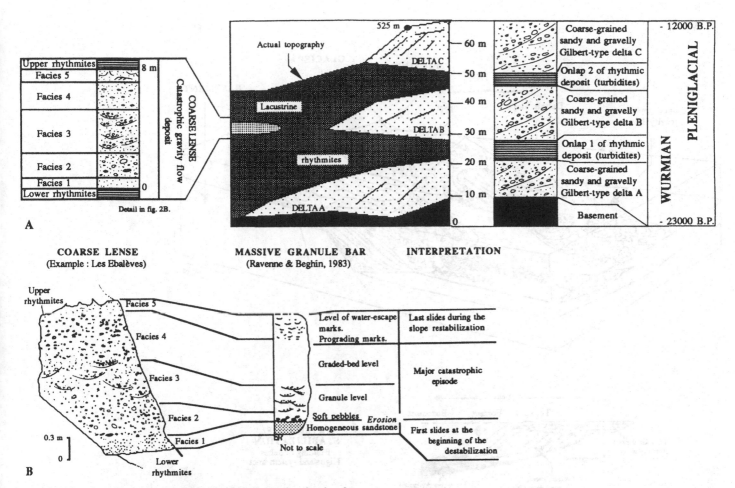

COARSE LENSE
(Example : Les Ebalèves)

MASSIVE GRANULE BAR
(Ravenne & Beghin, 1983)

INTERPRETATION

Fig. 2. A: Stratigraphic column of the Combe d'Ain Formation showing three superimposed deltas. Some gravelly lenses can be intercalated within the distal lacustrine rhythmic deposit. Sedimentation on deltas is dominated by gravity flow mechanisms.

B: Facies association of a coarse lens observed distally in the basin. Comparison with the massive granule bar described by Ravenne and Beghin (1983). *Facies 1*: Coarse homogeneous sand interpreted as a very dilute debris flow deposit. *Facies 2*: Structureless matrix- to clast-supported gravel bed with a very erosive base showing soft pebbles of lower rhythmite and olistoliths (several meters long) of basal deltaic facies. This facies corresponds to a debris flow deposit. *Facies 3*: Coarse to gravelly sand. Troughs are common and infilled by clast-supported gravel beds. Facies 3 corresponds to debris flow deposits. *Facies 4*: Matrix-supported gravel bed showing a crude alignment of clasts. This facies is interpreted as a debris flow deposit. *Facies 5*: Medium, homogeneous sand with water-escape structures and prograding marks interpreted as a liquefied flow deposit.

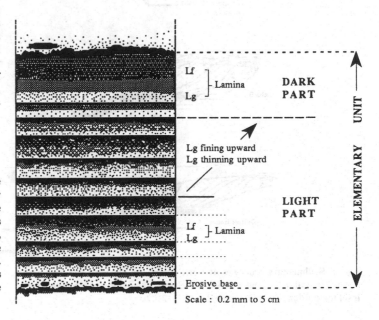

Fig. 3. Detail of one distal rhythmite. Dark and light 'beds' are observed on the field. One couplet (light + dark) is called an elementary unit. One elementary unit is composed of several laminae showing an erosive base (particularly the first lamina of one elementary unit) and normal graded bedding. One lamina is composed of one basal coarse layer, noted as Lg, and one upper fine layer, noted as Lf. The light part of one elementary unit is composed of many laminae which are organized into sets showing Lg layers fining and thinning upward. Number of laminae inside such set and number of sets in the light part is random. The dark part is always composed of 2 or 3 laminae (max.) and the last lamina shows a very developed Lf layer.

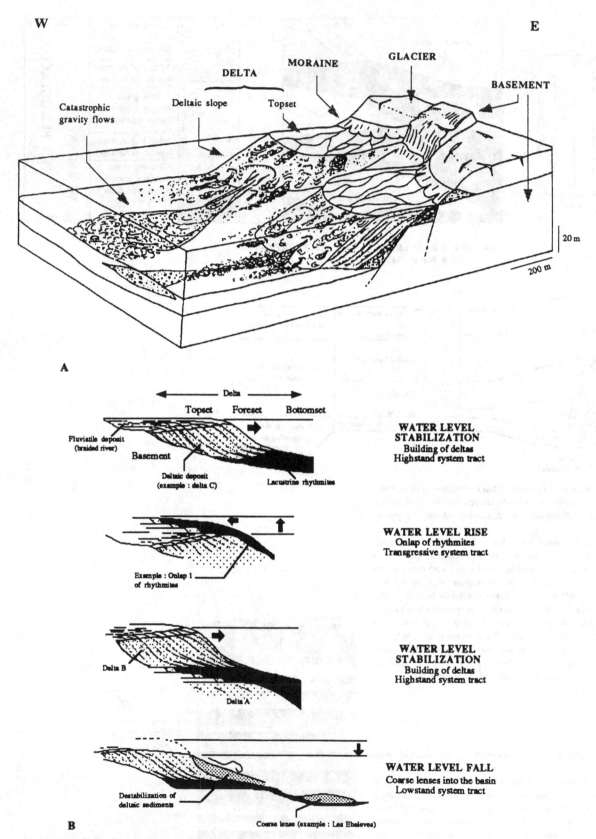

Fig. 4. Sedimentary processes at work during the deposition of the Combe d'Ain Formation (A) and sedimentary changes according to lacustrine water level fluctuations through time. (B).

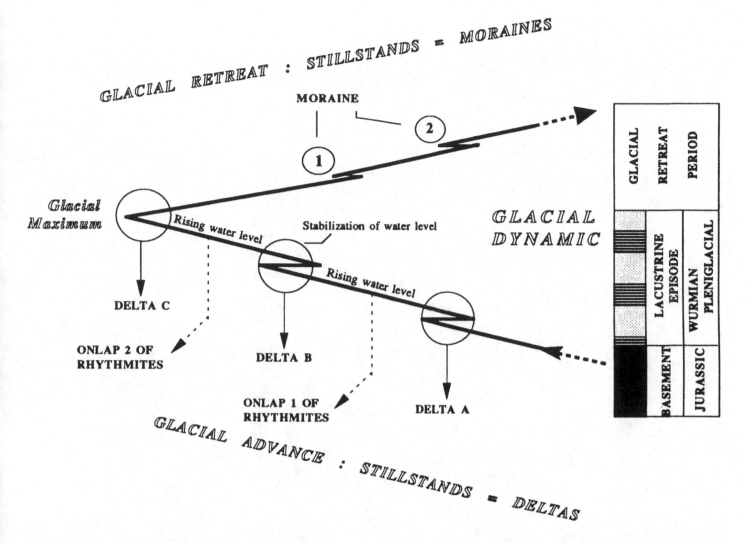

Fig. 5. Relations between sedimentation and glacial dynamics. The Combe d'Ain lake was dammed by a glacial tongue. During a glacial stillstand, the lacustrine water level was stabilized and deltas (A, B, C) could build up forming a high-stand system tract. During glacial advances, the water level rose and a transgressive system tract occurred with the onlap of rhythmites. Each delta corresponds to a stillstand during the glacial advance leading to the glacial maximum. Each moraine corresponds to a stillstand of the glacier during its retreat.

Bibliography

Campy, M., 1982. *Le quaternaire franc-comtois.* Thèse d'Etat, Univ. Besançon.

Dupis, A., Bossuet, G., Choquier, A., Campy, M., Lamy Au Rousseau, R. & Festeau, A., 1989. Reconnaissance géophysique de la formation de la Combe d'Ain. *Rapport inédit dactylographié Centre de Recherce Géophysique, C.N.R.S. – Centre des Sciences de la Terre de Dijon,* disponible au Centre de Recherche Géophysique de Garchy et au Centre des Sciences de la Terre, Univ. de Bourgogne, Dijon.

Lamy Au Rousseau, R., 1987. Les dépôts glacio-lacustres de la Combe d'Ain (Jura): approche sédimentologique et dynamique de dépôt. *Rapport de Dipl. Et. Approfondies,* Univ. Bourgogne, Dijon.

Lamy Au Rousseau, R., 1990. *Dynamique sédimentaire dans un lac progalciaire. Deltas, rhthmites, variations du niveau de l'eau, changements climatiques. Exemple du bassin de la Combe d'Ain (Jura, France) au Pléniglaciaire würwien.* Thèse de Doctorat de l'Univ. Bourgogne.

Ravenne, C. & Beghin, P., 1983. Apport des expériences en canal à l'interprétation sédimentologique des dépôts de cônes détritiques sous-marins. *Rev. Inst. français du Pétrole,* **38,** 279–97.

Plio-Pleistocene continental carbonates of the Marseilles Basin travertines (Provence, southern France)

PIERRE ARLHAC,[1] DENISE NURY[1] AND PHILIPPE BLOT[2]

[1]*Laboratoire de Géologique structurale et appliquée, Université de Provence, case postale 28, F-13331, Marseille cedex 3, France*
[2]*Laboratoire de Géologie, Université de Reims, F-51100, Reims, France*

Introduction

The Marseilles Tertiary sedimentary basin, which opens onto the Mediterranean Sea to the west, is surrounded by Mesozoic limestones. Most of the basin fill is Oligocene in age and belongs to a synrift environment (Nury, 1988; Guieu & Roussel, 1990). The travertine formation, commonly ascribed to the Calabrian (early Pleistocene, Villafranchian facies), unconformably overlies those Oligocene deposits (Fig. 1). This formation is currently found in the form of table-like structures scattered over the Marseilles Basin (Dupire, 1985). These subhorizontal or gently dipping forms range in altitude from 214 m above sea level at the Collet-Redon outlier hill, to only 20 m at Janet Cape (Arlhac *et al.*, 1988). Only the 'La Viste Area' travertines and tufas are described in this contribution. Here, tufa is defined as calcareous waterfall deposits and travertine as calcareous stratified deposits.

Description

The Marseilles travertine (and tufa) formation consists of soft, grey, heterogeneous limestone rocks of low density and high porosity. Where the formation is stratified, the stratification planes undulate very irregularly over the Oligocene substratum formation. A schematic lithologic sequence is presented in Fig. 2. The basal deposits (detrital pioneer stage) consist of clays and breccias interbedded with stromatolitic limestone. Above, vertically-oriented basal reed stems coated with carbonate (rhizoliths) interbed with transported stems and probable clays and sands (now missing). This stem sequence is overlain by micritic limestone beds alternating with porous, incrusted stromatolitic limestone beds. Identified algae include *Phormidium* (Cyanophycea – blue-green algae) and *Vaucheria* (Xanthophycea – Bryophyta). The micritic limestone contains moss mounds; some are identified as *Cratoneurum* and *Eucladia*. Contemporaneous with this travertine sequence are subvertically-oriented micritic limestone beds. These beds contain moss structures and abundant small burrows. Towards the top of the lithologic sequence, sapropel marlstones or siliciclastics interbed more commonly with the micritic limestones and stromatolitic limestones. A skeleton of *Elephas meridionalis* (Bonifay,

1967) was found in the siliciclastics of the northern sector of the study area. This entire sequence is topped by siliciclastics.

Interpretation

In the modern, calcareous tufa and travertine are mostly formed through the incrustation of moss. Calcite crystals precipitate as the result of the physiological activity of three organisms: Bryophyta, Cyanophycea, bacteria (Casanova, 1984). Also, Cyanophycea tufa associated with mosses mantles the beds of small creeks. The morphologies of travertine and tufa deposits are directly influenced by the direction of water flow (Fig. 3). Two major morphologies include: (1) subhorizontal calcareous layers containing gravels, leaves and wood fragments, covered by thick calcite layers or beds composed of the basal portions of reed stems or palustral sediment layers, interpreted as travertine forming in marshes, and (2) subvertical drapes, interpreted as waterfall tufas resulting from accretion and growth perpendicular to the vertical water currents.

The travertine and tufa formation of the Marseille Basin is interpreted as stream and marsh deposits which accumulated on the Oligocene substratum during the Plio-Pleistocene. Siliciclastics comprise the traction deposits and the various limestones are biochemical precipitates of travertine and tufa. The basal detrital pioneer layers are interpreted as deposits of alternating fast flow (siliciclastics) and calm (calcareous ponds) conditions. The reed stem layers are the preserved portions of *in situ* and transported reed marshes. The beds of micritic and stromatolitic limestones are interpreted as travertine deposits and marsh tufas in shallow water where reeds could not grow. The sub-vertical micrites are interpreted as waterfall tufas containing the living and feeding burrows of various insects in different life stages that were adapted to a stream environment. Palustrine edge sediments comprised the sapropel marlstones.

Paleogeographic reconstruction

Four successive sedimentary episodes can be recognized in the La Viste area (Figs 3 and 4). The sedimentary area of Episode I

357

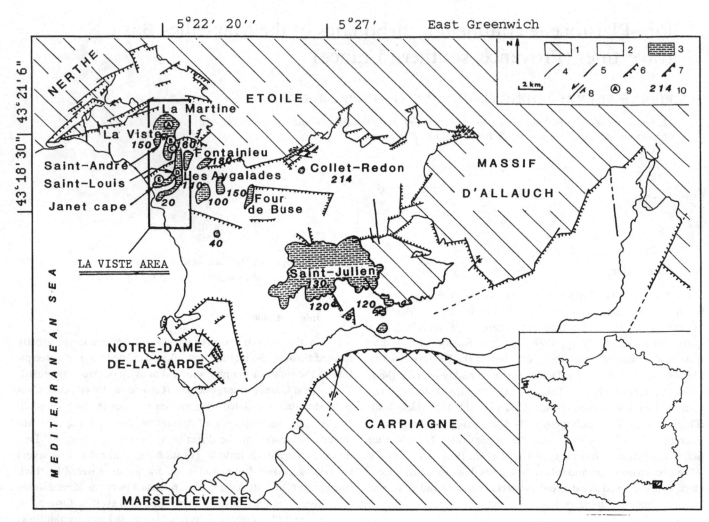

Fig. 1. Location of the two main travertine formations: 'La Viste area' and 'Saint Julien area' travertines in the Marseilles Basin. Legend: 1 = Mesozoic limestones; 2 = Oligocene deposits; 3 = tufa bioherms; 4 = outcrop boundaries; 5 = fractures; 6 = dip-slip faults; 7 = overthrusts; 8 = strike-slip faults; 9 = measured sections; 10 = altitude measurements.

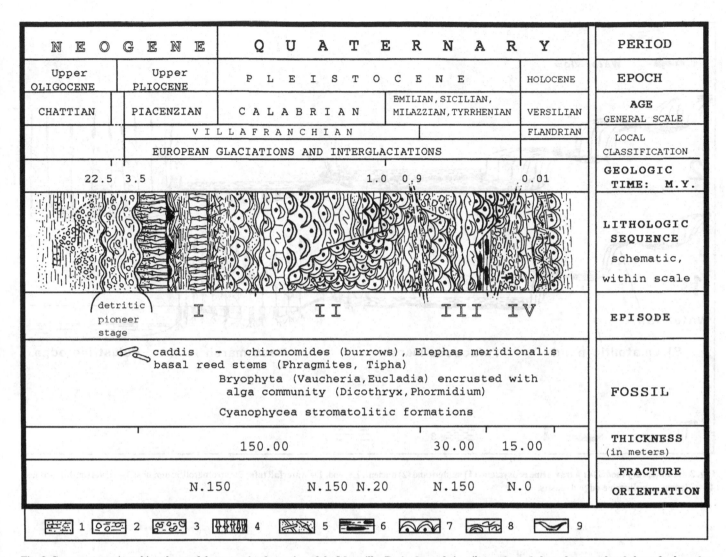

NEOGENE		QUATERNARY				PERIOD
Upper OLIGOCENE	Upper PLIOCENE	PLEISTOCENE			HOLOCENE	EPOCH
CHATTIAN	PIACENZIAN	CALABRIAN	EMILIAN,SICILIAN, MILAZZIAN,TYRRHENIAN		VERSILIAN	AGE GENERAL SCALE
	VILLAFRANCHIAN				FLANDRIAN	LOCAL CLASSIFICATION
	EUROPEAN GLACIATIONS AND INTERGLACIATIONS					
22.5 3.5			1.0 0.9		0.01	GEOLOGIC TIME: M.Y.
						LITHOLOGIC SEQUENCE schematic, within scale
detritic pioneer stage	I	II	III IV			EPISODE
caddis - chironomides (burrows), Elephas meridionalis basal reed stems (Phragmites, Tipha) Bryophyta (Vaucheria,Eucladia) encrusted with alga community (Dicothryx,Phormidium) Cyanophycea stromatolitic formations						FOSSIL
150.00			30.00	15.00		THICKNESS (in meters)
N.150		N.150 N.20	N.150	N.0		FRACTURE ORIENTATION

1 = silt, sands, and clays; 2 = gravel and clays; 3 = breccias; 4 = reed stems (rhizoliths); 5 = gravel and wood fragments; 6 = sapropel marlstones; 7 = tufa mounds; 8 = limestones (stromatolitic and micritic); 9 = stromatolitic limestone

Fig. 2. Summary stratigraphic column of the travertine formation of the Marseilles Basin. Legend: 1 = silt, sands, and clays; 2 = gravel and clays; 3 = breccias; 4 = reed stems (rhizoliths); 5 = gravel and wood fragments; 6 = sapropel marlstones; 7 = tufa mounds; 8 = limestones (stromatolitic and micritic); 9 = stromatolitic limestone. See text for more details.

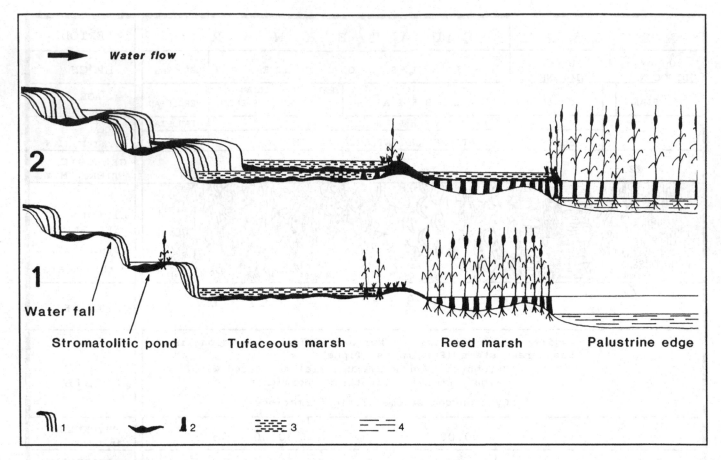

Fig. 3. Sedimentary model for a travertine ecosystem. (1) modern and (2) ancient. Legend: 1 = waterfall tufa; 2 = stromatolitic mounds; 3 = travertine limestones; 4 = calcareous palustrine edge deposits.

(Fig. 4B), containing travertine and tufa deposits, was bounded at the north (La Martine) by a 150° N-oriented fault. This fault was active during deposition; evidence includes breccia beds. A marsh covered the study area with reeds growing in the north and south with water flow in a southerly direction over the shallower, central portion of the marsh. During Episode II (Fig. 4C), escarpments were created due to new fault activity oriented 20° N to 150° N. Drainage was altered and waterfall tufas only formed in certain places. Subsequently, tufa marsh deposits grew above the waterfall deposits. Renewed faulting in the 150° N direction again altered the drainage pattern of the area during Episode III (Fig. 4D). Two sedimentary areas can be distinguished. The southern sector contained a deeper calcareous pond containing marly sapropels, tufas and travertines, while the northern sector was full of calcareous marshes dissected by streams. A new tectonic phase with faults oriented in the 0° N direction altered drainage once again (Episode IV, Fig. 4E) and the resulting deposits were stream siliciclastics and minor limestones.

Bibliography

Arlhac, P., Blot, P. & Nury, D., 1988. The use of Quaternary travertines in neotectonics. Example in the Marseilles graben (Provence, southern France). *C.N.F.-INQUA, Colloquium: Methodology and Applications of Neotectonics, Orleans, France, 3–5 October 1988.*

Bonifay, E., 1967. La tectonique récente du bassin de Marseille dans le cadre de l'évolution post-Miocène du littoral méditerranéen français. *Bull. Soc. Geol., France,* **7**, 549–60.

Casanova, J., 1984. *Genèse d'un carbonate d'un travertin pleistocène: interprétation palaeoécologique du sondage Peyre 1 (Compregnac, Aveyron, France),* pp. 219–25. Geobios Special Memoir No. 8.

Dupire, S., 1985. *Etude cartographique à 1/25,000 de la zone sud du bassin de Marseille. Aperçu géomorphologique et néotectonique.* Masters Thesis, Provence Univ.

Guieu, G. & Roussel, J., 1990. Arguments for the pre-rift uplift and rift propagation in the Ligurian-Provençal Basin (northwestern Mediterranean) in the light of Pyrenean-Provençal orogeny. *Tectonics,* **9** (5), 1113–42.

Nury, D., 1988. *L'Oligocène de Provence méridionale. Stratigraphie, dynamique sédimentaire, reconstitutions paléogéographiques.* Bureau Rech. Geol. & Min. Documents, no. 163.

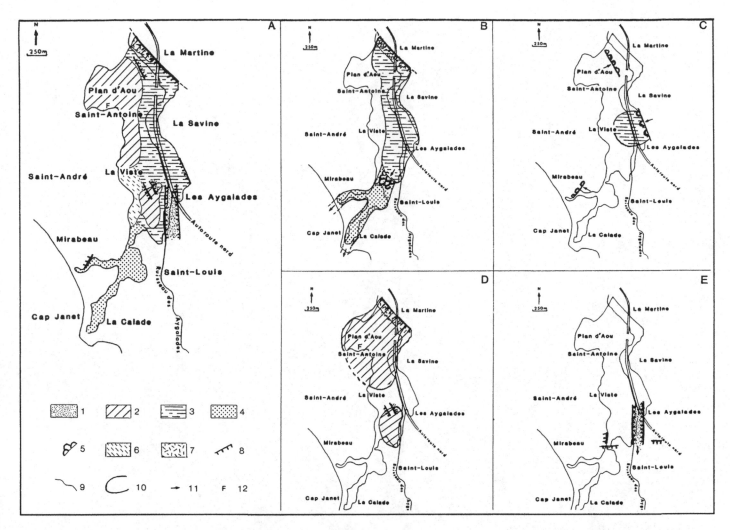

Fig. 4. Schematic maps showing the paleogeographic evolution of the 'La Viste area' travertine formation. Legend: A = general geologic map; B,C,D,E = progressive growth of the extent of the travertine formation, 1 = Episode IV deposits (detrital); 2 = Episode III deposits (tufa limestones and sapropels); 3 = mosses; 4 = basal reed stems; 5 = waterfall tufa = deposits of Episodes I and II; 6 = no data; 7 = breccias; 8 = normal faults; 9 = outcrop boundaries; 10 = boundaries of inferred sedimentation areas; 11 = waterflow direction; 12 = fossil bed. See text for details.

Lake Agassiz; Manitoba, Ontario, Saskatchewan, North Dakota and Minnesota (Canada and USA)

JAMES T. TELLER

Dept. Geological Sciences, University of Manitoba, Winnipeg, Manitoba, Canada, R3T 2N2

Location, age and areal extent

The areal extent of glacial Lake Agassiz sediment in Canada and the USA is shown in Fig. 1. Age of the lake and of its sediment is oldest in the south (ca. 11,500–9000 years BP) and youngest in the north (10,000–8000 years BP) as a result of the retreating and fluctuating glacial margin during the late Quaternary. The maximum area ever covered at one time by the lake was about 350,000 km², whereas the total area affected (shown in Fig. 1) was nearly 1,000,000 km² (Teller *et al.*, 1983).

General nature of sediment and stratigraphic column

The thickness of Lake Agassiz sediment in the main offshore basin of southern Manitoba is shown in Fig. 2. The Assiniboine delta (Fig. 2) is an underflow fan, as are all major deltas of the lake

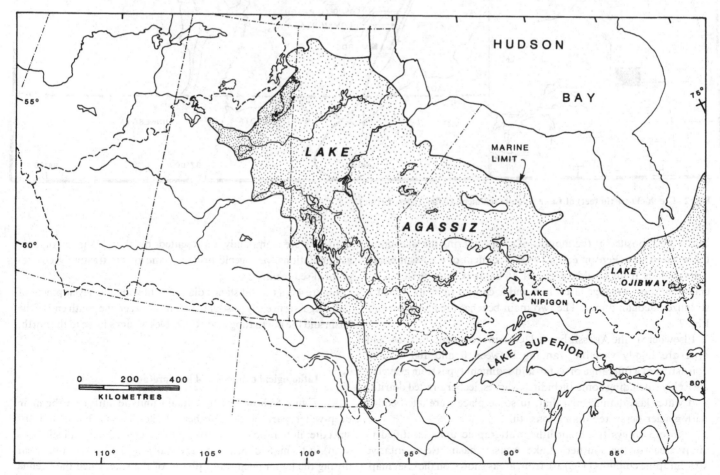

Fig. 1. Areal extent of glacial Lake Agassiz sediment cover in Canada and USA. (After Teller & Thorleifson, 1983.)

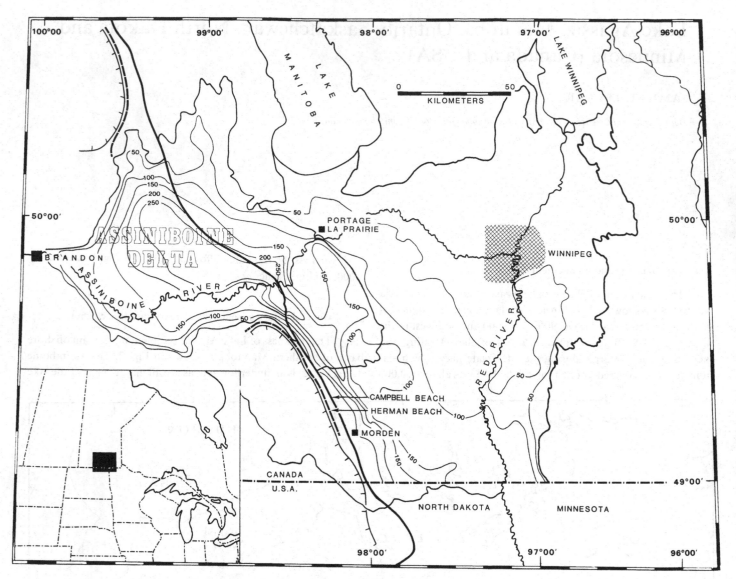

Fig. 2. The thickness (in feet) of Lake Agassiz sediment. (After Teller, 1976b.)

that were deposited at the mouths of rivers entering the lake along the western side (Fenton *et al.*, 1983). Sediment of the Assiniboine delta grades from bouldery gravel at its apex near Brandon to clayey silt near Portage la Prairie, and is capped in many areas by stabilized aeolian dunes. The two main beaches are also shown in Fig. 2.

Elsewhere in the Agassiz basin, lacustrine sediment thickness and type are highly variable, ranging between 0 and 100 m, and consisting of a full range of water-laid sediment types from gravel to silty clay (including some turbidites, varves, iceberg-rafted detritus and water-laid diamictons) that, in some places, are interbedded with glacier-deposited diamictons (tills).

Figure 3 shows the main lithostratigraphic units identified in eight areas of the southern Lake Agassiz basin. Representative stratigraphic columns (1–8) are from areas shown on the inset map. 'Dry' and 'Low-Water Sediment' phases within the Lake Agassiz

sequence are commonly represented by fluvial sands, organic accumulations, pedogenic horizons and/or contrastingly coarse-grained sediment.

Figure 4 is a time–distance diagram showing the development of Lake Agassiz in early post-glacial time between the south end of the basin and the fluctuating Red River–Des Moines Lobe to the north.

Lithological columns and interpretation

Figure 5 shows the three main units identified in the main deep-water part of the southern Lake Agassiz basin and the smectite: illite ratio and relative percentages of clay and silt (plus sand) in them. Sedimentation rates averaged 12–13 mm per year during the major deep-water phases of the lake, when the Brenna and Sherack Formations were deposited, and increased during the

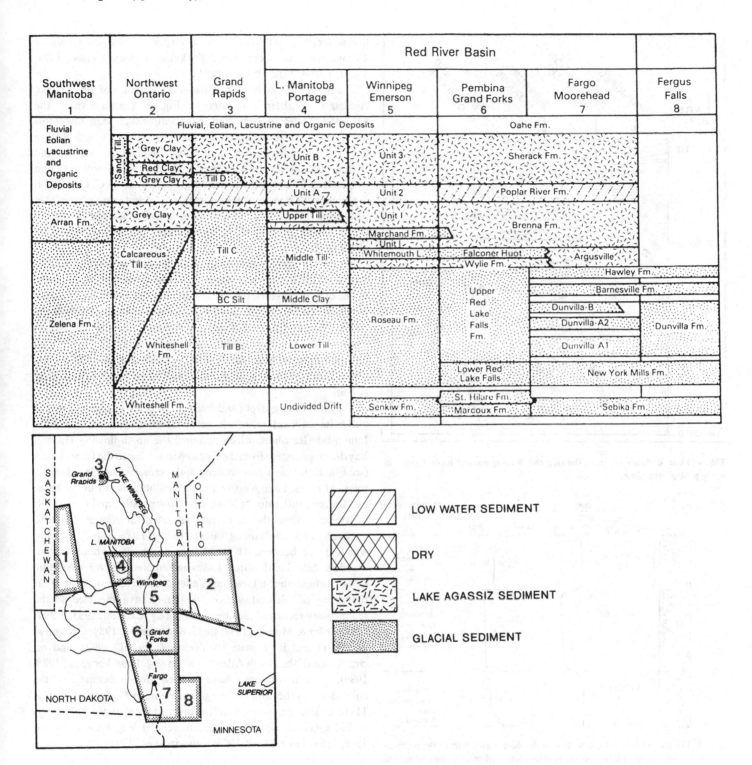

Fig. 3. Main lithostratigraphic units identified in eight areas of the southern Lake Agassiz basin. (After Fenton et al., 1983.)

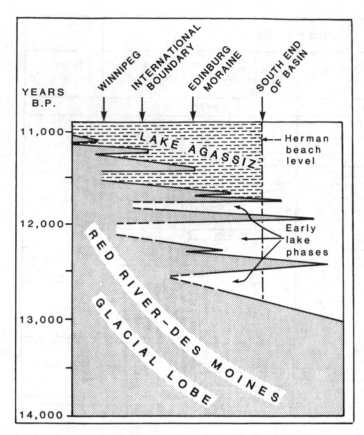

Fig. 4. Time–distance diagram showing the development of Lake Agassiz in early post-glacial time.

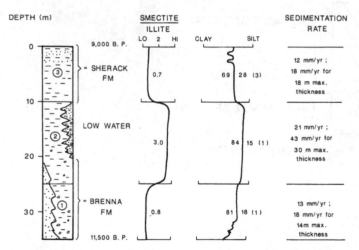

Fig. 5. The three main units identified in the deep-water part of the southern Lake Agassiz basin and the smectite: illite ratio and relative percentages of clay and silt (plus sand).

low-water phase to an average of 21 mm per year and a maximum of 43 mm per year where Unit 2 thickness is greatest (Last, 1974; Teller, 1976b; Teller & Last, 1988).

The modern Lake Manitoba Basin is a remnant of glacial Lake Agassiz in southern Manitoba (see Fig. 2). Figure 6 shows the variation of selected parameters in the sedimentary sequence of this basin and the lithostratigraphic units identified. A and B were deposited when the basin was still part of glacial Lake Agassiz, and correspond to Unit 1 and the Brenna Formation and to Unit 2 and the Sherack Fm, respectively, in Figs. 3 and 5. Units C, D, E and F are all post-9000 years BP (Holocene), and include a number of thin pedogenic zones – some of which are shown by abrupt changes in the 'moisture %' curve of Fig. 6 – that represent low lake levels (see also Fig. 7).

Figure 7 shows the chronology of fluctuations in the level of Lake Agassiz (11,500–9000 years BP) in the southern part of the basin and of one of its successors, Lake Manitoba (9000–0 years BP). Note the numerous pedogenic zones, which indicate low-water levels in the lake's history. Climatic history and other interpretive aspects shown in this figure are based largely on Elson (1967), Harris *et al.* (1974), Ritchie (1976), Arndt (1977), Teller & Last (1979, 1981, 1982), Last (1980), Teller & Fenton (1980) and Nambudiri & Shay (1986).

The areal extent, depth and overflow of glacial Lake Agassiz, as well as its sedimention, were largely controlled by the margin of the Laurentide Ice Sheet, which dammed the north-flowing Hudson Bay drainage system from the Rocky Mountains to the Great Lakes (see Fig. 1); isostatic rebound and outlet erosion also played major roles. At times, Lake Agassiz and its 2 million km² drainage basin overflowed south into the Mississippi River system and to the Gulf of Mexico. When the ice margin retreated far enough north to uncover eastward-draining outlets to the Lake Superior basin, which it did between 11–10 ka and again after about 9.5 ka, overflow shifted to the Great Lakes and entered the North Atlantic Ocean through the St. Lawrence Valley. Studies have suggested that the timing of this eastward overflow may have impacted on the sedimentary record of the Great Lakes (e.g. Clayton, 1983; Teller, 1985; Teller & Mahnic, 1988; Lewis & Anderson, 1989; Colman *et al.*, 1991) and its climate (Anderson & Lewis, 1992), and on circulation of the North Atlantic Ocean (e.g. Broecker *et al.*, 1988, 1989). In turn, this may have affected global climate and the coincidence of the timing of the global Younger Dryas cold event at 11–10 ka and the eastward shift in the routing of about 500 km³/yr of Lake Agassiz water appear to be related (e.g. Broecker *et al.*, 1988, 1989, 1990; Teller, 1990; Keigwin *et al.*, 1991).

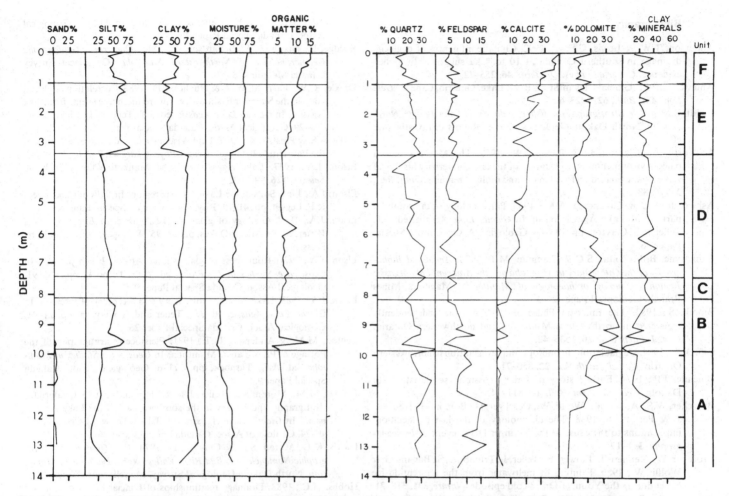

Fig. 6. The variation of selected parameters and identified lithostratigraphic units of the Lake Agassiz basin. (After Teller & Last, 1981.)

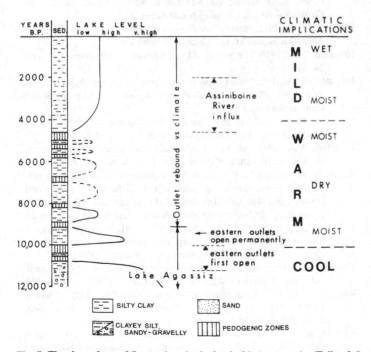

Fig. 7. The chronology of fluctuations in the level of Lake Agassiz. (Teller & Last, 1981; Fig. 12.)

Bibliography

Anderson, T.W. & Lewis, C.F.M., 1992. Climatic influences of deglacial drainage in southern Canada at 10 to 8 ka suggested by pollen evidence. *Geograph. phys. Quatern.*, **46**, 255–72.

Antevs, E., 1951. Glacial Clay in Steep Rock Lake, Ontario, Canada. *Geol Soc. Am. Bull.*, **62**, 1223–62.

Arndt, B.M., 1977. *Stratigraphy of offshore sediment of Lake Agassiz, North Dakota.* North Dakota Geological Survey Report of Investigation 60.

Ashworth, A., Clayton, L. & Bickley, W., 1972. The Mosbeck Site: a paleoenvironmental interpretation of the late Quaternary history of Lake Agassiz based on fossil insect and mollusc remains. *Quat. Res.*, **2**, 176–88.

Ashworth, A.C. & Cavancara, A.M., 1983. Paleoecology of the southern part of the Lake Agassiz basin. In *Glacial Lake Agassiz*, ed. J.T. Teller & L. Clayton, pp. 133–56. Geological Assoc. Canada Special Paper 26.

Bannatyne, B.B., Zoltai, S.C. & Tamplin, M.J., 1970. *Annotated Bibliography of the Quaternary in Manitoba and the Adjacent Lake Agassiz Region (including archaeology of Manitoba).* Manitoba Mines Branch, Geological Paper 2/70.

Björck, S., 1983. The Emerson Phase of Lake Agassiz, independently registered in northwestern Minnesota and northwestern Ontario: *Can. J. Earth Sci.*, **20**, 1536–42.

Björck, S., 1984. Deglaciation chronology and revegetation in northwestern Ontario. *Can. J. Earth Sci.*, **22**, 850–71.

Bluemle, J.P., 1974. Early history of Lake Agassiz in southeast North Dakota. *Geol. Soc. Am. Bull.*, **85**, 811–14.

Broecker, W.S., Andre, M., Wolfli, W., Oeschger, H., Bonani, G., Kennett, J. & Peteet, D., 1988. The chronology of the last deglaciation: implications to the cause of the Younger Dryas event. *Paleoceanography*, **3**, 1–19.

Broecker, W., Kennett, J., Flower, B., Teller, J., Trumbore, S., Bonani, G. & Wolfli, W., 1989. Routing of meltwater from the Laurentide Ice Sheet during the Younger Dryas cold episode. *Nature*, **341**, 318–21.

Broecker, W., Bond, G., Kas, M., Bonani, G. & Wolfli, W., 1990. A salt oscillator in the glacial Atlantic? I. The concept. *Paleoceanography*, **5**, 469–77.

Brophy, J.A. & Bluemle, J.P., 1983. The Sheyenne River: its geological history and effects on Lake Agassiz. In *Glacial Lake Agassiz*, ed. J.T. Teller & L. Clayton, pp. 173–86. Geological Assoc. Canada Special Paper 26.

Christiansen, E.A., 1979. The Wisconsinan deglaciation of southern Saskatchewan and adjacent areas. *Can. J. Earth Sci.*, **16**, 913–38.

Clayton, L., 1983. Chronology of Lake Agassiz drainage to Lake Superior. In *Glacial Lake Agassiz*, ed. J.T. Teller & L. Clayton, pp. 291–307. Geological Assoc. Canada Special Paper 26.

Clayton, L., Laird, W., Klassen, R. & Kupsch, W., 1965. Intersecting minor lineations on Lake Agassiz plain. *J. Geol.*, **73**, 652–6.

Clayton, L. & Moran, S.R., 1982. Chronology of Late Wisconsinan Glaciation in Middle North America. *Quat. Sci. Rev.*, **1**, 55–82.

Clayton, L., Teller, J.T. & Attig, J.W., 1985. Surging of the southwestern part of the Larentide Ice Sheet. *Boreas*, **14**, 235–41.

Colman, S.M., Jones, G., Forester, R. & Foster, D., 1990. Holocene paleoclimatic evidence and sedimentation rates from a core in southwestern Lake Michigan. *J. Paleolimnol.*, **4**, 269–84.

David, P.P., 1971. The Brookdale Road section and its significances in the chronological studies of dune activities in the Brandon Hills of Manitoba. In *Geoscience Studies in Manitoba*, ed. A.C. Turnock, pp. 293–9. Geological Assoc. Canada Special Paper 9.

Dredge, L.A., 1982. Relict ice-scour marks and late phases of Lake Agassiz in northernmost Manitoba. *Can. J. Earth Sci.*, **19**, 1079–87.

Dredge, L.A., 1983. Character and development of Northern Lake Agassiz and its relation to Keewatin and Hudsonian ice regimes. In *Glacial Lake Agassiz*, ed. J.T. Teller & L. Clayton, pp. 117–31. Geological Assoc. Canada Special Paper 26.

Dredge, L., Nixon, F. & Richardson, R., 1986. *Quaternary Geology and Geomorphology of Northwestern Manitoba.* Geological Survey Canada Memoir 418.

Drexler, C.W., Farrand, W.R. & Hughes, J.D., 1983. Correlation of glacial lakes in the Superior basin with eastward discharge events from lake Agassiz. In *Glacial Lake Agassiz*. ed. J.T. Teller & L. Clayton, pp. 309–29. Geological Assoc. Canada Special Paper 26.

Dyke, A.S. & Prest, V.K., 1987. Late Wisconsinan and Holocene history of the Laurentide Ice Sheet. *Geogr. Phys. Quat.*, **41**, 237–63.

Elson, J.A., 1957. Lake Agassiz and the Mankato-Valders problem. *Science*, **126**, 99–1002.

Elson, J.A., 1961. Soils of the Lake Agassiz region. In *Soils of Canada*, ed. R.F. Legget, pp. 41–79. Royal Soc. Canada Special Publ. 3.

Elson, J.A., 1967. Geology of glacial Lake Agassiz. In *Life, Land, and Water*, ed. W. Mayer-Oakes, pp. 36–95. Winnipeg, Univ. Manitoba Press.

Elson, J.A., 1971. Roundness of glacial Lake Agassiz beach pebbles. In *Geoscience Studies in Manitoba*, ed. A.C. Turnock, pp. 285–91. Geological Assoc. Canada Special Paper 9.

Elson, J.A., 1983. Lake Agassiz – discovery and a century of research. In *Glacial Lake Agassiz*, ed. J.T. Teller and L. Clayton, pp. 21–41. Geological Assoc. Canada Special Paper 26.

Fenton, M.M. & Anderson, D.T., 1971. Pleistocene stratigraphy of the Portage la Prairie area, Manitoba. In *Geoscience Studies in Manitoba*, ed. A.C. Turnock, pp. 271–6. Geological Assoc. Canada Special Paper 9.

Fenton, M.M., Moran, S.R, Teller, J.T. & Clayton, L., 1983. Quaternary stratigraphy and history in the southern part of the Lake Agassiz basin. In *Glacial Lake Agassiz*, ed. J.T. Teller & L. Clayton, pp. 49–74. Geological Assoc. Canada Special Paper 26.

Harris, K.L., Moran, S.R. & Clayton, L., 1974. *Late Quaternary Stratigraphic Nomenclature, Red River Valley, North Dakota and Minnesota.* North Dakota Geological Survey, Miscellaneous Series 52.

Hobbs, H.C., 1983. Drainage relationships of Glacial Lakes Aitkin and Upham and Early Lake Agassiz in northeastern Minnesota. In *Glacial Lake Agassiz*, ed. J.T. Teller and L. Clayton, pp. 245–59. Geological Assoc. Canada Special Paper 26.

Johnston, W.A., 1915. *Rainy River District, Ontario; Surficial Geology and Soils.* Geological Survey Canada Memoir 82.

Johnston, W.A., 1916. The genesis of Lake Agassiz: a confirmation. *J. Geol.*, **24**, 625–38.

Johnston, W.A., 1921. *Winnipegosis and Upper Whitemouth River areas, Manitoba; Pleistocene and Recent deposits.* Geological Survey Canada Memoir 128.

Johnston, W.A., 1934. *Surface Deposits and Groundwater Supply of the Winnigeg Map Area.* Geological Survey Canada Memoir 174.

Johnston, W.A., 1946. Glacial Lake Agassiz, with special reference to the mode of deformation of the beaches. *Geol. Surv. Can. Bull.*, **7**.

Kehew, A. & Clayton, L., 1983. Late Wisconsinan floods and development of the Souris-Pembina spillway system in Saskatchewan, North Dakota, and Manitoba. In *Glacial Lake Agassiz*, eds. J.T. Teller & L. Clayton, pp. 187–209. Geological Assoc. Canada Special Paper 26.

Keigwin, L.D., Jones, G.A., & Lehman, S.J., 1991. Deglacial meltwater discharge, North Atlantic deep circulation, and abrupt climate change. *J. Geophysical Res.*, **96**, no. C9, 16,811–16,826.

Klassen, R.W., 1972. Wisconsin events and the Assiniboine and Qu'Appelle valleys of Manitoba and Saskatchewan. *Can. J. Earth Sci.*, **9**, 544–60.

Klassen, R.W., 1975. Quaternary geology and geomorphology of Assiniboine and Qu'Appelle valleys of Manitoba and Saskatchewan. *Geol. Surv. Can. Bull.*, **228**.

Klassen, R.W., 1983a. Assiniboine delta and the Assiniboine-Qu'Appelle valley system – implications concerning the history of Lake Agassiz in southwestern Manitoba. In *Glacial Lake Agassiz*, ed. J.T. Teller &

L. Clayton, pp. 211–29. Geological Assoc. Canada Special Paper 26.

Klassen, R.W., 1983b. Lake Agassiz and the late glacial history of northern Manitoba. In *Glacial Lake Agassiz*, ed. J.T. Teller and L. Clayton, pp. 97–115. Geological Assoc. Canada Special Paper 26.

Klassen, R.W., 1986. *Surficial Geology of North-Central Manitoba*. Geological Survey of Canada Memoir 419.

Last, W.M., 1974. *Clay Mineralogy and Stratigraphy of Offshore Lake Agassiz Sediments in Southern Manitoba*. MSc Thesis, Univ. Manitoba.

Last, W.M., 1980. *Sedimentology and Post-Glacial History of Lake Manitoba*. PhD Dissertation, Univ. Manitoba.

Last, W.M., 1982. Holocene carbonate sedimentation in Lake Manitoba, Canada. *Sedimentology*, 29, 691–704.

Last, W.M. & Teller, J.T., 1983. Holocene climate and hydrology of the Lake Manitoba basin. In *Glacial Lake Agassiz*, ed. J.T. Teller & L. Clayton, pp. 333–53. Geological Assoc. Canada Special Paper 26.

Leverett, F., 1932. *Quaternary Geology of Minnesota and Parts of Adjacent States*. US Geol. Survey Prof. Paper 161.

Lewis, C.F.M. & Anderson, T.W., 1989. Oscillations of levels and cool phases of the Laurentide Great Lakes caused by inflows from glacial Lakes Agassiz and Barlow-Ojibway. *J. Paleolimnol.*, 2, 99–146.

Matsch, C.L., 1983. River Warren, the southern outlet of glacial Lake Agassiz, ed. J.T. Teller & L. Clayton, pp. 231–44. Geological Assoc. Canada Special Paper 26.

Mayer-Oakes, W.J. (Ed.), 1967. *Life, Land, and Water*. Proceedings of conference on environmental studies of glacial Lake Agassiz region: University of Manitoba Press, Winnipeg.

McAndrews, J.H., 1967. Paleoecology of the Seminary and Mirror Pool peat deposits. In *Life, Land, and Water*, ed. W. Mayer-Oakes, pp. 253–69. University of Manitoba Press, Winnipeg.

McPherson, R.A., Leith, E.I. & Anderson, D.T., 1971. Pleistocene stratigraphy of a portion of southeastern Manitoba. In *Geoscience Studies in Manitoba*, ed. A.C. Turnock, pp. 277–83. Geological Assoc. Canada Special Paper 9.

Mollard, J.D., 1983. The origin of reticulate and orbicular patterns on the floor of the Lake Agassiz basin. In *Glacial Lake Agassiz*, ed. J.T. Teller & L. Clayton, pp. 355–74. Geological Assoc. Canada Special Paper 26.

Moran, S.R., 1972. *Subsurface geology and foundation conditions in Grand Forks, North Dakota*. North Dakota Geological Survey, Miscellaneous Series 44.

Moran, S., Arndt, M., Bluemle, J., Camara, M., Clayton, L., Fenton, M., Harris, K., Hobbs, H., Keatinge, R., Sackreiter, D., Salomon, N. & Teller, J., 1976. Quaternary stratigraphy and history of North Dakota, southern Manitoba, and northwestern Minnesota. In *Quaternary Stratigraphy of North America*, ed. W.C. Mahaney, pp. 133–58. Dowden, Hutchison, & Ross, Stroudsburg.

Moran, S., Clayton, L. & Cvancara, A., 1971. New sedimentological and paleontological evidence for history of Lake Agassiz; Snake River section, Red Lake County, Minnesota. *Proc. North Dakota Acad. Sciences*, 24, 61–73.

Nambudiri, E.M.V. & Shay, C.T., 1986. Late Pleistocene and Holocene pollen stratigraphy of the Lake Manitoba basin, Canada. *Palaeontographica, Abt. B.* 202, 155–77.

Nambudiri, E.M.V., Teller, J.T. & Last, W., 1980. Pre-Quaternary microfossils – a guide to errors in radiocarbon dating. *Geology*, 8, 123–6.

Nielsen, E., Gryba, E. & Wilson, M., 1984. Bison remains from a Lake Agassiz spit complex in the Swan River valley, Manitoba: depositional environment and paleoecological implications. *Can. J. Earth Sci.*, 21, 829–42.

Nielson, E., McKillop, W. & McCoy, J., 1982. The age of the Hartman moraine and the Campbell beach of Lake Agassiz in northwestern Ontario. *Can. J. Earth Sci.*, 19, 1933–7.

Nikiforoff, C., 1945. The life history of Lake Agassiz; alternative interpretation. *Am. J. Sci.*, 245, 205–39.

Prest, V.K., 1963. *Red Lake-Lansdowne House Area, Northwestern Ontario. Surficial Geology*. Geological Survey Canada Paper 63–6.

Ritchie, J.C., 1976. The late Quaternary vegetation history of the western interior of Canada. *Can. J. Botany*, 54, 1793–818.

Ritchie, J.C., 1983. Paleoecology of the central and northern parts of the glacial Lake Agassiz basin. In *Glacial Lake Agassiz*, ed. J.T. Teller & L. Clayton, pp. 157–70. Geological Assoc. Canada Special Paper 26.

Ritchie, J.C. & Koivo, L.K., 1975. Postglacial diatom stratigraphy in relation to the recession of glacial Lake Agassiz. *Quat. Res.*, 5, 529–40.

Rittenhouse, G., 1934. A laboratory study of an unusual series of varved clays from northern Ontario. *Am. J. Sci.*, 28, 110–20.

Rominger, J. & Rutledge, P., 1952. Use of soil mechanics data in correlating and interpretation of Lake Agassiz sediments. *J. Geol.*, 60, 160–80.

Ross, J. & Karner, F., 1967. Petrography of core and well samples from Lake Agassiz and associated sediments, Grand Forks, North Dakota. *Proc North Dakota Acad. Sci.*, 21, 147–61.

Schreiner, B.T., 1983. Glacial Lake Agassiz in Saskatchewan. In *Glacial Lake Agassiz*, ed. J.T. Teller & L. Clayton, pp. 75–96. Geological Assoc. Canada Special Paper 26.

Shay, C.T., 1967. Vegetation history of the southern Lake Agassiz basin during the past 12,000 years. In *Life, Land, and Water*, ed. W. Mayer-Oakes, pp. 231–52. University of Manitoba Press, Winnipeg.

Teller, J.T., 1976a. *Thickness of Fine-Grained Sediment (clay, silt, sand) in the Main Lake Agassiz Basin (Red and Assiniboine River Valleys) of Southern Manitoba*. Manitoba Mineral Resources Division, Surficial Map 76–2.

Teller, J.T., 1976b. Lake Agassiz deposits in the main offshore basin of southern Manitoba. *Can. J. Earth Sci.*, 13, 27–43.

Teller, J.T., 1985. Glacial Lake Agassiz and its influence on the Great Lakes. In *Quaternary Evolution of the Great Lakes*. ed. P.F. Karrow & P.E. Calkin, pp. 1–16. Geological Assoc. Canada Special Paper 30.

Teller, J.T., 1987. Proglacial lakes and the southern margin of the Laurentide Ice Sheet. In *The Last Deglaciation in North America and adjacent oceans. Responses and mechanisms*, ed. W. Ruddiman & H.E. Wright, pp. 39–69. Geological Society of America, The Geology of North America, K-3.

Teller, J.T., 1989. Importance of the Rossendale site in establishing a deglacial chronology along the southwestern margin of the Laurentide Ice Sheet. *Quatern. Res.*, 32, 12–23.

Teller, J.T., 1990. Volume and routing of late-glacial runoff from the southern Laurentide Ice Sheet. *Quaternary Res.*, 34, 12–23.

Teller, J.T. & Clayton, L. (Eds.), 1983. *Glacial Lake Agassiz*, Geological Assoc. Canada Special Paper 26.

Teller, J.T. & Fenton, M.M., 1980. Late Wisconsinan glacial stratigraphy and history of southeastern Manitoba. *Can. J. Earth Sci.*, 17, 19–35.

Teller, J.T. & Last, W.M., 1979. *Post-Glacial Sedimentation and History in Lake Manitoba*. Manitoba Dept. Mines, Natural Resources and Environment, Report 79–41.

Teller, J.T. & Last, W.M., 1981. Late Quaternary history of Lake Manitoba, Canada. *Quatern. Res.*, 16, 97–116.

Teller, J.T. & Last, W.M., 1982. Pedogenic horizons in lake sediments. *Earth Surf. Proc. Landforms*, 7, 367–79.

Teller, J.T. & Last, W.M., 1988. Lake Agassiz – a giant Pleistocene freshwater lake. *Internat. Assoc. Sedimentologists Symposium on Sedimentology Related to Mineral Deposits, Beijing, Abstracts*, pp. 255–6.

Teller, J.T. & Mahnic, P., 1988. History of sedimentation in the northwestern Lake Superior basin and its relation to Lake Agassiz overflow. *Can. J. Earth Sci.*, 25, 1660–73.

Teller, J.T., Moran, S.R. & Clayton, L., 1980. The Wisconsinan deglaciation of southern Saskatchewan and adjacent areas: Discussion. *Can. J. Earth Sci.*, 17, 539–41.

Teller, J.T. & Thorleifson, L.H., 1983. The Lake Agassiz–Lake Superior Connection. In *Glacial Lake Agassiz*, eds. J.T. Teller & L. Clayton, pp. 261–90. Geological Assoc. Canada Special Paper 26.

Teller, J.T., Thorleifson, L.H., Dredge, L.A., Hobbs, H.C. & Schreiner, B.T., 1983. Maximum extent and major features of Lake Agassiz. In *Glacial Lake Agassiz*, ed. J.T. Teller & L. Clayton, pp. 43–5. Geological Assoc. Canada Special Paper 26.

Tuthill, S.J., 1967. Paleo-zoology and molluscan paleontology of the glacial Lake Agassiz region. In *Life, Land, and Water*, ed. W. Mayer-Oakes, pp. 299–312. University of Manitoba Press, Winnipeg.

Upham, W., 1895. *The Glacial Lake Agassiz.* US Geological Survey Monograph 25.

Winter, T.C., 1967. Linear sand and gravel deposits in the subsurface of glacial Lake Agassiz. In *Life, Land, and Water*, ed. W. Mayer-Oakes, pp. 141–54. University of Manitoba Press, Winnipeg.

Wright, H.E. & Glaser, P.H., 1983. Postglacial peatlands of the Lake Agassiz plain, northern Minnesota. In *Glacial Lake Agassiz*, ed. J.T. Teller & L. Clayton, pp. 375–89. Geological Assoc. Canada Special Paper 26.

Wyder, J.E., 1971. Subsurface stratigraphy of the Lake Agassiz basin in central Manitoba. In *Geoscience Studies in Manitoba*, ed. A.C. Turnock, pp. 263–9. Geological Assoc. Canada Special Paper 9.

Zoltai, S.C., 1961. Glacial history of part of northwestern Ontario. *Proc. Geol. Assoc. Can.*, **13**, 61–83.

Zoltai, S.C., 1967. Eastern outlets of Lake Agassiz. In *Life, Land, and Water*, ed. W. Mayer-Oakes, pp. 107–20. University of Manitoba Press, Winnipeg.

The Bonneville Basin, Quaternary, western United States

CHARLES G. OVIATT,[1] DOROTHY SACK[2] AND DONALD R. CURREY[3]

[1] *Dept. Geology, Kansas State University, Manhattan, KS, 66506 USA*
[2] *Dept. Geography, University of Wisconsin, Madison, WI, 53706 USA*
[3] *Dept. Geography, University of Utah, Salt Lake City, UT, 84112 USA*

The Bonneville Basin is located in the eastern Great Basin of the western United States (Fig. 1). It is a closed topographic basin (e.g. Eardley *et al.*, 1957; Gwynn, 1980; Currey, 1980, 1990; Currey *et al.*, 1984a; Arnow & Stephens, 1986) created by Basin and Range extensional tectonism beginning about 15 Ma (Gwynn, 1980; Miller, 1990). Although Tertiary lakes existed in and near the Bonneville basin, the details of their stratigraphic records are not well known, and the boundaries of the Tertiary basins may have been considerably different than those of the Quaternary basin. Only the Quaternary part of the record is treated here. For more information and references on the Tertiary rocks, see Machette (1985) and Miller (1990).

Closed-basin lakes have expanded and contracted in the Bonneville Basin many times during the Quaternary (Figs 2 and 3) largely in response to alternating periods of increased and decreased effective moisture (e.g. Eardley *et al.*, 1973; Currey & James, 1982; Scott *et al.*, 1983; Machette, 1985, 1988; Oviatt & Currey, 1987; Benson *et al.*, 1990; Currey, 1990; Oviatt *et al.*, 1990). Late Pleistocene and Holocene shorelines are well preserved in the basin (Currey, 1980, 1982, 1990; Currey *et al.*, 1983, 1984a, b; Scott *et al.*, 1983; Oviatt *et al.*, 1990). Lake Bonneville (28,000–13,000 years BP; Currey, 1990) is the only Quaternary lake in the basin known to have overflowed its external threshold to achieve open-basin status.

Researchers have described lacustrine sediments in the Bonneville Basin dating from most of the Quaternary Period. The best documented deposits, however, are of late Pleistocene and Holocene age. Named stratigraphic units include the Little Valley, Cutler Dam and Bonneville Alloformations (late Pleistocene; Figs 3 and 4), but many other lacustrine, alluvial and eolian units remain unnamed. Geochronometric control has been obtained using the following techniques: tephrochronology, U-series, radiocarbon, amino-acid, thermoluminescence, soil stratigraphy and geologic correlation (reliability indices D, E, and F; e.g. Broecker & Kaufman, 1965; Scott *et al.*, 1983; Spencer *et al.*, 1984; McCalpin, 1986; McCoy, 1987; Oviatt & Nash, 1989; Benson, 1990).

Figure 4 shows an example of the kinds of stratigraphic information that are accessible in outcrops in the Bonneville Basin. In this section, the Little Valley Alloformation is represented by a single facies (marl), but nearby sections include a fine-grained deltaic facies. The Bonneville Alloformation is represented by three facies: marl, fine-grained deltaic sand and fluvial sand. Available chronometric data are listed in the figure caption.

The total thickness of Quaternary sediments varies around the basin, but ranges up to about 300 m (Eardley *et al.*, 1973). The sediments deposited in the succession of lakes in the Bonneville Basin are diverse, reflecting the changing physical environment through time and the numerous depositional environments (facies) in the large, intricate basin. They include marls, fine-grained and coarse-grained deltas, and barrier/spit complexes of sand and gravel (e.g., Eardley *et al.*, 1973; Gwynn, 1980). Fossils include pollen, plant macrofossils, diatoms, algal tufa, sponges, ostracodes, molluscs, fish and land mammals (Gilbert, 1890; Eardley *et al.*, 1957; Eardley & Gvodetsky, 1960; Bissell, 1963; Bright, 1963; Eardley *et al.*, 1973; Gwynn, 1980; Currey & James, 1982; Scott *et al.*, 1983; Spencer *et al.*, 1984; McCoy, 1987; Oviatt, 1987a; Oviatt *et al.*, 1990.

Bonneville Basin lacustrine sediments are composed of fine to coarse siliciclastics, carbonates, and saline minerals (Eardley & Gvosdetsky, 1960; Eardley *et al.*, 1973; Spencer *et al.*, 1984; McKenzie & Eberli, 1987; Currey, 1990). Geochemically the lake water has varied from highly saline to alkaline to freshwater (Gwynn, 1980; Spencer *et al.*, 1984, 1985a, b; McKenzie & Eberli, 1987). Carbon, oxygen and sulfur isotopes have been studied in late Quaternary to Holocene sediments (Spencer *et al.*, 1984; McKenzie & Eberli, 1987).

Many researchers have investigated the isostatic response of the crust to the Lake Bonneville water load (Gilbert, 1890; Crittenden, 1963; Currey, 1982; Currey *et al.*, 1983, 1984b; Scott *et al.*, 1983; Bills & May, 1987). Studies of the neotectonics of the basin are continuing (Machette, 1988).

371

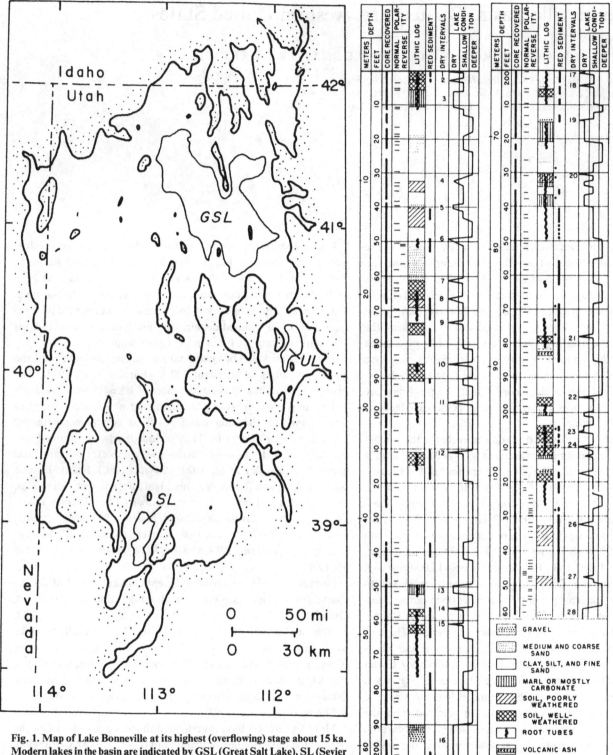

Fig. 1. Map of Lake Bonneville at its highest (overflowing) stage about 15 ka. Modern lakes in the basin are indicated by GSL (Great Salt Lake), SL (Sevier Lake), and UL (Utah Lake).

Fig. 2. Log of the Burmester core, taken by Eardley and colleagues on the south shore of Great Salt Lake. The Bishop volcanic ash (740 ka) is shown at a depth of 100 m (from Eardley *et al.*, 1973, Fig. 1, p. 212). Reproduced with permission of the authors and the Geological Society of America.

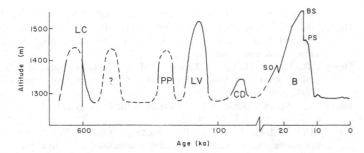

Fig. 3. Major lake cycles in the Bonneville Basin during the middle and late Quaternary (modified from McCoy, 1987, Fig. 7). Named lake cycles are: Pokes Point (PP), Little Valley (LV), Cutler Dam (CD), and Bonneville (B). The Lava Creek ash (LC) is shown, as are the Stansbury oscillation (SO), Bonneville shoreline (BS), and Provo shoreline (PS).

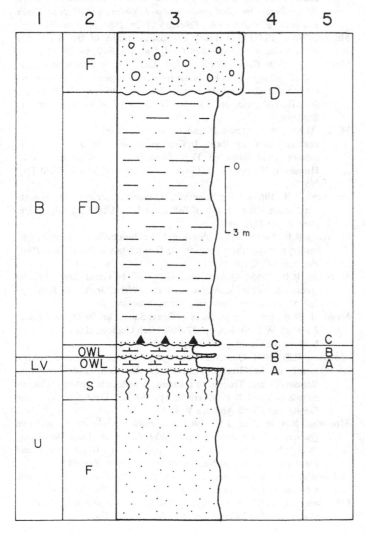

Fig. 4. Stratigraphic section exposed along the Sevier River, northeast of Delta, Utah (39° 25.7′ N., 112° 26.4′ W), showing two lacustrine stratigraphic units. Column 1 indicates stratigraphic units: U = unnamed alluvial deposits; LV = Little Valley Alloformation; B = Bonneville Alloformation. Column 2 indicates depositional environments: F = fluvial; S = soil; OWL = open-water lacustrine (littoral to pelagic marl deposition); FD = fine-grained deltaic (underflow fan). Column 3 indicates lithology: circles and dots = pebbly sand; dots and dashes = thinly bedded silt and fine sand; triangles = reworked basaltic volcanic ash; hachured dashes = marl; dots = sand. Column 4 indicates unconformities (wavy contacts): A = pre-Little Valley unconformity; B = Little Valley/Bonneville unconformity caused by regression and transgression of lake; C = intra-Bonneville unconformity caused by scour of bottom currents during the Bonneville flood; D = intra-Bonneville unconformity caused by river entrenchment near the end of the Bonneville cycle. Column 5 indicates geochronology determined at this and nearby sections: A = amino acid ratios in fossil snail shells (alloisoleucine/isoleucine [aIle/Ile], hydrolysate) averaging 0.4 ± 0.06 (*Amnicola*), [230]Th ages of 13.6 and 14.7 ka, and ^{14}C ages ranging from about 18 to 12 ka; C = Pavant Butte basaltic ash (15.5 ka) reworked from Bonneville OWL unit (this ash is found in-situ within the Bonneville marl elsewhere in the area). The top of the section is at an altitude of 1448 m. Data are from an unpublished manuscript by Oviatt, McCoy, and Nash (1990), and Varnes & Van Horn (1984).

Bibliography

Arnow, T. & Stephens, D., 1986. *Hydrologic characteristics of the Great Salt Lake, Utah: 1847–1986.* US Geological Survey Water-Supply Paper 2332.

Benson, L.V., Currey, D.R., Dorn, R.I., Lajoie, K.R., Oviatt, C.G., Robinson, S.W., Smith, G.I. & Stine, S., 1990. Chronology of expansion and contraction of four Great Basin lake systems during the past 35,000 years. *Palaeogeogr., Palaeoclimatol., Palaeoecol.,* **78,** 241–86.

Benson, L., Currey, D.R., Lao, Y. & Hostetler, S., 1992. Lake-size variations in the Lahontan and Bonneville basins between 13,000 and 9000 ^{14}C yr BP. *Palaeogeogr., Palaeoclimatol., Palaeoecol.,* **95,** 19–32.

Bills, B.G. & May, G.M., 1987. Lake Bonneville: Constraints on lithospheric thickness and upper mantle viscosity from isostatic warping of Bonneville, Provo, and Gilbert stages shorelines. *J. Geophys. Res.,* **92** (no. B11), 11,493–11,508.

Bissell, H.J., 1963. Lake Bonneville: geology of southern Utah valley. *US Geol. Surv. Prof. Paper* 257-B, 101–30.

Bright, R., 1963. *Pleistocene lakes Thatcher and Bonneville, southeastern Idaho.* PhD Dissertation, University of Minnesota, Minneapolis, MN.

Broecker, W.S., & Kaufman, A., 1965. Radiocarbon chronology of Lake Lahontan and Lake Bonneville II, Great Basin. *Geol. Soc. Am. Bull.,* **76,** 537–66.

Crittenden, M.D., 1963. New data on the isostatic deformation of Lake Bonneville. *US Geol. Surv. Prof. Paper* 454-E, pp. E1–E31.

Currey, D.R., 1980. Coastal geomorphology of Great Salt Lake and vicinity: *Utah Geol. Mineral Surv. Bull.,* **116,** 69–82.

Currey, D.R., 1982. *Lake Bonneville: selected features of relevance to neotectonic analysis.* US Geol. Survey Open-File Report 82–1070.

Currey, D.R., 1990. Quaternary paleolakes in the evolution of semidesert basins, with special emphasis on Lake Bonneville and the Great Basin, USA. *Palaeogeogr., Palaeoclimatol., Palaeoecol.,* **76,** 189–214.

Currey, D.R., Atwood, G. & Mabey, D.R., 1984a. *Major levels of Great Salt Lake and Lake Bonneville.* Utah Geol. Mineral Surv. Map 73.

Currey, D.R. & James, S.R., 1982. Paleoenvironments of the northeastern Great Basin and northeastern Basin rim region: a review of geological and biological evidence. *Soc. Am. Archaeol Papers* No. 2, 27–52.

Currey, D.R. & Oviatt, C.G., 1985. Durations, average rates, and probable causes of Lake Bonneville expansions, stillstands, and contractions during the last deep-lake cycle, 32,000 to 10,000 years ago. In *Problems of and prospects for predicting Great Salt Lake levels;* papers from a conference held in Salt Lake City, March 26–28, 1985, ed. P.A. Kay & H.F. Diaz, pp. 9–24. Center for Public Affairs and Administration, University of Utah. Also in *Geograph. J. Korea,* **10,** 1085–99.

Currey, D.R., Oviatt, C.G. & Plyler, G.B., 1983. Lake Bonneville stratigraphy, geomorphology, and isostatic deformation in west-central Utah. *Utah Geol. Mineral Surv. Spec. Studies* 62, 63–82.

Currey, D.R., Oviatt, C.G. & Czarnomski, J.E., 1984b. Late Quaternary geology of Lake Bonneville and Lake Waring. *Utah Geol. Assoc. Publ.,* **13,** 227–37.

Eardley, A.J. & Gvosdetsky, V., 1960. Analysis of Pleistocene core from Great Salt Lake, Utah. *Geol. Soc. Am. Bull.,* **71,** 1323–44.

Eardley, A.J., Gvosdetsky, V. & Marsell, R.E., 1957. Hydrology of Lake Bonneville and sediments and soils of its basin. *Geol. Soc. Am. Bull.,* **68,** 1141–201.

Eardley, A.J., Shuey, R.T., Gvosdetsky, V., Nash, W.P., Picard, M.D., Grey, D.C. & Kukla, G.J., 1973. Lake cycles in the Bonneville basin, Utah. *Geol. Soc. Am. Bull.,* **84,** 211–16.

Gilbert, G.K., 1890. *Lake Bonneville.* US Geol. Survey Monograph 1.

Gwynn, J.W. (Ed.) 1980. Great Salt Lake: a scientific, historical, and economic overview. *Utah Geol. Mineral Surv. Bull.,* **116,** 1–400.

Hunt, C.B., 1982. Pleistocene Lake Bonneville, ancestral Great Salt Lake, as described in the notebooks of G.K. Gilbert, 1875–1880. *Brigham Young Univ. Geol. Studies* 29 (pt. 1), 1–225.

Hunt, C.B., Varnes, H.D., and Thomas, H.E., 1953. Lake Bonneville: geology of northern Utah Valley, Utah. *US Geol. Surv. Prof. Paper* 257-A, 1–99.

Ives, R.L., 1951. Pleistocene valley sediments of the Dugway area, Utah. *Geol. Soc. Am. Bull.,* **62,** 781–97.

Jarrett, R.D. & Malde, H.E., 1987. Paleodischarge of the late Pleistocene Bonneville Flood, Snake River, Idaho, computed from new evidence. *Geol. Soc. Am. Bull.,* **99,** 127–34.

Machette, M.N., 1985. Late Cenozoic geology of the Beaver Basin, southwestern Utah. *Brigham Young Univ. Geol. Studies* 32 (pt. 1), 19–37.

Machette, M.N. (Ed.) 1988. *In the Footsteps of G.K. Gilbert – Lake Bonneville and neotectonics of the eastern basin and range province.* Utah Geological and Mineral Survey Miscellaneous Publication 88–1.

Madsen, D.B. & Currey, D.R., 1979. Late Quaternary glacial and vegetation changes, Little Cottonwood Canyon area, Wasatch Mountains, Utah. *Quat. Res.,* **12,** 254–70.

Malde, H.E., 1968. The catastrophic late Pleistocene Bonneville Flood in the Snake River Plain, Idaho. US Geol. Survey Prof. Paper 596, 1–52.

McCalpin, J., 1986. Thermoluminescence (TL) dating in seismic hazard evaluations: An example from the Bonneville basin, Utah. In *Proceedings of the 22nd Symposium on Engineering Geology and Soils Engineering: Boise, Idaho, February 24–26, 1986,* pp. 156–76.

McCoy, W.D., 1987. Quaternary aminostratigraphy of the Bonneville basin, western United States. *Geol. Soc. Am. Bull.,* **98,** 99–112.

McKenzie, J.A. & Eberli, G.P., 1987. Indications for abrupt Holocene climatic change: Late Holocene oxygen isotope stratigraphy of the Great Salt Lake, Utah. In *Abrupt Climatic Change,* eds. W.H. Berger & L.D. Labeyrie, pp. 127–36. D. Reidel Publishing Company, Stuttgart.

Miller, D.M., 1990. Mesozoic and Cenozoic tectonic evolution of the northeastern Great Basin. In *Geology and ore deposits of the northeastern Great Basin,* ed. D.R. Shaddrick, J.A. Kizis Jr. & E.L. Hunsaker, III, pp. 43–73. Geological Society of Nevada Field Trip No. 5.

Morrison, R.B., 1965a. Lake Bonneville: Quaternary stratigraphy of eastern Jordan Valley, south of Salt Lake City, Utah. *US Geol. Surv. Prof. Paper* 477, 1–80.

Morrison, R.B., 1965b. New evidence on Lake Bonneville stratigraphy and history from southern Promontory Point, Utah. *US Geol. Surv. Prof. Paper* 525-C, C110–19.

Morrison, R.B., 1965c. Quaternary geology of the Great Basin. In *The Quaternary of the United States,* ed. H.E. Wright Jr. & D.G. Frey, pp. 265–85. Princeton University Press, Princeton.

Morrison, R.B., 1966. Predecessors of Great Salt Lake. In *The Great Salt Lake,* ed. W.L. Stokes, pp. 77–104. Utah Geological Society Guidebook to the Geology of Utah, no. 20.

Morrison, R.B., 1991. Quaternary stratigraphic, hydrologic, and climatic history of the Great Basin, with emphasis on Lakes Lahontan, Bonneville, and Tecopa. In *Quaternary nonglacial geology: Conterminous U.S.,* ed. R.B. Morrison, pp. 283–320. Geol. Soc. Am., The Geology of North America V. K-2.

Morrison, R.B. & Frye, J.C., 1965. *Correlation of the middle and late Quaternary successions of the Lake Lahontan, Lake Bonneville, Rocky Mountain (Wasatch Range), southern Great Plains, and eastern midwest areas.* Nevada Bureau of Mines Report 9.

Nakiboglu, S.M. & Lambeck, K., 1983. A reevaluation of the isostatic rebound of Lake Bonneville. *J. Geophys. Res.,* **88,** 10439–47.

O'Connor, J.E., 1993. *Hydrology, hydraulics, and geomorphology of the Bonneville Flood.* Geol. Soc. Am. Spec. Paper 274.

Oviatt, C.G., 1987a. Lake Bonneville stratigraphy at the Old River Bed, Utah. *Am. J. Sci.*, **287**, 383–98.

Oviatt, C.G., 1987b. Late Cenozoic capture of the Sevier River into the Sevier Desert basin, Utah. *Utah Geol.* Assoc. Publication 16, 265–269.

Oviatt, C.G., 1988. Late Pleistocene and Holocene lake fluctuations in the Sevier Lake basin, Utah, USA. *J. Paleolimnol.*, **1**, 9–21.

Oviatt, C.G., 1989. *Quaternary geology of part of the Sevier Desert, Millard County, Utah.* Utah Geological and Mineral Survey Special Studies 70.

Oviatt, C.G., & Currey, D.R., 1987. *Pre-Bonneville Quaternary lakes in the Bonneville basin, Utah.* Utah Geological Association Publication 16, pp. 257–63.

Oviatt, C.G., Currey, D.R., Miller, D.M., 1990. Age and paleoclimatic significance of the Stansbury shoreline of Lake Bonneville, Utah. *Quat. Res.*, **33**, 291–305.

Oviatt, C.G., Currey, D.R. & Sack, D., 1992. Radiocarbon chronology of Lake Bonneville, eastern Great Britain, U.S.A. *Palaeogegr., Palaeoclimatol., Palaeoecol.*, **99**, 225–41.

Oviatt, C.G., McCoy, W.D. & Nash, W.P., 1994. *Sequence stratigraphy of lacustrine deposits: A Quaternary example from the Bonneville basin, Utah.* Geol. Soc. Am. Bull. 106. (In press.)

Oviatt, C.G., McCoy, W.D. & Reider, R.G., 1987. Evidence for a shallow early or middle Wisconsin-age lake in the Bonneville basin, Utah. *Quat. Res.*, **27**, 248–62.

Oviatt, C.G. & Nash, W.P., 1989. Late Pleistocene basaltic ash and volcanic eruptions in the Bonneville Basin, Utah. *Geol. Soc. Am. Bull.*, **101**, 292–303.

Passey, Q.R., 1981. Upper mantle viscosity derived from the difference in rebound of the Provo and Bonneville shorelines: Lake Bonneville Basin, Utah. *J. Geophys. Res.*, **86**, 11701–8.

Sack, D., 1989. Reconstructing the chronology of Lake Bonneville: An historical review. In *History of geomorphology. 19th annual Binghamton geomorphology symposium volume*, ed. K.J. Tinkler, pp. 223–56. Unwin Hyman, London.

Sack, D., 1990. *Quaternary geology of Tule Valley, west-central Utah.* Utah Geological and Mineral Survey Map 124 [1:100,000].

Sack, D., 1992. Obliteration of surficial paleolake evidence in the Tule Valley subbasin of Lake Bonneville, In Quaternary coasts of the United States: Marine and lacustrine systems. *SEPM (Soc. for Sedimentary Geology) Special Pub.*, **48**, 427–33.

Scott, W.E., McCoy, W.D., Shroba, R.R. & Rubin, M., 1983. Reinterpretation of the exposed record of the last two cycles of Lake Bonneville, western United States. *Quat. Res.*, **20**, 261–85.

Spencer, R.J., Baedecker, M.J., Eugster, H.P., Forester, R.M., Goldhaber, M.B., Jones, B.F., Kelts, K., McKenzie, J., Madsen, D.B., Rettig, S.L., Rubin, M. & Bowser, C.J., 1984. Great Salt Lake and precursors, Utah: the last 30,000 years. *Contrib. Mineral. Petrol.*, **86**, 321–34.

Spencer, R.J., Eugster, H.P., Jones, B.F. & Rettig, S.L., 1985a. Geochemistry of Great Salt Lake, Utah I. Hydrochemistry since 1850: *Geochim. Cosmochim. Acta*, **49**, 727–37.

Spencer, R.J., Eugster, H.P. & Jones, B.F., 1985b. Geochemistry of Great Salt Lake, Utah II. Pleistocene-Holocene evolution. *Geochim. Cosmochim. Acta*, **49**, 739–47.

Thompson, R.S., Toolin, L.J., Forester, R.M. & Spencer, R.J., 1990. Accelerator-mass spectrometer (AMS) radiocarbon dating of Pleistocene lake sediments in the Great Basin. *Palaeogeogr., Palaeoclimatol., Palaeoecol.*, **78**, 301–13.

Varnes, D.J. & Van Horn, R., 1961. A reinterpretation of two of G.K. Gilbert's Lake Bonneville sections, Utah. *US Geol. Surv. Prof. Paper* 424-C, C98–9.

Varnes, D.J. & Van Horn, R., 1984. *Surficial geologic map of the Oak City area, Millard County, Utah.* US Geological Survey Open-File Report 84–115, scale 1:31,680.

Willett, H.C., 1977. The prediction of Great Salt Lake levels on the basis of recent solar-climatic cycles. *Utah Geol.*, **4**, 113–23.

Williams, J.S., 1962. Lake Bonneville: Geology of Southern Cache Valley, Utah. *US Geol. Surv. Prof. Paper* 257-C, 131–52.

Lakes Salpeten and Quexil, Peten, Guatemala, Central America

MARK BRENNER

Florida Museum of Natural History, Gainesville, Florida 32611, USA

Lakes Salpeten (104 m above mean sea level) and Quexil (110 m above mean sea level) lie in the tropical, karst lowlands of Peten, Guatemala (Fig. 1). The lakes are sufficiently deep ($z_{max} > 30$ m) to have held water during the late Pleistocene. Long cores collected in

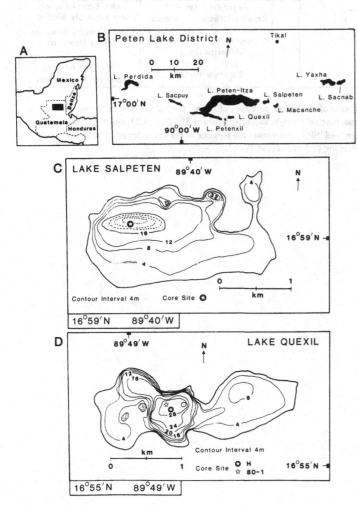

Fig. 1. A: Regional map showing the location of the Peten Lake District. B: Map of the Peten Lake District showing the locations of Lakes Salpeten and Quexil. C: Bathymetric map of Lake Salpeten. D: Bathymetric map of Lake Quexil.

1978 (Quexil core H) and 1980 (Salpeten 80–1 and Quexil 80–1) contain a record of paleoecological conditions from the late Pleistocene to present (Figs. 2 and 3). Radiocarbon measurements on terrestrial wood from Lake Quexil core 80–1 accurately date the Pleistocene–Holocene climatic transition (SI-5257: 8.94–9.01 m = $10,750 \pm 460$ BP and AA-3062: 8.95 m = $10,300 \pm 110$ BP. A *Biomphalaria* sp. shell from 15.96 m in Quexil 80–1 yielded a date of $27,450 \pm 500$ BP (AA-3064), but may be subject to hard-water-lake error. The Pleistocene–Holocene boundary in Salpeten core 80–1 was assigned by palynological correlation with the dated Quexil stratigraphy.

Pleistocene–Holocene climatic change was marked in Peten by a shift from very arid to more mesic conditions. Pollen data show an abrupt shift from sparse, dry vegetation to a semi-evergreen forest, which developed rapidly under conditions more mesic than at present. The climatic boundary is also marked by a shift from deposition of mineral-rich (calcite, gypsum, quartz, clay) sediments, to organic-rich, lacustrine gyttja. Laminae in the early Holocene deposits from Quexil suggest an anoxic hypolimnion under deep water.

Radiocarbon dates on bulk sediments from Peten lakes are questionable because of hard-water-lake error and because mid-Holocene deposits are comprised, in part, of re-deposited colluvium with unknown isotopic content. Holocene stratigraphies from the area have been reliably dated by correlation of the palynological or lithological stratigraphies with the archaeologically documented prehistory of the region. Mid- to late-Holocene climatic changes are masked by the impact of Maya activities on the local landscape. Maya settlement in Peten lasted from 3000 to 400 BP, and is detected in the cores by evidence of regional deforestation and soil erosion in the Quexil and Salpeten watersheds. Forest clearance destabilized drainage basin soils and led to rapid sedimentation rates. Maya-period deposits are clay-rich and poor in pollen and lacustrine microfossils.

Although the Classic Maya civilization endured a major collapse in the ninth century AD, it is thought that Postclassic (850–1650 AD) population densities in the Peten were sufficient to keep the region largely deforested. Re-establishment of high forest vegetation about 400 years ago is correlated with regional Maya depopulation

LAKE SALPETEN

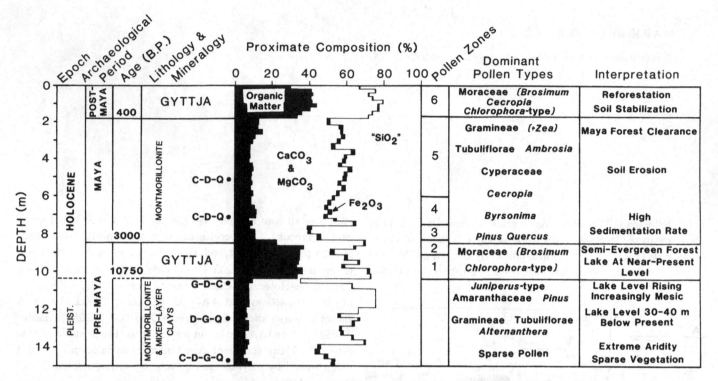

Fig. 2. Stratigraphy of Lake Salpeten Core 80–1. Non-clay minerals are listed in order of abundance and are as follows: C = calcite, D = dolomite, G = gypsum, Q = quartz. Proximate composition was calculated in the following manner: Organic matter is weight loss on ignition at 550°C, $CaCO_3$ and $MgCO_3$ is the sum of the carbonate equivalents of Ca and Mg, Fe_2O_3 is computed from Fe concentration, and 'SiO$_2$' represents the balance of the dry sediment. Carbonate content is overestimated at levels where much Ca is bound in gypsum. Pollen zones are according to Leyden (1987).

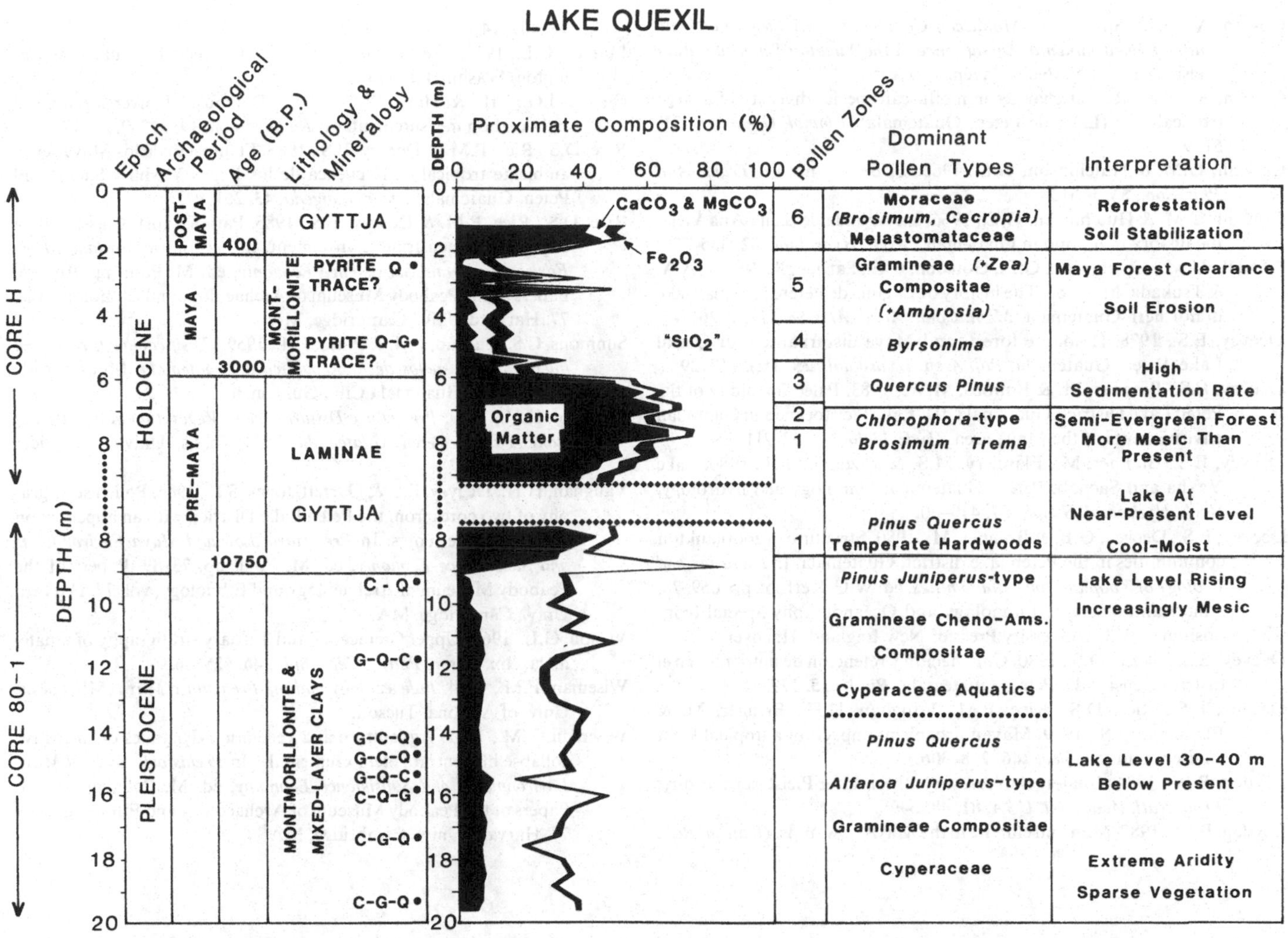

LAKE QUEXIL

Fig. 3. Stratigraphy of Lake Quexil Cores H (0–9.24 m) and 80–1 (7.49–19.68 m). Non-clay minerals are listed in order of abundance and are as follows: C = calcite, G = gypsum, Q = quartz. Proximate composition was calculated in the following manner: Organic matter is 2.5 times organic carbon (Core H) or weight loss on ignition at 550°C (Core 80–1), $CaCO_3$ and $MgCO_3$ is the sum of the carbonate equivalents of Ca and Mg, Fe_2O_3 is computed from Fe concentration, and 'SiO_2' represents the balance of the dry sediment. Topmost (0–1.0 m) sediments were not analyzed chemically. Carbonate content is overestimated at levels where much Ca is bound in gypsum. Pollen zones are according to Vaughan et al. (1985).

that coincided with European intrusion in the region. Forest cover restabilized basin soils and prevented rapid erosion and consequent high sedimentation rates. Reforestation is detectable in the pollen record and is coincident with a return to deposition of organic gyttja. At present the Quexil and Salpeten watersheds are largely uninhabited, though there has been some localized slash-and-burn agriculture in recent years.

Bibliography

Binford, M.W., 1983. Paleolimnology of the Peten Lake District, Guatemala, I. Erosion and deposition of inorganic sediment as inferred from granulometry. *Hydrobiologia*, **103**, 199–203.

Binford, M.W., Brenner, M., Whitmore, T.J., Higuera-Gundy, A., Deevey, E.S. & Leyden, B.W., 1987. Ecosystems, paleoecology and human disturbance in subtropical and tropical America. *Quatern. Sci. Rev.*, **6**, 115–28.

Brenner, M., 1978. *Paleolimnological Assessment of Human Disturbance in the Drainage Basins of Three Northern Guatemalan Lakes.* MS Thesis, Univ. of Florida, Gainesville.

Brenner, M., 1983a. *Paleolimnology of the Maya Region.* PhD Thesis, Univ. of Florida, Gainesville.

Brenner, M., 1983b. Paleolimnology of the Peten Lake District, Guatemala, II. Mayan population density and nutrient loading of Lake Quexil. *Hydrobiologia*, **103**, 205–10.

Brenner, M., Leyden, B.W. & Binford, M.W., 1990. Recent sedimentary histories of shallow lakes in the Guatemalan savannas. *J. Paleolimnol*, **4**, 239–52.

Covich, A., 1970. *Stability of Molluscan Communities: A Paleolimnologic Study of Environmental Disturbance in the Yucatan Peninsula.* PhD Thesis, Yale Univ., New Haven.

Covich, A., 1976. Recent changes in molluscan species diversity of a large tropical lake (Lago de Peten, Guatemala). *Limnol. Oceanogr.*, **21**, 51–9.

Cowgill, U.M. & Hutchinson, G.E., 1963. El Bajo Sante Fe. *Trans. Am. Phil. Soc.*, **53**, 1–51.

Cowgill, U.M. & Hutchinson, G.E., 1966. La Aguada de Santa Ana Vieja: the history of a pond in Guatemala. *Arch. Hydrobiol.*, **62**, 335–72.

Cowgill, U.M., Hutchinson, G.E., Goulden, C.E., Patrick, R., Racek, A.A. & Tsukada, M., 1966. The history of Laguna de Petenxil: a small lake in northern Guatemala. *Mem. Conn. Acad. Arts Sci.*, **17**, 1–26.

Deevey, E.S., 1978. Holocene forests and Maya disturbance near Quexil Lake, Peten, Guatemala. *Pol. Arch. Hydrobiol.*, **25** (1/2), 117–29.

Deevey, E.S., Brenner, M. & Binford, M.W., 1983. Paleolimnology of the Peten Lake District, Guatemala, III. Lake Pleistocene and Gamblian environments of the Maya area. *Hydrobiologia*, **103**, 211–16.

Deevey, E.S., Brenner, M., Flannery, M.S. & Yezdani, G.H., 1980. Lake Yaxha and Sacnab, Peten, Guatemala: limnology and hydrology. *Arch. Hydrobiol. Suppl.*, **57**, 419–60.

Deevey, E.S., Deevey, G.B. & Brenner, M., 1980. Structure of zooplankton communities in the Peten lake district, Guatemala. In *Evolution and Ecology of Zooplankton Communities*, ed. W.C. Kerfoot, pp. 669–78. American Society of Limnology and Oceanography Special Symposium, vol. 3, University Press of New England, Hanover.

Deevey, E.S. & Rice, D.S., 1980. Coluviación y retención de nutrientes en el distrito lacustre del Peten, Guatemala. *Biotica*, **5**, 129–44.

Deevey, E.S., Rice, D.S., Rice, P.M., Vaughan, H.H., Brenner, M. & Flannery, M.S., 1979. Mayan urbanism: impact on a tropical karst environment. *Science*, **206**, 298–306.

Leyden, B.W., 1984. Guatemalan forest synthesis after Pleistocene aridity. *Proc. Natl. Acad. Sci. USA*, **81**, 4856–9.

Leyden, B.W., 1987. Man and climate in the Maya Lowlands. *Quatern. Res.*, **28**, 407–14.

Lundell, C.L., 1937. *The Vegetation of Peten.* Carnegie Institute of Washington, Washington, D.C.

Ogden, J.G., III, & Hart, W.C., 1977. Dalhousie University natural radiocarbon measurements II. *Radiocarbon*, **19**, 392–9.

Rice, D.S., Rice, P.M. & Deevey, E.S., 1983. El impacto de los Mayas en el ambiente tropical de la cuenca de los Lagos Yaxha y Sacnab, El Peten, Guatemala. *Amer. Indigena*, **43**, 261–97.

Rice, D.S., Rice, P.M. & Deevey, E.S., 1985. Paradise lost: Classic Maya impact on a lacustrine environment. In *Prehistoric Lowland Maya Environments and Subsistence Economy*, ed. M. Pohl, pp. 91–105. Papers of the Peabody Museum of Archaeology and Ethnology, vol. 77. Harvard Univ., Cambridge, MA.

Simmons, C.S., Tarano, J.M. & Pinto, J.H., 1959. *Clasificación de reconocimiento de los suelos de la República de Guatemala.* Ministerio de Agricultura, Guatemala City, Guatemala.

Vaughan, H.H., 1979. *Prehistoric Disturbance of Vegetation in the Area of Lake Yaxha, Peten, Guatemala.* PhD Thesis, Univ. of Florida, Gainesville.

Vaughan, H.H., Deevey, E.S. & Garrett-Jones, S.E., 1985. Pollen stratigraphy of two cores from the Peten Lake District, with an appendix on two deep-water cores. In *Prehistoric Lowland Maya Environments and Subsistence Economy*, ed. M. Pohl, pp. 73–89. Papers of the Peabody Museum of Archaeology and Ethnology, vol. 77. Harvard Univ., Cambridge, MA.

Vinson, G.L., 1962. Upper Cretaceous and Tertiary stratigraphy of Guatemala. *Am. Assoc. Petrol. Geol. Bull.*, **46**, 425–56.

Wiseman, F.M., 1974. *Paleoecology and the Prehistoric Maya.* MS Thesis, Univ. of Arizona, Tucson.

Wiseman, F.M., 1985. Agriculture and vegetation dynamics of the Maya collapse in Central Peten, Guatemala. In *Prehistoric lowland Maya Environments and Subsistence Economy*, ed. M. Pohl, pp. 63–71. Papers of the Peabody Museum of Archaeology and Ethnology, vol. 77. Harvard Univ., Cambridge, MA.

Salinas del Bebedero Basin (República Argentina)

MIGUEL ANGEL GONZÁLEZ

CONICET and Carl C:zon Caldenius Foundation C.C.289 -Sucursal 13 (B)-1413 Buenos Aires, República Argentina

Introduction

Salinas del Bebedero is of tectonic origin, a fault-block basin (González, 1981, 1982, 1983) in a region with intense neotectonic activity (Polanski, 1963; Regairaz & Videla Leaniz, 1968). The region's present-day annual rainfall is less than 500 mm, with potential evapotranspiration greater than 1000 mm (annual average). Thus, the local rainfall only permits the existence of a temporary and extremely shallow lake with a depth less than 0.5 m, allowing economic exploitation of halite.

During global cooling episodes, however, Salinas del Bebedero received great amounts of seasonal meltwaters from the Cordillera de los Andes (Fig. 1a; González, 1981, 1982, 1983; González & Minetti, 1989; Maidana & González, 1990), by means of the Desaguadero fluvial system (Fig. 1A) and the Bebedero inflow creek. Each high lacustrine stage reflected an important glacial advance in the Central Argentinian Andes, which, as yet, has not been directly dated. Thus, these lacustrine chronologies are the first to show glacial episodes in that region.

Pleistocene lacustrine stages

Some evidence does exist for Pre-Wisconsinian (= Pre-Eemian) lacustrine stages (Fig. 2A). On the downwind margins of the Salinas del Bebedero (NW, N and NE; Fig. 1B), there are at least three morphologically well-developed paleo-shorelines, 100–200 m wide and 2–3 m high. They are composed of coarse sand and gravel (lacustrine fore-shore facies), indicating at least four lacustrine stage (L.S. A, B, C and D). These stages are well-dated by means of ^{14}C on *Chilina parchappi* d'Orbigny shells (a freshwater gastropod). On the southeastern, southern and west-southwestern sides of the lake, these similar paleo-shorelines are masked by eolian deposits (paleo-dunes). Only the fourth and fifth lacustrine stages (L.S. D and E (?)) are exposed as pelithic sediments of the lacustrine floor (Figs 2B, 2C).

L.S. A: oldest buried lacustrine fore-shore facies. One age of $20,140 \pm 370$ years BP would indicate that a Cordilleran cold stage occurred prior to the glacial stage Llanquihue II of the Southern Chilean Andes, as identified by Mercer

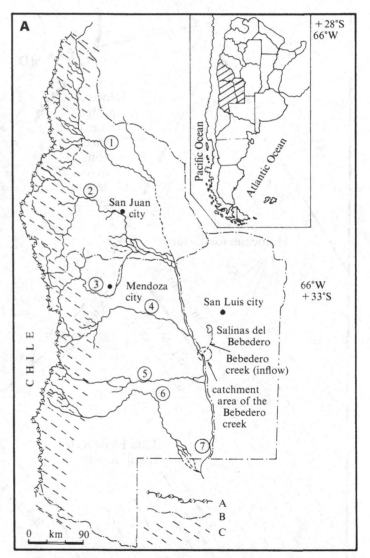

Fig. 1.A: Catchment area of the Desaguadero fluvial system, and catchment area of Bebedero Creek. 1. Bermejo River; 2. San Juan River; 3. Mendoza River; 4. Tunuyán River; 5. Diamante River; 6. Atuel River; 7. Desaguadero River. (A) International boundary; (B) Interprovincial boundary; (C) Cordillera de los Andes.

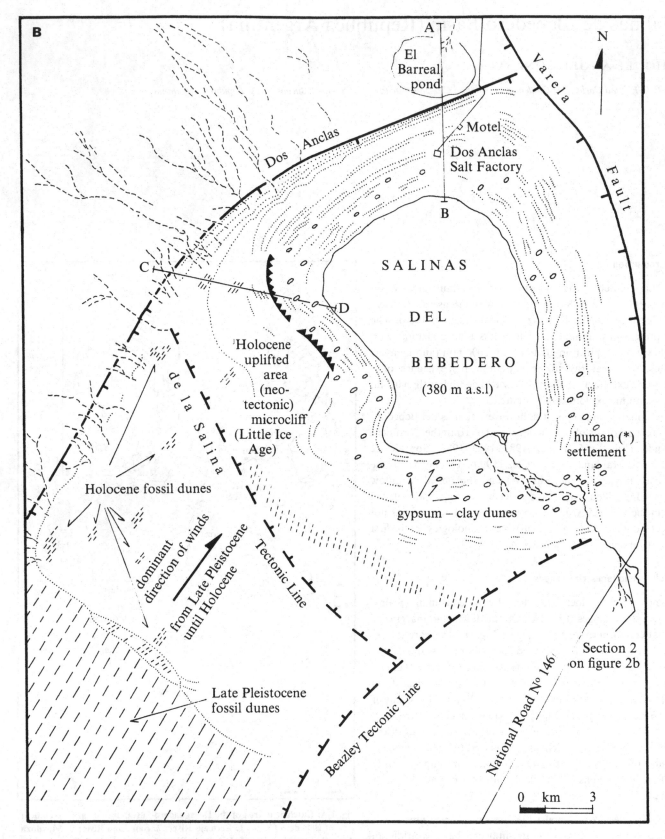

Fig. 1B: Detail of Salinas de Bebedero basin and its main features. Location of the cross sections presented in Figs 2A, 2B and 2C. (*) Section 1 on Fig. 2B.

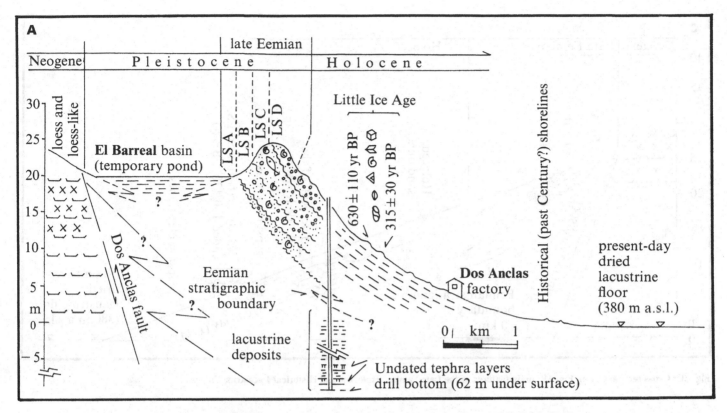

Fig. 2A: Cross section AB on Fig. 1B, showing the shorelines of the Late Eemian and the Lacustrine Stages of the Little Ice Age.

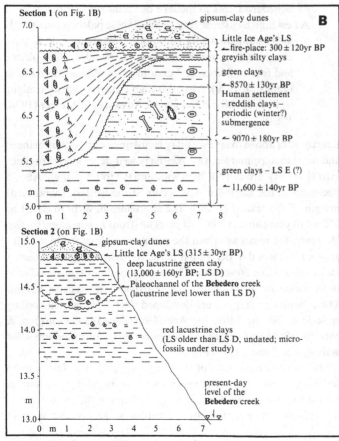

Fig. 2B: Main stratigraphic features in the southeastern part of the basin (Sections 1 and 2 on Fig. 1B).

(1984); or perhaps, this horizon denotes the start of the same Glacial Stage Llanquihue II.

L.S. B: lacustrine fore-shore facies. Two dates with a mean age of 17,410 ± 130 years BP (Table 1) indicates possible correlation with the Glacial Stage Llanquihue II of Mercer (1984).

L.S. C: lacustrine fore-shore facies. One age of 14,700 ± 180 years BP would indicate a possible correlation with the glacial stage Llanquihue III of Mercer (1984) and, perhaps, with the first of two glacial 'pulses' of Southern Chilean Andes, as reported by Porter (1981). Other fossils include fish (*Percichtys* sp., Cr. A. Cione, pers. comm.) and the first appearance of a saline tolerant gastropod (*Littoridina parchappi* d'Orbigny).

L.S. D: lacustrine fore-shore facies. A set of two dates (Table 1) with a mean age of 13,775 ± 105 years BP, and another set of nine ages (some from the N, NW and NE fore-shore facies; other ones from the SE paleolacustrine pelithic deposits), with a mean age of 12,860 ± 380 years BP. This indicates at least two sub-stages correlated with the second glacial pulse of Porter (1981), and with the culmination of the glacial stage Llanquihue III of Mercer (1984). Furthermore, it is possible to correlate the L.S. D with a cool and wet episode from the Bolivian Andes and Altiplano (see Kessler, 1963, 1984; Servant & Fontes, 1978; Lauer & Frankenberger, 1984) as an early 'Younger Dryas-like' episode in the southern hemisphere.

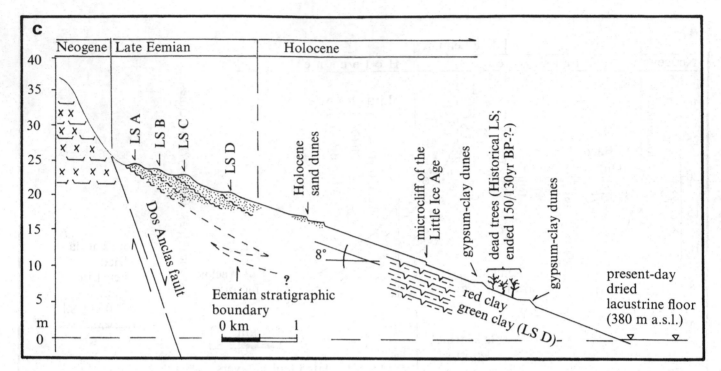

Fig. 2C: Cross section CD on Fig. 1B; main features of the Late Eemian, Holocene and Historical Lacustrine Stages.

L.S. E(?): clay sediments interpreted as deep lacustrine, situated on the southern side of the basin (Fig. 2B). One age of 11,600 ± 140 years BP. Also, *L. Parchappi* and *Biomphalaria peregrina* d'Orb. shells are present. It is not yet well established whether these clay layers correspond to another sub-stage of the L.S. D, or if they correspond to a younger, posthumous Late Pleistocene L.S. E(?). They could perhaps be connected to a possible 'Younger Dryas-like' glacial stage in the Argentinean Central Andes.

Holocene lacustrine stages

These lacustrine stages are represented by several, undated shorelines, which are smaller than the Late Pleistocene ones and are mainly 4–5 m wide and 0.5 m high. Also, toward the southwest (windward side of lake), a microcliff occurs (Figs 2A, B, C), recording the Little Ice Age.

A seasonal human settlement on the southern side of the lake (Fig. 2B) was developed during the spring every year, from 9070 ± 180 years BP until 8570 ± 130 years BP (^{14}C ages on *Rhea* sp. eggshells, as the main food remains; Balbuena *et al.*, 1982; Table 1) around 6 m above the present-day lacustrine floor.

After 8570 years BP, a new high lacustrine stage buried the human settlement with clay sediments. This L.S. is still undated, but it is possible that it is related to some of the Neoglacial Advances, defined by Mercer (1976, 1984) in the southern Andean regions.

During a low lacustrine episode, paleochannels were cut by Bebedero Creek in the lake bed sediments (Fig. 2B).

A new high lacustrine episode filled these paleochannels with finely laminated silty carbonate. Fossils include *L. parchappi*, calcareous microfossils (ostracodes, charophytes and foraminiferas) and diatom assemblages (Maidana & González, 1990). Also, this high lacustrine episode might be linked with the Neoglacial Advances of Mercer (1976, 1984).

Exactly 10 m above the present-day lacustrine floor, to the northern and northeastern margin, two relatively well-developed shorelines exist (Fig. 2A), around 10–20 m wide and 0.5 to 1.5 m high. Also, a microcliff of the Little Ice Age occurs at the western (windward) margin of the lake (Fig. 2C). The shorelines have been dated as 630 ± 110 years BP and 350 ± 70 years BP (from *L. parchappi* shells). However, the mean age from the second shoreline combined with an age from wood underlying it in the southern margin suggests a mean age of 315 ± 30 years BP. These two shoreline deposits indicate two episodes or 'pulses' of meltwater arrival from Andean regions. These lacustrine stages were developed during the global cooling episode called the Little Ice Age (González, 1983; González & Minetti, 1989) and correlated to secular cycles of minimum solar activity (see Siscoe, 1978, Fig. 3).

On the southern side of the basin (Fig. 2B) a last human settlement (300 ± 120 years BP on wood) is overlain by lacustrine deposits of the second Little Ice Age, indicating the existence of a dry episode between these two lacustrine stages. A thin layer of

Table 1. *Carbon-14 ages and correlations of lacustrine stages of Salines del Bedero basin*

Sample N°	Age (years BP)	mean age ($\bar{X} \pm s$)	Stages Lacustrine	Stages Glacial	depth of water	lacustrine evidence
—	undated	18th/19th century	high	Late Little Ice Age	+5/+6 m	dead forest; historical documents
AC-0706	300 ± 120	315 ± 30	high	Little Ice Age	+10 m	Shorelines, micro cliff; human settlements
AC-0379	350 ± 70					
AC-0492	630 ± 110	630 ± 110	high			
—	undated	—	high	Neoglacial Advances(?)	?	Shorelines
—	undated	—	low	warm	shallow (dried?)	paleochannels
—	undated	—	high	Neoglacial Advances(?)	?	shorelines
AC-0906	8570 ± 130	top	low (episodic inundation)	Hypsithermal	shallow dried(?)	Archaeological site
AC-0904	9070 ± 130	basal				
AC-0711	11,600 ± 140	11,600 ± 140	E (?)	Younger Dryas Like (?)	?	deep sediments
AC-1183	12,270 ± 240	12,300 ± 60	D	Late Llanquihue III[1] and 2nd 'pulse' of Porter[2]		deep sediments
AC-1180	12,355 ± 205					
AC-0708	13,200 ± 150				+25 m	shorelines and deep sediments
AC-0375	12,700 ± 190					
AC-0374	12,700 ± 150					
AC-0373	13,000 ± 160	13,000 ± 250	D	(Possibly early Younger – Dryas like – ? –)		
AC-0371	13,290 ± 190					
AC-0369	13,260 ± 200					
AC-0105	13,000 ± 140					
AC-0710	13,850 ± 160	13,775 ± 105	D			shorelines
AC-0709	13,700 ± 160					
AC-0707	14,700 ± 180	14,700 ± 180	C	Llanquihue III[1] 1st 'pulse'[2]	+25 m	shorelines
AC-0994	17,320 ± 270	17,410 ± 130	B	Llanquihue II	+25 m	shorelines
AC-0368	17,500 ± 300					
AC-0993	20,140 ± 370	20,140 ± 370	A	Early Llanquihue II (?)	+25 m	shorelines
—	undated	—	—	Pre-Eemian	>25 m	deep sediments

(bracketed value spanning AC-1183 through AC-0105 block: 12,860 ± 380)

Notes:
[1] Mercer, 1984.   [2] Porter, 1981.

yellowish gypsum crystals and some pelithic sediments of this second lacustrine stage are found across the basin; they are commonly wind-transported, forming abundant small gypsum-clay dunes. Fossils of this dry episode include ostracodes tolerant of high salinity (*Iliocypris, Cyprinotus, Limnocythere, Cyprideis, Darwinula* and *Cypridopsis*: González *et al.*, 1980, 1981; García, 1987), foraminiferas (*Streblus parkinsonianus, Elphidium gunteri*, and *Discorbis*

sp.; González *et al.*, 1980, 1981), charophytes (*Lamprothamnium, Chara*, and some others; González *et al.*, 1980, 1981; García, 1987), and diatoms (Maidana & González, 1990).

Finally, some historical documents indicate a high lacustrine stage from at least the eighteenth century to the lower half of the nineteenth century. On the western side of the basin, lower than the microcliff of the Little Ice Age, abundant *in situ* small trees occur

(Fig. 2C), killed by an inundation episode. This last high and very saline inundation episode, as yet undated, may correspond to the historically documented high lacustrine stage.

Bibliography

Balbuena, J.L., González, Angiolini, F.E. & Albero, M.C., 1982. Presencia humana en la Salina de Bebedero (provincia de San Luis, República Argentina) durante el Holoceno temprano. Su significado paleoclimático. In *Primer Reunión Nacional de Ciencias de Hombre en Zonas Aridas.* Univ. Nacional Cuyo and CRICYT-ME, Mendoza.

Garcia, A., 1987. *El gametangio femenino de Charophytae actuales de Argentina. Análisis comparado con el registro fósil correspondiente.* PhD Thesis, Univ. Nacional de la Plata.

González, M.A., 1981. Evidencias paleoclimáticas en la Salina del Bebedero (San Luis, Argentina). *VIII Congreso Geol. Argentino, Actas 3,* pp. 411–38.

González, M.A., 1982. Oscilaciones pleistocénicas del nivel lacustre en la actual Salina de Bebedero (San Luis, Argentina). Su relación con la última glaciación en la Cordillera de Mendoza y San Juan. In *Commission Genesis and Lithology of Quaternary Deposits, South American Regional Meeting, INQUA and Univ. Nacional del Comahue, Neuquén, Abstracts,* pp. 41–3.

González, M.A., 1983. Pleistocene and Holocene lake levels in the actual Salina del Bebedero, Argentina. [14]C dates. Relations with the latest Pleistocenic glaciation. In *Symp. on Desert Encroachment, Fast Tropical Erosion, and Coastal Subsidence and Submergence, INQUA – IGCP, Hamburg, Abstracts,* pp. 88–9.

González, M.A. & Minetti, J.L., 1989. *Present day lacustrine bodies in the Argentine Republic. Their Possibilities Concerning to Quaternary Paleoclimatic Records.* (Unpublished manuscript.)

González, M.A., Musacchio, E.A., Garcia, A., Pascual, R., & Corte, A.E., 1980. Sobre la presencia de foraminíferos en sedimentos holocenos de la Salina del Bebedero (San Luis, Argentina). In *Primer Simp. Sobre Problemas del Litoral Atlántico Bonaerense, Com. Invest. Cient. Bs. Asoc., Mar del Plata, Resúmenes,* pp. 253–69.

González, M.A., Musacchio, E.A., Garcia, A., Pascual, R. & Corte, A.E., 1981. Las líneas de costa (Holoceno) de la Salina del Bebedero (San Luis, Argentina). Implicancias paleoambientales de sus microfósiles. *VIII Congreso Geol. Argentino, Actas 3,* pp. 617–28.

Kessler, A., 1963. Über Klima and Wasserhaushalt des Altiplano (Bolivien-Perú) während des Hochstandes der letzen Vereisung. *Erkunde,* **17** (3/4).

Kessler, A., 1984. The paleohydrology of the Late Pleistocene Lake Tanka, on the Bolivian Altiplano and recent climatic fluctuations. In *Late Cainozoic Paleoclimates of the Southern Hemisphere,* ed. J.C. Vogel, pp. 115–22. SASQUA International Symposium, Swaziland, Proceedings.

Lauer, W. & Frankenberger, P., 1984. Late glacial glaciation and the development of climate in southern South America. In *Late Cainozoic Paleoclimates of the Southern Hemisphere,* ed. J.C. Vogel, pp. 103–14. SASQUA International Symposium, Swaziland, Proceedings.

Maidana, M.I. & González, M.A., 1990. Diatom assemblages from Late Pleistocene and Holocene Lacustrine sediments of Salinas de Bebedero, (San Luis, República Argentina). *11th International Diatom Symposium, San Francisco, Abstracts.*

Mercer, J.H., 1976. Glacial History of Southernmost South America. *Quatern. Res.,* **6** (2), 125–66.

Mercer, J.H., 1984. Late Cainozoic glacial variations in South America South of the Equator. In *Late Cainozoic Paleoclimates of the Southern Hemisphere,* ed. J.C. Vogel, pp. 45–58. SASQUA International Symposium, Swaziland, Proceedings.

Polanski, J., 1963. Estratigrafía, neotectónica y geomorfología del Pleistoceno pedemontano, entre los ríos Diamante y Mendoza (Provincia de Mendoza). *Asoc. Geol. Argentina Rev.,* **17** (3–4), 127–349.

Porter, St. C., 1981. Pleistocene glaciation in the southern Lake District of Chile. *Quatern. Res.,* **16**, 263–92.

Regairaz, C.A. & Videla Leaniz, J.R., 1968. Fenómenos de neotectónica y su influencia sobre la morfología actual del Piedemonte Mendocino. *Ter. J. Geol. Argentinas, Actas* **2**, 21–34.

Servant, M. & Fontes, J. Ch., 1978. Les Lacs Quaternaires des hauts plateaux des Andes Boliviennes. Premieres interpretations paleoclimatiques. *Cahiers ORSTOM, sér. géol.,* **X** (1), 9–23.

Siscoe, G.L., 1978. Solar-terrestrial influences on weather and climate. *Nature,* **276**, 348–52.

Holocene Coorong Lakes, South Australia

JOHN K. WARREN

School of Applied Geology, Curtin University, GPO Box U1987, Perth WA 6001, Australia

General geology

The word Coorong is the European form of an aboriginal word 'Karanj' meaning long neck of water, a description of the 120 km-long Coorong Lagoon (Fig. 1A, B). The Coorong coastal plain was documented as an area of 'primary' dolomite deposition by Mawson (1929), but the Holocene geology was not studied in any detail until the work of Alderman & Skinner (1957), followed by Skinner (1963), von der Borch (1965, 1976), von der Borch and Jones (1976), von der Borch and Lock (1979), Muir, Lock & von der Borch (1980), Lock (1982), Bolz and von der Borch, (1984), Rosen *et al.* (1988a, b), Rosen *et al.* (1989), and Warren (1989). Lakes in the Salt Creek region contain the most abundant and thickest dolomites in the Coorong coastal plain (Fig. 1A, B).

Although the Coorong Dolomites take their name from the lagoon, dolomite is not forming in the bottom sediments of the perennial Coorong Lagoon which is floored by laminated aragonite and Mg-calcite muds. The Coorong Lagoon is an estuary more than 200 km long and a few kilometers wide. In the early Holocene it extended as a single water body from Kingston to the Murray River mouth (Fig 1A, B). Today, only the northern end of the Coorong Lagoon, from the Murray River mouth to Salt Creek, is a continuous water body. The southern part of the lagoon from Salt Creek to Kingston is now known as the ephemeral Coorong Lagoon. Like the area to the north, it was once a continuous lagoonal water body, but was broken up into a series of ephemeral lakes by the migration across the lagoon of the Holocene coastal dunes of Younghusband Peninsula. It is now an area of elongate temporary lakes that are separated by marginal flats. Lake waters are brackish in winter and hypersaline in summer with thin layers of muddy dolomites precipitating in some of the lakes (Fig. 1A, B; Alderman & Skinner, 1957).

The area behind the Coorong Lagoon is characterized by a series of near-parallel Pleistocene beach-dune ridges that are separated by interdunal furrows filled with lagoonal and lacustrine carbonate sediments (Schwebel, 1983). A calcarenite beach-dune ridge was deposited along the Quaternary coast of South Australia each time the ice caps melted and sea level reached a high stand (Schwebel, 1978; Warren, 1983). There have been 21 high stillstands in the last 700,000 years, shaping the 13 distinct coastal ranges of the South-

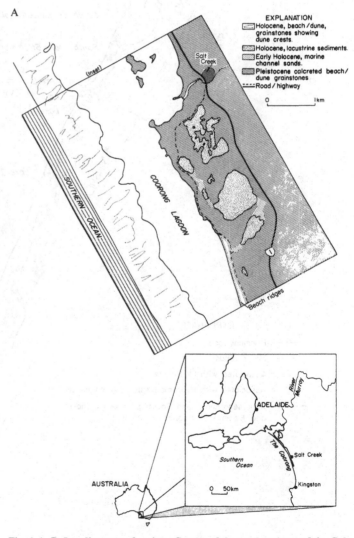

Fig. 1 A, B. Locality map of various Coorong lakes and an inset of the Salt Creek Lake Chain (after Warren, 1988, 1990).

387

B

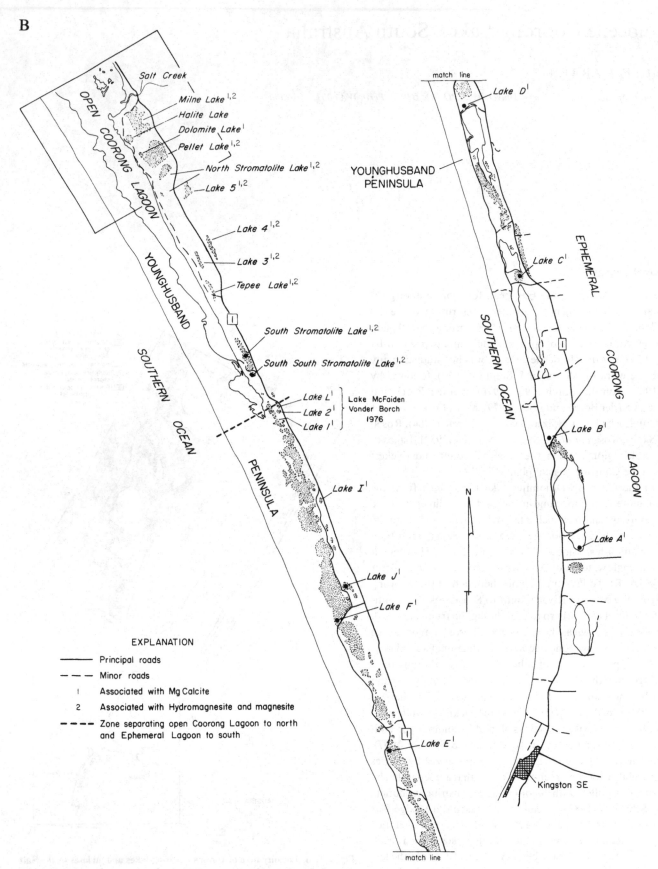

EXPLANATION

——— Principal roads

– – – Minor roads

1   Associated with Mg Calcite

2   Associated with Hydromagnesite and magnesite

– – – – Zone separating open Coorong Lagoon to north
        and Ephemeral Lagoon to south

Fig. 1 *(cont.)*

east Province of South Australia (Schwebel, 1983). The purest and thickest modern dolomites are found in the ephemeral lakes of the interdunal furrow immediately behind the 120,000 years BP dune ridge on the mainland side of the open Coorong Lagoon. It is the lacustrine sediments that occur in these interdunal corridors that are the subject of this contribution (Fig. 1A, B).

Hydrology

Holocene lacustrine carbonate units precipitate in interdunal corridors where sluggish, seaward-flowing, Mg-enriched continental groundwaters, resurface and evaporate in the interdunal depressions (von der Borch et al., 1975). Hence, thicker dolomitic units are typically found in the near-coastal zone lakes immediately landward of the Coorong Lagoon where slow-moving meteoric groundwater is forced to the surface as it rides up over an underlying wedge of saline marine-derived water.

O'Driscoll (1960) estimated the seaward flow rate of groundwater in the Coorong coastal plain to be 0.1 cm/day. O'Driscoll & Shepherd (1960) noted that in an area 40 km northeast of Salt Creek there was little or no recharge to the saline water table (10,000–28,000 TDS) and that across a very flat hydraulic gradient of 1 in 5000 there was a slow migration of saline groundwater toward the Coorong Lagoon. Other references relevant to the hydrology of the area are O'Driscoll (1960), O'Driscoll & Shepherd (1960) and Holmes & Waterhouse (1983).

Climate

The climate in the Coorong region is Mediterranean-like, with hot, dry summers (Dec. to Mar.), and cool wet winters (June to Sept.). Prevailing winds in the area are westerlies and southerlies and afternoon sea breezes dominate throughout the study area. Maximum rainfall is 800 mm in the far southeast of South Australia, 600 mm around Kingston, and 400 mm in the area about Salt Creek (Fig. 1A, B). Von der Borch (1976) observed that dolomite in the Coorong region formed in areas of the coastal plain where rainfall was less than 700 mm. This encompasses the area between Kingston and Salt Creek (Fig. 1A, B). He suggested that south of the 700 mm isohyet, the reduced evaporation rates and higher rainfall prevented the concentration of lake waters to salinities where dolomite could precipitate. Regional reconnaissance confirms this observation and goes further by showing that more magnesian-rich dolomites tend to be found at the more arid northwestern portion of the coastal plain near Salt Creek (Warren, 1990).

In the Salt Creek region, evaporation rates exceed precipitation except for a few weeks in June and July (South Australian Bureau of Meteorology, public data). To the southeast, near Kingston, precipitation exceeds evaporation for three to four months each year (May to August). Air temperatures range from $-1\,°C$ to $38\,°C$ with a mean annual air temperature of $13.5\,°C$. Water temperatures vary between $10\,°C$ and $28\,°C$ and the pH from 8–10. During the period 1982–85 the summer temperatures within subaerially exposed lake sediments were as high as $50\,°C$, winter temperatures were as low as $5\,°C$. Detailed analyses of some Coorong lake waters can be found in von der Borch (1965), and the hydrogeochemistry of the regional groundwaters in O'Driscoll (1960). Mg:Ca ratios in different lakes varies from 1 to 20. Sulfate concentrations measured by Skinner (1963) in several lakes in the coastal plain range from 6.1 to 8.2% and are close to the expected value for seawater sulfate (7.68%).

Coorong Lake Stratigraphy

Coorong Lakes can be subdivided into two types:

Type 1 lakes have no Holocene marine–estuarine biota in their basal sedimentary unit and so were not connected to the Coorong Lagoon in the early Holocene. They have been fed by rainwater and continental groundwaters throughout their Holocene history (Fig. 1B; Milne Lake, Tepee Lake, Lakes 3–5).

Type 2 lakes have a Holocene basal sedimentary unit that contains a marine–estuarine biota and so were connected to the Coorong lagoon in the early Holocene (e.g. Salt Creek lake chain). They are further subdivided into Type 2a and 2b lakes. *Type 2a lakes* are found within calcrete-floored depressions surrounded by Pleistocene dunes. Type 2a lakes sampled in this study typically occupied the 120,000 year interdunal corridor located immediately landward of the modern Coorong lagoon. Thickening of the basal sedimentary unit in the corridor regions separating two adjacent Type 2a lakes in the Salt Creek lake chain gives these lakes their characteristic cross-section whereby lakes are separated by sills composed of a thickened basal unit (see Fig. 3). *Type 2b lakes* are the lakes of the ephemeral Coorong Lagoon (Lakes A–F) located southeast of the open Coorong Lagoon. They lack a calcreted Pleistocene margin on their seaward side. They were part of the open Coorong Lagoon until isolated by Holocene beach-dune migration and spit accretion by calcareous sediments derived mainly from the Younghusband Peninsula.

All lacustrine sequences cored in the various Coorong lakes are shallowing-upward cycles exhibiting kindred up-core transitions of textures that are independent of mineralogy (Fig. 2).

The *basal unit* of the Holocene lacustrine succession is a massive siliciclastic-carbonate grainstone to wackestone. Cores through the unit frequently contain whole pelecypod and gastropod shells as well as minor dolomite. Mineralogy is variable, depending on the relative proportions of quartz sand, bioclasts and micrite. Typical mineralogies are calcite, aragonite, magnesian calcite, frequently there is a quartz sand in the lower parts of the unit often with minor dolomite. The dolomite is micritic, commonly calcian-rich, and occurs either as pore fills or nodular cement patches; it is always a minor constituent of the basal unit. The basal unit now fills the narrow corridors that once joined type 2a lakes to each other or to the Coorong Lagoon. In these corridor areas, the basal unit can be as thick as 2–3 m (Figs 1A, B and 3). Cores taken within interdunal corridors typically fine upward, reflecting a lessening of bottom currents and supply of coarser sediment as the lake's tie to the open Coorong lagoon was blocked by laterally accreting of beach ridges (Warren, 1988). Beneath the lacustrine fill, and away from the corridors, the basal unit thins to tens of centimeters (Fig. 3). In the

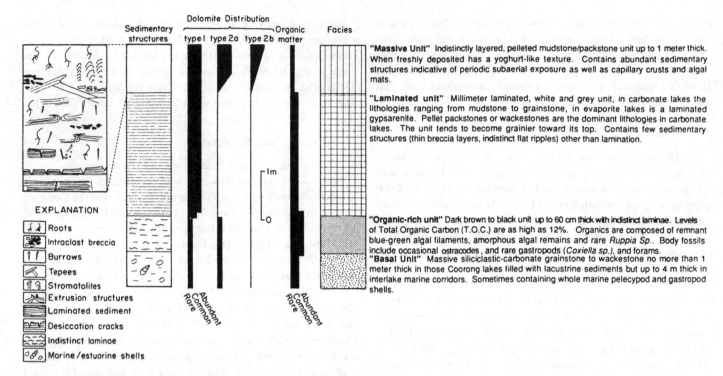

Fig. 2. Vertical sequence from a Coorong Lake (after Warren, 1988).

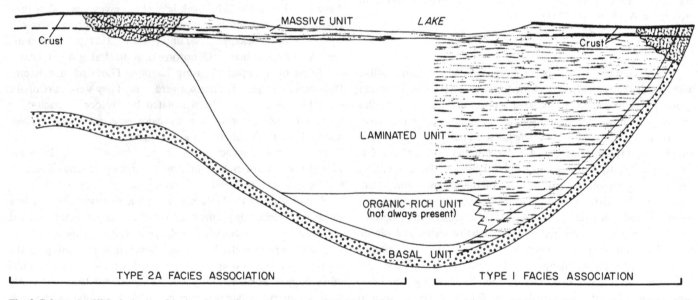

Fig. 3. Schematic of lithofacies in a Coorong Lake. Shows the separate subdivision of facies between type 1 and type 2 lakes and different position of dolomite in the stratigraphy. Dolomite in basal unit (open circles), dolomite in marginal position, typically calcian dolomite (stippled), dolomite as lake fill (horizontal line pattern). (After Warren, 1988.)

central areas of type 2a lakes of the Salt Creek lake chain this thinned basal unit contains up to 4% organic matter (TOC av. = 1.41 ± 1.21%). In type 2a lakes, most of the basal unit was deposited in the early Holocene when the interdunal depressions were consistently filled with lagoonal seawater replenished by a surface connection with the oceanic–estuarine waters of the Coorong Lagoon (e.g., Salt Creek lake chain and South Stromatolite Lake; Fig. 1A, B).

In type 2b lakes the basal unit comprises the greater volume of the Holocene sediment fill. In marginward parts of these 2b lakes, the unit is predominantly a grainstone–packstone with muddy interbeds a few cm thick. In the more central parts of larger type 2b lakes, the unit is muddier and can sometimes be composed of layered to laminated wackestone–mudstones containing articulated pelecypod and ostracode shells. Like the marine/estuarine sediments of the open Coorong lagoon, this layered portion of the

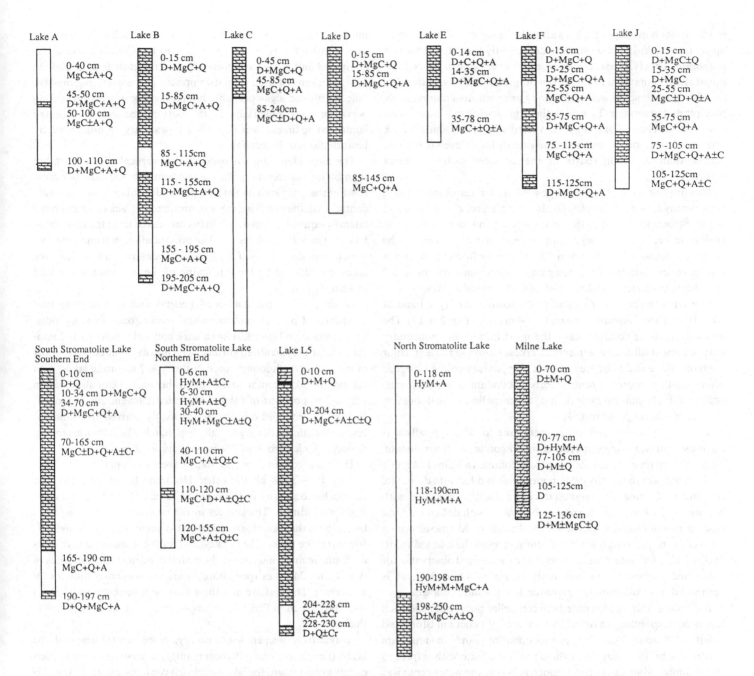

Fig. 4. Core mineralogies of single cores from selected Coorong lakes. Correlation lines between lakes are not shown as each lake was sampled by a single core. See figure 1 for localities of various lakes. A = aragonite; C = calcite; D = dolomite; HyM = hydromagnesite; M = magnesite; MgC = magnesian calcite; Q = quartz. (after Warren, 1990).

basal unit contains the skeletal remains of a marine–estuarine biota, and so can be distinguished from the laminated lacustrine unit described in the next section.

Beneath the central portions of type 1 lakes the basal unit is up to 70 cm thick and composed of siliciclastic or reworked carbonate dune sands mixed with up to 40% mud. In type 1 lakes the unit is predominantly a quartzose sand with minor calcian dolomite.

In the type 2a lakes of the Salt Creek lake chain and South

Stromatolite Lake there is a black, to dark brown unit immediately above the basal unit; in core it is defined by its obvious dark color. This *organic-rich unit* is a pelleted and sometimes laminated mudstone containing degraded algal remains, diatoms and degraded high plant material, as well as occasional small ostracods and gastropods. The percentage of organic material in this unit varies but ranges up to 12%, locally the unit includes sufficient organic matter to classify it as a sapropel (Fig. 4; Warren, 1988). The unit is most common in deeper hollows atop the basal unit where it can be up to 1 m thick. Its mineralogy is dominantly low Mg-calcite and aragonite. This is in part a reflection of mineralogy of the comminuted skeletal material in the sediment, predominantly ostracodes, and dwarf molluscs as well as a varying proportion of inorganically precipitated aragonite. Typically, the organic-rich unit contains no dolomite. In Halite Lake, a gypsum-filled salina, this unit is a dark brown to brown packstone with consistently lower levels of organic matter than its counterparts in the adjacent carbonate lakes (Warren, 1988).

The organic-rich unit was deposited under restricted, anoxic, periodically lacustrine conditions. In the Salt Creek lake chain and South Stromatolite Lake, the deposition of this unit in the early Holocene heralded the beginning of restricted conditions in the lakes and defined the time when the lakes were first cut off from a surface connection to the Coorong Lagoon; an event that converted interdunal estuaries into density-stratified perennial lakes.

Overlying the organic-rich unit or the basal unit of type 1 and 2a lakes is a *pelletal laminated unit* up to 3–4 m thick (Figs 2 and 3). The laminated unit is volumetrically the most important sedimentary unit in almost all the type 1 and 2a lakes. Laminae range in color from off-white and light grey to dark grey; darker colored laminae often contain more organic debris (abundant algal filaments, occasional *Ruppia* rhizomes) and fewer pellets. Total organic carbon values range up to 6%.

Mineralogy of the unit varies from lake to lake; typically it is composed of varying proportions of aragonite, hydromagnesite, magnesite, magnesian calcite and rare dolomite. In Milne Lake, one of the few lakes in the Coorong region that is filled with dolomite, the unit is dominantly magnesian-rich dolomite associated with magnesite. In Tepee Lake, another lake filled with dolomite, it is a calcian-rich dolomite associated with Mg-calcite. Magnesite occurs only in the uppermost few tens of centimeters of the lake sediment. Halite Lake is an interesting exception to the general observation of laminated carbonates in this unit; in Halite Lake the unit is composed of mm-laminated gypsarenite–aragonite couplets.

Individual lamina alternate between pellet-poor and pellet-rich layers corresponding to respective textures of pellet mudstone and pellet packstone. Relative proportions of sand to mud are controlled by the location of the depositional site with respect to water depth, wind and current velocity. Where the water depth was deeper or the wave fetch shorter, the sequence is dominated by mudstones. Where the water depth was shallower, or the wave energy higher, the sequence is dominated by pelletal packstones. Prevailing southerly and westerly winds deposited grainier packstone–wackestone sequences in the northern and easterly portions of most type 1 and 2a lakes. In the deeper and/or more southwesterly areas of the same lakes, the sediments are pelletal mudstones. Many type 2b lakes are adjacent to the Holocene dunes of Younghusband Peninsula, hence the proportion of skeletal sand in the laminated unit is typically higher on the southwesterly side of these lakes.

The laminated unit was deposited under subaqueous lacustrine conditions. Laminae show no evidence of doming, bioerosion or mud cracks, nor do they contain evidence of subaerial exposure seen in present day surface sediments in the Coorong lakes. The laminated unit lacks the breccias, the extrusion structures, and the lesser amounts of organic matter present in the overlying massive unit. Benthonic algae probably bound these sediments preventing pervasive wave reworking of the bottom sediments. Once the laminated sediment was deposited it was largely unaffected by bioturbation or desiccation.

The laminated unit was deposited by vertical accretion rather than by lateral accretion. This assessment is based on: the occurrence of the laminated unit in all type 1 and 2a lakes, a flat lake-wide contact with the overlying massive unit, and the lack of an extensive laterally-equivalent massive unit in cores taken from the margins of lakes in the Salt Creek region. The progradation that characterizes present-day deposition of the overlying 'massive unit' in the same lakes is evidenced by the thickening of the unit toward the lake margin (Fig. 3).

Directly above the laminated pelletal unit is a massive unit composed of pelletal mudstones and wackestones showing indistinct laminae and layers. It has a dark grey color below a 1–2 mm-thick, light-grey oxidized surface. The massive unit contains most of the Holocene dolomite cored in the type 2 Coorong lakes. And yet, not all lakes contain dolomite in this unit; it can also contain varying proportions of calcite, Mg-calcite, magnesite, aragonite, hydromagnesite and cristobalite (probably derived from diatom tests). Organic matter is generally less than 0.5%, notwithstanding a total organic carbon of 2% in one sample (Fig. 4).

The massive unit forms in areas covered by ephemeral surface waters. It overlies all the other Holocene lacustrine units and encroaches on many of the marginal areas of the lake including Pleistocene dunes. Thicknesses in the massive unit vary across a lake. About the lake edge it can be up to a meter thick, although 40–60 cm is more usual. The thickest and most extensive massive units are found in marginal flats on the easterly and northeasterly sides of type 1 and 2a lakes (prevailing winds are westerlies and southwesterlies). The massive unit thins into the lake center to where in some lakes such as Pellet Lake it is no more than a few centimeters thick.

In relatively deeper, lower-energy, more central areas of the lakes, the massive unit is predominantly a mudstone; in the higher-energy areas toward the lake margins, it contains more pellets and is sometimes a pelletal packstone–wackestone covered by a surface layer of wave ripples. In both areas the unit has a transitional, but abrupt, contact 1–3 cm thick into the underlying laminated unit.

The massive unit is highly-bioturbated and crosscut by burrows, plant rhizomes and rootlets. About the lake edges it contains capillary crusts, tepees and breccias (Warren, 1988, for full description and interpretation of sedimentary structures). This unit is the capstone to all Coorong vertical sequences and is the last stage of lacustrine infill in all lakes cored during this study. It was laid down under conditions of ephemeral surface waters that also characterize all the present-day Coorong lakes.

The massive unit was not intersected at depths of more than 1.1 m in any of the lakes sampled, and no core from the central regions of any Coorong lake contained the massive unit in direct contact with the basal unit (Fig. 3; Warren, 1988). This implies that in these areas the transition from subaqueous sedimentation of the basal unit to that of the laminated unit was direct and not separated by a drying out phase. It also implies that the Coorong interdunal lakes were filled with water throughout most of their Holocene history. The bulk of the massive unit formed in the latter stages of a lake's infill when the sediment surface had accreted to where it approached the surface level of the perennial lake waters. Prior to this time, the lakes were permanently water-filled with water surfaces within a meter of the present water table and the bulk of the precipitated mud was accreting as laminated sediment atop the subaqueous lake floor. Once the top surface of the laminated unit approached the water level of the lake, sediment for the massive unit was readily supplied by lateral accretion and the massive unit began to rapidly prograde across the lake.

Dolomite in the Coorong Coastal Plain

Coorong dolomites are true 'primary dolomites'; dolomite is precipitating as dolomite and is not replacing an earlier carbonate mineral. Dolomites in the various lakes of the Coorong coastal plain in South Australia occur in three mineralogical associations (von der Borch, 1965; Warren, 1990):

Dolomite ± Mg-calcite (widespread).
Dolomite ± Magnesite (common in northwest).
Dolomite ± aragonite ± hydromagnesite (rare and only in northwest).

Most Holocene dolomite occurrences in the near coastal lakes of the Coorong region are associated with Mg-calcite. Only in the more arid settings found at the northwestern end of the Coorong region, near Salt Creek, is dolomite commonly found in association with magnesite and occasionally with hydromagnesite and aragonite. Dolomite in the near surface sediments of lakes near Salt Creek such as Pellet Lake and Milne Lake is often associated with magnesite in the lake center and Mg-calcite about the lake margin. The dolomites associated with magnesite tend to be more magnesian-rich than dolomites associated with Mg-calcite, a direct reflection of the chemistry of the mother waters. Waters which precipitate magnesite are typically more concentrated and more magnesium-rich than waters precipitating Mg-calcite, hence co-precipitated dolomites will also be more magnesium-rich.

Dolomite has been the main lacustrine mineral of concern in geological studies of this area, yet it makes up no more than 10% of the carbonate minerals forming surficial Holocene deposits across the coastal plain.

Whilst most dolomite-dominated units in the Coorong coastal plain are less than a meter thick and usually only a few tens of centimeters thick, dolomitic units as thick as 3 m occur in small interdunal depressions near Salt Creek and immediately landward of the Coorong Lagoon (Rosen et al., 1988b; Warren, 1988). In lakes in the southern reaches of the ephemeral Coorong Lagoon near Kingston, thin modern dolomites occur sporadically,

especially toward the lake margins. Thin Holocene dolomites also occur at the surface of interdunal lakes located tens of kilometers inland where the associated sediments were continental lacustrine not lagoonal estuarine (e.g., Tilley Swamp, near Naracoorte, located 100 km from coast; for other localities see von der Borch & Lock, 1979).

Salt Creek lake chain is located at the northern end of the Coorong coastal plain (Figs 1A, B and 4). It embraces Halite Lake, Dolomite Lake, Pellet Lake and North Stromatolite Lake and is the area of thickest Holocene dolomite in the Coorong coastal plain with Holocene lake-wide dolomitic units up to 3 meters thick (Rosen et al., 1988a, 1989; Warren, 1988). In the early Holocene, before the depression was filled with sediment, the interdunal corridor of the Salt Creek lake chain (Fig. 1A) was a narrow estuarine lagoon, a perennial narrow embayment open to the Coorong Lagoon via the southern end of North Stromatolite Lake (Warren, 1988). At that time the basal unit of estuarine–marine carbonates was laid down on the lagoon floor. Surface exchange of waters between the Coorong Lagoon and the Salt Creek waterway slowed and then stopped as the entrance to the then lagoon chain was blocked by the build-up of beach ridges at the southern end of North Stromatolite Lake (Fig. 1A). As the entrance to the embayment began to clog, the narrow confined channels connecting North Stromatolite Lake, Pellet Lake, Dolomite Lake and Halite lake were also choking with sandy estuarine sediments. Blockage converted the previously restricted estuarine–marine lagoon into a chain of isolated lakes; the same isolated lakes that can be seen today although they are now filled with lacustrine carbonates. Isolation of the lakes led to lake water chemistries which evolved separately from one lake to another, and so different mineral assemblages were laid down in different lakes (Fig. 4). In other nearby lakes, such as Milne Lake (Fig. 1A), there was no Holocene connection to the marine lagoon, but the floor of the interdunal corridor sat well below the regional water table, and so the depression was a perennial lake gradually filling with Holocene dolomite. As the various interdunal depressions filled with lacustrine sediment, the lake water-surface became a lake water-table.

Bibliography

Ahmad, R. & Hoestetler, P.B., 1988. Recent advances in the study of Holocene dolomitic carbonate sedimentation in the Coorong area of South Australia. *Abstracts of Aust. Geol. Soc. Ann. Meeting*, pp. 40–1.

Alderman, A.R. & Skinner, C.W., 1957. Dolomite sedimentation in the South-East of South Australia. *Am. J. Sci.*, **255**, 561–7.

Bolz, R.W. & von der Borch, C.C., 1984. Stable isotope study of the Coorong area, South Australia. *Sedimentology*, **31**, 837–49.

Brown, R.G., 1965. *Sedimentation in the Coorong Lagoon, South Australia*. PhD Thesis, Department of Geology, Univ. Adelaide.

De Deckker, P. & Geddes, M.C., 1980. Seasonal fauna of ephemeral saline lakes near the Coorong Lagoon, South Australia. *Aust. J. Marine Freshwater Res.*, **31**, 677–99.

Eriksson, K.A. & Warren, J.K., 1983. A paleohydrological model for early Proterozoic dolomitization and silicification. *Precambrian Res.*, **21**, 299–321.

Hayball, A.J., McKirdy, D.M., Warren, J.K., von der Borsch, C.C. & Padley, D., 1991. Organic facies of Holocene carbonates in North Stromatolite Lake, Coorong, Region, South Australia. *15th Int. Mtg. Organic Geochemistry, Manchester, UK, Programme and Abstracts*, pp. 19–20.

Holmes, J.W. & Waterhouse, J.D., 1983. Hydrology. In *Natural History of the South East*, ed. M.J. Tyler, C.R. Twidale, J.K. Ling & J.W. Holmes, pp. 49–59. Royal Soc. South Australia.

Kendall, G.C. St. C. & Warren, J.K., 1987. Hardgrounds in carbonate platforms and the development of tepee structures. *Sedimentology*, **84**, 1007–27.

Lock, D.E., 1982. *Groundwater Controls on Dolomite Formation in the Coorong Region of South Australia and its Ancient Analogues*. PhD Thesis, The Flinders University of South Australia.

Mawson, Sir D., 1929. Some South Australian limestones in the process of formation. *Quart. J. Geol. Soc.*, **85**, 613–23.

McKirdy, D.M., Padley, D., Macdonald, F., Warren, J.K., Hayball, A.J. & von der Borch, C.C., 1992. Coorong lacustrine carbonates as Holocene analogues of ancient evaporite-associated hydrocarbon source rocks. *Amer. Ass. Petrol. Geol. Int. Mtg., Sydney, Abstracts*, pp. 63–4.

Miser, D.E., 1987. *Microstructures in natural and synthetic dolomites*. PhD Thesis, Univ. Texas, Austin.

Muir, M., Lock, D. & von der Borch, C.C., 1980. The Coorong model for penecontemporaneous dolomite formation in the Middle Proterozoic McArthur Group, Northern Territory, Australia: In *Concepts and Models of Dolomitization*, ed D.H. Zenger, J.B. Dunham & R.L. Ethington, pp. 51–67. Society of Economic Paleontologists and Mineralogists, Spec. Pub. No. 28.

O'Driscoll, E.P.D., 1960. *The Hydrology of the Murray Basin Province*. Bull. Geol. Surv. South Aust., No. 35.

O'Driscoll, E.P.D. & Shepherd, R.G., 1960. *The Hydrology of Part County Cardwell in the Upper Southeast of South Australia*. Geol. Survey of South Australia, Report of Investigation, No. 15.

Peterson, M.N.A. & von der Borch, C.C., 1965. Modern chert in a carbonate precipitating locality. *Science*, **149**, 1501–3.

Rosen, M.R., Miser, D.E. & Warren, J.K., 1988a. Sedimentology, mineralogy, and isotopic analysis of Pellet Lake, Coorong region, South Australia. *Sedimentology*, **35**, 105–22.

Rosen, M.R., Miser, D.E. & Warren, J.K. 1988b. Compositional variations of dolomite from a chain of ephemeral lakes in the Coorong region, South Australia (abs): *Am. Assoc. Petrol. Geol. Bull.*, **72**, 241.

Rosen, M.R., Miser, D.E., Starcher, M.A. & Warren, J.K., 1989. Formation of dolomite in the Coorong region, South Australia. *Geochim. Cosmochim. Acta*, **53**, 661–9.

Schwebel, D.A., 1978. *Quaternary Stratigraphy of the South-East of South Australia*. PhD Thesis, The Flinders Univ. South Australia.

Schwebel, D.A., 1983. Quaternary dune systems: In *Natural History of the South East*, ed. M.J. Tyler, C.R. Twidale, J.K. Ling & J.W. Holmes, pp. 15–24. Royal Soc. South Australia.

Skinner, H.C.W., 1963. Precipitation of calcian dolomites and magnesian calcites in the southeast of Australia. *Am. J. Sci.*, **261**, 449–72.

Von der Borch, C.C., 1965. The distribution and preliminary geochemistry of modern carbonate sediments of the Coorong area, South Australia. *Geochim. Cosmochim. Acta*, **29**, 781–99.

Von der Borch, C.C., 1976. Stratigraphy and formation of Holocene dolomitic carbonate deposits of the Coorong area, South Australia. *J. Sed. Petrol.*, **46**, 952–6.

Von der Borch, C.C. & Jones, J.B., 1976. Spherular modern dolomite from the Coorong area, South Australia. *Sedimentology*, **23**, 587–91.

Von der Borch, C.C. & Lock, D., 1979. Geological significance of Coorong dolomites. *Sedimentology*, **26**, 813–24.

Von der Borch, C.C., Lock, D. & Schwebel, D., 1975. Groundwater formation of dolomite in the Coorong region of South Australia. *Geology*, **3**, 283–5.

Walter, M.R., Golubic, S. & Preiss, W.V., 1973. Recent stromatolites from hydromagnesite and aragonite depositing lakes near Coorong Lagoon, South Australia. *J. Sed. Petrol.*, **43**, 1021–30.

Warren, J.K., 1983. On pedogenic calcrete as it occurs in the vadose zone of Quaternary calcareous dunes in coastal South Australia. *J. Sed. Petrol.*, **53**, 787–96.

Warren, J.K., 1986. Source rock potential of shallow water evaporitic settings. *J. Sed. Petrol.*, **56**, 442–54.

Warren, J.K., 1988. Sedimentology of Coorong Dolomite in the Salt Creek region, South Australia. *Carbonates and Evaporites*, **3**, 175–99.

Warren, J.K., 1989. *Evaporite Sedimentology: Importance in Hydrocarbon Accumulation*. Prentice Hall, Englewood Cliffs.

Warren, J.K., 1990. Sedimentology and mineralogy of dolomitic Coorong lakes, South Australia. *J. Sed. Petrol.*, **60** (6), 843–58.

Lake Balaton, Hungary

TIBOR CSERNY

Hungarian Geological Institute, Népstadion út. 14, H-1142 Budapest, Hungary

Lake Balaton is the largest lake in central Europe, its environment being one of the most valuable and important recreation areas in Hungary. Lake Balaton is situated in the western part of Hungary (Fig. 1), its mean water level is 104.8 m above that of the Adriatic Sea. At mean level, its surface area is 596 km², average depth is 3.25 m and water volume is at 1.9 km³; Lake Balaton is shallow with a large surface area (Fig. 2). The water level has fluctuated within wide limits in accordance with climate changes (Fig. 3). Over the past decades, eutrophication has continued at an exponential rate.

Lake Balaton and its surroundings have been studied in several phases since the turn of the century. In 1981, a new study program was initiated to understand better the thickness and properties of the recent sediments and reconstruct the geological history of the lake. Thirty-three boreholes up to 240 m in depth were drilled between 1981–89; 370 km of seismo-acoustic and echographic profiles were recorded in 1987 (Fig. 4).

The results can be summarized as follows:

The Holocene sediments comprise clayey silt, silt and sand. Molluscan shell debris and peat occur near the base of some cores. A few cm-thick layers of polygenetic pebbles (derived from the northern shore) cover the Pannonian basement; these vary in dimensions and have been weathered. DTA and X-ray analyses indicate that the Holocene sediments contain carbonate (40–80%), quartz, muscovite, chlorite and clay minerals (illite, montmorillonite) in decreasing abundance. Magnesium calcite decreases in abundance towards the lower parts of the cores. Organic carbon contents are not a function of depth. Natural sulfur was found in extracts from 2.5–3.5 m depth intervals.

The isopach map of loose mud (Fig. 5) and the seismostratigraphic-tectonic map of the basement (Fig. 6) have been compiled on a scale of 1:50,000. The muddy sediments are generally 5 m thick. There are elevations in the basement covered with 1 to 1.5 m thick mud layers, while in troughs, mud layers may reach 8 m in thickness. By applying radiocarbon and palynological dating, mud accumulation rates could be calculated as 0.4–0.6 mm/year. On the basis of seismic time sections (Figs 6, 7 and 8), the lake basement contains seven 'sequence-types' which are distinguishable by their characteristic geophysical signals. Sequence-types 1–3, 6 and 7 represent horizontal sequences and cannot be easily divided into layers. Sequence-types 4 and 5 represent dipping strata and can be subdivided into four layers (P,O,A,B; see Fig. 8) in some parts of the lake. The geological identification of the sequence-types and their layers based on new boreholes is a task for the near future.

Diatoms, pollen, ostracodes and molluscs are the most useful in establishing geological history and changes in the ecological conditions of Lake Balaton (Figs 9 and 10). Ten ecozones were derived from relative quantative faunal changes through time. Radiocarbon dating of the peat indicates that Lake Balaton has been in existence since 12,000 years BP.

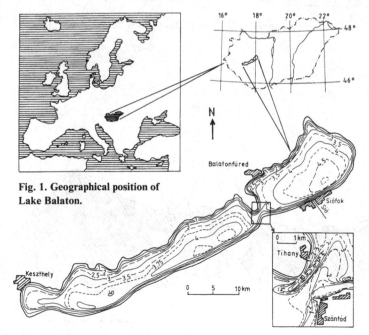

Fig. 1. Geographical position of Lake Balaton.

Fig. 2. The isobathic map of Lake Balaton.

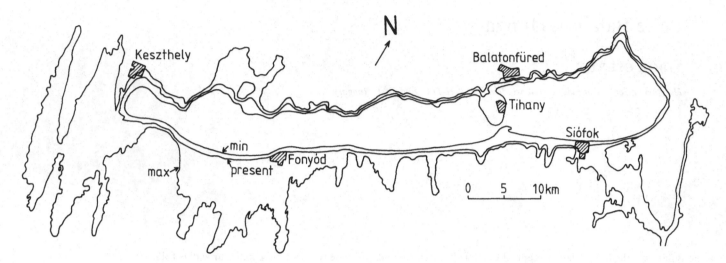

Fig. 3. Min./max. and present extension of Lake Balaton.

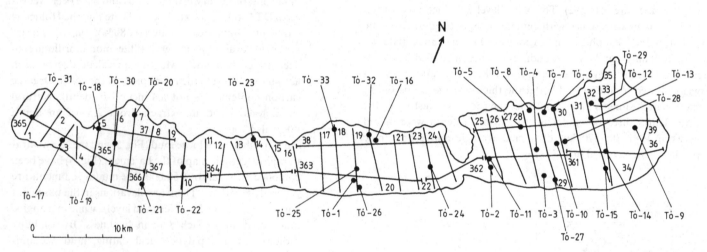

Fig. 4. Layout of boreholes and geophysical logs.

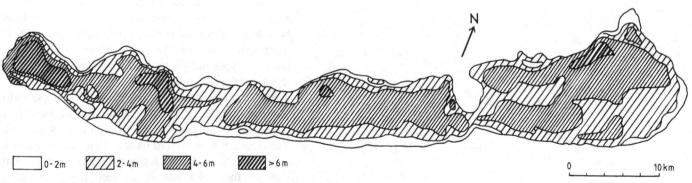

0-2m	2-4m	4-6m	>6m

Fig. 5. Isopach map of loose mud.

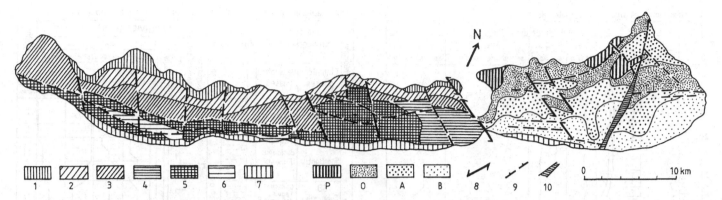

Fig. 6. Seismostratigraphic-tectonic map of the basement. Explanation of legend: sequence-types 1–7 designate sedimentary beds of different lithology (grain size and physical properties) distinguished on seismo-acoustic logs based on amplitude and frequency of reflected signals. 1–3, 6, and 7 identify specific horizontal beds while 4 and 5 indicate beds dipping a few degrees to the S-SE. Layers P, O, A, and B: layers differentiated within sequence-types 4 and 5 in some portions of the lake. A transcurrent fault is defined as a horizontal displacement.

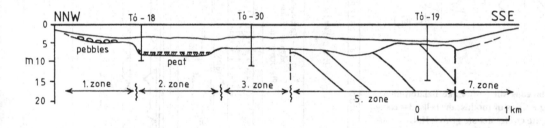

Fig. 7. Profile No. 5. See Fig. 4 for exact location of profile no. 5 in western Lake Balaton. Zones refer to areas where similarly numbered sequence-types occur (see Fig. 6) in which horizontal layers cannot as yet be differentiated.

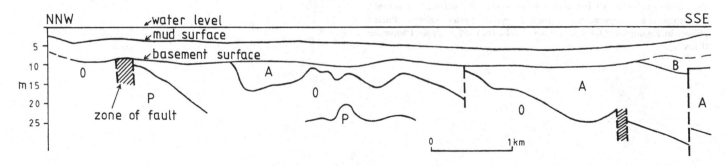

Fig. 8. Profile No. 30. See Fig. 4 for exact location of profile no. 30 in eastern Lake Balaton. Layers P,O,A, and B refer to layers differentiated within sequence-types 4 and 5.

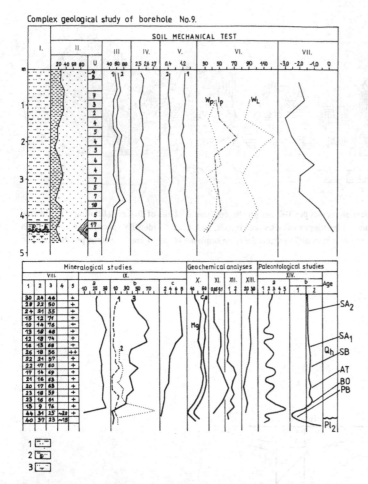

Complex geological study of borehole No.9.

Stratigraphical and ecological characteristics of the borehole Tó-24

Fig. 9. I: Geological column: 1 = argillaceous silt (soft Lake Balaton mud); 2 = argillaceous silt containing large quantities of molluscan shells; 3 = compact argillaceous silt. In addition, elongate circles denote gravel. II: Granulometric distribution (mud, silt, sand, and gravel) (%): U = coefficient of inequality. III: Water content: 1 = adsorptive; 2 = structural and adsorptive water. IV: Bulk density (g/cm³). V: Body density (g/cm³): 1 = natural state; 2 = dry state. VI: Plasticity (%): W_p = Plastic limit; I_p = Plasticity index; W_L = Liquid limit. VII: Consistency index (relative measurement). VIII: Mineralogical analyses (%): 1 = detrital minerals; 2 = clay minerals; 3 = carbonates; 4 = goethite, 5 = organic matter. IX: Carbonate content; a = CO_2 (%), b = detrital (epigenetic); 1 = dolomite, 2 = calcite; chemogenetic (syngenetic); 3 = Mg-calcite; c = $CaCO_3/MgCO_3$ ratio in Mg = calcite. X: Relative value of exchangeable Ca-Mg ions. XI: Bitumen (%). XII: Organic carbon (%). XIII: Bitumen/organic carbon ratio. XIV: Palynological analyses: a = quantity of pollen: 1 = in traces, 2 = very scarce; 3 = scarce; 4 = moderate; 5 = frequent; b = temperature: 1 = cool; 2 = warm. Letters refer to climatic phases in Figure 10. Q_h = Quaternary (Holocene), Pl_2 = Upper Pannonian (Liassic).

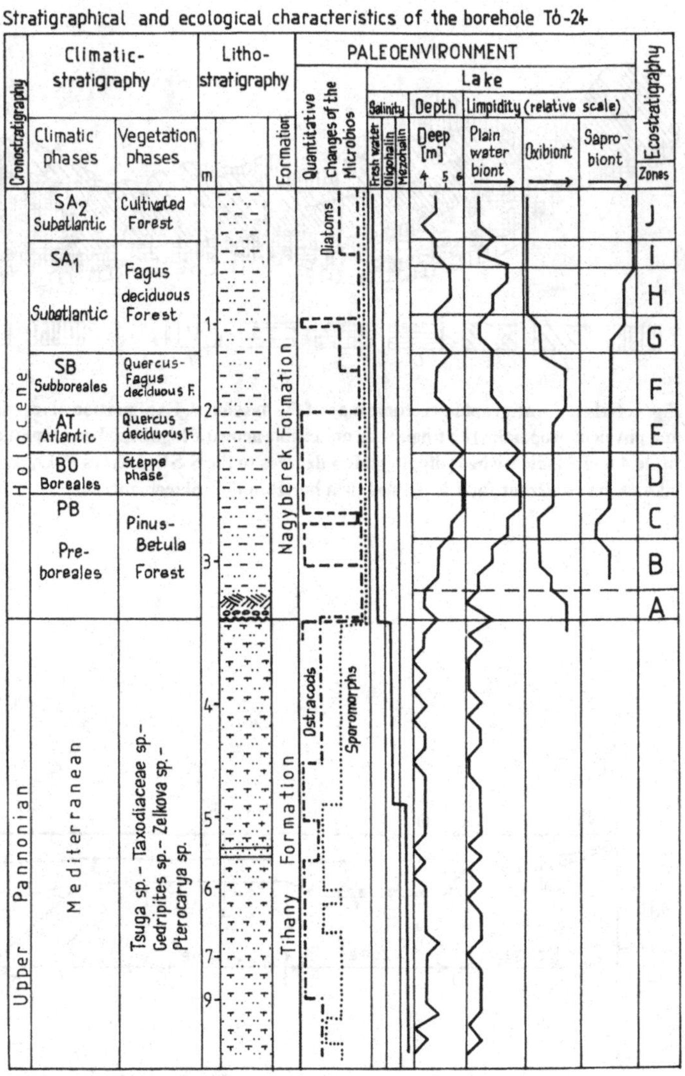

Fig. 10. Lithological legend same as in Fig. 9. In addition, cross-hatching and elongate circles at the bottom of the Holocene denote gravel overlain by peat and the stippled layer in the middle portion of the Pannonian denotes sandy silt.

Bibliography

Baranyi, S. (Ed.), 1979. *Summary of Investigations of Lake Balaton*. Vízügyi Műszaki Gazdasági Tájékoztató 112 sz., VIZDOK, Budapest. (In Hungarian.)

Bauer, I. & Sárdi, A., 1984. Mapping of the bottom sediments of Lake Balaton through the use of subsurface radar. *Vízügyi Közlemények*, **LXVI** (3), 456–66. (In Hungarian.)

Bendefy, L. & Nagy, I.V., 1969. *Secular Changes in Lake Balaton's shoreline*. Műszaki Könyvkiadó, Budapest. (In Hungarian.)

Bodor, E., 1987. Formation of Lake Balaton palynological aspects – Holocene environment in Hungary. *Contributions to the INQUA Congress, Canada*, pp. 77–80.

Brukner-Wein, A., 1988. Actuogeological drilling in Lake Balaton in 1982: organic geochemistry. *Földt Int. Évi, Jel.*, 1986-ról, 581–609. (In Hungarian.)

Cserny, T., 1987. Results of recent investigations of the Lake Balaton deposits – Holocene environment in Hungary. *Contributions to the INQUA Congress, Canada*, pp. 67–76.

Cserny, T. & Corrada, R., 1989. Complex geological investigation of Lake Balaton (Hungary) and its result. *Acta Geol. Hung.*, **32** (1–2), 117–30.

Entz, G. & Sebestyén, O., 1942. *The File of Lake Balaton*. Természettudományi Társulat Könyvkiadó Vállalat, Budapest. (In Hungarian.)

Herodek, S. & Máté, F., 1984. Eutrophication and its reversibility in Lake Balaton (Hungary). *Proceedings of SHIGA Conference 1984 on conservation and Management of World Lake Environment, Japan*, pp. 95–102.

Herodek, S., Laczkó, L. & Virág, Á., 1988. Lake Balaton research and management. *Contributions of the Third International Conference on the Conservation and Management of Lakes, 'Balaton 88', Keszthely*, p. 111.

Illés, I., 1981. *Balaton*. Natura, Budapest. (In Hungarian.)

Lóczy, L., 1913. *Results of the Scientific Investigations of the Balaton Region I/1/1. Geological Formations of the Balaton Area and their Regional Distribution*. Budapest. (In Hungarian.)

Máté, F., 1987. Mapping of the recent sediments of the bottom of Lake Balaton. *Földt Int. Évi. Jel.*, 1985-ról, 366–79. (In Hungarian.)

Miháltz-Fragó, M., 1982. Palynological examination of bottom samples from Lake Balaton. *Földt Int. Évi. Jel.*, 1981-ról, 439–48. (In Hungarian.)

Müller, G., 1970. High-magnesian calcite and protodolomite in Lake Balaton (Hungary) sediments. *Nature*, **226** (5247), 749–50.

Müller, G. & Wagner, F., 1978. Holocene carbonate evolution in Lake Balaton (Hungary): a response to climate and impact of man. In *Modern and Ancient Lake Sediments*, ed. A. Matter & M.E. Tucker, pp. 57–81. IAS Special Publication 2.

Nagy-Bodor, E., 1988. Palynological study of Pannonian and Holocene deposits from Lake Balaton. *Földt Int. Évi. Jel.* 1986-ról, pp. 568–80. (In Hungarian.)

Raincsák-Kosáry, Z. & Cserny, T., 1984. Results of the engineering geological mapping of the Balaton region. *Földt Int. Évi. Jel.*, 1982-ról, 49–57. (In Hungarian.)

Rónai, A., 1969. The geology of Lake Balaton and surroundings. Mitt. Internat. *Verein, Limnologie*, **17**, 275–81.

Series of Hydrological Maps, 1976. *21. Balaton 1. Hydrography and geomorphology*. Vituki, Budapest. (In Hungarian.)

Somlyódi, L., 1983. *Eutrophication of Lake Balaton*. Vituki Kózlemények 38. (In Hungarian.)

Szesztay, K. (Ed.), 1966. *Investigation of the filling up of Lake Balaton (1963–1964)*. (In Hungarian.)

Zólyomi, B., 1953. Die Entwicklungsgeschichte der Vegetation Ungarns seit dem letzten interglatzial. *Acta Biol. Acad. Sci. Hungary*, **4**, 367–430.

Zólyomi, B., 1987. Degree and rate of sedimentation in Lake Balaton – Pleistocene environment in Hungary. *Contributions to the INQUA Congress, Canada*, pp. 57–79.

Lake Miragoane, Haiti (Caribbean)

MARK BRENNER[1], JASON H. CURTIS[2], ANTONIA HIGUERA-GUNDY[3], DAVID A. HODELL[4],

GLENN A. JONES[5], MICHAEL W. BINFORD[6] AND KATHLEEN T. DORSEY[7]

[1] *Dept. Fisheries & Aquaculture, Univ. of Florida, 7922 NW 71st St., Gainesville, FL 32606, USA*
[2] *Dept. Geology, Univ. of Florida, Gainesville, FL 32611, USA*
[3] *Florida Museum of Natural History, Univ. of Florida, Gainesville, FL 32611, USA*
[4] *Dept. Geology, Univ. of Florida, Gainesville, FL 32611, USA*
[5] *Woods Hole Oceanographic Institution, Woods Hole, MA 02543, USA*
[6] *Graduate School of Design, Harvard University, Cambridge, MA 02138, USA*
[7] *Kinnetic Laboratories, Inc., Santa Cruz, CA 95060, USA*

Lake Miragoane is one of the largest, deepest, freshwater lakes in the Caribbean (area = 7.06 km², maximum depth = 41 m, conductivity = 350 μS cm^{-1}). The basin lies in limestone terrain and is situated in a tectonic rift system on the north shore of Haiti's southern peninsula (Fig. 1). The lake surface is 20 m above mean sea level. Chronology for a 7.7 m sediment core is based on ^{210}Pb dating, conventional ^{14}C dates on bulk organic matter, and AMS ^{14}C dates on ostracode shells and terrestrial wood. The core spans the sedimentary record from ≈ 10.5 ka BP to present. Geochemical, palynological and stable isotope data elucidate the paleoecology and paleoclimate of this island site.

Carbonates dominate the lacustrine deposits in the core, and can constitute > 90% of the dry mass at some levels (Fig. 2). Organic matter content in the sediments is generally low (< 10%), but reaches a maximum of 35.6% at 40 cm. Ostracode shells (*Candona* sp.) are abundant throughout the section and indicate freshwater conditions spanning the period of deposition. δ^{18}O measurements on *Candona* sp. shells (Fig. 3) provide a high-resolution record of changing evaporation:precipitation (E:P) ratio for the last ten millennia. Pollen data corroborate the variations in E:P ratio based on the stable isotope and provide evidence of changes in temperature and seasonality from the late Pleistocene to Present (Fig. 4).

High δ^{18}O values at the base of the section indicate a high E:P ratio and low lake level during the latter part of the Younger Dryas chronozone (i.e., ≈ 10.5–10.0 ka BP). The inferred aridity is substantiated by the dominance of xeric palms and a montane shrub community. The latter suggests temperatures colder than today. Progressively depleted δ^{18}O values between 7.4 and 6.0 m depth indicate increasingly mesic conditions. During this early Holocene

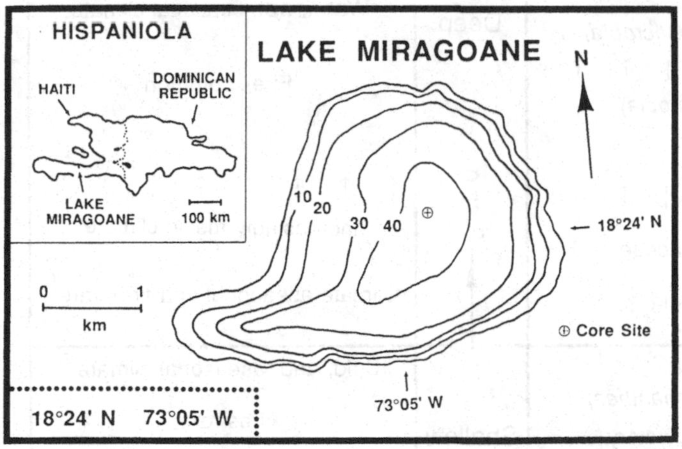

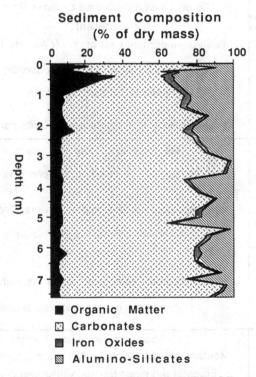

Fig. 2. Composition of sediments in the Lake Miragoane core, showing the relative amounts of organic matter, carbonates, iron oxides and alumino-silicates. Organic matter is loss on ignition at 550°C. Carbonates are the $CaCO_3$ and $MgCO_3$ equivalents of Ca and Mg. Iron oxides represent the Fe_2O_3 equivalents of Fe. Alumino-silicates are the balance of the matrix and represent clays, quartz, and biogenic silica.

Fig. 1. Bathymetric map of Lake Miragoane, Haiti, showing 10-m contours and the site where the 7.7-m sediment core was collected in 41 m of water. Inset map shows Haiti and the Dominican Republic, the two nations that occupy the Caribbean island of Hispaniola. Lake Miragoane lies about 5 km from the coast, on the north shore of Haiti's southern peninsula.

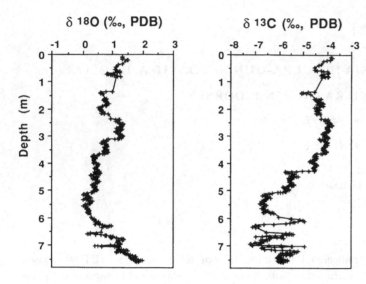

δ 18O (‰, PDB)

δ 13C (‰, PDB)

Fig. 3. δ¹⁸O and δ¹³C measurements on *Candona* sp. ostracod shells. Stable isotope values are relative to PDB standard and are a five-point running mean of data from 515 samples taken at approximately 1-cm intervals over the length of the section.

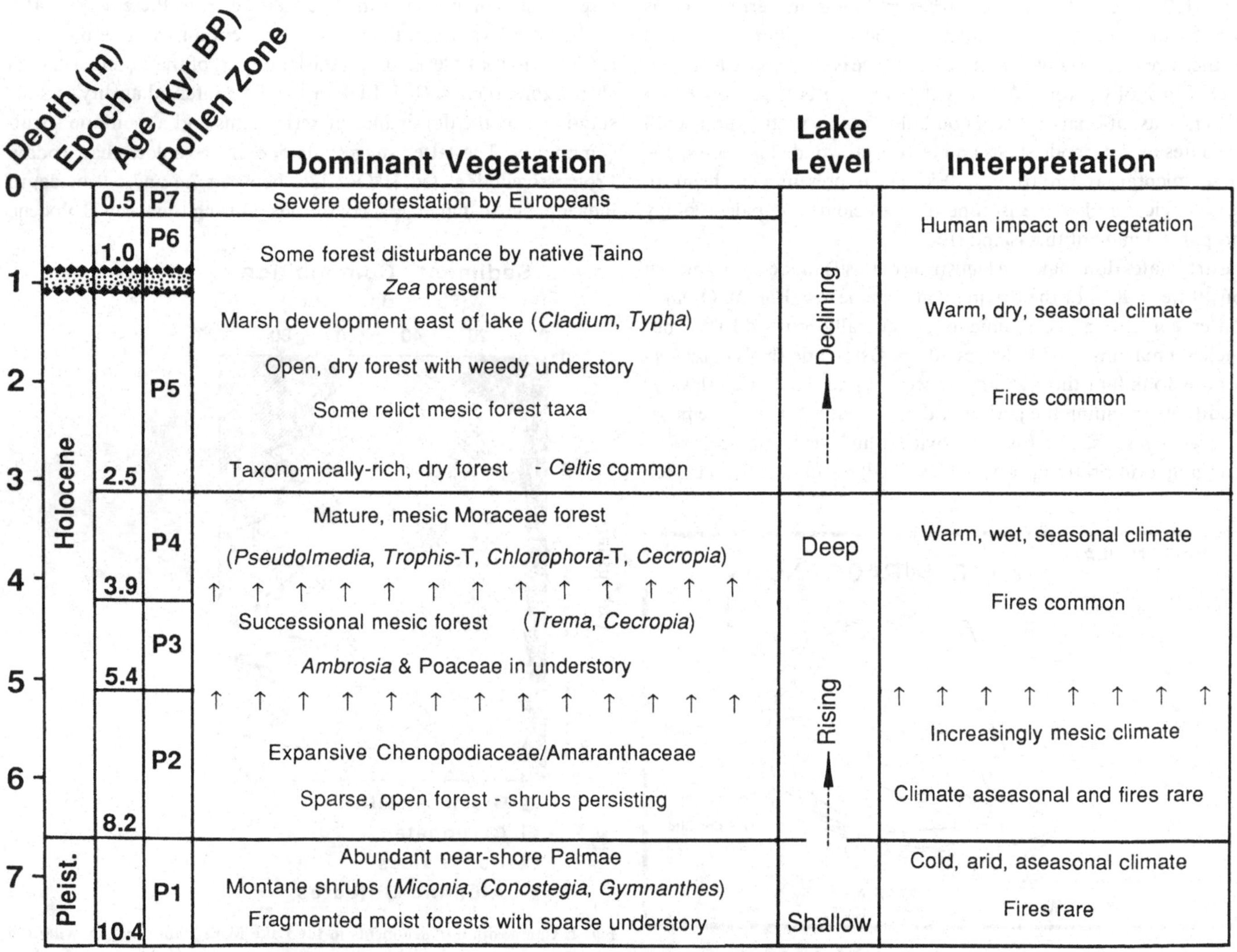

Depth (m)	Epoch	Age (kyr BP)	Pollen Zone	Dominant Vegetation	Lake Level	Interpretation
0		0.5	P7	Severe deforestation by Europeans		Human impact on vegetation
	Holocene	1.0	P6	Some forest disturbance by native Taino	Declining	
1				*Zea* present		Warm, dry, seasonal climate
			P5	Marsh development east of lake (*Cladium*, *Typha*)		
2				Open, dry forest with weedy understory		Fires common
				Some relict mesic forest taxa		
3		2.5		Taxonomically-rich, dry forest - *Celtis* common		
			P4	Mature, mesic Moraceae forest	Deep	Warm, wet, seasonal climate
4		3.9		(*Pseudolmedia*, *Trophis*-T, *Chlorophora*-T, *Cecropia*) ↑ ↑ ↑ ↑ ↑ ↑ ↑ ↑ ↑ ↑ ↑ ↑ ↑ ↑		Fires common
			P3	Successional mesic forest (*Trema*, *Cecropia*)		
5		5.4		*Ambrosia* & Poaceae in understory ↑ ↑ ↑ ↑ ↑ ↑ ↑ ↑ ↑ ↑ ↑ ↑ ↑	Rising	↑ ↑ ↑ ↑ ↑ ↑ ↑ ↑
			P2	Expansive Chenopodiaceae/Amaranthaceae		Increasingly mesic climate
6				Sparse, open forest - shrubs persisting		Climate aseasonal and fires rare
		8.2				
7	Pleist.		P1	Abundant near-shore Palmae		Cold, arid, aseasonal climate
				Montane shrubs (*Miconia*, *Conostegia*, *Gymnanthes*)		Fires rare
		10.4		Fragmented moist forests with sparse understory	Shallow	

Fig. 4. Paleoenvironmental interpretations based on stable isotopes, geochemistry, pollen and charcoal remains. Ages for the base of pollen zones P6 and P7 were estimated by extrapolation of the depth vs. cumulative mass relation derived from ²¹⁰Pb dating. Ages for the base of pollen zones P1–P5 were computed according to the second-order polynomial that best describes the age/depth curve for the complete section: Age [years] = 47 + 395 (depth [m]) + 127 (depth [m])². Shaded area (0.72–1.1 m) indicates unrecovered section.

time span (10.0–7.0 ka BP), lake level rose appreciably and Chenopodiaceae–Amaranthaceae dominated the pollen assemblage. High lake levels were maintained throughout the mid-Holocene moist period (7.0–3.2 ka BP) and mature Moraceae forest was established by 3.9 ka BP under warm, wet, seasonal climatic conditions.

After 3.2 ka BP (3.6 m), stepwise $\delta^{18}O$ enrichment indicates progressively drier conditions (high E:P ratio), interrupted by moister episodes, none of which was as wet as the early-to-mid-Holocene period. Evidence of climatic drying is also found in the palynological record. Pollen data indicate substantial loss of Moraceae-dominated mesic vegetation and expansion of open, dry forests possessing a weedy understory.

Paleoclimatic interpretation for the last millennium relies on the $\delta^{18}O$ record because regional vegetation, perhaps subjected to human impact as early as 3–4 ka BP, was increasingly disturbed by native Taino (1.0–0.5 ka BP), and thereafter (0.5 ka BP) by European settlers. Surficial sediments contain few arboreal pollen grains and reflect the acute deforestation that the watershed has experienced in recent decades. The stable isotope record should be unaffected by human activities, although dramatic anthropogenic changes in watershed hydrology might produce shifts in the $\delta^{18}O$ signal. Assuming that the oxygen isotope ratio reflects changes in E:P only, the record indicates a gradual trend toward increasing dryness over the last ≈ 1000 years. $\delta^{18}O$ values in near-surface deposits (0–19 cm) are high, with 5-point running means ranging between 1.30 and 1.50 ‰. The period from 2.5 to 1.7 ka BP (3.10–2.37 m) was also characterized by high E:P, with $\delta^{18}O$ values ranging from 1.05 to 1.34 ‰, but oxygen isotope values in modern sediments suggest conditions as dry as any since the early Holocene. In order to test the regional significance of the recent drying trend detected in Lake Miragoane, it will be necessary to carry out high-resolution paleoclimatic studies in other low-elevation circum-Caribbean lakes.

Bibliography

Binford, M.W., Brenner, M., Whitmore, T.J., Higuera-Gundy, A., Deevey, E.S. & Leyden, B., 1987. Ecosystems, paleoecology and human disturbance in subtropical and tropical America. *Quatern. Sci. Rev.*, **6**, 115–28.

Bond, R.M., 1935. Investigations of some Hispaniolan lakes (Dr. R.M. Bond's expedition) II. Hydrology and hydrography. *Arch. Hydrobiol.*, **28**, 137–61.

Brenner, M. & Binford, M.W., 1988. A sedimentary record of human disturbance from Lake Miragoane, Haiti. *J. Paleolimnol.*, **1**, 85–97.

Eyerdam, W.J., 1961. An excursion to Lake Miragoane, Haiti. *Nautilus*, **75**, 71–4.

Higuera-Gundy, A., 1989. Recent vegetation changes in southern Haiti. In *Biogeography of the West Indies: Past, Present, and Future*, ed. C.A. Woods, pp. 191–200. Sandhill Crane Press Inc., Gainesville.

Higuera-Gundy, A., 1991. *Antillean Vegetational History and Paleoclimate Reconstructed from the Paleolimnological Record of Lake Miragoane, Haiti*. PhD Dissertation, Univ. Florida, Gainesville.

Hodell, D.A., Curtis, J.H., Jones, G.A., Higuera-Gundy, A., Brenner, M., Binford, M.W. & Dorsey, K.T., 1991. Reconstruction of Caribbean climate change over the past 10,500 years. *Nature*, **352**, 790–3.

Holdridge, L.R., 1945. A brief sketch of the flora of Hispaniola. In *Plants and plant science in Latin America*, ed. F. Verdoorn, pp. 76–8. Chronica Botanica Co., Waltham, MA.

Mann, P., Hempton, M.R., Bradley, D.C. & Burke, K., 1983. Development of pull-apart basins. *J. Geol.*, **91**, 529–54.

Mann, P., Taylor, F.W., Burke, K. & Kulstad, R., 1984. Subaerially exposed Holocene coral reef, Enriquillo Valley, Dominican Republic. *Geol. Soc. Am. Bull.*, **95**, 1084–92.

Morgan, G.S. & Woods, C.A., 1986. Extinction and the biogeography of West Indian land mammals. *Biol. J. Linn. Soc.*, **28**, 167–203.

Rouse, I., 1989. Peopling and repeopling of the West Indies. In *Biogeography of the West Indies: Past, Present, and Future*, ed. C.A. Woods, pp. 119–36. Sandhill Crane Press Inc., Gainesville.

Rouse, I. & Moore, C., 1984. Cultural sequence in southwestern Haiti. *Bull. Bur. Natl. Ethnol.*, **1**, 25–38.

Taylor, F.W., Mann, P., Valastro, S. Jr & Burke, K., 1985. Stratigraphy and radiocarbon chronology of a subaerially exposed Holocene coral reef, Dominican Republic. *J. Geol.*, **93**, 311–32.

Woodring, W.P., Brown, J.S. & Burbank, W.S., 1924. *Geology of the Republic of Haiti*. Republic of Haiti Department of Public Works, Geological Survey of the Republic of Haiti, Port-Au-Prince.

Lake Titicaca, Bolivia–Peru

D. WIRRMANN

Agreement ORSTOM-Universidad Mayor de San Andrés, La Paz, Bolivia, ORSTOM, 72 Route d'Aulnay, 93143 BONDY Cédex, France

Main physical characteristics

South latitude: 15° 13′ 19″ to 16° 35′ 37″

West longitude: 68° 33′ 36″ to 70° 02′ 13″

Altitude above sea level: 3809 m

Total area: 8562 km²

Total water volume: 903 km³

Mean dissolved salt concentration: 0.9–1.2 g/l with a sodium chloride facies (Carmouze *et al.*, 1981).

The origin of Lake Titicaca (Fig. 1) is still not well established and subject to discussion. Lavenu (1981) speaks of the filling of a tectonic depression produced by a distension phase during the Late Tertiary. Later on, during the Pleistocene, various lacustrine transgressions occured (paleolakes Mataro, Cabana, Ballivian, Minchin and Tauca: Ahlfeld (1972); Lavenu *et al.* (1984); Servant & Fontes (1978)).

Detailed climatological, geological, hydrophysical and hydrochemical parameters can be found in: Monheim (1956); Fontes *et al.* (1979); Boulangé & Aquize Jaen (1981); Carmouze & Aquize Jaen (1981); Carmouze *et al.* (1983); Kirkish & Taylor (1984); Powell *et al.* (1984); Kunzell & Kessler (1986); Richerson *et al.* (1986); Iltis (1987); Liberman (1987); Quintanilla *et al.* (1987); Wirrmann (1992).

Present limnological conditions

Bathymetry and local bottom relief control the distribution of the lacustrine macrophytes (Collot *et al.*, 1983) and the ostracode fauna (Mourguiart, 1987). The superficial sediments are mainly autochthonous and biogenic. Their distribution is also controlled by bathymetry (Fig. 2, adapted after Boulangé *et al.*, 1981). The total thickness of the lacustrine deposits is unknown. Lago Grande is classified as a monomictic warm lake and Lago Huiñaimarca (also known as Lago Titicaca Menor or Lago Pequeño) is classified as a polymictic warm lake, except for the Chua depression which is similar to the Lago Grande (Lazzaro, 1981).

More detailed limnological data can be found in: Haas (1955); Richerson *et al.* (1977); Guerlesquin (1981); Iltis (1984, 1988); Loubens *et al.* (1984); Vincent *et al.* (1986); Liberman & Miranda (1987); Loubens & Osorio (1988); Repelin *et al.* (1988); Richerson & Carney (1988).

Late Quaternary evolution

Core TD1 (Fig. 3), the longest one obtained, was taken with a 6 m-Mackereth corer in 19 m of water in the western central part of Lago Huiñaimarca.

The 20,000 years of record (¹⁴C chronology established on bulk sediment or organic matter according to the samples) have been

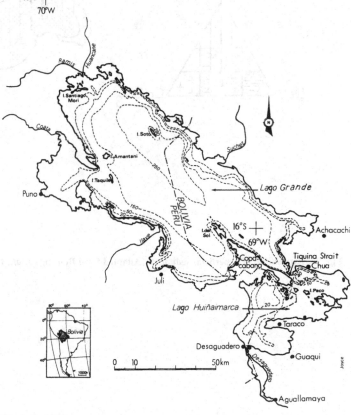

Fig. 1. Locality map.

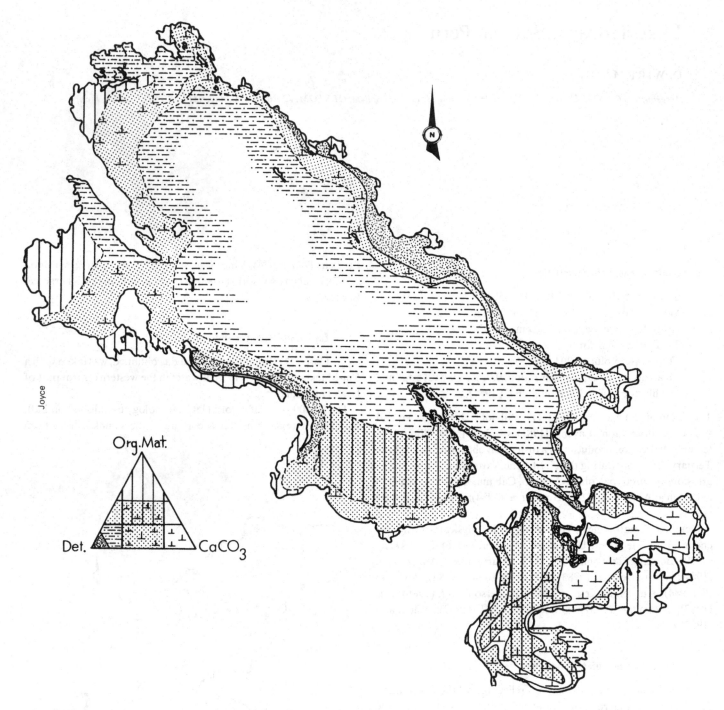

Joyce

Fig. 2. Distribution of superficial sediment. (Adapted from Boulangé *et al.*, 1981.)

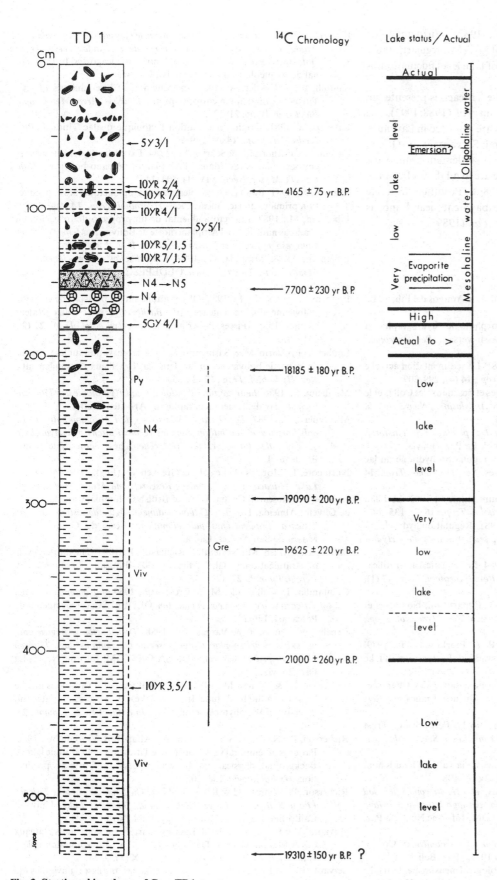

Fig. 3. Stratigraphic column of Core TD1.

analyzed by sedimentology (Wirrmann, 1987; Wirrman & de Oliveira Almeida, 1987) micropaleontology (Mourguiart, 1987) and palynology (Ybert, 1988). The curve of lake level fluctuations is deduced from these data.

The Late Quaternary evolution of Lake Titicaca is presented in de Oliveira Almeida (1986) and in Wirrmann *et al.* (1988; 1992). The Late Quaternary paleoclimatic variations are reported in Lavenu *et al.* (1984); Servant & Fontes (1984) and Servant *et al.* (1989). Preliminary results concerning the human settlement around the lake basin and during the Tiwanaku stage are available in Bouysse-Cassagne (1987, 1988) and Kolata (1989). Pollution in Lake Titicaca, particularly in the shallow embayment near Puno, is discussed in a recent book by Northcote *et al.* (1989).

Bibliography

Ahlfeld, F., 1972. *Geología de Boliva*. Editorial 'Los Amigos del Libro', La Paz-Cochabamba, Bolivia.

Boulangé, B. & Aquize Jaen, E., 1981. Morphologie, hydrographie et climatologie du lac Titicaca et de son bassin versant. *Rev. Hydrobiol. Trop.*, 14 (4), 269–87.

Boulangé B., Vargas, C. & Rodrigo, L.A., 1981. La sédimentation actuelle dans le lac Titicaca. *Rev. Hydrobiol. Trop.*, 14 (4), 299–309.

Bouysse-Cassagne, T., 1987. Le jeu des hommes et des dieux: les Collas et le contrôle de l'île de Titicaca. *Cah. des Am. latines, Nouv., sér.*, 6, 61–92.

Bouysse-Cassagne, T., 1988. *Lluvias y cenizas. Dos pachacuti en la historia*. Bibl. Andina, ser. Historia, No 4, Hisbol, La Paz, Bolivia.

Carmouze, J.-P. & Aquize Jaen, E., 1981. La régulation hydrique du lac Titicaca et l'hydrologie de ses tributaires. *Rev. Hydrobiol. Trop.*, 14 (4), 311–28.

Carmouze, J.-P., Aquize, E., Arze C., & Quintanilla, J., 1983. Le bilan énergétique du lac Titicaca. *Rev. Hydrobiol. Trop.*, 16 (2), 135–44.

Carmouze, J.-P., Arce C. & Quintanilla, J., 1981. Régulation hydrochimique du lac Titicaca et l'hydrochimie de ses tributaires. *Rev. Hydrobiol. Trop.*, 14 (4), 329–48.

Carmouze, J.-P., Arce C. & Quintanilla, J., 1984. Le lac Titicaca: stratification physique et métabolisme associé. *Rev. Hydrobiol. Trop.*, 17 (1), 3–11.

Collot, D., Koriyama, F. & Garcia, E., 1983. Répartition, biomasses et production des macrophytes du lac Titicaca. *Rev. Hydrobiol. Trop.*, 16 (3), 241–62.

Fontes, J.-C., Boulangé, B., Carmouze, J.-P. & Florkowski, T., 1979. Preliminary oxygen-18 and deuterium study of the dynamics of lake Titicaca. In *Isotopes in Lake Studies* I.A.E.A., Vienna, 145–50.

Guerlesquin, M., 1981. Contribution à la connaissance des Characées d'Amérique du Sud (Bolivie, Equateur, Guyane Française). *Rev. Hydrobiol. Trop.*, 14 (4), 381–404.

Haas, F., 1955. *Mollusca: Gastropoda*. Report No 17, Percy Sladen Trust Expedition to Lake Titicaca in 1937. *Trans. Limn. Soc. London*, ser. 3, 1 (3), 275–308.

Iltis, A., 1984. Algues du lac Titicaca et des lacs de la vallée d'Ichu Khota (Bolivie). *Cryptogamie Algologie*, V (2–3), 85–108.

Iltis, A., 1987. *Datos sobre las temperaturas, el pH, la conductibilidad electrica y la transparencia de las aguas del lago Titicaca boliviano (1985–1986)*. Convenio UMSA-ORSTOM, Informe No 3, La Paz, Bolivia.

Iltis, A., 1988. *Biomasas fitoplanctonicas del lago Titicaca boliviano*. Convenio UMSA–ORSTROM, Informe No 10, La Paz, Bolivia.

Kirkish, M. & Taylor, M., 1984. Micrometeorological measurements at lake Titicaca (Peru-Bolivia). *Verh. Internat. Verein. Limnol.*, 22, 1232–6.

Kolata, A., 1989. *La tecnología y organización de la producción agrícola en el estado de Tiwanaku. Primer informe de resultados del proyecto Wilajawira*. Proyecto Wilajawira, Univ. de Chicago and Inst. Nacional de Arqueología de Bolivia (eds), La Paz.

Kunzell, F., & Kessler, A., 1986. Investigation of level changes of lake Titicaca by maximum entropy spectral analysis. *Arch. Met. Geoph. Biocl.*, ser. B, 36, 219–27.

Lavenu, A., 1981. Origine et évolution tectonique du lac Titicaca. *Rev. Hydrobiol. Trop.*, 14 (4), 289–97.

Lavenu, A., Fornari, M. & Sébrier, M., 1984. Existence de deux nouveaux épisodes lacustres dans l'Altiplano péruano-bolivien. *Cah. ORSTOM, sér. Géol.*, XIV (1), 103–14.

Lazzaro, X., 1981. Biomasses, peuplements phytoplanctoniques et production primaire du lac Titicaca. *Rev. Hydrobiol. Trop.*, 14 (4), 349–80.

Liberman, M., 1987. Impacto ambiental de un proyecto de irrigación en praderas naturales del Altiplano norte de Bolivia. In *Memoria del 1er Congreso de Praderas Nativas de Boliva*. P.M.P.R., Oruro, Bolivia.

Liberman, M. & Miranda, C., 1987. *Contribución al conocimiento del fitoplanctón del lago Titicaca*. OLDEPESCA, Documento de Pesca 003, Lima.

Loubens, G. & Osorio, F., 1988. Observations sur les poissons de la partie bolivienne du lac Titicaca. III. *Basilichthys bonariensis* (Valenciennes, 1835) (Pisces, Atherinidae). *Rev. Hydrobiol. Trop.*, 21 (2), 151–77.

Loubens, G., Osorio, M. & Sarmentio, J., 1984. Observation sur les poissons de la partie bolivienne du lac Titicaca. I. Milieux et peuplements. *Rev. Hydrobiol. Trop.*, 17 (2), 153–61.

Monheim, F., 1956. *Beiträge zur Klimatologie und Hydrologie des Titicacabeckens*. Heidelberger Geographische Arbeiten, Heft 1.

Mourguiart, P., 1987. *Les Ostracodes lacustres de l'Altiplano bolivien. Le polymorphisme, son intérêt dans les reconstitutions paléohydrologiques et paléoclimatiques de l'Holocène*. Thèse 3ème cycle, Univ. Bordeaux I.

Northcote, T., Morales, P., Levy D. & Greaven M., (Eds), 1989. *Pollution in Lake Titicaca, Peru: Training, Research and Management*. Westwater Research Center, Univ. of British Columbia, Canada.

de Oliveira Almeida, L., 1986. *Estudio sedimentológico de testigos del lago Titicaca. Implicaciones paleoclimáticas*. Tesis de Grado, Univ. Mayor de San Andrés, La Paz.

Powell, T., Kirkish, M., Neale, P. & Richerson, P., 1984. The diurnal cycle of stratification in lake Titicaca: eddy diffusion. *Verh. Internat. Verein. Limnol.*, 22, 1237–43.

Quintanilla, J., Calliconde, M. & Crespo, P., 1987. *La quimica del lago Titicaca y su relación con el plancton*. OLDEPESCA, Documento de Pesca 004, Lima.

Répelin, R., Pinto, J. & Vargas, M., 1988. *Distribución y migraciones nictimerales del zooplancton en el sector boliviano del lago Titicaca (Logo Pequeño)*. Convenio UMSA-ORSTOM, Informe No 11, La Paz, Bolivia.

Richerson, P., & Carney, M., 1988. Patterns of temporal variations in lake Titicaca, a high altitude tropical lake. II. Succession rate and diversity of the phytoplancton. *Verh. Internat. Verein. Limnol.*, 23, 734–8.

Richerson, P., Neale, P., Wurtsbaugh, W., Alfar R. & Vincent, W., 1986. Patterns of temporal variation in lake Titicaca, a high altitude lake. I. Background, physical and chemical processes and primary production. *Hydrobiologia*, 138, 205–20.

Richerson, P., Widmer, C. & Kittel, T., 1977. *The limnology of lake Titicaca (Peru–Bolivia), a Large High Altitude Tropical Lake*. Univ. of California, Davis, Inst. of Ecology, Publ. No 14.

Servant, M. & Fontes, J.-C. 1978. Les lacs quarternaires des hauts plateaux des Andes boliviennes. Datations par le ^{14}C. Interprétation paléoclimatique. *Cah. ORSTOM, sér. Géol.*, X (1), 9–23.

Servant, M. & Fontes, J.-C., 1984. Les basses terrasses fluviatiles du Quaternaire récent des Andes boliviennes. Datations par le ^{14}C.

Interprétation paléoclimatique. *Cah. ORSTOM., sér. Geol.*, **XIV** (1), 15–28.

Servant, M., Argollo, J., de Oliveira Almeida, L., Servant-Vildary, S. & Wirrmann, D., 1989. Paleohydrology in the bolivian Andes during the last 15,000 years: paleoclimatic scenarios. *Int. Symposium on Global Changes in South America during the Quaternary – Past-Present-Future, São Paulo, Brasil, Special Publ.* No. 1, pp. 182–3.

Vincent, W., Wurtsbaugh, W., Neale, P. & Richerson, P. 1986. Polymixis and algal production in a tropical lake: latitudinal effects on the seasonality of photosynthesis. *Freshwater Biol.*, **16**, 781–803.

Wirrmann, D., 1987. *El lago Titicaca: sedimentológia y paleohidrológia durante el Holoceno (10,000 años B.P.-Actual)*. Convenio UMSA-ORSTOM, Informe No 6, La Paz, Bolivia.

Wirrmann, D., 1992. Morphology and bathymetry. In *Lake Titicaca*, ed. C.

Dejoux & A. Altis, pp. 14–20. Kluwer Academic Publishers.

Wirrmann, D., Mourguiart, P. & de Oliveira Almeida, L., 1988. Holocene sedimentology and Ostracods repartition in Lake Titicaca. Paleohydrological interpretations. *Quaternary of South America and Antarctic Peninsula*, **6**, 89–127.

Wirrman, D., & de Oliveira Almeida, L., 1987. Low Holocene level (7700 to 3650 years ago) of Lake Titicaca. (Bolivia). *Paleogeogr., Palaeoclimatol., Palaeoecol.*, **59**, 315–23.

Wirrmann, D., Ybert, J.-P., & Mourguiart, P., 1992. A 20,000 paleohydrological record from Lake Titicaca. In *Lake Titicaca*, ed. C. Dejoux & A. Iltis, pp. 37–45. Kluwer Academic Publishers.

Ybert, J.-P., 1988. Apports de la palynologie à la connaissance de l'histoire du lac Titicaca (Bolivie-Pérou) au cours du Quaternaire récent. *Inst. fr. Pondichéry, trav. sec. sci. tech.*, **XXV**, 139–50.

The Finger Lakes of New York State, USA

HENRY T. MULLINS AND ROBERT W. WELLNER

Department of Geology, Syracuse University, Syracuse, N.Y. 13244, USA

General stratigraphy

The Finger Lakes of New York State consist of 11, glacially scoured lake basins (Fig. 1). They vary considerably in maximum length (5–61 km), lake level elevation (116–334 m) and maximum water depth (9–186 m), which results in variable limnology (Bloom-field, 1978). Uniboom seismic reflection surveys from eight of the 11

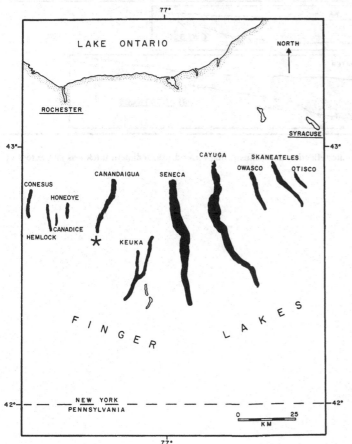

Fig. 1. Location map of Finger Lakes of central New York State, USA. Star highlights drillcore location. Approximate midpoint is 42° 40′ N/77° 00′ W. From Mullins *et al.* (1989).

lakes indicate that bedrock has been eroded as much as 304 m *below* sea-level and that maximum sediment fill is as much as 275 m thick (Fig. 2; Mullins & Hinchey, 1989; Mullins *et al.*, 1991). Morpho-stratigraphy and radiocarbon dates constrain the age of these deposits as less than 14,000 years. The bulk of these deposits were deposited rapidly between 14,000 and 13,500 years ago. Buried eskers beneath Owasco, Seneca and Canandaigua lakes argue for sediment input by subglacial meltwaters (Mullins & Hinchey, 1989).

Detailed stratigraphy

The subsurface stratigraphy of the Finger Lakes has been defined by two drillcores collected in July 1990. The drillsites are located in a dry valley 3 km south of Canandaigua Lake (Fig. 1). On-land multichannel seismic reflection data (Fig. 3) indicate approximately 155 m of sediment fill over bedrock at the drill sites. Drillcore no 1 penetrated 119 m of section; drillcore no 2 only 13 m (Fig. 3; Mullins *et al.*, 1991). Although core recovery was less than 50%, gamma-ray and resistivity logs provide details of subsurface stratigraphy (Fig. 4).

Four, first-order stratigraphic units have been defined (Figs 3 and 4): (1) a basal sand and gravel unit (that terminated drillcoring) which we correlate with *Valley Heads moraine* (~ 14 ka) outcrops to the south, (2) a fine-grained *glaciolacustrine* unit (90 m thick) which grades upward from massive clays with sand, to dropstone mud, to rhythmically bedded silt and sand, (3) a coarsening-upward (sand to gravel) *alluvial* facies with artesian water (15 m thick), capped by, (4) a cyclic sequence of *peat and lake clay* (12 m thick); basal peat at the contact with underlying alluvium has a radiocarbon date of 13,600 ± 200 years.

The Valley Heads moraine and outwash material is interpreted to have been deposited by pressurized, subglacial meltwaters asso-ciated with collapse of the Laurentide ice sheet and rapid ice flow ~ 14 ka (Mullins & Hinchey, 1989). Fine-grained glaciolacustrine sediments are also believed to have been in part pumped into the Finger Lakes basins by subglacial meltwaters and vertical facies variations likely reflect the rapid (~ 500 years) retreat of the ice margin. The coarsening-upward sands and gravels possibly record

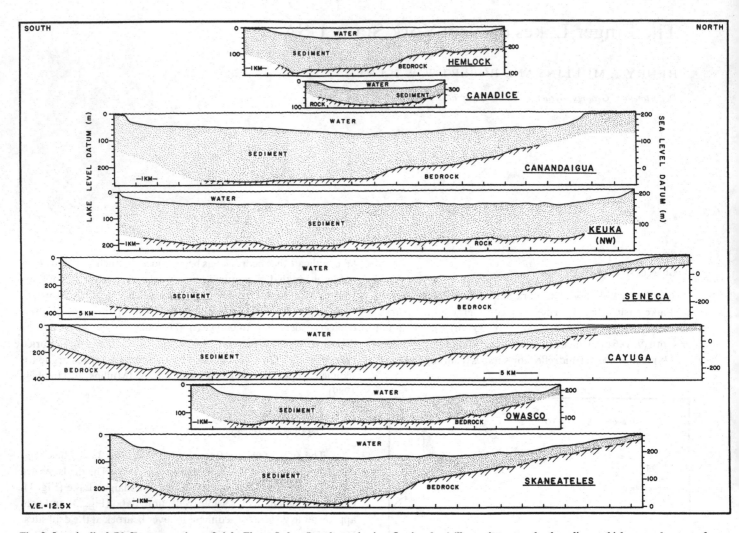

Fig. 2. Longitudinal (N–S) cross-sections of eight Finger Lakes (based on seismic reflection data) illustrating water depth, sediment thickness and extent of bedrock erosion (from Mullins & Hinchey, 1989).

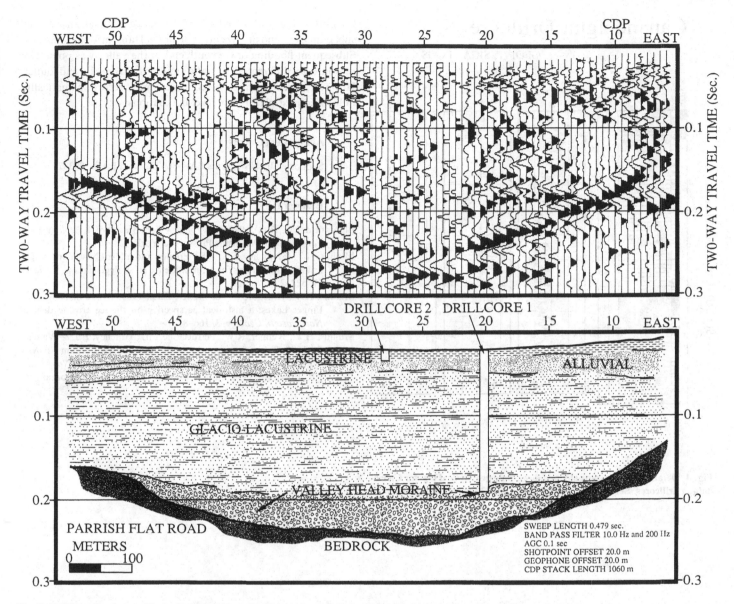

Fig. 3. Seismic reflection profile (top) and interpretation (bottom) across Canandaigua Lake dry valley where drillcores were recovered.

Canandaigua Drillcore

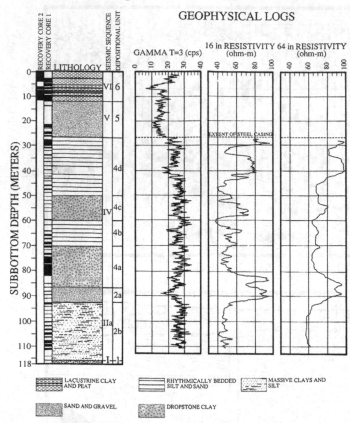

GEOPHYSICAL LOGS

Fig. 4. Detailed lithostratigraphy and down-hole geophysical logs for Canandaigua drillcores. See Fig. 1 for location and Fig. 3 for reflection profile.

the abrupt lowering of lake level, drainage reversal, incision of subaerial glens and progradation of alluvial fans, prior to 13,600 ka. Subsequent flooding of the south end of the valley was possibly a response to differential isostatic uplift of the north end outlet. However, cyclic peat-clay sequences may record climatically-influenced post-glacial lake-level fluctuations.

Acknowledgements

Supported by NSF grants EAR-8607326 and EAR-8903870.

Bibliography

Bloomfield, J.A., (Ed.), 1978. *Lakes of New York State*, vol. 1, *Ecology of the Finger Lakes*. Academic Press, New York,

Mullins, H.T. & Hinchey, E.J., 1989. Erosion and infill of New York Finger Lakes: Implications for Laurentide ice sheet deglaciation. *Geology*, **17**, 622–25.

Mullins, H.T., Hinchey, E.J. & Muller, E.H., 1989. Origin of New York Finger Lakes: a historical perspective on the ice erosion debate. *Northeastern Geology*, **3**, 166–81.

Mullins, H.T., Wellner, R.W., Petruccione, J.L., Hinchey, E.J. & Wanzer, S., 1991. Subsurface geology of the Finger Lakes region. In *New York State Geological Association Fieldtrip Guidebook No. 63*, ed. J.R. Ebert, pp. 1–54. State University of New York, Oneonta.

The saline lakes of southern British Columbia, Canada

ROBIN W. RENAUT AND DOUGLAS STEAD

Dept. of Geological Sciences, University of Saskatchewan, Saskatoon, Saskatchewan S7N 0W0, Canada

General summary

There are three main groups of saline lakes in the semi-arid, south-central Interior of British Columbia – Basque, Kamloops and Osoyoos (Fig. 1). These lakes are fed predominantly by groundwater, and occupy small closed depressions in glaciated hilly terrain above an adjacent large, glacial valley. Most of the lake waters are of Mg-Na-SO$_4$ or Na-Mg-SO$_4$ composition.

The climate is semi-arid, and both hotter and drier than the Fraser Plateau to the north where another major saline lake group is present (Renaut, Stead and Owen, this volume). Consequently, permanent bedded evaporites exist in each of the southern lake groups. Most lakes are bounded by open grassland and sagebrush scrub. Broken woodland surrounds the Basque lakes.

The Basque Group

The seven saline lakes at Basque (Fig. 2) are small (< 3 ha) and lie in very small catchments (mostly < 0.5 km^2) in rugged mountainous terrain 3 km west of the Thompson Valley, near Ashcroft (Cole, 1924). The local bedrock comprises greenstones, argillites and cherts of the Cache Creek Terrane, and is mantled discontinuously by glacial deposits. All the lakes are ephemeral and predominantly groundwater-fed. Except for Lakes 5 and 7, they are covered by shallow brines during spring and early summer. The steep, vegetated, valley sides have limited peripheral mudflat development to the up-valley margins.

Lakes 1 to 4 lie along a narrow faulted valley, each dammed by glacial and colluvial debris. They have strong Mg-Na-SO$_4$ brines (100 to 320 g l^{-1} TDS). The floors of Lakes 1, 2 and 3 consist of numerous mud-rimmed brine-pools (1 to 25 m diameter) lying within muds. Lakes 6 and 7 lie in small, steep depressions in a parallel valley at ~30 m higher elevation and have Na-Mg-SO$_4$ brines. Lake 5, a small ephemeral carbonate playa, occupies a faulted valley converging from the north. On the flat valley floor to the north of the road, there is a zone of groundwater discharge and extensive Mg-carbonate precipitation.

Soap Lake, a small (25 ha) outlying saline, alkaline playa-lake, lies in a catchment composed predominantly of volcanic rocks, 5 km south of Spences Bridge (Fig. 1) (Cole, 1926; Cummings, 1940).

The Kamloops Group

More than 20 saline lakes occur within a 20 km radius of Kamloops (Fig. 3). About 10 ephemeral lakes lie in rolling hills along the margins of a 12 km stretch of Highway 1, west of the city. Most are ephemeral with Na-Mg-SO$_4$ or Mg-Na-SO$_4$ waters (Cummings, 1940; Topping & Scudder, 1977; Hudec & Sonnenfeld, 1980). Salsola Pond contains strong Na-CO$_3$-SO$_4$ brines, as do Buse Lake and Barnes Lake, east of Kamloops. An outlying group, including several perennial saline lakes with Na-CO$_3$-SO$_4$ and Mg-Na-SO$_4$ brines, occurs on the Tranquile Plateau, 8–10 km to the north of the city.

The Osoyoos Group

A small group of saline lakes is present in the dry scrub-covered hills ~12 km west of Osoyoos (Fig. 1). The most notable is Spotted Lake, a groundwater-fed playa lake which lies in a broad, shallow depression. Approximately 60% of the lake bed consists of shallow pools within mud, similar to those at the Basque lakes. The pools are from 6–25 m in diameter and up to 2 m deep. The brine is of Mg-Na-SO$_4$ composition (Cole, 1924; Goudge, 1926; Nesbitt, 1974).

Several minor closed-basin lakes with Mg-Na-SO$_4$ brines lie in the hills north of Osoyoos, including Mahoney and Green lakes, both of which are perennial. Mahoney Lake is meromictic (Northcote & Halsey, 1969; Northcote & Hall, 1983; Hall & Northcote, 1990).

Sediments and mineralogy

Except for the Basque group (Nesbitt, 1974, 1990; Eugster & Hardie, 1978), there have been no systematic studies of the lake sediments. At Basque, epsomite and bloedite are the main salts to form in the brine pools of lakes 1 to 4 (Fig. 4). Excavations of Lakes

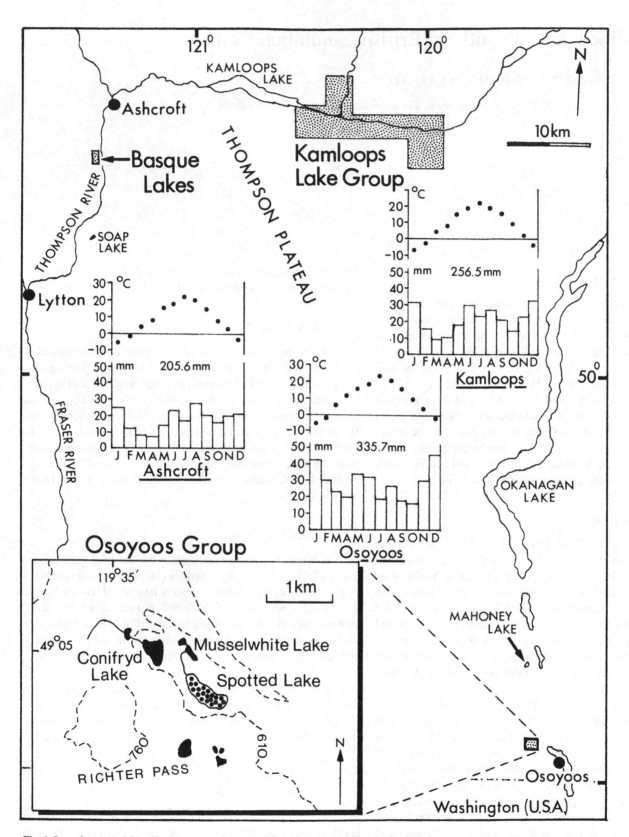

Fig. 1. Location map of the saline lake groups in southern British Columbia. The graphs show the mean monthly and annual precipitation (mm), and temperature data (monthly means) for selected stations near the saline lake group (data from Atmospheric Environment Service, 1982).

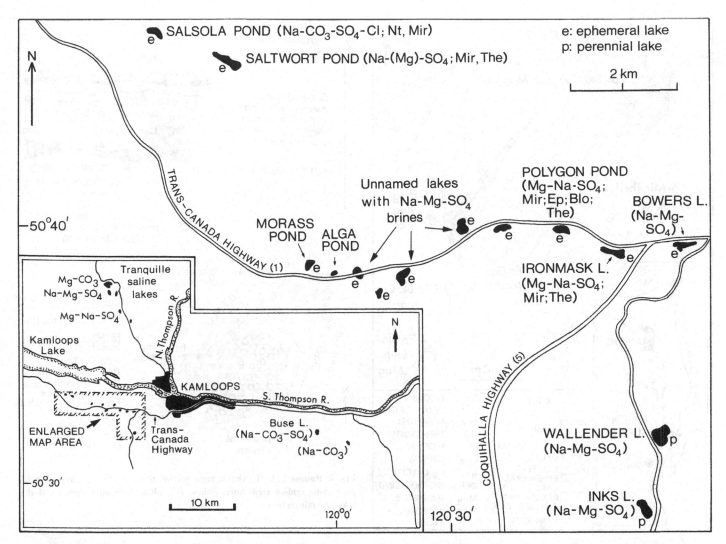

Fig. 2. The saline lakes of the Kamloops region. Brine compositions are noted where data are available. Mineralogical data are based on a few grab samples only: Blo: bloedite; Ep: epsomite; Mir: mirabilite; Nt: natron; The: thenardite.

1 and 2 by Goudge (1926) showed that the larger brine pools contain at least 5 m of evaporites, principally epsomite and bloedite, the latter increasing in abundance with depth. Greenish black interstitial lacustrine muds within and between the pools contain gypsum, magnesite, protodolomite and from ∼5 to 25% siliciclastic detritus. Up to 8% organic matter is present. Nesbitt (1990) has provided a detailed account of the mineralogy and geochemical evolution of Basque lake No. 2.

Although the mineralogy of the Kamloops lakes has not been investigated, preliminary X-ray analyses have indicated similarities with the other sulfate and carbonate lakes in British Columbia. Salsola Pond reputedly has at least 12 m of bedded evaporites; several of the sodium sulfate lakes have > 5 m of bedded salts.

Brief accounts of the Spotted Lake sediments were given by Jenkins (1918), Goudge (1926) and Cummings (1940). Goudge found crystalline salts extending to the bottom of the brine pools. X-ray diffraction analyses of nearshore mud collected in September 1987 (Renaut, unpub. data) revealed gypsum, calcite and Mg-calcite. Clear salts from a nearshore pool were mainly epsomite, underlain by soft, muddy epsomite and bloedite. Efflorescent gypsum, thenardite and hexahydrite crusts occur along the shore zone and on interpool muds. Mirabilite probably freezes out during autumn and winter.

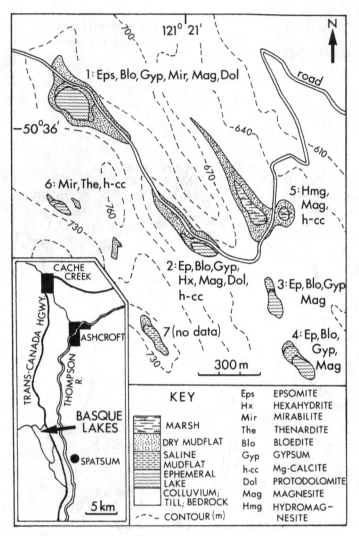

Fig. 3. Location map and depositional subenvironments of the Basque Lakes.

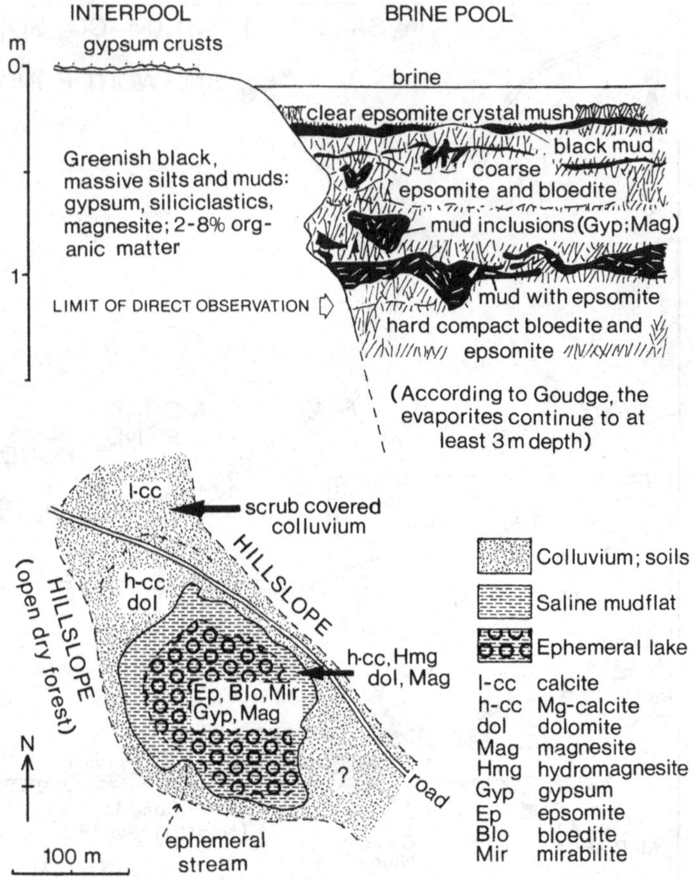

Fig. 4. Basque Lake 1. Above: cross-section through small excavated brine pool along mid-eastern shore. Below: Depositional subenvironments of the lake and mineralogy.

Bibliography

Atmospheric Environment Service, 1982. *Canadian Climate Normals 1951–1980. Temperature and precipitation (British Columbia)*. Environment Canada, Ottawa, Ontario.

Cole, L.H., 1924. Sodium and magnesium salts of Western Canada. *Trans. Can. Inst. Min. Metall.*, **27**, 209–47.

Cole, L.H., 1926. *Sodium Carbonate at Soap Lake, B.C.* Rep. Mines Branch, Can. Dept. Mines No. 687.

Cummings, J.M., 1940. *Saline and Hydromagnesite Deposits of British Columbia*. Bull. B.C. Dept. Mines No. 4.

Eugster, H.P. & Hardie, L.A., 1978. Saline lakes. In *Lakes: Chemistry, Geology, Physics*, ed. A. Lerman, pp. 237–93. Springer, New York.

Goudge, M.F., 1926. *Magnesium Sulphate in British Columbia*. Rep. Mines Branch, Can. Dept. Mines No. 642, 62–80.

Hall, K.J. & Northcote, T.G., 1990. Production and decomposition processes in a saline meromictic lake. *Hydrobiologia*, **197**, 115–28.

Hudec, P. & Sonnenfeld, P., 1980. Comparison of Caribbean solar ponds and inland solar lakes of British Columbia. In *Hypersaline Brines and Evaporitic Environments*, ed. A. Nissenbaum, pp. 101–14. Elsevier, Amsterdam.

Jenkins, O.P., 1918. Spotted lakes of epsomite in Washington and British Columbia. *Am. J. Sci.*, **46**, 638–44.

Nesbitt, H.W., 1974. *The Study of Some Mineral-Aqueous Solution Interactions*. PhD Thesis, Johns Hopkins Univ., Baltimore.

Nesbitt, H.W., 1990. Groundwater evolution, authigenic carbonates and sulphates of the Basque Lake No. 2 Basin, Canada. In *Fluid-mineral interactions: a tribute to H.P. Eugster*, ed. R.J. Spencer & I. Ming-Chou, pp. 355–71. Spec. Publ. Geochem. Soc., vol. 2.

Northcote, T.G. & Hall, K.J., 1983. Limnological contrasts and anomalies in two adjacent saline lakes. In *Proc. II Int. Symp. Athalassic (Inland) Saline Lakes* (Develop. Hydrobiol., 16), ed. U.T. Hammer pp. 179–94. Junk, The Hague.

Northcote, T.G. & Halsey, T.G., 1969. Seasonal changes in the limnology of some meromictic lakes in southern British Columbia. *J. Fish. Res. Bd. Can.*, **26**, 1763–87.

Topping, M.S. & Scudder, C.G.E., 1977. Some physical and chemical features of saline lakes in central British Columbia. *Syesis*, **10**, 145–66.

The saline lakes of the Fraser Plateau, British Columbia, Canada

ROBIN W. RENAUT[1], DOUGLAS STEAD[1] AND R. BERNHART OWEN[2]

[1] Dept. of Geological Sciences, University of Saskatchewan, Saskatoon, Saskatchewan S7N 0W0, Canada
[2] Dept. of Geography, Hong Kong Baptist College, 224 Waterloo Road, Kowloon, Hong Kong

General summary

There are more than 1000 perennial and ephemeral saline lakes on the intermontane Fraser Plateau of Interior British Columbia. They are clustered in two groups – the Chilcotin Plateau Group, and the larger Cariboo Plateau Group (Figs 1 and 2). They range from small ephemeral playa-lakes, some with permanent evaporites, through to larger meromictic perennial lakes. Ice retreated from the area ~ 10,000 years ago and the lake basins developed in a former proglacial setting. Most of the saline lakes lie in narrow, shallow, closed depressions within hummocky till, or between till mounds and eskers. They are commonly clustered along paleomeltwater channels. A few saline lakes occur in mountainous terrain along the plateau margins. Most lakes are small (0.1

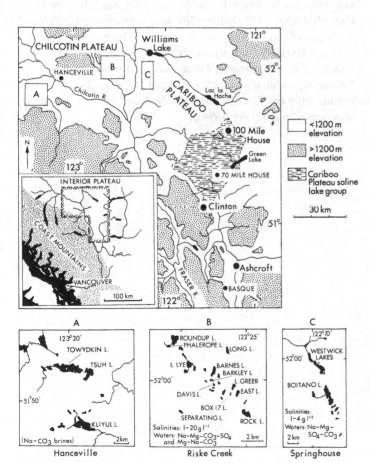

Fig. 1. Location map of the Cariboo and Chilcotin saline lake groups. Insets show the principal saline lakes of the Chilcotin Plateau.

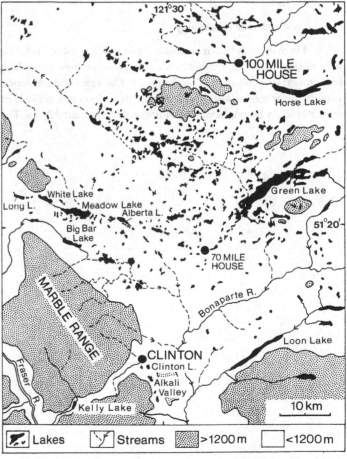

Fig. 2. Location map of the saline lakes of the Cariboo Plateau. Only the larger lakes are shown.

419

to 2.0 km diameter) and shallow (<2 m), but there are a few perennial, meromictic lakes that are up to 5 km across and up to 15 m deep (Northcote & Halsey, 1969). Faunal studies have been conducted by Scudder (1969) and Hammer & Forró (1992).

Most lakes are fed by direct precipitation (including snowmelt) and groundwater discharge, catchments being very small. The waters have exceptional compositional diversity. The dominant brines are of Na-CO$_3$-(SO$_4$)-Cl, Mg-Na-SO$_4$, and Na-Mg-SO$_4$ types but with numerous subtypes (Reinecke, 1920; Goudge, 1926a, b; Walker & Parsons, 1927; Cummings, 1940; Topping & Scudder, 1977; Hudec & Sonnenfeld, 1980; Hammer, 1986; Renaut & Long, 1987). Most alkaline brines occur where groundwaters are in contact with the basalts that underlie the plateau; sulfate brines are common where groundwaters are in contact with sulfide-bearing Paleozoic and Mesozoic metasediments. Carbonate precipitation acts as the major chemical divide in the evolution of the main types of brine (Renaut, 1990).

The Cariboo region ranges from semi-arid to sub-humid with extremes of temperature from −45 °C to > +35 °C. Less than 90 days annually are frost-free and most lakes remain ice-covered for four to six months. Most of the Cariboo Plateau has dense coniferous forest cover, trees commonly extending almost to the shoreline.

Sediments and mineralogy

Three main subenvironments are found in most of the saline lake basins: hillslope, mudflat (dry and/or saline) and saline lake (ephemeral or perennial) (Figs 3 and 4). The vegetated hillslopes, usually linear mounds of till or esker sands and gravels, are sites of minor erosion by unchannelled runoff and minor earthflows. The sediments of the mudflats range from entirely siliciclastic through mixed siliciclastic-carbonate to almost pure carbonate muds. They include minerals precipitated interstitially from shallow groundwater and those formed within the water-column when the lakes cover the mudflats. Stromatolites are common at sites of groundwater discharge (Renaut, 1991; 1994a). The lacustrine sediments are variously dominated by carbonate muds (both laminated and unlaminated), organic-rich (sapropelic) muds and subaqueous evaporites (Renaut & Long, 1987, 1989; Renaut & Stead, this volume). Subaerial and sublacustrine springs, some with small travertine deposits, occur along the margins of several lakes.

A large suite of salts and authigenic minerals is present, reflecting the diverse brine compositions. Alkaline earth carbonates, including abundant hydromagnesite, dolomite and magnesite muds (Cummings, 1940; Grant, 1987; Renaut & Stead, 1991) dominate many saline mudflats and playa floors. In the saline, alkaline lakes natron is the principal evaporite mineral, forming during autumn and winter from cooling brines concentrated by evaporation. In the sulfate lakes, epsomite, mirabilite and other sulfate minerals form from the most concentrated brines (Gonzales, 1987; Renaut & Stead, 1990; Renaut, 1994b).

Most modern sedimentation in the saline lakes is autochthonous and results from: (1) small, low-relief catchments, (2) the dense vegetative cover which has stabilized soils and (3) a ratio of Ca + Mg:HCO$_3$ + CO$_3$ in the inflow waters near to unity, allowing formation of abundant carbonate sediment. The sediment record in the playa lakes is commonly poorly preserved due to disruption by interstitial mineral growth, freeze–thaw and seasonal ground-ice formation, volume changes associated with hydration–rehydration, sediment flowage due to differential loading and other processes (Renaut, 1989; 1994a).

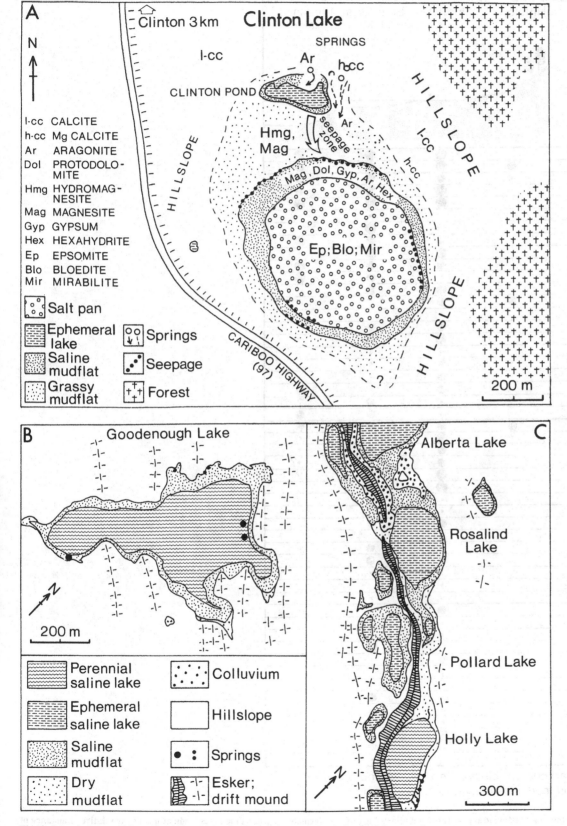

Fig. 3. Location map of the saline lakes of the Cariboo Plateau. Only the larger lakes are shown.

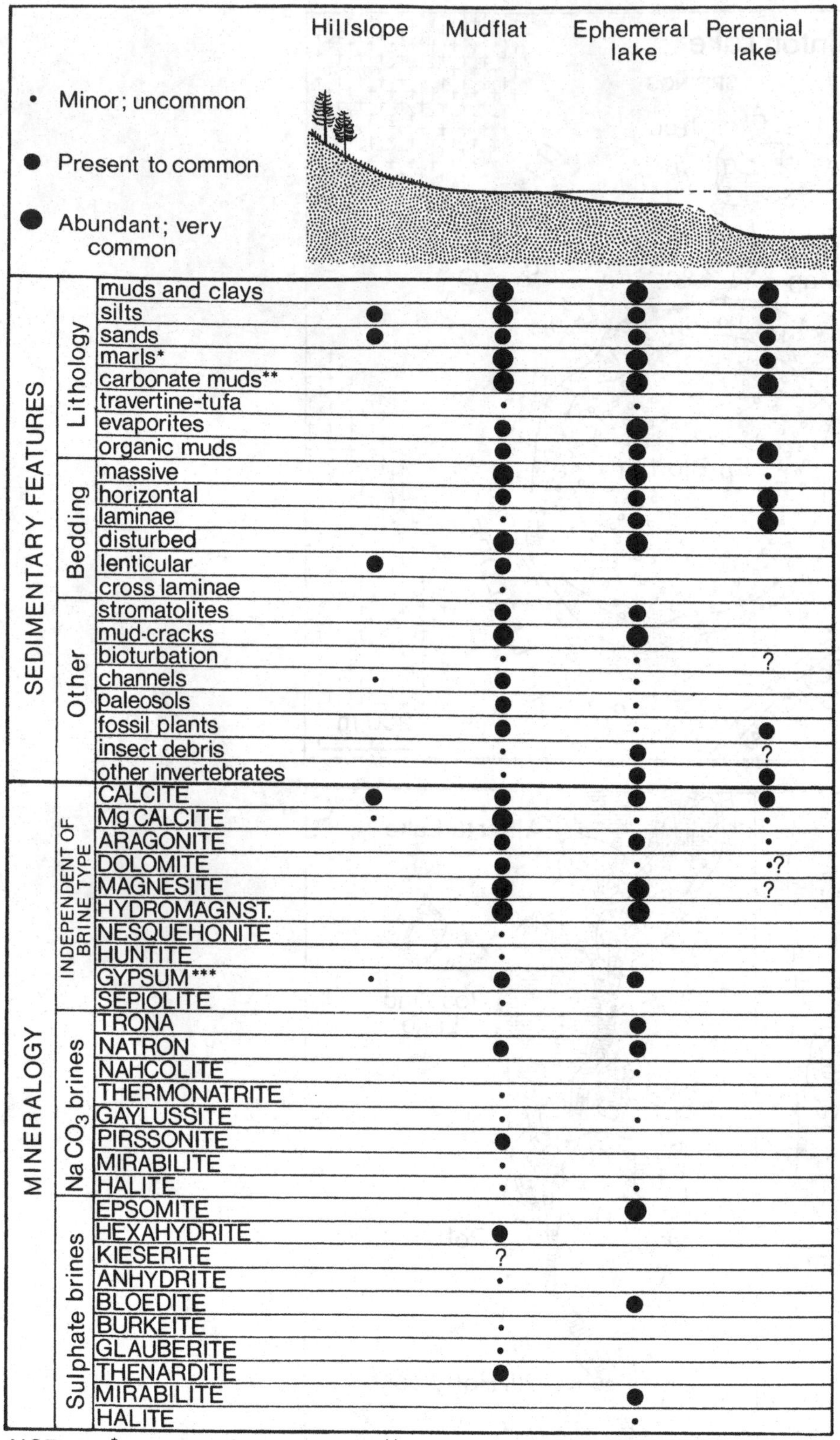

Legend:
- • Minor; uncommon
- ● Present to common
- ⬤ Abundant; very common

			Hillslope	Mudflat	Ephemeral lake	Perennial lake
SEDIMENTARY FEATURES	Lithology	muds and clays		⬤	⬤	⬤
		silts	●	⬤	⬤	⬤
		sands	●	⬤	⬤	●
		marls*		⬤	⬤	●
		carbonate muds**		⬤	⬤	⬤
		travertine-tufa		•	•	
		evaporites		●	⬤	
		organic muds		●	⬤	⬤
	Bedding	massive		⬤	⬤	•
		horizontal		⬤	⬤	●
		laminae		•	•	⬤
		disturbed		⬤	⬤	
		lenticular	●	⬤		
		cross laminae		•		
	Other	stromatolites		⬤	⬤	
		mud-cracks		⬤	⬤	
		bioturbation		•	•	?
		channels	•	⬤	•	
		paleosols		⬤		
		fossil plants		●	•	●
		insect debris		•	●	?
		other invertebrates		•	●	●
MINERALOGY	INDEPENDENT OF BRINE TYPE	CALCITE	●	⬤	●	●
		Mg CALCITE	•	⬤	●	•
		ARAGONITE		⬤	●	•
		DOLOMITE		⬤	•	•?
		MAGNESITE		⬤	⬤	?
		HYDROMAGNST.		⬤	⬤	
		NESQUEHONITE		•		
		HUNTITE		•		
		GYPSUM***	•	●	●	
		SEPIOLITE		•		
	Na CO₃ brines	TRONA			●	
		NATRON		●	●	
		NAHCOLITE			•	
		THERMONATRITE		•		
		GAYLUSSITE		•	•	
		PIRSSONITE		●		
		MIRABILITE		•		
		HALITE		•	•	
	Sulphate brines	EPSOMITE			●	
		HEXAHYDRITE		●		
		KIESERITE		?		
		ANHYDRITE		•		
		BLOEDITE			●	
		BURKEITE		•		
		GLAUBERITE		•		
		THENARDITE		●		
		MIRABILITE			●	
		HALITE			•	

NOTES: *marl: 20-75% carbonate; **carbonate mud: >75% carbonate; ***gypsum uncommon in saline, alkaline lake basins

Fig. 4. Summary chart showing the main sedimentary features and mineralogy of the Cariboo Plateau saline lakes. Indications of the relative abundance of features and minerals are based on observations of more than 30 basins. Magnesite and hydromagnesite are common in lakes with sulfate brines; dolomite is most common in sodium carbonate lake basins.

Bibliography

Cummings, J.M., 1940. *Saline and Hydromagnesite Deposits of British Columbia*. Bull. B.C. Dept. Mines No. 4.

Gonzales, A., 1987. Sedimentology and hydrochemistry of Clinton Lake, Interior British Columbia. *Prog. Abstr. Joint Annu. Meet. Geol. Assoc. Can. Miner. Assoc. Can.* 12, p. 47.

Goudge, M.F., 1926a. *Sodium Carbonate in British Columbia*, pp. 81–102. Rep. Mines Branch, Can. Dept. Mines No. 642.

Goudge, M.F., 1926b. *Magnesium Sulphate in British Columbia*, pp. 62–80. Rep. Mines Branch, Can. Dept. Mines No. 642.

Grant, B., 1987. *Magnesite, Brucite and Hydromagnesite Occurrences in British Columbia*. Open File Rep. B.C. Geol. Surv. Branch No. 1987–13.

Hammer, U.T., 1986. *Saline Lake ecosystems of the world*. Junk, Dordrecht.

Hammer, U.T. & Forró, L., 1992. Zooplankton distribution and abundance in saline lakes of British Columbia, Canada. *Int. J. Salt Lake Res.*, 1, 65–80.

Hudec, P. & Sonnenfeld, P., 1980. Comparison of Caribbean solar ponds and inland solar lakes of British Columbia. In *Hypersaline Brines and Evaporitic Environments*, ed. A. Nissenbaum, pp. 101–14. Elsevier, Amsterdam.

Northcote, T.G. & Halsey, T.G., 1969. Seasonal changes in the limnology of some meromictic lakes in southern British Columbia. *J. Fish. Res. Bd. Can.*, **26**, 1763–87.

Reinecke, L., 1920. Mineral deposits between Lillooet and Prince George, British Columbia. *Mem. Geol. Surv. Can.*, p. 118.

Renaut, R.W., 1989. Paleolimnological implications of Holocene sedimentation in the playa lakes of British Columbia, Canada. *Abstracts V Int. Symp. Palaeolimnol. Ambleside, Cumbria, UK.*, p. 88.

Renaut, R.W., 1990. Recent carbonate sedimentation and brine evolution in the saline lake basins of the Cariboo Plateau, British Columbia, Canada. *Hydrobiologia*, **197**, 67–81.

Renaut, R.W., 1991. Stromatolite development in the playa lakes of British Columbia, Canada: morphology, distribution and paleoenvironmental implications. *Abstr. V Int. Symp. Athalassic (Inland) Saline Lakes, Lake Titicaca, Bolivia.*

Renaut, R.W. 1994a. Morphology, distribution and preservation potential of microbial mats in the hydromagnesite-magnesite playas of the Cariboo Plateau, British Columbia, Canada. *Hydrobiologia*. (In press.)

Renaut, R.W. 1994b. Carbonate and evaporite sedimentation at Clinton Lake, British Columbia, Canada. In *Paleoclimate and Basin Evolution of Playa Systems*, ed. M.R. Rosen. Special Papers, Geol. Soc. Am. (In press.)

Renaut, R.W. & Long, P.R., 1987. Freeze-out precipitation of salts in saline lakes – examples from Western Canada. In *Crystallization and Precipitation*, ed. G. Strathdee, M.O. Klein & L.A. Melis, pp. 33–42. Pergamon, Oxford.

Renaut, R.W. & Long, P.R., 1989. Sedimentology of the saline lakes of the Cariboo Plateau, Interior British Columbia, Canada. *Sed. Geol.*, **64**, 239–64.

Renaut, R.W. & Stead, D., 1990. Sedimentology, mineralogy and hydrochemistry of the magnesium sulphate playa-lakes of Alkali Valley, British Columbia, Canada. *Abstracts (Posters) 13 Int. Sed. Congress, Nottingham, UK.*, pp. 187–8.

Renaut, R.W. & Stead, D., 1991. Recent magnesite-hydromagnesite sedimentation in playa basins of the Cariboo Plateau, British Columbia. *Pap. B.C. Geol. Surv. Branch*, 1991–1, pp. 279–88.

Scudder, C.G.E., 1969. The fauna of saline lakes on the Fraser Plateau in British Columbia. *Verh. Int. Verein. Limnol.*, **17**, 430–9.

Topping, M.S. & Scudder, C.G.E., 1977. Some physical and chemical features of saline lakes in central British Columbia. *Syesis*, **10**, 145–66.

Walker, T.L. & Parsons, A.L., 1927. Notes on Canadian minerals, tremolite, clinohumite, stromeyerite, natron and hexahydrite. *Univ. Toronto Stud. Geol.*, **24**, 21–3.

Last Chance Lake, a natric playa-lake in interior British Columbia, Canada

ROBIN W. RENAUT AND DOUGLAS STEAD

Dept. of Geological Sciences, University of Saskatchewan, Saskatoon, Saskatchewan S7N 0W0, Canada

General summary

Last Chance Lake (Fig. 1) is a small, ephemeral saline lake with highly alkaline (pH 9.0–10.5), Na-CO_3-SO_4-Cl brines (> 350 g l^{-1} TDS). The lake lies in a forested catchment on the intermontane Cariboo Plateau in Interior British Columbia at an altitude of ~ 1083 m (Renaut, Stead and Owen, this volume). It is significant for two main reasons – it is one of the few modern saline lakes in North America with permanent bedded natron salts; it exhibits a distinctive 'spotted' morphology, characterized by hundreds of individual brine pools within lacustrine muds. Natron was harvested unsuccessfully from the lake periodically between 1924 and about 1940 (Davis, 1924; Cummings, 1940).

The lake, which has a surface area of ~ 13 ha, lies between linear mounds of glacial till that rise up to ~ 5 m above the lake surface. It is fed principally by direct groundwater discharge, seasonal coldwater (6–7 °C) spring seepages and snowmelt. In common with many ephemeral hypersaline lakes in the Pacific Northwest (e.g. Jenkins, 1918; Reinecke, 1920; Allison & Mason, 1947; Bennett, 1962), the lake consists of several hundred brine pools enclosed within mixed siliciclastic-carbonate muds (Fig. 2). The bowl-shaped pools, which range from < 1 m to ~ 3 m deep, are small (1–2 m) and circular towards the margins, becoming larger (> 25 m across), irregularly lobate, and composite toward the center. Brine pools cover ~ 75% of the lake floor, decreasing in abundance towards the southeast where the basin floor shallows. At a few m depth, the lake sediments rest upon a thin mantle (1–5 m) of glacial till that in turn overlies Tertiary basalt lavas. Corroded gravels derived from the till have been brought to the surface within the muds, forming stone-polygons around many of the shallower peripheral pools. Mud-dykes, mud-volcanoes and similar diapiric structures are common within and between pools (Renaut & Long, 1989).

Sediments and mineralogy

Following spring snowmelt, the whole lake surface is flooded with shallow (< 50 cm; salinity: 50–60 g l^{-1} TDS) brine (Fig. 3). Algae, cyanobacteria and *Artemia* are abundant. Progressively, with summer evaporative concentration, the soft interpool muds become exposed and encrusted with several efflorescent salts (trona, natron and alkaline earth carbonates), gradually isolating the brine pools.

The dark greenish interpool muds are massive, rich in organic matter (mostly microbial; up to 17 wt%), and composed predominantly of carbonate minerals (calcite, protodolomite), clay minerals, feldspars, quartz and mafic minerals. Angular and corroded basalt clasts and rounded metamorphic rocks are locally supported within the muddy matrix.

Crystal rafts form periodically within the isolated pools during summer and may include both trona and natron, the latter crystallizing during cool nights. During autumn and winter, natron crystallizes from cooling brines, initially as crystal rafts, but also as rigid crusts of clear crystals up to 7 cm thick at the base of the pools. Minor mirabilite crystallizes from the sulfate-enriched residual waters. Coarse, black inclusion-rich natron crystals occur in the muds that underlie most of the larger brine pools (Goudge, 1926). During winter the entire lake surface is covered by ice several centimeters thick. With rising temperatures and snowmelt in the spring (April), most of the evaporites redissolve annually, but the black pool-base crystals remain throughout the year (Renaut & Long, 1987; Renaut, 1990).

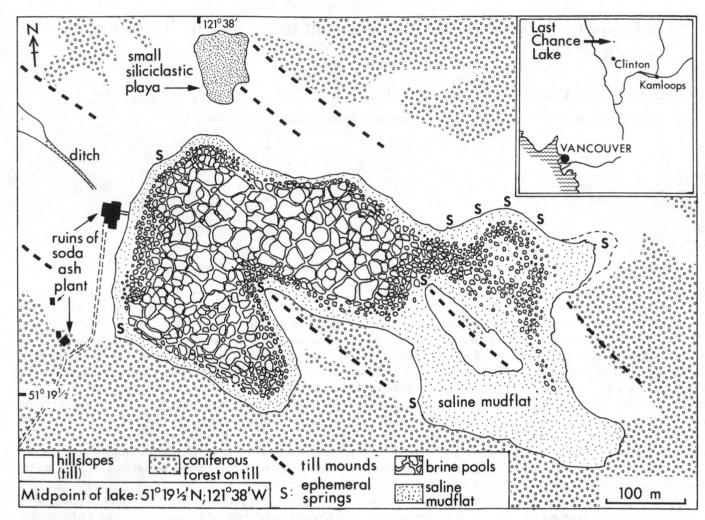

Fig. 1. Last Chance Lake, showing setting and the distribution of brine pools.

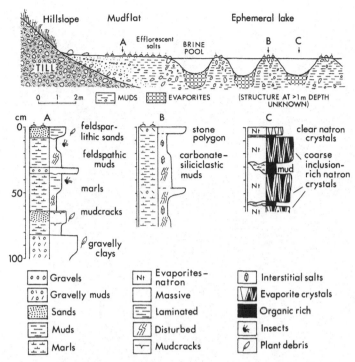

Fig. 2. Sedimentary profiles of shallow cores from the western margin of Last Chance Lake.

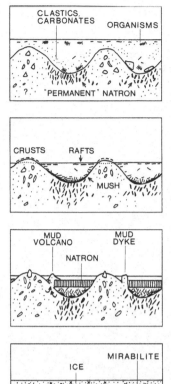

SPRING (April–early June)

Following snowmelt, lake is covered by shallow (30–50 cm) brine as winter salts are dissolved. Intense biological activity with algae, cyanobacteria and Artemia. Minor clastics washed in from marginal hillslopes. Spring discharge active.

SUMMER (mid June–early September)

Evaporative concentration leads to falling lake level and exposure of pools. Carbonates precipitated interstitially and as crusts at interpool sites. Trona and natron crystallize as rafts and pool-base crystal mush.

AUTUMN (mid-September–November)

Falling temperatures lead to cooling of the brine and crystallization of a clear natron crust (3–5 cm). Gravitational instability produces small mud dykes and mud volcanoes within crusts.

WINTER (November–March)

Minor mirabilite freezes out from residual brine. Formation of snow and ice cover (>5 cm) across entire lake surface. Groundwater discharge still active locally below ice cover.

Fig. 3. Seasonal changes in Last Chance Lake.

Bibliography

Allison, I.S. & Mason, R.S., 1947. *Sodium Salts of Lake County, Oregon.* Short Pap. Oreg. Dept. Geol. Miner. Ind. No. 17.

Bennett, W.A.G., 1962. *Saline Lake Deposits in Washington.* Wash. Div. Geol. Bull., No. 44.

Cummings, J.M., 1940. *Saline and Hydromagnesite Deposits of British Columbia.* Bull. B.C. Dept. Mines No. 4.

Davis, A.W., 1924. Central District (No. 3), Non-metallic section. *Annu. Rep. Minist. Mines B.C.* (1923), pp. 169–72.

Goudge, M.F., 1926. *Sodium Carbonate in British Columbia.* pp. 81–102. Rep. Mines Branch, Can. Dept. Mines No. 642.

Jenkins, O.P., 1918. Spotted lakes of epsomite in Washington and British Columbia. *Am. J. Sci.*, **46**, 638–44.

Reinecke, L., 1920. *Mineral deposits between Lillooet and Prince George, British Columbia. Mem. Geol. Surv. Can.*

Renaut, R.W., 1990. Recent carbonate sedimentation and brine evolution in the saline lake basins of the Cariboo Plateau, British Columbia, Canada. *Hydrobiologia*, **197**, 67–81.

Renaut, R.W. & Long, P.R., 1987. Freeze-out precipitation of salts in saline lakes – examples from Western Canada. In *Crystallization and Precipitation*, ed. G. Strathdee, M.O. Klein & L.A. Melis, pp. 33–42. Pergamon, Oxford.

Renaut, R.W., & Long, P.R., 1989. Sedimentology of the saline lakes of the Cariboo Plateau, Interior British Columbia, Canada. *Sed. Geol.*, **64**, 239–64.